R. A. Aliev, B. Fazlollahi, R. R. Aliev

Soft Computing and its Applications in Business and Economics

Studies in Fuzziness and Soft Computing, Volume 157

Further volumes of this series can be found on our homepage: springeronline.com

Vol. 141. G.C. Onwubolu, B.V. Babu
New Optimzation Techniques in Engineering, 2004
ISBN 3-540-20167-X

Vol. 142. M. Nikravesh, L.A. Zadeh, V. Korotkikh (Eds.)
Fuzzy Partial Differential Equations and Relational Equations, 2004
ISBN 3-540-20322-2

Vol. 143. L. Rutkowski
New Soft Computing Techniques for System Modelling, Pattern Classification and Image Processing, 2004
ISBN 3-540-20584-5

Vol. 144. Z. Sun, G.R. Finnie
Intelligent Techniques in E-Commerce, 2004
ISBN 3-540-20518-7

Vol. 145. J. Gil-Aluja
Fuzzy Sets in the Management of Uncertainty, 2004
ISBN 3-540-20341-9

Vol. 146. J.A. Gámez, S. Moral, A. Salmerón (Eds.)
Advances in Bayesian Networks, 2004
ISBN 3-540-20876-3

Vol. 147. K. Watanabe, M.M.A. Hashem
New Algorithms and their Applications to Evolutionary Robots, 2004
ISBN 3-540-20901-8

Vol. 148. C. Martin-Vide, V. Mitrana, G. Păun (Eds.)
Formal Languages and Applications, 2004
ISBN 3-540-20907-7

Vol. 149. J.J. Buckley
Fuzzy Statistics, 2004
ISBN 3-540-21084-9

Vol. 150. L. Bull (Ed.)
Applications of Learning Classifier Systems, 2004
ISBN 3-540-21109-8

Vol. 151. T. Kowalczyk, E. Pleszczyńska, F. Ruland (Eds.)
Grade Models and Methods for Data Analysis, 2004
ISBN 3-540-21120-9

Vol. 152. J. Rajapakse, L. Wang (Eds.)
Neural Information Processing: Research and Development, 2004
ISBN 3-540-21123-3

Vol. 153. J. Fulcher, L.C. Jain (Eds.)
Applied Intelligent Systems, 2004
ISBN 3-540-21153-5

Vol. 154. B. Liu
Uncertainty Theory, 2004
ISBN 3-540-21333-3

Vol. 155. G. Resconi, J.L. Jain
Intelligent Agents, 2004
ISBN 3-540-22003-8

Vol. 156. R. Tadeusiewicz, M.R. Ogiela
Medical Image Understanding Technology, 2004
ISBN 3-540-21985-4

Rafik A. Aliev
Bijan Fazlollahi
Rashad R. Aliev

Soft Computing and its Applications in Business and Economics

Springer

Professor Rafik Aziz Aliev
Department of Control Systems
Azerbaijan State Oil Academy
Joint MBA Program (USA, Azerbaijan)
20 Azadlyg Avenue
Baku az1010
Azerbaijan
E-mail: raliev@iatp.az

Professor Rashad Rafik Aliev
Department of Mathematics
Eastern Mediterranean University
Gazimagusa
Turkish Republic of North Cyprus
Mersin 10
Turkey
E-mail: rashad.aliyev@emu.edu.tr

Professor Bijan Fazlollahi
Department of Computer Information Systems &
Institute of International Business
Robinson College of Business
Georgia State University
Atlanta, GA 30303
USA
E-mail: DSCBBF@langate.gsu.edu

ISSN 1434-9922
ISBN 3-540-22138-7 Springer Berlin Heidelberg New York

Library of Congress Control Number: 2004106462

Springer is a part of Springer Science+Business Media
springeronline.com

Printed in Germany

Typesetting: camera-ready by authors
Cover design: E. Kirchner, Springer-Verlag, Heidelberg
Printed on acid free paper 62/3020/M - 5 4 3 2 1 0

Foreword

"Soft Computing and its Applications in Business and Economics," or SC-BE for short, is a work whose importance is hard to exaggerate. Authored by leading contributors to soft computing and its applications, SC-BE is a sequel to an earlier book by Professors R.A. Aliev and R.R. Aliev, "Soft Computing and Its Applications," World Scientific, 2001. SC-BE is a self-contained exposition of the foundations of soft computing, and presents a vast compendium of its applications to business, finance, decision analysis and economics. One cannot but be greatly impressed by the wide variety of applications — applications ranging from use of fuzzy logic in transportation and health case systems, to use of a neuro-fuzzy approach to modeling of credit risk in trading, and application of soft computing to e-commerce.

To view the contents of SC-BE in a clearer perspective, a bit of history is in order. In science, as in other realms of human activity, there is a tendency to be nationalistic — to commit oneself to a particular methodology and relegate to a position of inferiority or irrelevance all alternative methodologies.

As we move further into the age of machine intelligence and automated reasoning, we run into more and more problems which do not lend themselves to solution through the use of our favorite methodology. In this event, it may be expedient to explore the possibility of using an alternative methodology, or a combination of our favorite methodology with an alternative methodology, or a combination of alternative methodologies. This is the key idea that was the genesis of soft computing.

The idea became a reality in 1991, when the Berkeley Initiative in Soft Computing (BISC) was launched at UC Berkeley. Today, the so-called BISC Group is a worldwide internet-linked community with over 5000 members.

Basically, soft computing is a coalition of methodologies whose principal constituents are: fuzzy logic, neurocomputing, evolutionary computing, proba-bilistic computing, chaotic computing, rough set theory and parts of machine learning. The central thesis of soft computing is that, in general, best results are obtained when the constituent methodologies of soft computing are used in combination rather than in a stand-alone mode.

SC-BE is an outstanding treatise on soft computing and its applications, especially in the realms of business and economics. One of the many outstanding features of SC-BE is its excellence of organization. More concretely, the first two chapters are devoted to exposition of the basics of the principal constituents of soft computing: fuzzy set theory and fuzzy logic; neurocomputing; probabilistic computing; evolutionary computing; and chaotic computing. Following Chapter 2, Chapter 3 describes how the constituent methodologies may be used in combination to achieve superior performance. The principal modes of combination are neurofuzzy, neuro-genetic, fuzzy-genetic and neuro-fuzzy-genetic. These modes of combination underlie the recently developed (a) fuzzy-logic-based methodology of computing with words and perceptions; and (b) multi-agent distributed intelligence systems, which are described at the end of the chapter. The material in this chapter attests to the authors mastery of the theory of soft computing and extensive experience in the realm of its applications.

At this juncture, the most widely used combination is that of neuro-fuzzy systems. A goal example relates to what may be called the Arabshahi idea — a basic idea which was described by Arabshahi in the context of the backpropagation algorithm, but, in fact, has a much broader applicability. Specifically, most algorithms contain some parameters which can be adjusted to fit a particular problem. In many cases, the choice of values of such parameters is a matter of judgment based on experience. What fuzzy logic — and only fuzzy logic provides — is a language for representing judgment and perceptions derived from experience in the form of fuzzy if-then rules. Addition of such rules to the main algorithm leads to a combined algorithm — an algorithm which reduces the need for experience and im-proves performance.

The remaining six chapters provide an exceptionally complete, insightful and up-to-date coverage of a wide variety of applications of soft computing to business, economics, finance, forecasting, decision analysis and related fields. Among the numerous applications which are discussed are those relating to fuzzy linear programming, probabilistic scheduling for oil refinery, decision support, medicine, credit assessment, inventory control, stock market prediction, portfolio selection, risk management and e-commerce. These chapters place on the table a vast compendium of hard-to-get information about the ways in which soft computing can be employed to formulate and solve important real-world problems in business, economics and related fields.

What is obvious and yet not widely recognized, is that soft computing, taken as a whole, is much more powerful than any of its constituents taken in isolation. What this implies is that a strong case can be made for the thesis that a course on soft computing should be a standard course in every engineering curriculum. This is not the case at present, but is likely to become a reality in the not distant future. In this context, SC-BE could play an important role both as a reference and a textbook.

In summary, SC-BE is an outstanding work — a source of many new ideas and techniques, and a vast resource of illuminating information about soft computing and its numerous applications. It is a must reading for anyone who is interested in the conception, design and utilization of systems which are called upon to perform complex tasks requiring a high level of machine intelligence. The authors, Professors R.A. Aliev, B. Fazlollahi and R.R. Aliev; the publisher, Springer-Verlag; the series editor, Professor J. Kacprzyk; and the book editor, T. Ditzinger, deserve our thanks and congratulations for producing an outstanding book that is certain to contribute to a wider acceptance and use of soft computing as a powerful tool for constructing systems with superior MIQ (Machine IQ).

April 15, 2004 *Lotfi A. Zadeh*

Dedication

Dedicated to Professor Lotfi Zadeh, the founder of Soft Computing

Preface

Soft Computing is of central importance among postmodern information technologies for the creation of hybrid intelligent systems with high MIQ (Machine IQ). It was born about a decade ago, when there was intense competition between various emerging technologies. The founder of Soft Computing theory, Professor Lotfi Zadeh had perceived that more could be gained by cooperation than by claims and counter-claims of superiority of these new technologies. Soft Computing concept is based on the idea that in general, better results can be achieved by employing the constituent methodologies in combination than in a stand-alone mode. Soft Computing is consortium of such intelligent paradigms as Fuzzy Logic (FL), Neurocomputing (NC), Evolutionary Computing (EC), Probabilistic Computing (PC), and Chaotic Computing (CC) that enable one to solve many important real-world problems, which could not have been solved by mentioned and other existing technologies.

"Soft Computing and Its Applications" by R.A.Aliev and R.R.Aliev, World Scientific, 2001, a best-selling textbook, presents a unified framework and was the first to provide a systematic account of the major concepts and methodologies of Soft Computing. It introduces the general theory, foundations, and design principles of Soft Computing based systems and covers mainly applications in engineering problems.

Significant progress in Soft Computing technology and the need for results that rely on more realistic assumptions inspired new researchers to revisit business problems, i.e. the problems that have been traditionally tackled by introducing simplifying assumptions in the past. Today businesses are undergoing a major paradigm shift, moving from traditional management into a world of agile and smart organizations and processes. An agile and smart corporation is able to rapidly and intelligently respond to market changes. For this reason, corporations have been seeking to develop and adopt new technologies, such as soft computing technology, in management of business processes that can assist in developing more realistic solutions quickly and intelligently.

In recent years, a great number of papers and some books have explored the use of fuzzy logic ("Fuzzy Logic for Business, Finance, and Management" by G.Bojadziev and M.Bojadziev, World Scientific, 1997), Neural Networks ("Business Applications of Neural Networks" by P.Lisboa, A.Vellido, and B.Edinbury, World Scientific, 2000), Genetic Algorithms ("Evolutionary Computation in Economics and Finance" by Shu-Heng Chen, Physica-Verlag, 2002) and Soft Computing ("Soft Computing for Risk Evaluation and Management: Applications in

Technology, Environment and Finance", D. Ruan, M. Fedrizzi, J. Kacprzyk (eds.), Springer-Verlag, 2001) as a tool for designing intelligent systems in business, finance, management and economics.

Books on applications of emerging technologies in business and economics, including the books mentioned, present recent progress in the application of constituent methodologies, in particular neural networks, fuzzy logic, chaos etc. This book highlights some of the recent developments in practical applications of soft computing in business and eccnomics. It is the first book on application of SC based hybrid methods combining fuzzy logic, neuro-computing, evolutionary computing, probabilistic computing and chaotic computing in functional areas of business and economics.

This book is organized into 9 chapters. The first chapter introduces the general concept of Soft Computing technology and answers to the question "what is Soft Computing?" It also introduces the comparative features of the constituents of SC and their intelligent synergy. In order to make the book self-contained, a review of the theories of constituents of SC: fuzzy logic, neuro-computing, evolutionary computing, probabilistic computing and chaotic computing is given in chapter 2. The reader is expected to have a basic knowledge of these emerging technologies. The integrations of those constituents methodologies formed the core of SC is subject of chapter 3. A distinguishing feature of this book is that it deals with all main synergy of different constituent paradigms. Much attention is given to very widely-used in practice, neuro-fuzzy technology. Fuzzy logic and genetic algorithms, together with neurocomputing technologies are recognized as one of major parts of SC, consequently a special place in this chapter is given to hybridization between these constituents. This chapter also introduces the reader to neuro-genetic, fuzzy-genetic and other SC technologies. It also includes a new aspect of Distributed Intelligence, namely, Soft Computing based multi-agent systems.

Soft Computing in this book is presented not only with the theoretical development but also with a large variety of realistic applications to business and economics problems, which are considered in six consequent chapters – chapters 4-9.

Chapter 4 deals with Soft Computing based qualitative and quantitative forecasting methods.

It represents neuro and fuzzy computing based time series forecasting, fuzzy Delphi method, Soft Computing based prediction in chaotic time series. Chapter 5 covers a wide range of methods and procedures for creation of new generation of Decision making systems and DSS based on SC technology. It introduces the evolutionary and fuzzy chaos approaches to fuzzy linear programming, fuzzy decision making, and multi-agent distributed intelligent systems. The emphasis is on development of Soft Computing based multi-agent DSS, hybrid DSS based on genetic algorithms and simulation. Application of Soft Computing technology in marketing is subject of chapter 6. This chapter includes soft computing based marketing analysis of a customer's behavior, credit evaluation, fraud detection and service quality evaluation.

Chapter 7 highlights some of the recent practical applications of soft computing in operations management. It covers scheduling fuzzy-probabilistic expert system for oil refineries, fuzzy regression based quality evaluation and neuro-fuzzy pattern recognition in manufacturing, soft computing based inventory control and project scheduling and other related problems.

Soft computing provides effective tools for dealing with complex problems in finance characterized almost always with uncertainty, vagueness and imprecision. Such tools are considered in chapter 8 for creation of intelligent stock market predicting, loan assessment, and risk management systems. This chapter also covers Soft Computing approaches to portfolio selection, trading DSS and solution of other important problems in finance.

A new way of conducting business over the Internet, Electronic Commerce (EC) is growing exponentially. Because EC is relatively new and available information regarding this kind of business strategy and environment is often inexact and vague, the application of Soft Computing technology to solve EC problems seems very appropriate. Chapter 9 includes Soft Computing based methods for creation of a multi-agent system for EC decisions, personalization of EC, and solution of a very important problem in EC, namely, risk analysis.

This book will be valuable aid to anyone considering the application of Soft Computing theory and technology to real problems of business and economics, because it contains a number of detailed accounts of such applications. It brings in a systematic way Soft Computing into the university and college educational systems and may be basic text for introducing business managers, teachers, and scientists from various fields of business and economics to the Soft Computing technology, enabling them to initiate projects and make applications.

We would like to express our thanks to professor L. Zadeh, founder of Soft Computing theory for his constant and invaluable support of our research and for his help in the publication of this book. Special thanks are due to Professors Mo. Jamshidi, I. Turksen, J. Kacprzyk, T. Whalen, S.Ulyanov and V. Loia for helpful discussions on various topics of Soft Computing and its applications.

We are grateful to our colleagues dr. R. Vahidov and B. Guirimov for many enjoyable and productive conversations and collaborations.

R. A. Aliev
B. Fazollahi
R. R. Aliev

Contents

1 Introduction to Soft Computing

1.1 Basic concept of Soft Computing

Artificial intelligence as a science has been existing for about 40 years now. The main problem of this science is replication of human reasoning processes and behavior with the aid of computers and other artificial devices as well as construction of machines simulating decision making by humans in imprecise and uncertain environments. In most cases these various areas, where precise models, methods, and algorithms for solving problems characterized by uncertainty are not available, are attributed to the field of artificial intelligence. Methods of artificial intelligence are based on two characteristic features:

1. Use of information in symbolic form i.e. letters, words, phrases, signs, figures;
2. Search with the aid of symbolic logic. When processing symbolic information, the computer converts the words and phrases to the form of binary digits. Then the computer recognizes or compares the sequences of such symbols (converted to digital form).

The classics of artificial intelligence stated that the ability of computers to manipulate symbols as easily as numbers, compare sequences of symbols, and then, depending on the results of comparison perform subsequent operations, would allow realization of the functions typical for the human mind, i.e. functions of deductive logical reasoning, in machines. It may seem that the potential abilities of a computer in creation of artificial intelligence based on the symbolic information processing are unlimited. Despite the significant success of artificial intelligence (in the classical sense) in developing a wide range of systems for solving problems, automatically proving theorems, recognizing patterns as well as in constructing expert systems and natural language understanding systems, the expectations have not been achieved to a full degree. Traditional artificial intelligence is not capable of solving problems which require the use of common sense, and it does not accept procedures, which are similar to human abilities of understanding and reasoning. Traditional artificial intelligence has not succeeded in solving problems for intelligent robotics, computer vision, recognition of speech and hand-written graphics, machine translation, learning through experience and many other important real-world problems. The above problems, as well as many others have intrinsic imprecision and uncertainty that cannot be neglected. As Prof. L.Zadeh noted, the traditional artificial intelligence could achieve more success in pursuing its goals if it did not limit itself to processing symbolic information only and using the first order logic. All traditional artificial intelligence systems have been implemented using the Hard Computing technology, which restricts considerably the abilities of those systems. Moreover, the traditional artificial intelligence, due to the features shown above does not accept the numerical methods, which are important for accounting for uncertainty and imprecision. Due to the above limitations, the MIQ (Machine Intelligence Quotient) for traditional artificial intelligence systems is not sufficiently high. There is a strong need to increase MIQ for intelligent systems. Soft Computing methodology implies cooperative activity rather than autonomous one for such new computing paradigms as fuzzy logic, neural networks, evolutionary computation and others. This approach allows solving many important real-world problems, which were impossible to solve using traditional artificial intelligence methods [2-4, 9, 11-13].

The combination of such intelligent paradigms (used as computing techniques) as Fuzzy Logic (FL), Neural Networks (NN), Probabilistic Reasoning (PR), Genetic Algorithms (GA), and Chaos Theory (ChT) dealing with pervasive imprecision and uncertainty of the real-world problems is named Soft Computing (SC). Unlike traditional Hard Computing (HC), SC can tolerate imprecision, uncertainty and partial truth without loss of performance and effectiveness for the end use. In no more than a decade we will see reorientation of Artificial Intelligence towards Soft Computing from the traditional Hard Computing. L.Zadeh noted that, unlike the traditional Hard Computing, Soft Computing aims at accommodation with the pervasive imprecision of the real world. The guiding principle of Soft Computing is: exploit the tolerance for imprecision, uncertainty, and partial truth to achieve tractability, robustness and low solution cost [14]. We can easily come to the conclusion that precision has a cost (unfortunately, this obvious principle often is neglected). Therefore, in order to solve a problem with an acceptable cost we need to aim at a decision with only the necessary degree of precision, not exceeding the requirements. The impressive examples of the aforesaid are problems of landing a helicopter or parking a car. Let's consider the second case. One can park a car without taking any distance and angle measurements, because the final position of the car is not specified clearly. If it is specified, though, then the measurements are necessary, say, in the range of fractions of millimeter or several seconds of arc. This will require many hours of manoeuvres and measurements from the devices for solving the problem. Moreover, the cost of decision will increase exponentially as the precision increases.

Soft Computing technology is of great importance for data compression, especially in HDTV, audio recording, speech recognition, image understanding, and related fields. Actually, soft-computing-based concepts and techniques have already been playing an essential role in the conception, design and manufacturing of high MIQ products and systems. As noted by Zadeh, the perfect model of SC is human brain.

As it was mentioned above, all traditional artificial intelligence systems, including expert systems, widely used in various areas of human activity, have been implemented on the base of Hard Computing, often using computers. However, this base obviously limits the effectiveness and, generally, the possibility of creating artificial intelligence systems for different purposes. Currently, the significant increase can be noticed in the number of applied artificial intelligence systems based not on numerical (symbolic) computation and traditional Hard Computing, but on neural networks, fuzzy computing, evolutionary programming, and chaotic computing. There is also an increase in the number of publications in proceedings of scientific conferences which are devoted to fuzzy logic, genetic algorithms, artificial life, biological computing, neural computing etc. This increase provides the evidence that the focus of the investigations and implementations of real artificial intelligence systems makes a shift towards Soft Computing.

Fig. 1.1 shows the structure of Soft Computing technology forming the basis for computational intelligence.

The following main components of Soft Computing are known by now: Fuzzy Logic (FL), Neural Networks (NN), Probabilistic Reasoning (PR), Genetic Algorithms (GA), and Chaos Theory (ChT) (Fig. 1.1). In our (and not only our) framework of SC, FL is the kernel of SC. FL's main characteristic is the robustness of its interpolative reasoning mechanism. Within Soft Computing, Fuzzy Logic occupies a special place bacause it can be used as a springboard for generalization of any theory, including its partners in SC consortium. In SC FL is mainly concerned with imprecision and approximate reasoning, NN with learning, PR with uncertainty and propagation of belief, GA with global optimization and search and ChT with nonlinear dynamics. Each of these computational paradigms (emerging reasoning technologies) provide us with complementary reasoning and searching methods to solve complex, real-world problems. In large scope, FL, NN, PR, and GA are complementary rather that competitive [2-4, 14].

The interrelations between the components of SC, shown in Fig.1.1, make the

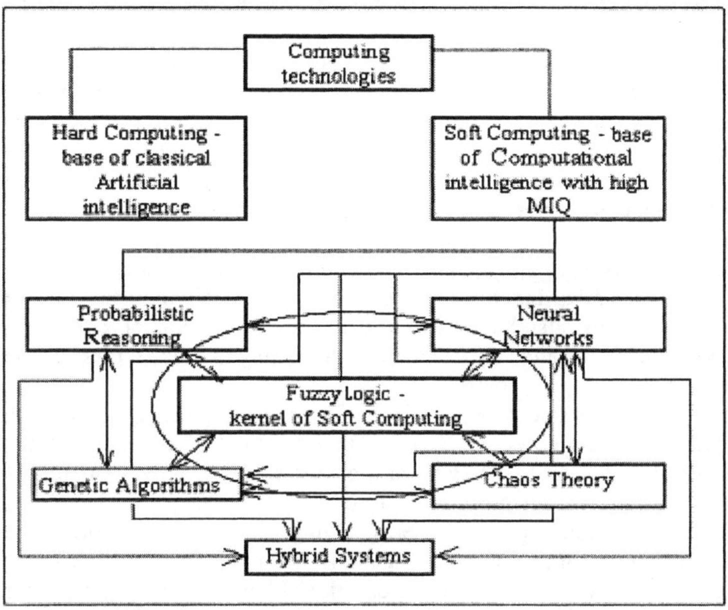

Fig. 1.1. The main components of Soft Computing

theoretical foundation of Hybrid Intelligent Systems. As noted by L. Zadeh "currently the term hybrid intelligent systems is gaining currency as a descriptor of systems in which FL, NC, and PR are used in combination. In my view, hybrid intelligent systems are the wave of the future" [16]. The use of Hybrid Intelligent Systems are leading to the development of numerous manufacturing system, multimedia system, intelligent robots, trading systems, which exhibits a high level of MIQ.

1.2 Combination of Constituents of Soft Computing

The constituents of SC can be used independently (fuzzy computing, neural computing, evolutionary computing etc.), and more often in combination [1,4,8,10-12,15]. Based on independent use of the constituents of Soft Computing, fuzzy technology, neural technology, chaos technology and others have been recently applied as emerging technologies to both industrial and non-industrial areas. Fuzzy logic is the leading constituent of Soft Computing. In Soft Computing, fuzzy logic plays a unique role. FL serves to provide a methodology for computing with words [12,17]. It has been successfully applied to many industrial spheres, robotics, complex decision making and diagnosis, data compression, and many other areas. To design a system processor for handling knowledge represented in a linguistic or uncertain numerical form we need a fuzzy model of the system.

Fuzzy sets can be used as a universal approximator, which is very important for modeling unknown objects. If an operator cannot tell linguistically what kind of action he or she takes in a specific situation, then it is quite useful to model his/her control actions using numerical data. However, fuzzy logic in its so-called pure form is not always useful for easily constructing intelligent systems. For example, when a designer does not have sufficient prior information (knowledge) about the system, development of acceptable fuzzy rule base becomes impossible. As the complexity of the system increases, it becomes difficult to specify a correct set of rules and membership functions for describing adequately the behavior of the system. Fuzzy systems also have the disadvantage of not being able to extract additional knowledge from the experience and correcting the fuzzy rules for improving the performance of the system.

Another important component of Soft Computing is neural networks. Artificial neural networks viewed as parallel computational models, are parallel fine-grained implementation of non-linear static or dynamic systems. A very important feature of these networks is their adaptive nature, where "learning by example" replaces traditional "programming" in problems solving. Another key feature is the intrinsic parallelism that allows fast computations. Artificial neural networks are viable computational models for a wide variety of problems including pattern classification, speech synthesis and recognition, curve fitting, approximation capability, image data compression, associative memory, and modeling and control of non-linear unknown systems [6,7]. Neural networks are favorably distinguished for efficiency of their computations and hardware implementations. Another advantage of neural networks is generalization ability, which is the ability to classify correctly new patterns.

A significant disadvantage of neural networks is their poor interpretability. One of the main criticisms addressed to neural networks concerns their black box nature [3].

Evolutionary Computing (EC) is a revolutionary approach to optimization. One part of EC—genetic algorithms—are algorithms for global optimization. Genetic algorithms are based on the mechanisms of natural selection and genetics [5]. One advantage of genetic algorithms is that they effectively implement parallel multicriteria search. The mechanism of genetic algorithms is simple. Simplicity of operations and powerful computational effect are the two main advantages of genetic algorithms. The disadvantages are the problem of convergence and absence of strong theoretical foundation. The requirement of coding the domain of the real variables into bit strings also seems to be a drawback of genetic algorithms. It should be also noted that the computational speed of genetic algorithms is low.

Probabilistic Reasoning (PR) offers the mechanism to evaluate the outcome of systems affected by probabilistic uncertainty. PR uses the operation of conditioning to update the probability values and perform a probabilistic inference. The probabilistic approach (objective or subjective) provides a rigorous framework for representation of a probabilistic knowledge, modeling random phenomena and for analyzing them. Moreover, PR approach does not distinguish between ambiguity of the knowledge and uncertainty generating errors and lack of complete knowledge. Fuzzy Logic is able to make this distinction.

Here we should also note that human reasoning does not follow the axioms of probability theory and in case when the evidence is uncertain, the computational complexity significantly increases.

At present Chaotic computing is a very important field of scientific research and treated as a basis for a new technology. A chaotic system is a deterministic system that exhibits random behavior. Chaos Theory deals with the non-linear dynamical systems that exhibit extreme sensitivity to initial conditions. Behavior of chaotic systems is characterized by a strange attractor, which has a fractal dimension bounded by the topological and the Euclidean dimensions.

Table 1.1 presents the comparative characteristics of the components of Soft Computing.

Using chaotic analysis we determine the predictability and formulate prediction strategies of system's behavior. Chaotic computing deals also with nonlinear systems with unknown functional form and, possibly, noise. Chaotic computing gives a tool to determine a new perspective of nonlinear data analysis. No assumption is made about the behavior of the data.

In addition to the aforesaid it should be noted that identifying chaos in real-world problems is a complex task. Another weakness of chaotic computing is that practical numerical analysis of chaotic systems in most cases is connected with computational difficulty.

For each component of Soft Computing there is a specific class of problems, where the use of other components is inadequate. For example, the well-known problem of parking a car can be solved successfully by using only Fuzzy Logic, and not by using Neural Networks, GA etc.

As it was shown above, the components of Soft Computing—Fuzzy Logic, Neuro Computing, Probabilistic Reasoning etc. complement each other, rather than compete. It becomes clear that FL, NC, PR, and GA are more effective when used in combinations.

Table 1. 1. Comparative characteristics of the components of Soft Computing

	Fuzzy Sets	Artificial Neural Networks	Evoltionary Computting, GA	Probabilistic Reasoning	Chaotic computing
Weaknesses	□ Knowledge acquisition □ Learning	□ Black Box Interpretability	□ Coding □ Computational speed	□ Limitation of the axioms of Probabiity □ Lack of complete knowledge □ Computational complexity	□ Computational complexity □ Chaos identification complexity
Strengths	□ Interpretability □ Transparency □ Plausibility □ Graduallity □ Modeling □ Reasoning □ Tolerance to imprecision	□ Learning □ Adaptation □ Faul tolerance □ Curve fitting □ Generalization ability □ Approximation ability	□ Computational efficiency □ Global optimiz ation	□ Rigorous framework □ Well understanding	□ Nonlinear dynamics simulation □ Discovering chaos in observed data (with noise) □ Determining the predictability □ Prediction strategies

The following are known principal combinations of the components of Soft Computing [4]:

Neuro Computing + Fuzzy Logic (Neuro-Fuzzy: NF)
Fuzzy Logic + Genetic Algorithms (FG),
Fuzzy Logic + Chaos Theory (FCh),
Neural Networks + Genetic Algorithms (NG);
Neural Networks + Chaos Theory (NCh);
Fuzzy Logic + Neural Networks + Genetic Algorithms (FNG),
Neural Networks + Fuzzy Logic + Genetic Algorithms (NFG).
Fuzzy Logic+Probabilistic Reasoning (FP)
Other combinations of constituents of SC are possible as well.

Lack of interpretability of neural networks on one hand and poor learning capability of fuzzy systems on the other hand are similar problems that limit the application of these tools. Neural-Fuzzy systems are hybrid systems which try to solve this problem by combining the learning capability of connectionist models with the interpretability property of fuzzy systems. As it was noted above, in case of dynamic work environment, the automatic knowledge base correction in fuzzy systems becomes necessary. On the other hand, artificial neural networks are successfully used in problems connected to knowledge acquisition using learning by examples with the required degree of precision. Incorporating neural networks in fuzzy systems for fuzzification, construction of fuzzy rules, optimization and adaptation of fuzzy knowledge base, implementation of fuzzy reasoning, and defuzzification is the essence of the Neuro-Fuzzy approach. Section 3.1 of the book is devoted to Neuro-Fuzzy systems.

The combination of rule-based fuzzy systems employing "rule of thumb" strategy used by humans in decision making, with genetic algorithms, which allow to perform global search, enables creation of effective, robust, and adaptive systems. Often, the membership functions used in fuzzy rules in knowledge bases of fuzzy systems and fuzzy performance indices are represented as non-differentiable fuzzy numbers, e.g. trapezoids, triangles etc. Use of gradient-based methods for development of such systems becomes infeasible. An effective technique in this case is GA.

The combination of FL and GA allows optimization of fuzzy knowledge base of fuzzy logic control system by defining optimal number of rules in knowledge base and optimal values for centers and shapes of membership functions. Here GA is used for constructing relational matrix and membership functions of the designed fuzzy systems. In turn, in the combination of FL with GA, theory of fuzzy systems can be used for improving the behavior of genetic operators or genetic algorithms on whole, i.e. it is possible to create fuzzy tools for improving effectiveness of GA via developing fuzzy genetic algorithms.

The combination of genetic algorithms with neural networks yields promising results as well. It is known that one of main problems in development of artificial neural systems is selection of a suitable learning method for tuning the parameters of a neural network (weights, thresholds, and structure). The most known algorithm is the "error backpropagation" algorithm. Unfortunately, there are some difficulties with "backpropagation". First, the effectiveness of the learning considerably depends on initial set of weights, which are generated randomly. Second, the "backpropagation", like any other gradient-based method, does not avoid local minima. Third, if the learning rate is too slow, it requires too much time to find the solution. If, on the other hand, the learning rate is too high it can generate oscillations around the desired point in the weight space. Fourth, "backpropagation" requires the activation functions to be differentiable. This condition does not hold for many types of neural networks.

Genetic algorithms used for solving many optimization problems when the "strong" methods fail to find appropriate solution, can be successfully applied for learning neural networks, because they are free of the above drawbacks.

The models of artificial neurons, which use linear, threshold, sigmoidal and other transfer functions, are effective for neural computing. However, it should be noted that such models are very simplified. For example, reaction of a biological axon is chaotic even if the input is periodical. In this aspect the more adequate model of neurons seems to be chaotic. Model of a chaotic neuron can be used as an element of chaotic neural networks. The more adequate results can be obtained if using fuzzy chaotic neural networks, which are closer to biological computation.

Fuzzy systems with If-Then rules can model non-linear dynamic systems and capture chaotic attractors easily and accurately. Combination of Fuzzy Logic and Chaos Theory gives us useful tool for building system's chaotic behavior into rule structure.

Identification of chaos allows us to determine predicting strategies. If we use a Neural Network Predictor for predicting the system's behavior, the parameters of the strange attractor (in particular fractal dimension) tell us how much data are necessary to train the neural network. The combination of Neurocomputing and Chaotic computing technologies can be very helpful for prediction and control.

Different methods exist for learning fuzzy neural networks, i.e. neural networks with fuzzy signals and/or fuzzy weights. In particular, direct fuzzification of the ordinary delta-rule is used. Another method is based on alpha-cuts of fuzzy sets, interval arithmetic, and consequent application of back propagation. In the latter case, the algorithm can fail to converge to correct values of weights. In either case it is necessary to calculate the gradient of the fuzzy error measure. The corresponding derivatives are too complex, especially in case of more general fuzzy sets for input and output signals and weights.

Neuro-genetic algorithms for learning fuzzy neural networks based on a combination of FL, NC, and GA are free of the above drawbacks and produce effective results.

It is necessary to note that Probabilistic Reasoning and Fuzzy Reasoning are complementary rather than competitive. The cooperation between these formalisms gives a useful tool for modeling and reasoning under uncertainty in complicated real-world problems. Such cooperation is of particular importance for constructing perception-based intelligent information systems.

We hope that the mentioned intelligent combinations will develop further, and the new ones will be proposed. These SC paradigms will form the basis for creation and development of Computational Intelligence.

References

1. Aliev RA (1994) Fuzzy Expert Systems. İn: Aminzadeh F, Jamshidi M (eds) SOFT COMPUTING: Fuzzy Logic, Neural Networks and Distributed Artificial Intelligence. NJ: PTR Prentice Hall, pp 99-108
2. Aliev RA, Aliev RR (1997-1998) Soft Computing (in Russian), vol I, II, III. Baku: ASOA Press
3. Aliev R, Bonfig K, Aliew F (2000) Soft Computing. Berlin: Verlag Technic

4. Aliev RA, Aliev RR (2001) Soft Computing and its Application. World Scientific, New Jersey, London, Singapure, Hong Kong
5. Goldberg DE (1989) Genetic algorithms in search, optimization and machine learning. Reading, MA: Addison-Wesley
6. Haykin S (1994) Neural Networks: A Comprehensive Foundation. Marmillau and IEEE Computer Society
7. Mohamad HH (1995) Fundamentals of artificial neural networks.Cambridge. MIT Press
8. Nauck D, Klawonn F, Kruse R (1997) Foundations of Neuro-Fuzzy Systems.NY: John Wiley and Sons
9. Pearson DW, Steele NC, Albrecht RF (1995) Artificial Neural Nets and Genetic Algorithms. In: Inter. Conf. in Ales, France
10. Welstead ST(ed) (1994) Neural Networks and Fuzzy Logic Applications in C/C++, Professional Computing. NY: John Wiley
11. Yager RR, Zadeh LA(eds) (1994) Fuzzy sets, neural networks and Soft Computing. NY: VAN Nostrand Reinhold
12. Zadeh LA (1994) Soft Computing and Fuzzy Logic. IEEE Software 11(6): 48-58
13. Zurada YM, Marks RJ, Robinson CY (1994) Computational Imitating life. J Piscataway, IEEE Press
14. Zadeh LA (1994) The roles of fuzzy logic and soft computing in the conception, design and deployment of intelligent systems. J BT Technol 14(4): 32-36
15. Zadeh LA (1994) Fuzzy logic, Neural Networks and Soft Computing . J Comm of ACM 37(3): 77-84
16. Zadeh LA (1995) Foreword. In: First European Congress on Intelligent Techniques and Soft Computing – EUFIT'95, p VII
17. Zadeh LA (1996) Fuzzy Logic=Computing with words. J IEEE Transactions on fuzzy systems 2, vol 4

2 Constituent Methodologies of Soft Computing

2.1 Elements of Fuzzy Sets Theory

2.1.1 Fuzzy sets and Operations on them [14]

Definition. Fuzzy sets

Let X be a classical set of objects, called the universe, whose generic elements are denoted x. Membership in a classical subset A of X is often viewed as a characteristic function $\propto_A$ from X to {0,1} such that

$$\propto_A(x) = \begin{cases} 1 & \text{iff } x \in A \\ 0 & \text{iff } x \notin A \end{cases}$$

where {0,1} is called a valuation set; 1 indicates membership while 0 - non-membership.

If the valuation set is allowed to be in the real interval [0,1], then A is called a fuzzy set [13,16,52,53,113,117]. $\propto_A(x)$ is the grade of membership of x in A

$$\propto_A : X \to [0,1]$$

As closer the value of $\propto_A(x)$ is to 1, so much x belongs to A.

A is completely characterized by the set of pair.

$$A = \{(x, \propto_A(x)),\ \ x \in X\}\ .$$

Example

$$A=0.1/1+0.3/2+0.5/3+0.7/4+0.8/5+0.9/6+0.95/7+1.0/8+0.97/9+$$

$$+0.9/10+0.8/11+0.7/12+0.5/13+0.3/14+0.1/15.$$

A is a fuzzy set. A classical version of this fuzzy set is

$$A=\{1,2,3,4,5,6,7,8,9,10,11,12,13,14,15\}$$

or

$$A=1/1+1/2+1/3+1/4+1/5+1/6+1/7+1/8+1/9+1/10+1/11+1/12+1/13+1/14+1/15.$$

For graphical representation see Fig. 2.1.

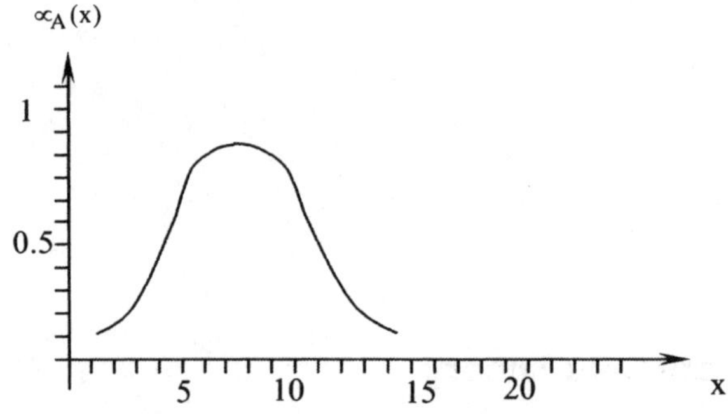

Fig. 2.1. Fuzzy set

Example

The representation of temperature within a range $[T_1, T_2]$ by fuzzy and crisp sets is shown in Fig. 2.2a and 2.2b, respectively. In the first case we use membership function $[T_1, T_2] \rightarrow [0,1]$ for describing linguistic concepts *"cold", "normal", "warm"*. In the second case right - open intervals are used for describing of traditional variable by crisp sets.

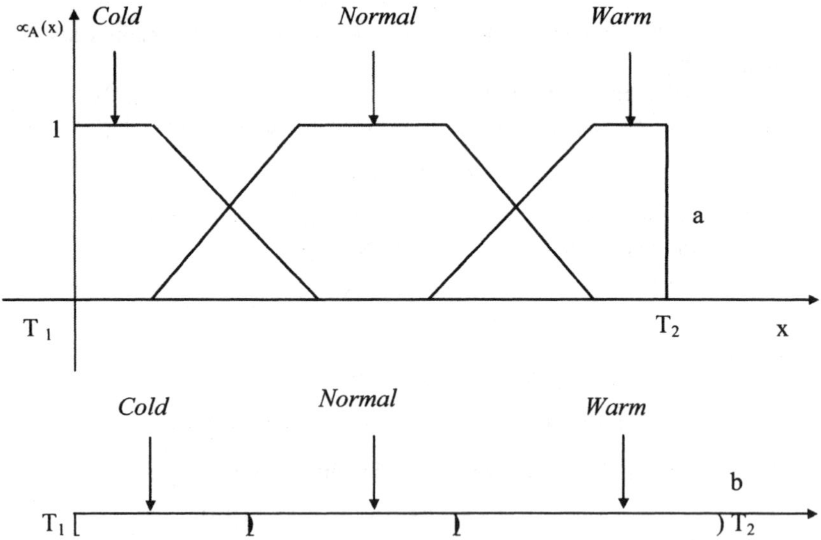

Fig. 2.2. Representation of temperature by fuzzy (Fig. 2.2a) and crisp (Fig. 2.2b) sets

Fuzzy sets with crisply defined membership functions are called ordinary fuzzy sets.

If membership function of a fuzzy set A assigns to each element of the universal set X a closed interval of real numbers, then this type of fuzzy sets are called interval-valued fuzzy sets

$$\propto_A : X \to \varepsilon\,[0,1],$$

where $\varepsilon[0,1]$ denotes the family of all closed intervals of real members in $[0,1]$.

Example

$$A = \propto_A (x_1^{\alpha_1}, x_1^{\alpha_2})/x_1 + \square + \propto_A (x_n^{\alpha_1}, x_n^{\alpha_2})/x_n$$

Membership function of interval-valued fuzzy set A is given in Fig. 2.3.

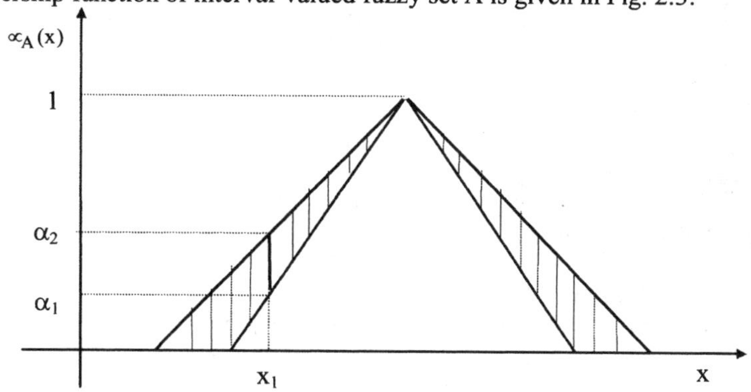

Fig. 2.3. Interval-valued fuzzy set A

Fuzzy sets whose membership function itself is an ordinary fuzzy set are called fuzzy sets of type 2. Their membership functions are

$$\propto_A : X \to F([0,1]), \tag{2.1}$$

where F([0,1]) is the set of all ordinary fuzzy sets that can be defined in [0,1], i.e. fuzzy power set of [0,1].

Example

Fuzzy set of type 2 is shown in Fig. 2.4.

If $x = x_1$ four numbers α_1, α_2, α_3, α_4 are produced, by which the ordinary fuzzy set defined with trapezoidal membership function assigned A to x_1 is determined.

The fuzzy sets of still higher types could be obtained recursively in similar way.

If fuzzy set in X has membership function as type m-1, m>1 fuzzy set on [0,1], then that set is referred to as fuzzy set type m.

Fuzzy set whose membership function value is a random variable is called probabilistic set.

In a more general case the valuation set of fuzzy sets may be a partially ordered set L

$$\propto_A : X \to L .$$

Such generalized type of fuzzy sets are referred to as L-fuzzy set. All the types of considered above fuzzy sets are special L-fuzzy sets.

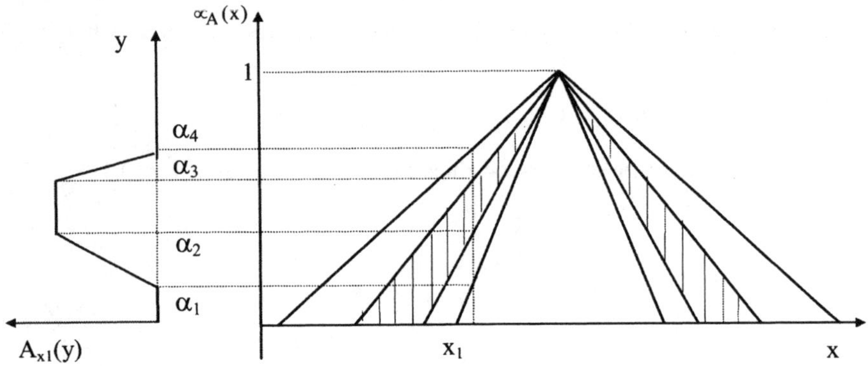

Fig. 2.4. Fuzzy set of type 2

Definition. Equality of fuzzy sets
Two fuzzy sets A and B are said to be equal if and only if

$$\forall x \in X \quad \propto_A(x) = \propto_B(x) \quad A = B .$$

Definition. The support and the crossover point of a fuzzy set. The singleton
The support of a fuzzy set A is the ordinary subset of X that has nonzero membership in A:

$$\sup A = A^{+0} = \{x \in X, \propto_A(x) > 0\}$$

The elements of x such as $\propto_A(x)=1/2$ are the crossover points of A.

A fuzzy set that has only one point in X with $\propto_A=1$ as its support is called a singleton.

Definition. The height of a fuzzy set. Normal and subnormal sets
The height of A is

$$\operatorname{hgt}(A) = \sup_{x \in X} \propto_A(x)$$

i.e., the least upper bound of $\propto_A(x)$.

A is said to be normalized iff $\exists\, x \in X, \propto_A(x)=1$. This definition implies $\operatorname{hgt}(A) = 1$. Otherwise A is called subnormal fuzzy set.

The empty set $\varnothing$ is defined as

$$x \in X,\ \propto_{\varnothing}(x) = 0, \text{ of course } \forall x \in X,\ \propto_X(x) = 1$$

Example

$$A=0.3/20+0.5/22+1.0/25+0.8/27+0.4/30.$$

Here the universe X is $X=\{15,20,22,25,27,30,33,35\}$

$$\text{supp } A=\{20,22,25,27,30\}.$$

A is normalized, i.e. $25\in X$, $\propto_A(25)=1$,

$$\text{hgt}(A)=\sup_{x\in X} \propto_A(x)=\sup_{x\in 15,\ldots,35}\{0,0.3,0.5,1.0,0.8,0.4\}=1.$$

Definition. α-level fuzzy sets

One of important way of representation of fuzzy sets is α-cut method. Such type of representation allows us to use properties of crisp sets and operations on crisp sets in fuzzy set theory.

The (crisp) set of elements that belongs to the fuzzy set A at least to the degree α is called the α-level set:

$$A^\alpha=\{x\in X, \propto_A(x)\geq\alpha\}.$$

$A^\alpha=\{x\in X, \propto_A(x)>\alpha\}$ is called "strong α-level set" or "strong α-cut".

Let us consider a simple example.

$$A=0.2/x_1+0.5/x_2+0.8/x_3+1.0/x_4.$$

The level set of A is $\{0.2,0.5,0.8,1.0\}$. Then the α-level sets are:

$A^{0.2}=\{x_1,x_2,x_3,x_4\}$

$A^{0.5}=\{x_2,x_3,x_4\}$

$A^{0.8}=\{x_3,x_4\}$

$A^{1.0}=\{x_4\}$.

Now we introduce fuzzy set A_α, defined as

$$A_\alpha(x)=\alpha A^\alpha(x). \tag{2.2}$$

Then the original fuzzy set A may be defined as. $A=\bigcup_{[0,1]} A_\alpha$. $\cup$ denotes the standard fuzzy union.

In our example

$$A_{0.2}=0.2/x_1+0.2/x_2+0.2/x_3+0.2/x_4$$

$$A_{0.5}=0/x_1+0.5/x_2+0.5/x_3+0.5/x_4$$

$$A_{0.8}=0/x_1+0/x_2+0.8/x_3+0.8/x_4$$

$$A_{1.0}=0/x_1+0/x_2+0/x_3+1.0/x_4.$$

The union of these four fuzzy sets gives

$$A=A_{0.2}\cup A_{0.5}\cup A_{0.8}\cup A_{1.0}.$$

The representation of A by union of fuzzy sets A_α is usually referred to as a decomposition of A. Below we give decomposition theorem of fuzzy sets.

Theorem 2.1
For every $A \in F(X)$

$$A = \bigcup_{\alpha \in [0,1]} A_\alpha \tag{2.3}$$

where A_α is defined by (2.2), F(X) denotes the fuzzy power set of X. Proof of this theorem is given in [52].

Definition. Convexity of fuzzy sets
A fuzzy set A is convex iff

$$\propto_A(\lambda x_1 + (1-\lambda)x_2) > \min(\propto_A(x_1), \propto_A(x_2)) \tag{2.4}$$

for all $x_1, x_2 \in R, \lambda \in [0,1]$, min denotes the minimum operator.

Alternatively, a fuzzy set A on R is convex iff all its α-level sets are convex in the classical sense. A typical convex fuzzy set is given in Fig. 2.5.

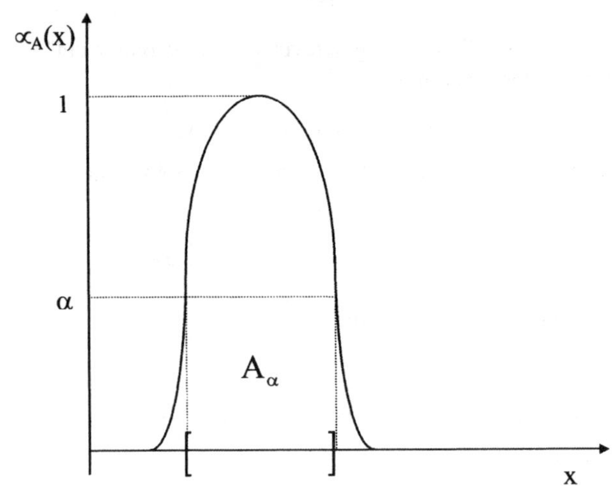

Fig. 2.5. Convex fuzzy set

A fuzzy set shown in Fig. 2.6 is not convex because it does not satisfy condition (2.4).

The α-cut given in Fig. 2.6 illustrates that some α-cuts are not convex.

Definition. The cardinality of a fuzzy set
When X is a finite set, the scalar cardinality |A| of a fuzzy set A on X is defined as

$$|A| = \sum_{x \in A} \propto_A(x).$$

Sometimes $|A|$ is called the power of A.

$\|A\| = |A|/|X|$ is the relative cardinality.

When X is infinite, |A| is defined as $|A| = \int_X \propto_A(x)\,dx$.

Example

$A = 0/1 + 0.3/2 + 0.6/3 + 0.9/4 + 1.0/5 + 0.7/6 + 0.4/7 + 0.2/8 + 0.1/9 + 0/10$

For the fuzzy set A the cardinality is

$|A| = 0.3+0.6+0.9+1.0+0.7+0.4+0.2+0.1=4.2$

Its relative cardinality is

$$\|A\| = \frac{4.2}{10} = 0.42 .$$

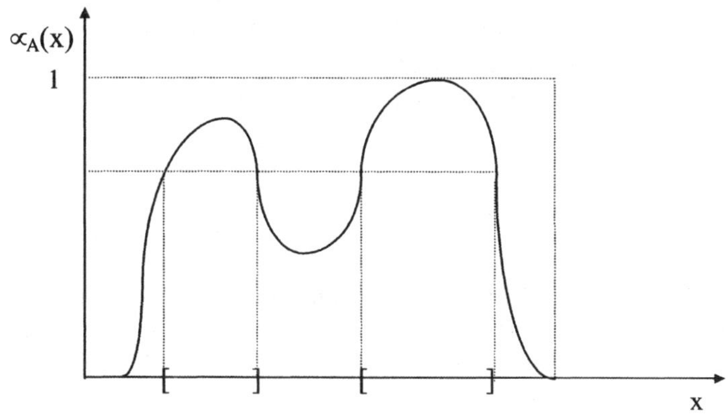

Fig. 2.6. Nonconvex fuzzy set

Definition. Fuzzy set inclusion

Given fuzzy sets A,B∈F(X) A is said to be included in B (A⊆B) or A is a subset of B if ∀x∈X, $\propto_A(x) \le \propto_B(x)$.

When the inequality is strict, the inclusion is said to be strict and is denoted as A<B.

It was mentioned above that each fuzzy set is uniquely defined by a membership function.

In the literature one can find different ways in which membership functions are represented.

List representation

If universal set $X = \{x_1, x_2, \ldots, x_n\}$ is a finite set, membership function of a fuzzy set A on X $\propto_A(x)$ can be represented as table. Such table lists all elements in the universe X and the corresponding membership grades as shown below

$$A = \propto_A(x_1)/x_1 + \ldots + \propto_A(x_n)/x_n = \sum_{i=1}^{n} \propto_A(x_i)/x_i .$$

Here symbol / (slash) does not denote division, it is used for correspondence between an element in the universe X (after slash) and its membership grade in the fuzzy set A (before slash). The symbol + connects the elements (does not denote summation).

If X is a finite set then

$$A = \int_X \propto_A(x)/x \ .$$

Here symbol $\int_X$ is used for denoting a union of elements of set X.

Example

An operator in an enterprise wants to classify the quality of fruits production. The values at the variable "Quality" and elements of the universe $X = \{0,1,2,\ldots,10\}$ are given in Table 2.1.

Fuzzy set $A \cong$ highest from the Table 2.1 may be represented as

$$A \cong \text{highest} = 0.1/0 + 0.1/1 + 0.2/2 + 0.3/3 + 0.4/4 + 0.5/5 + \\ + 0.6/6 + 0.7/7 + 0.8/8 + 0.9/9 + 1.0/10 \ .$$

Graphical representation

Graphical description of a fuzzy set A on the universe X is suitable in case when X is one or two-dimensional Euclidean space. Simple typical shapes of membership functions are usually used in fuzzy set theory and practice (see Table 2.2).

Table 2.1. The values of the variable "Quality"

Quality	$x \in X$
Very low	0
Almost low	1
Low	2
Better than low	3
Almost average	4
Average	5
Better than average	6
Almost high	7
High	8
Better than high	9
Highest	10

Example

Assume, for instance, a group of women aged near 30. In this case we may define intuitively fuzzy sets A_1=*"very young"*, A_2=*"young"*, A_3=*"yet young"*, A_4=*"young for somebody"* as shown in Fig. 2.7.

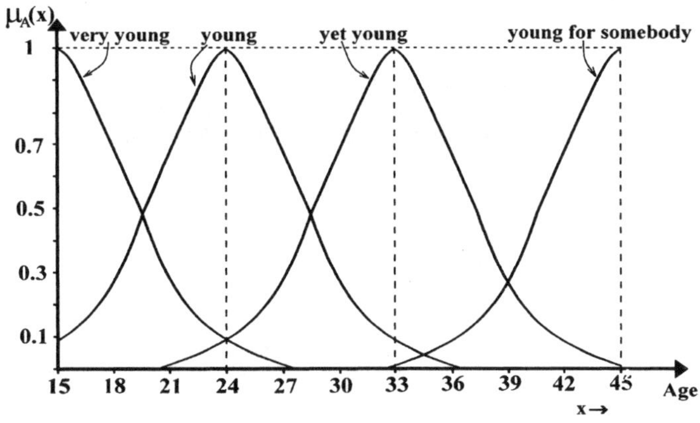

Fig. 2.7. Graphical representation of membership function

For comparison tabular representations of same fuzzy sets $A_i, i=\overline{1,4}$ are given below

$A_1 \cong \{1.0/15+0.7/18+0.3/21+0.1/24+0.0/30+...+0.0/45\}$

$A_2 \cong \{0.1/15+0.3/18+0.7/21+1.0/24+0.7/27+0.3/30+0.1/33+$
$+0.0/36+...+0.0/45\}$

$A_3 \cong \{0.0/15+\ldots+0.0/21+0.1/24+0.3/27+0.7/30+1.0/33+$
$+0.7/36+0.3/39+0.1/42+0.0/45\}$

$A_4 \cong \{0.0/15+\ldots+0.0/33+0.1/36+0.3/39+0.7/42+1.0/45\}$

where the numbers before the slashes are values of the corresponding membership function; the numbers after the slashes are arguments of the corresponding membership function which makes up a discrete universe, i.e.

$$\propto_{A_1}(x)=\{1.0,0.7,0.3,0.1,0.0,\ldots,0.0\}$$
$$\propto_{A_2}(x)=\{0.1,0.3,0.7,1.0,0.7,0.3,0.1,\ldots,0.0\}$$
$$\propto_{A_3}(x)=\{0.0,\ldots,0.0,0.1,0.3,0.7,1.0,0.7,0.3,0.1,0.0\}$$
$$\propto_{A_4}(x)=\{0.0,\ldots,0.0,0.1,0.3,0.7,1.0\}$$

and the age intervals,

$$X=\{15,18,31,24,27,30,33,36,39,42,45\}.$$

Fuzzy n - cube representation

All fuzzy sets on universe X with n elements can be represented by points in the n-dimensional unit cube – n-cube.

Assume that universe X contains n elements $X=\{x_1,x_2,...,x_n\}$. Each element $x_i, i=\overline{1,n}$ can be viewed as a coordinate in the n dimensional Euclidean space. A subset of this space for which values of each coordinate are restricted in [0,1] is called n-cube. Vertices of the cube, i.e. bit list (0,1,…,0) represent crisp sets. The points inside the cube define fuzzy subsets.

Example

Assume that n=2. Fig. 2.8 shows the fuzzy 2-cube or square. There are four crisp sets: (0,0); (1,1);(1,0);(0,1). The point A=(0.6,0.7) denotes the fuzzy subset.

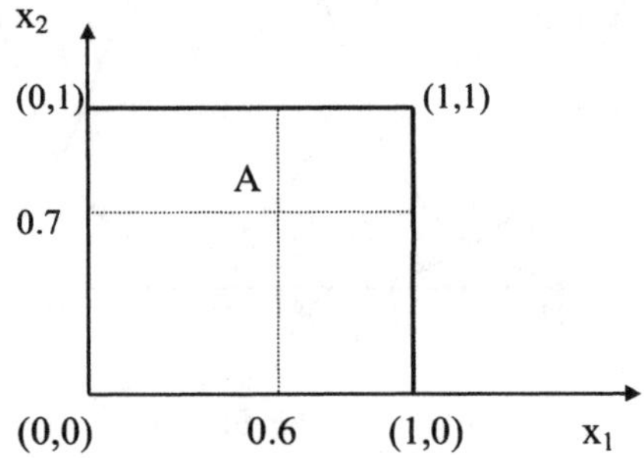

Fig. 2.8. Fuzzy 2-cube

Analytical representation

In case if universe X is infinite, it is not effective to use the above considered methods for representation of membership functions of a fuzzy sets. In this case it is suitable to represent fuzzy sets in an analytical form, which describes the shape of membership functions.

There are some typical formulas describing frequently used membership functions in fuzzy set theory and practice.

For example, bell-shaped membership functions often are used for representation of fuzzy sets. These functions are described by the formula:

$$\propto_A(x) = c \cdot \exp\left(-\frac{(x-a)^2}{b}\right)$$

which is defined by three parameters, a, b and c (Fig. 2.9).

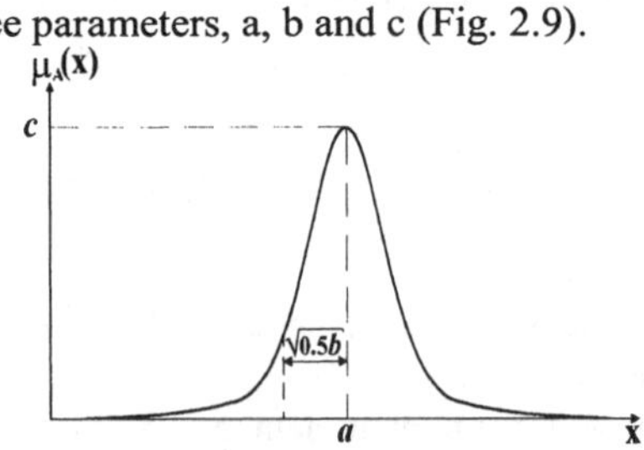

Fig. 2.9. Bell - shaped membership function

The triangular membership functions are frequently used in fuzzy arithmetic's for describing fuzzy numbers. These membership functions can be represented by the formula:

$$\propto_A(x) = \begin{cases} \frac{x-a_1}{a_2-a_1} r, & \text{if } a_1 \le x \le a_2 \\ \frac{a_3-x}{a_3-a_2} r, & \text{if } a_2 \le x \le a_3 \\ 0, & \text{otherwise} \end{cases}$$

which is fully defined by the four parameters a_1, a_2, a_3 and r (Fig. 2.10).

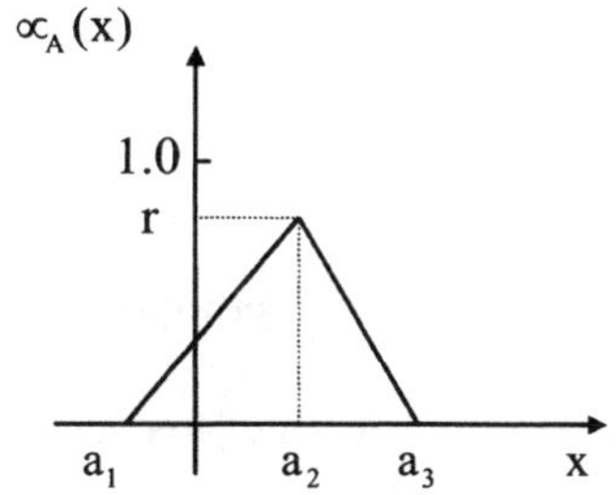

Fig. 2.10. A triangular membership functions

Trapezoidal membership functions are also quite useful. These functions may be described by the formula:

$$\propto_A(x) = \begin{cases} \frac{x-a_1}{a_2-a_1} r, & \text{if } a_1 \le x \le a_2 \\ r, & \text{if } a_2 \le x \le a_3 \\ \frac{a_4-x}{a_4-a_3} r, & \text{if } a_3 \le x \le a_4 \\ 0, & \text{otherwise} \end{cases}$$

which is fully defined by the five parameters, a_1, a_2, a_3, a_4 and r (Fig. 2.11).

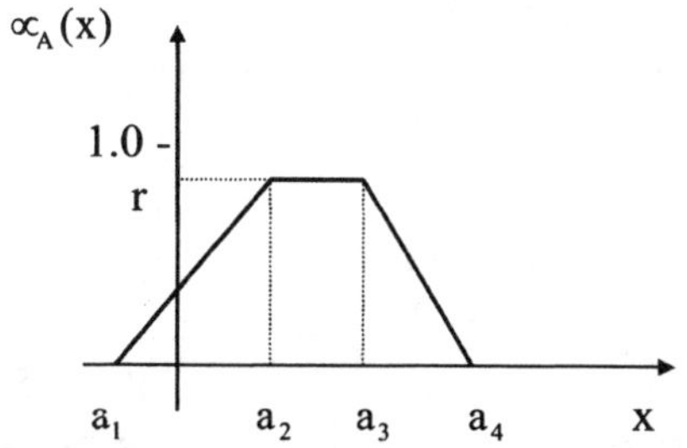

Fig. 2.11. A trapezoidal membership function

In general it is effective to represent the important typical membership functions by a parametrized family of functions. The following are formulas for describing the 6 families of membership functions

$$\propto_A(x, c_1) = [1 + c_1(x-a)^2]^{-1} \tag{2.5}$$

$$\propto_A(x, c_2) = [1 + c_2|x-a|]^{-1} \tag{2.6}$$

$$\propto_A(x, c_3, d) = [1 + c_3|x-a|^d]^{-1} \tag{2.7}$$

$$\propto_A(x, c_4, d) = \exp[-c_4|x-a|^d] \tag{2.8}$$

$$\propto_A(x, c_5) = \max\{0, [1 - c_5|x-a|]\} \tag{2.9}$$

$$\propto_A(x, c_6) = c_6 \exp\left[-\frac{(x-a)^2}{b}\right] \tag{2.10}$$

Here $c_i > 0, i = \overline{1,6}, d > 1$ are parameters, a denotes the elements of corresponding fuzzy sets with the membership grade equal to unity.

Table 2.2 summarizes the graphical and analytical representations of frequently used membership functions (MF).

The problem of constructing membership functions is problem of knowledge engineering.

There are many methods for estimation of membership functions. They can be classified as follows:

1. Membership functions based on heuristics.
2. Membership functions based on reliability concepts with respect to the particular problem.
3. Membership functions based on more theoretical demand.
4. Membership functions as a model for human concepts.
5. Neural networks based construction of membership functions.

The heuristic methods for estimation of the membership use partial information about it, such as the values that membership function takes at a finite number of samples in the universe of discourse.

For example, in order to build the membership function of the fuzzy set A="young" or "old" the following formula is used:

$$\propto_{young}(x) = \begin{cases} 0, & \text{if } x \le 25 \\ 1/(1+((x-25)/5)^{-2}), & \text{if } x > 25 \end{cases}$$

$$\propto_{old}(x) = \begin{cases} 0, & \text{if } x < 50 \\ 1/(1+((x-50)/5)^{-2}), & \text{if } x \ge 50 \end{cases}$$

Table 2.2. Typical membership functions

Type of Membership function	Craphical Representation	Analytical Representation
Triangular MF	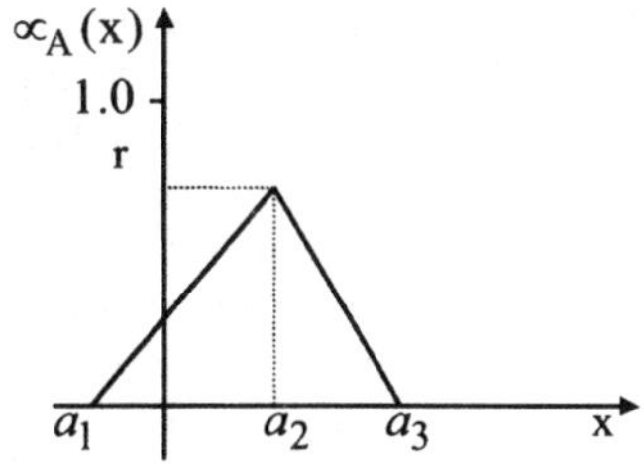	$\propto_A(x)=\begin{cases}\frac{x-a_1}{a_2-a_1}r, & \text{if } a_1\le x\le a_2,\\ \frac{a_3-x}{a_3-a_2}r, & \text{if } a_2\le x\le a_3,\\ 0, & \text{otherwise}\end{cases}$
Trapezoidal MF	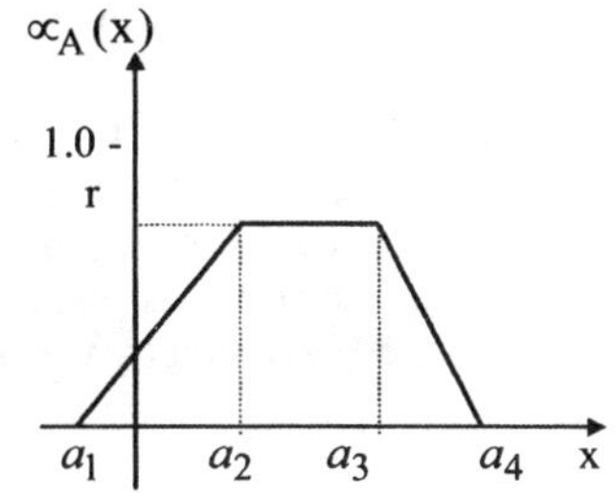	$\propto_A(x)=\begin{cases}\frac{x-a_1}{a_2-a_1}r, & \text{if } a_1\le x\le a_2,\\ r, & \text{if } a_2\le x\le a_3,\\ \frac{a_4-x}{a_4-a_3}r, & \text{if } a_3\le x\le a_4,\\ 0, & \text{otherwise}\end{cases}$
S-shaped MF	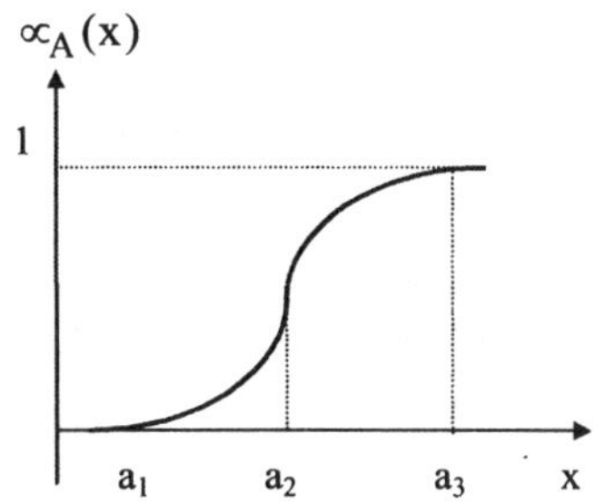	$\propto_A(x)=\begin{cases}0, & \text{if } x\le a_1,\\ 2\left(\frac{x-a_1}{a_3-a_1}\right)^2, & \text{if } a_1<x<a_2,\\ 1-2\left(\frac{x-a_1}{a_3-a_1}\right)^2, & \text{if } a_2\le x<a_3,\\ 1, & \text{if } a_3\le x\end{cases}$
Bell-shaped MF	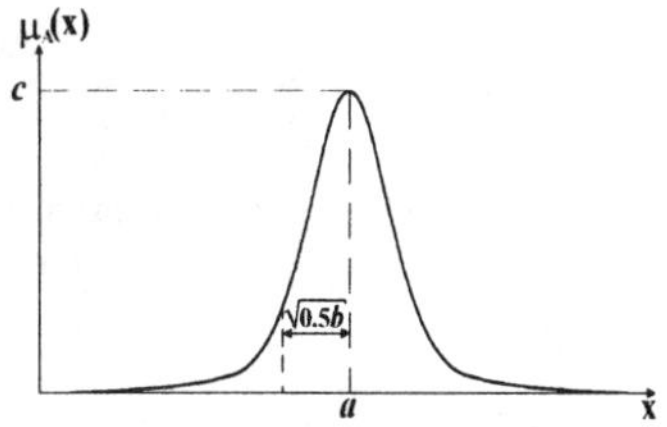	$\propto_A(x)=c\cdot\exp\left(-\frac{(x-a)^2}{b}\right)$

When membership function is estimated with respect to the particular problem, i.e. for fuzzy mathematical programming problem some of its parameters are chosen relative to particular problem. Most useful membership function in this area is linear or piecewise linear function with relativity parameters.

For example, $\alpha(x) = 1 - x/a,\ \ x \in [0,a]$, where the parameter a is chosen with respect to given problem.

The estimation methods of membership functions based on more theoretical demand use axioms, probability density functions and so on.

The construction of membership functions is of high relevance and importance to the field of control systems. Two different directions may be considered: the first defines the functions and identifies the system parameters; the other works with a given system and identifies the membership function under the control process.

Saaty method is referred to as estimation methods which considered membership functions as a model for human concepts. Its content consists of the following. Let $w_i>0$, i=1,..,n be the membership degrees of the n numbers. Then the matrix of relative weights whose ij-th element is w_i/w_j, is formed. The vector $(w_1,w_2,...,w_n)^T$ is an eigenvector corresponding to the largest eigenvalue. All the other eigenvalues are zero.

The idea is to estimate the matrix of relative weights and then obtain an estimate of the vector $(w_1,w_2,...,w_n)^T$ as an eigenvector corresponding to the largest eigenvalue of the relative weight matrix. To compare a set of n object in pairs according to their relative weights the objects are denoted as $A_1,...,A_n$ and their weights by $w_1,...,w_n$. The pair wise comparison is represented by the matrix

$$
\begin{array}{cccccc}
 & A_1 & A_2 & .. & A_n & \\
A = & \left\| \begin{array}{c} w_1/w_1 \\ w_2/w_1 \\ \cdots \\ w_n/w_1 \end{array} \right. & \begin{array}{c} w_1/w_2 \\ w_2/w_2 \\ \cdots \\ w_n/w_2 \end{array} & \begin{array}{c} \cdots \\ \cdots \\ \\ \cdots \end{array} & \left. \begin{array}{c} w_1/w_n \\ w_2/w_n \\ \cdots \\ w_n/w_n \end{array} \right\| & \begin{array}{c} A_1 \\ A_2 \\ \cdots \\ A_n \end{array}
\end{array}
$$

This matrix, called a reciprocal matrix, has positive entries everywhere and satisfies the reciprocal property $a_{ji} = 1/a_{ij}$. Multiplying this matrix by the vector $W=(w_1,...,w_n)^T$, we have

$$AW = nW \tag{2.11}$$

or

$$(A - nI)W = 0$$

This is a system of homogeneous linear equations which has a non-trivial solution iff the determinant of (A-nI) vanishes, that is, n is an eigenvalue of A. The matrix A is also consistent, that is,

$$a_{ik} = a_{ik} / a_{ij} .$$

In general case, the precise values of $w_i\ /w_j$ are not known and must be estimated. Since the eigenvalues are perturbed by a small change in coefficients, equation (2.11) becomes

$$A'W' = \lambda^{max} W' \tag{2.12}$$

where λ^{max} is the largest eigenvalue of A'. To simplify the notation, equation (2.12) is expressed in the form of

$$AW = \lambda^{max} W$$

where A is Saaty's matrix of pair wise comparisons. The eigenvector associated with the largest eigenvalue is the desired vector of weights.

Very effective way for constructing membership functions is to use neural networks by learning from examples. This problem will be discussed in section 2.2.

The extension principle of Zadeh is an important tool by which it is possible to fuzzify classical mathematical theories.

Definition. Extension principle
A principle for fuzzifying crisp function is called extention principle.

Let f be a mapping from X to a universe $Y, f: X \to Y$, i.e.

$$y = f(x), x \in X, y \in Y.$$

Let instead $x \in X$ a fuzzy quantity A = about x on X is given. It is necessary to find the fuzzy image by f of the fuzzy argument A, i.e. $B = f(A)$. The membership values of B should be determined by the membership values of A. If f is a one-to-one mapping we can take

$$\propto_B(y) = \propto_A(f^{-1}(y)) = \propto_A(x) \qquad (2.13)$$

where f^{-1} is the inverse of f.

If f is not one-to-one mapping, then in accordance to Zadeh

$$\propto_B(y) = \sup_{x|y=f(x)} \propto_A(x). \qquad (2.14)$$

Graphical representation of extension principle is shown in Fig. 2.12.

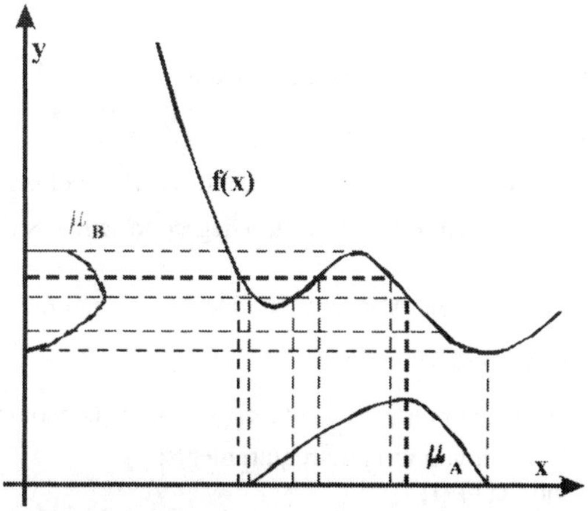

Fig. 2.12. Graphical representation of extension principle

Example
Let

$$f(x) = x^4$$

$$A=\{(-2,0.3),(-1,0.6),(0,0.8),(1,1),(2,0.5)\} .$$

By applying the (2.14) we obtain

$$\propto_B(y) = \{(0,0.8),(1,1),(16,0.5) \quad .$$

In general for n-ary function f, i.e. when $X = X_1 \cdot X_2 \cdot ...X_n$ and the argument A is Cartesian product $A = A_1 \cdot A_2 \cdot ...A_n$ in accordance with the extension principle of Zadeh we can write

$$\propto_B(y) = \sup_{\substack{(x_1x_2,...,x_n)\in X_1\cdot X_2\cdot ...X_n \\ y=f(x_1x_2,...,x_n)}} \min(\propto_{A_1}(x_1),...,\propto_{A_n}(x_n)) \tag{2.15}$$

There exist three standard fuzzy operations: fuzzy intersection, union and complement which are generalization of the corresponding classical set operations.

Let's A and B be two fuzzy sets in X with the membership functions $\propto_A$ and $\propto_B$, respectively. Then the operations of intersection, union and complement are defined as given below.

Definition. Fuzzy standard intersection and union
The intersection ($\cap$) and union ($\cup$) of fuzzy sets A and B can be calculated by the following formulas:

$$\forall x \in X \quad \propto_{A\cap B}(x) = \min(\propto_A(x), \propto_B(x))$$

$$\forall x \in X \quad \propto_{A\cup B}(x) = \max(\propto_A(x), \propto_B(x)),$$

where $\propto_{A\cap B}(x)$ and $\propto_{A\cup B}(x)$ are the membership functions of $A\cap B$ and $A\cup B$, respectively.

Definition. Standard fuzzy complement
The complement $\overline{A}$ of A is defined by the membership function:

$$\forall x \in X \ \propto_{\overline{A}}(x) = 1 - \propto_A(x).$$

As already mentioned $\propto_A(x)$ is interpreted as the degree to which x belongs to A. Then by the definition $\propto_{\overline{A}}(x)$ can be interpreted as the degree to which x does not belong to A.

In general, fuzzy complement of A type c A^c may be defined by a function

$$c:[0,1] \to [0,1],$$

which assigns a value $A^c(x)$ to each membership grade $\propto_A(x)$. As it is shown in [52] function c must satisfy the following axiomatic requirements:

c(0)=1 and c(1)=1;
for all a,b$\in$[0,1], if a$\leq$b, then c(a)$\geq$c(b);
c is a continuous function;

$$c(c(a))=a \quad \text{for each } a \in [0,1].$$

For example the threshold - type complements

$$c(a) = \begin{cases} 1, & \text{if } a \le t \\ 0, & \text{if } a > t \end{cases}$$

are general fuzzy complements. Here $t \in [0,1]$ is the threshold of c.

The standard fuzzy operations do not satisfy the law of excluded middle $A \cup \overline{A} = X$ and the law of contradiction $A \cap \overline{A} = \varnothing$ of classical set theory. But commutativity, associativity, idempotency, distributivity, and De Morgan laws are held for standard fuzzy operations.

For fuzzy union, intersection and complement operations there exist a broad class of functions. Function that qualify as fuzzy intersections and fuzzy unions are defined as t-norms and t-conorms.

Definition. t-norms

t-norm is a binary operation in [0,1], i.e. a binary function t from [0,1] into [0,1] that satisfies the following axioms

$$t(\propto_A(x),1) = \propto_A(x); \tag{2.16}$$

$$\text{if } \propto_A(x) \le \propto_C(x) \text{ and } \propto_B(x) \le \propto_D(x) \tag{2.17}$$

$$\text{then } t(\propto_A(x), \propto_B(x)) \le t(\propto_C(x), \propto_D(x));$$

$$t(\propto_A(x), \propto_B(x)) = t(\propto_B(x), \propto_A(x)); \tag{2.18}$$

$$t(\propto_A(x), t(\propto_B(x), \propto_C(x))) = t(t(\propto_A(x), \propto_B(x), \propto_C(x))). \tag{2.19}$$

Here (2.16) is boundary condition, (2.17)-(2.19) are conditions of monotonicity, commutativity and associativity, respectively.

The function t takes as its arguments the pair consisting of the element membership grades in set A and in set B, and yields membership grades of the element in the $A \cap B$

$$A \cap B(x) = t[A(x), B(x)] \quad \forall x \in X$$

The following are frequently used t-norm-based fuzzy intersection operations:

Standard intersection

$$t_0(\propto_A(x), \propto_B(x)) = \min\{\propto_A(x), \propto_B(x)\} \tag{2.20}$$

Algebraic product

$$t_1(\propto_A(x), \propto_B(x)) = \propto_A(x) \cdot \propto_B(x) \tag{2.21}$$

Bounded difference

$$t_2(\propto_A(x), \propto_B(x)) = \propto_{A \cap B}(x) = \max(0, \propto_A(x) + \propto_B(x) - 1) \quad (2.22)$$

The intersection of A and B in accordance to (2.22) is

$A \cap B$=0/1+0/2+0/3+0/4+0/5+0.1/6+0.13/7+0.13/8+0.11/9+0.1/10+

+0.11/11+0.16/12+0.16/13+0.11/14+0/15+0/16+0/17+0/18+0/19

Drastic intersection

$$t_3(\propto_A(x), \propto_B(x)) = \begin{cases} \min\{\propto_A(x), \propto_B(x)\}, & \text{if } \propto_A(x) = 1 \\ & \text{or } \propto_B(x) = 1 \\ 0, & \text{otherwise} \end{cases} \quad (2.23)$$

For four fuzzy intersections the following is true

$$t_3(\propto_A(x), \propto_B(x)) \le t_2(\propto_A(x), \propto_B(x)) \le t_1(\propto_A(x), \propto_B(x)) \le t_0(\propto_A(x), \propto_B(x)) \quad (2.24)$$

Definition. t-conorms

t-conorm is a binary operation in [0,1], i.e. a binary function $S:[0,1]\cdot [0,1] \to [0,1]$ that satisfies the following axioms

$$S((\propto_A(x), 0) = \propto_A(x) \qquad \text{(boundary condition)} \quad (2.25)$$

If $\propto_A(x) \le \propto_C(x)$ and $\propto_B(x) \le \propto_D(x)$ then

$$S(\propto_A(x), \propto_B(x)) \le S(\propto_C(x), \propto_D(x)); \qquad \text{(monotonicity)} \quad (2.26)$$

$$S(\propto_A(x), \propto_B(x)) = S(\propto_B(x), \propto_A(x)); \qquad \text{(commutativity)} \quad (2.27)$$

$$S(\propto_A(x), S(\propto_B(x), \propto_C(x))) = S(S(\propto_A(x), \propto_B(x), \propto_C(x))) \qquad \text{(associativity).} \quad (2.28)$$

The function S yields membership grade of the element in the set $A \cup B$ on the argument which is pair consisting of the same elements membership grades in set A and B

$$(A \cup B)(X) = S[A(x), B(x)] \quad (2.29)$$

The following are frequently used t-conorm based fuzzy union operations.

Standard union

$$S_0(\propto_A(x), \propto_B(x)) = \max\{\propto_A(x), \propto_B(x)\} \quad (2.30)$$

Algebraic sum

$$S_1(\propto_A(x), \propto_B(x)) = \propto_A(x) + \propto_B(x) - \propto_A(x) \cdot \propto_B(x) \quad (2.31)$$

Bounded sum

$$S_2(\propto_A(x),\propto_B(x)) = \min\{1,\propto_A(x)+\propto_B(x)\} \tag{2.32}$$

Drastic union

$$S_3(\propto_A(x),\propto_B(x)) = \begin{cases} \max\{\propto_A(x),\propto_B(x)\}, & \text{if } \propto_A(x)=0 \\ & \text{or } \propto_B(x)=0 \\ 1, & \text{otherwise} \end{cases} \tag{2.33}$$

For four fuzzy union operations the following is true

$$S_0(\propto_A(x),\propto_B(x)) \le S_1(\propto_A(x),\propto_B(x)) \le S_2(\propto_A(x),\propto_B(x)) \le \\ \le S_3(\propto_A(x),\propto_B(x)) \tag{2.34}$$

Parametrized families of t-norms and t-conorms

Different authors suggested parametrized families of t-norms and t-conorms in order to adapt them to context in which they are used. The following are examples of some parametrized t-norms and t-conorms.

Yager operation

$$\propto_{A\cap B}(x) = 1-\min\left\{1,((1-\propto_A(x))^\gamma+(1-\propto_B(x))^\gamma)^{\frac{1}{\gamma}}\right\} \quad \gamma\ge 1 \tag{2.35}$$

$$\propto_{A\cup B}(x) = \min\left\{1,((\propto_A(x))^\gamma+(\propto_B(x))^\gamma)^{\frac{1}{\gamma}}\right\} \quad \gamma\ge 1 \tag{2.36}$$

Dubois-Prade operator

$$\propto_{A\cap B}(x) = \frac{\propto_A(x)\cdot\propto_B(x)}{\max\{\propto_A(x),\propto_B(x),\gamma\}} \tag{2.37}$$

$$\propto_{A\cup B}(x) = \frac{\propto_A(x)+\propto_B(x)-\propto_A(x)\cdot\propto_B(x)-\min\{\propto_A(x),\propto_B(x),1-\gamma\}}{\max\{(1-\propto_A(x)),(1-\propto_B(x)),\gamma\}} \tag{2.38}$$

Aggregation operations

By this type of operations n $(n \ge 2)$ fuzzy sets are combined to produce a single fuzzy set.

Definition. Aggregation operation

Aggregation operation on n fuzzy sets is defined by a function

$$1:[0,1]^n \to [0,1].$$

Let's given fuzzy sets $A_1, A_2, ..., A_n$ defined on X. Function l produces an aggregate fuzzy set A by corresponding operating on membership functions of these sets for each $x \in X$

$$A(x) = l(A_1(x),\square\ , A_n(x)).$$

Function l must satisfy boundary conditions, be monotonic increasing and continuous function. The following are examples of aggregation operations.

Generalized means

$$l_\gamma(\propto_{A_1}, \propto_{A_2}, ..., \propto_{A_n}) = \left(\frac{\propto_{A_1}^\gamma + \propto_{A_2}^\gamma + ... + \propto_{A_n}^\gamma}{n} \right)^{\frac{1}{\gamma}} \tag{2.39}$$

where $\gamma \in R$ $(\gamma \neq 0)$ is the parameter that indicates the type of different means. When γ=1 (arithmetic mean)

$$l_1(\propto_{A_1}, \propto_{A_2}, ..., \propto_{A_n}) = \frac{1}{n}(\propto_{A_1} + \propto_{A_2} + ... + \propto_{A_n}). \tag{2.40}$$

Ordered weighted averaging operations

$$l_N(\propto_{A_1}, \propto_{A_2}, ..., \propto_{A_n}) = w_1 \propto_{B_1} + w_1 \propto_{B_2} + ... + w_n \propto_{B_n}), \tag{2.41}$$

where $\sum_{i=1}^{n} w_i = 1$, $\propto_{B_i}$ is the largest element in $\propto_{A_1}, \propto_{A_2}, ..., \propto_{A_n}$.

Other operations on fuzzy sets

Definition. Cartesian product of fuzzy sets

The Cartesian product of fuzzy sets $A_1, A_2, ..., A_n$ on universes $X_1, X_2, ..., X_n$ respectively is a fuzzy set in the product space $X_1 \cdot X_2 \cdot ... \cdot X_n$ with the membership function

$$\propto_{A_1 \cdot A_2 \cdot ... A_n}(x) = \min\{\propto_{A_i}(x_i) | x = (x_1, x_2, ..., x_n), \ x_i \in X_i\} \tag{2.42}$$

Definition. Power of fuzzy sets

m-th power of a fuzzy set A^m is defined as

$$\propto_{A_m}(x) = [\propto_A(x)]^m, \ \ \forall x \in X, \ \ \forall m \in R^+ \tag{2.43}$$

where R^+ is positively defined set of real numbers.

Definition. Concentration and dilation of fuzzy sets

Let A be fuzzy set on the universe:

$$A = \{x, \propto_A(x) / x \in X\}$$

Then the operator $Con_m A = \{(x, [\propto_A(x)]^m)/x \in X\}$ is called concentration of A and the operator $Dil_n A = \{(x, \sqrt{\propto_A(x)})/x \in X\}$ is called dilation of A.

The examples of conversions of fuzzy sets using the concentration and dilation operations are shown below

$$A \quad = \int \propto_A(x)/x$$

$$\text{Very } A \quad = \int [\propto_A(x)]^2/x$$

$$\text{Very very A} \quad = \int [\propto_A(x)]^4 / x$$

$$\text{More or less A} \quad = \int \sqrt{\propto_A(x)} / x$$

$$\text{A little A} \quad = \int \sqrt[4]{\propto_A(x)} / x$$

$$\text{Not A} \quad = \int (1 - \propto_A(x) / x$$

$$\text{Not very much} \quad = \int (1 - (\propto_A(x))^2) / x$$

Definition. Difference of fuzzy sets
Difference of fuzzy sets is defined by the formula:

$$\forall x \in X, \propto_{A|-|B}(x) = \max(0, \propto_A(x) - \propto_B(x)). \tag{2.44}$$

A |−| B is the fuzzy set of elements that belong to A more than to B.

Symmetrical difference of fuzzy sets A and B is the fuzzy set $A \nabla B$ of elements that belong more to A than to B:

$$\forall x \in X \quad \propto_{A\nabla B}(x) = |\propto_A(x) - \propto_B(x)| \tag{2.45}$$

2.1.2 Mathematics of Fuzzy Computing

Definition. Fuzzy topology
Let X be non-empty set. It is said that in the set X is defined fuzzy topological structure or simply fuzzy topology, if in X is indicated the class $F\mathfrak{J}$ of subsets of the set X, which satisfies three conditions:

1. $\varnothing, X \in F\mathfrak{J}$ (or $\chi_\varnothing, \chi_X \in F\mathfrak{J}$);

2. $\bigcup_{i=1}^{n} A_i \in F\mathfrak{J}$, if $A_i \in F\mathfrak{J}$ for all i; χ

3. $\bigcap_{i=1}^{n} A_i \in F\mathfrak{J}$ if $A_i \in F\mathfrak{J}$.

If $A \subseteq X$, then χ_A denotes the characteristic function of A in X, that is for A is given $\chi_A : X \to \{0,1\}$.

If in the set X fuzzy topology $F\mathfrak{J}$ is given then set X is called fuzzy topological space and denoted $(X, F\mathfrak{J})$. The members of $F\mathfrak{J}$ is called a $F\mathfrak{J}$ - open fuzzy subset.

The next basic concept is neighborhood.

Definition. Neighborhood
Let $(X, F\mathfrak{J})$ be a fuzzy topological space. It is said that a fuzzy subset U of X is a neighborhood (nbhd) of a fuzzy subset A if there exists an open fuzzy subset A' of X such that $A \subseteq A' \subseteq U$, U is nhbd of A.

Now we consider a fuzzy continuous function (F-continuous functions).

Definition. F-continuous function

Assume we have fuzzy topological spaces (X, FJ) and (Y, FU). A function f from X into Y is called F -continuous, if pre-image of any open subset of Y is an open subset of X, in other words, if fuzzy subset f^{-1} (B) is open in X every time, when B is open in Y.

Fuzzy geometry

In general, fuzzy geometry may be considered as extension of conventional geometry to the fuzzy case [27,69,87,89-91,97]. Fuzzy geometry includes the topological concepts of area, perimeter, compactness, length, adjacency etc. These measures can be used to reflect the spatial ambiguity in image distribution and analysis. In [97] the concept of fuzzy geometry and its application to image processing is described.

Definition. Fuzzy point

Fuzzy point $\widetilde{x}_0$ is a convex fuzzy subset of R^i. Fuzzy points in R is characterized by kernel x_0 whose precise location is only approximately known .

A crisp point $x_0 \in R^i$ is the kernel, from which membership function decreases in all directions monotonically [21]. In Fig. 2.13 and Fig. 2.14 fuzzy points with hyperpyramidal (Fig. 2.13) and hyperparaboloidal (Fig. 2.14) membership functions are shown. In first case imprecision of location of fuzzy point is expressed by intervals for the components, in second case by definite matrix in all directions of the space.

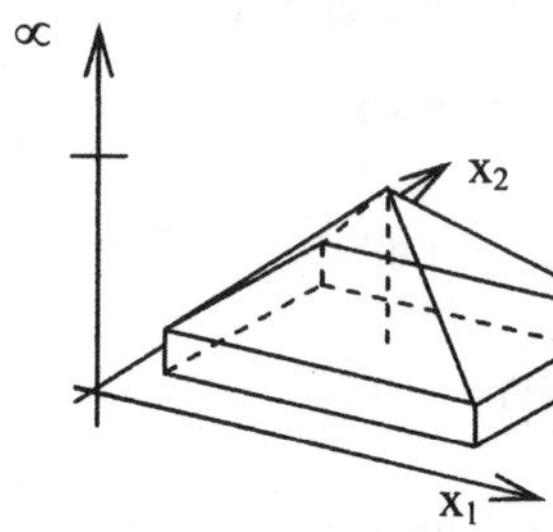

Fig. 2.13. Fuzzy points with hyperpyramidal membership

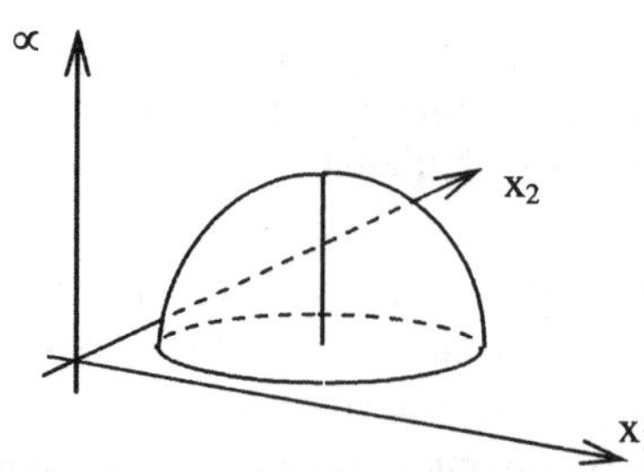

Fig. 2.14. Fuzzy points with hyperparaboloidal membership

Definition. Fuzzy interval
If fuzzy domain I of the real line R is bounded by two normalized convex fuzzy sets then it is called fuzzy interval. In Fig. 2.15 fuzzy interval with fuzzy ends $\propto_{\tilde{a}}(x)$ and $\propto_{\tilde{b}}(x)$ is given.

A crisp interval [a, b] is the kernel, from which the membership function decreases to zero [21].

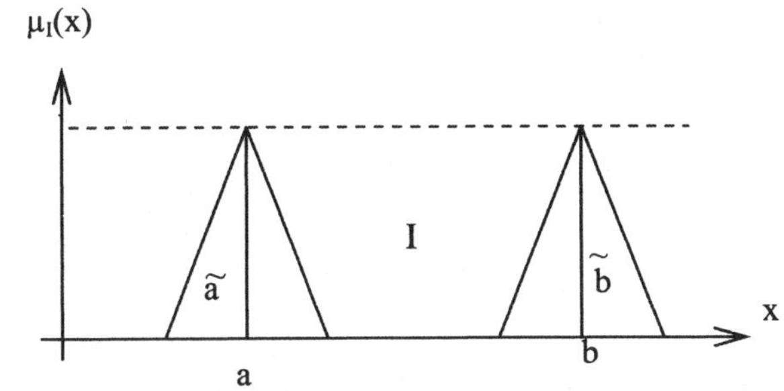

Fig. 2.15. Fuzzy interval

Analogously, fuzzy region in R^i is represented as a crisp region, which is surrounded by a fuzzy transition zone, in which the membership function decreases monotonically to zero [22].

Definition. Length of a fuzzy interval
Length of fuzzy interval I is defined as

$$L(I) = \int_{I_0} \propto_I(x)dx$$

Here $I_0 = \{x \mid \propto_I(x) > 0\}$ is support of fuzzy interval.

Definition. Distance between fuzzy points
A distance between two points $d(x_1, x_2)$, by using the extension principle translates to a fuzzy distance between fuzzy sets.

The fuzzy distance between two fuzzy sets A and B on X (X is metric space) is defined as [22,71].

$$\propto_{d(A,B)}(y) = \sup_{\substack{(x_1,x_2)\in X\cdot X \\ d(x_1,x_2)=y}} \min(\propto_A(x_1), \propto_B(x_2))$$

Example
Fuzzy distance in case when $X=R^1$, $d(x_1,x_2)=|x_1 - x_2|$ is shown in Fig. 2.16.

Some distances frequently used in practical problems are given below (for R^1):

$d(x_1, x_2) = (x_1 - x_2)^2$ Euclidean distance

$d(x_1,x_2)=(|x_1-x_2|^p)^{1/p}$ Minkowski metric

$d(x_1,x_2)=c|x_1-x_2|$ Tschebyscheff metric

$d(x_1,x_2)=|x_1-x_2|$ Hamming distance

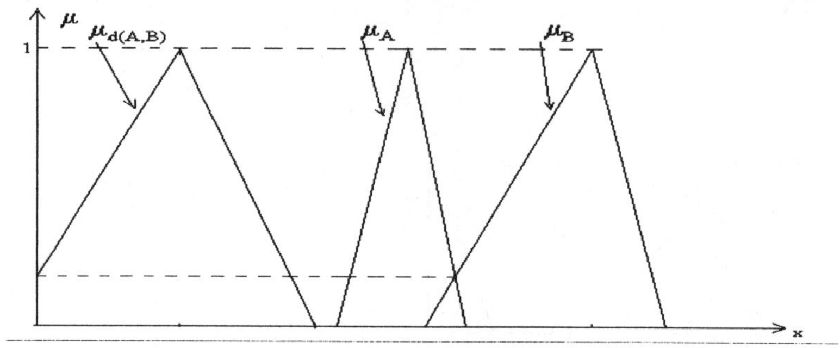

Fig. 2.16. Fuzzy distance between fuzzy sets

Definition. Fuzzy area

The area of fuzzy subset is defined as the area of fuzzy subset A given on R^2 is defined as

$$S(A)=\iint_{A_0} \propto_A(x,y)dxdy \tag{2.46}$$

Here $A_0=\{(x,y) \mid \propto_A(x,y)>0\}$ is support of fuzzy region A.

For fuzzy region, represented by piecewise membership function, the area is defined as [97]

$$S(A)=\sum_i \propto(i) \tag{2.47}$$

Definition. Perimeter

In case of fuzzy set A when $\propto_A$ is piecewise constant, the perimeter of fuzzy set A is defined as

$$P_A=\sum_{i,j,k} |\propto(i)-\propto(j)*L(i,j,k)| \tag{2.48}$$

Here $\propto(i)$ and $\propto(j)$ are the membership values of two adjacent regions, L(i,j,k) is length of a k-th arc of this regions.

Definition. Compactness

The compactness of a fuzzy set A with area S_A and perimeter P_A is defined as

$$C(A)=\frac{S_A}{P_A^2} \tag{2.49}$$

Definition. Length and Breadth of a fuzzy set
The length of a fuzzy set A is defined as

$$l(A) = \max_x \left\{ \int_x \propto_A (x, y) dy \right\}, \tag{2.50}$$

where the integral is taken over the region with $\propto_A (x, y) > 0$. For discrete case formula (2.50) takes form

$$l(A) = \max_x \left\{ \sum_y \propto_A (x, y) \right\} \tag{2.51}$$

The breadth of a fuzzy set A is defined as

$$b(A) = \max_y \left\{ \int_x \propto_A (x, y) dx \right\} \tag{2.52}$$

or

$$b(A) = \max_y \left\{ \sum_x \propto_A (x, y) \right\} \tag{2.53}$$

Definition. Adjacency
Let A and B be two piecewise constant fuzzy sets of X. Then adjacency between A and B is defined as

$$Ad(A, B) = \sum_{i,j,k} \propto_A (i) \propto_B (j) |L(i, j, k)| \tag{2.54}$$

Definition. Index of area coverage (IOAC)
IOAC of a fuzzy set A is defined as

$$IOAC(A) = \frac{S_A}{l(A) \cdot b(A)} \tag{2.55}$$

This index for fuzzy region represents the fraction of the maximum area (covered by the length and breadth of the region) actually covered by the region.

Fuzzy arithmetic

Fuzzy arithmetic is an important constituent of numerical analysis and main tool for dealing with uncertainly in a computationally effective way. The main concept and kernel of fuzzy arithmetic is fuzzy number [51].

Definition. Fuzzy number
A fuzzy number is a fuzzy set A on R which possesses the following properties:
a) A is a normal fuzzy set; b) A is a convex fuzzy set; c) α-cuts of A A^{α} are a closed interval for every $\alpha \in (0,1]$; d) the support of A A^{+0} is bounded.

In Fig. 2.17 some basic types of fuzzy numbers are shown. For comparison of a fuzzy number with a crisp number in Fig. 2.18 crisp number 2 is given.

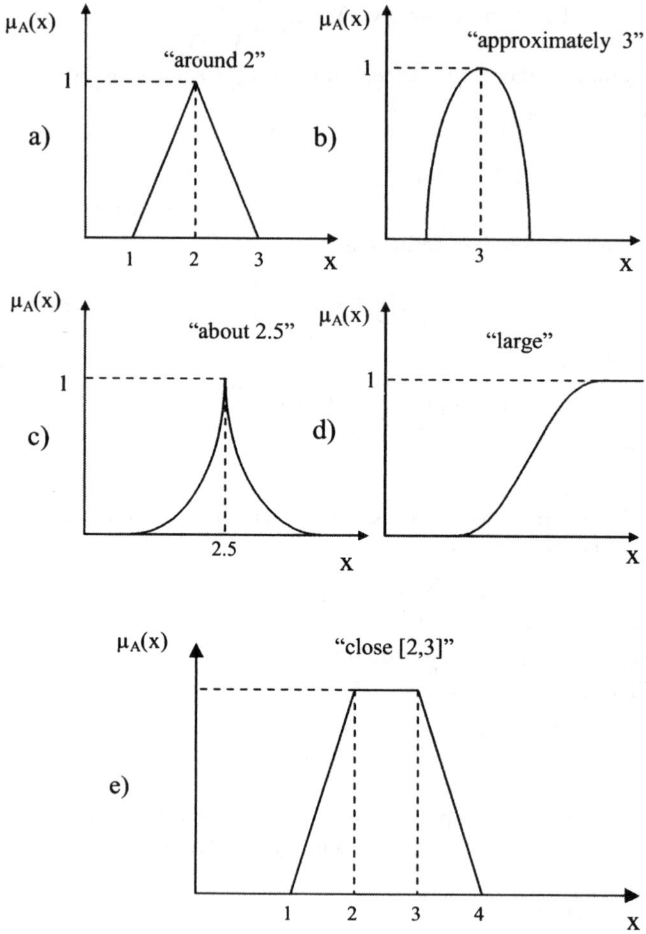

Fig. 2.17. Types of fuzzy numbers

Arithmetic operation on fuzzy numbers

There are different methods for developing fuzzy arithmetic. In this section we present three methods.

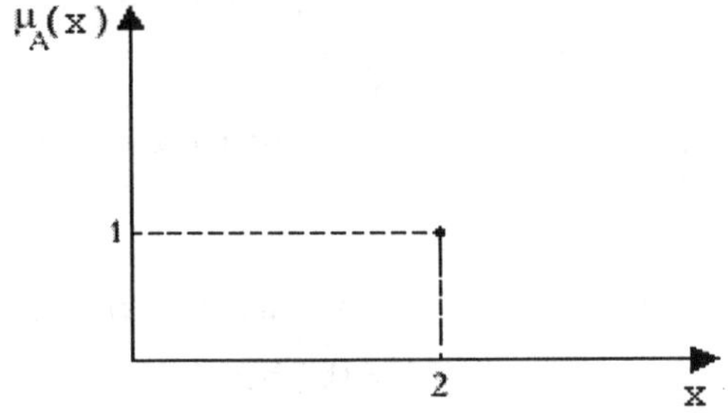

Fig. 2.18. Crisp number 2

Method based on the extension principle

By this method basic arithmetic operations on real numbers are extended to operations on fuzzy numbers. Let A and B be two fuzzy numbers and * denote any of four arithmetic operations $\{+, -, \cdot, :\}$.

A fuzzy set A*B on R can be defined by the equation

$$\forall z \in R \quad \mu_{(A*B)}(z) = \sup_{z=x*y} \min[\mu_A(x), \mu_B(y)] \tag{2.56}$$

It is shown in [52] that A*B is fuzzy number and the following theorem has been formulated and proved.

Theorem 2.2

Let $* \in \{+, -, \cdot, :\}$, and let A, B denote continuous fuzzy numbers. Then, the fuzzy set A*B defined by (2.56) is a continuous fuzzy number.

Then for four basic arithmetic operations on fuzzy numbers we can write

$$\mu_{(A+B)}(z) = \sup_{z=x+y} \min[\mu_A(x), \mu_B(y)] \tag{2.57}$$

$$\mu_{(A-B)}(z) = \sup_{z=x-y} \min[\mu_A(x), \mu_B(y)] \tag{2.58}$$

$$\mu_{(A \cdot B)}(z) = \sup_{z=x \cdot y} \min[\mu_A(x), \mu_B(y)] \tag{2.59}$$

$$\mu_{(A:B)}(z) = \sup_{z=x:y} \min[\mu_A(x), \mu_B(y)] \tag{2.60}$$

Method based on interval arithmetic and α-cuts

This method is based on representation of arbitrary fuzzy numbers by their α-cuts and use interval arithmetic to the α-cuts. Let $A, B \in R$ be fuzzy numbers and * denote any of four operations. For each $\alpha \in (0,1]$, the α-cut of $A * B$ is expressed as

$$(A * B)^\alpha = A^\alpha * B^\alpha \tag{2.61}$$

For $*$ we assume $0 \notin \text{supp}(B)$.
The resulting fuzzy number $A * B$ can be defined as (see section 2.1.1)

$$A * B = \bigcup_{\alpha \in [0,1]} \alpha (A * B)^{\alpha} \tag{2.62}$$

Next we using (2.61), (2.62) illustrate four arithmetic operations on fuzzy numbers.

Addition

Let A and B be two fuzzy numbers and A^{α} and B^{α} their α-cuts

$$A^{\alpha} = [a_1^{\alpha}, a_2^{\alpha}] \; ; \; B^{\alpha} = [b_1^{\alpha}, b_2^{\alpha}] \tag{2.63}$$

Then we can write

$$A^{\alpha} + B^{\alpha} = [a_1^{\alpha}, a_2^{\alpha}] + [b_1^{\alpha}, b_2^{\alpha}] = [a_1^{\alpha} + b_1^{\alpha}, a_2^{\alpha} + b_2^{\alpha}], \tag{2.64}$$

here

$$A^{\alpha} = \{x / \propto_A (x) \geq \alpha\} ; \quad B^{\alpha} = \{x / \propto_B (x) \geq \alpha\} \tag{2.65}$$

Subtraction

Subtraction of given fuzzy numbers A and B can be defined as

$$(A - B)^{\alpha} = A^{\alpha} - B^{\alpha} = [a_1^{\alpha} - b_2^{\alpha}, a_2^{\alpha} - b_1^{\alpha}], \forall \alpha \in [0,1] \tag{2.66}$$

We can determine (2.66) by addition of the image B^{-} to A

$$\forall \alpha \in [0,1], \qquad B^{\alpha^{-}} = [-b_2^{\alpha}, -b_1^{\alpha}] \tag{2.67}$$

Multiplication

Let two fuzzy numbers A and B be given. Multiplication $A \cdot B$ is defined as

$$(A \cdot B)^{\alpha} = A^{\alpha} \cdot B^{\alpha} = [a_1^{\alpha}, a_2^{\alpha}] \cdot [b_1^{\alpha}, b_2^{\alpha}] \qquad \forall \alpha \in [0,1] \tag{2.68}$$

Division

Division of two fuzzy numbers A and B is defined by

$$A^{\alpha} : B^{\alpha} = [a_1^{\alpha}, a_2^{\alpha}] : [b_1^{\alpha}, b_2^{\alpha}] \qquad \forall \alpha \in [0,1]$$

Fuzzy minimum and maximum of fuzzy numbers

The set R of real numbers has linear order. For $\forall x, y \in R$ the pair $(R, \leq)$ as mentioned above is a lattice. Fuzzy numbers do not have the structure of linear order, they can be ordered partially. In ordering of fuzzy numbers it is necessary to extend the lattice operation max and min to fuzzy minimum $\widetilde{\min}$ and fuzzy maximum $\widetilde{\max}$. Consider two fuzzy numbers $A, B \subset R$. Fuzzy minimum and fuzzy maximum we can define as:

$$\widetilde{\min}(A,B)(z) = \sup_{z=\min(x,y)} \min[A(x), B(y)], \tag{2.69}$$

$$\tilde{\max}(A,B)(z) = \sup_{z=\max(x,y)} \min[A(x),B(y)]. \qquad (2.70)$$

By using α-cuts we define

$$\forall\alpha \in [0,1] \quad \tilde{\min}(A^\alpha, B^\alpha) = [\min(a_1^\alpha, b_1^\alpha), \min(a_2^\alpha, b_2^\alpha)], \qquad (2.71)$$

$$\tilde{\max}(A^\alpha, B^\alpha) = [\max(a_1^\alpha, b_1^\alpha), \max(a_2^\alpha, b_2^\alpha)]. \qquad (2.72)$$

Definition. Absolute value of fuzzy number
Absolute value of fuzzy number is defined as:

$$\mathrm{abs}(A) = \begin{cases} \max(A,-A), & \text{for } R^+ \\ 0, & \text{for } R^- \end{cases}$$

LR - representation of fuzzy numbers
LR - representation of fuzzy numbers suggested by Dubois and Prade [35,36] allows considerably simplifying operations of fuzzy arithmetic and gaining computation speed. Let us consider some reference functions $L, R : R \to [0,1]$, which satisfy conditions :
a) $L(0) = R(0) = 1$, b) L (for left) is monotonically increasing, R (for right) is monotonically decreasing functions.
Using L and R it is possible to construct a wide class of fuzzy numbers.

Definition. LR-fuzzy number
A fuzzy number A is of LR-type if there exist reference functions L,R, and parameters $\alpha > 0$, $\beta > 0$ with

$$\propto_A(x) = \begin{cases} L\left(\dfrac{m-x}{\alpha}\right) & \text{if } x \le m, \\ R\left(\dfrac{x-m}{\beta}\right) & \text{if } x \ge m \end{cases} \qquad (2.73)$$

where m, α, β are called the mean value, left and right spreads of A, respectively. Symbolically we can write

$$A = (m, \alpha, \beta)_{LR} \qquad (2.74)$$

Addition
The addition of A and B is calculated by the formula:

$$A + B = (a, \alpha_l, \alpha_r)_{LR} + (b, \beta_l, \beta_r)_{LR} = (a+b, \alpha_l + \beta_l, \alpha_r + \beta_r)_{LR}$$

Subtraction

$$A - B = (a, \alpha_l, \alpha_r)_{LR} - (b, \beta_l, \beta_r)_{LR} = (a-b, \alpha_l + \beta_r, \alpha_r + \beta_l)_{LR}$$

Multiplication
Following formulas are defined for multiplication and division of fuzzy numbers when spreads are small as relative to the mean values:

$$A \cdot B = (a, \alpha_l, \alpha_r)_{LR} \cdot (b, \beta_l, \beta_r)_{LR} = (a \cdot b, a\,\beta_l + b\,\alpha_l, a\,\beta_r + b\,\alpha_r)_{LR}$$

(for a>0, b>0)

$$A \cdot B = (a, \alpha_l, \alpha_r)_{LR} \cdot (b, \beta_l, \beta_r)_{LR} = (a \cdot b, b \cdot \alpha_l - a \cdot \beta_r, b \cdot \alpha_r - a \cdot \beta_l)_{LR}$$

(for a<0, b>0)

$$A \cdot B = (a, \alpha_l, \alpha_r)_{LR} \cdot (b, \beta_l, \beta_r)_{LR} = (a \cdot b, -b \cdot \alpha_r - a \cdot \beta_r, -b \cdot \alpha_l - a \cdot \beta_l)_{LR}$$

(for a<0, b<0)

Scalar multiplication

$$\forall \lambda > 0, \lambda \in R, \quad \lambda \cdot (a, \alpha_l, \alpha_r)_{LR} = (\lambda \cdot a, \lambda \cdot \alpha_l, \lambda \cdot \alpha_r)_{LR}$$
$$\forall \lambda < 0, \lambda \in R, \lambda \cdot (m, \alpha, \beta)_{LR} = (\lambda m, -\lambda\beta, -\lambda\alpha)_{LR}$$

Inverse

$$A^{-1} = (1/a, \alpha_r/a^2, \alpha_l/a^2)_{LR}$$

Division

$$A : B = A \cdot B^{-1} = (a, \alpha_l, \alpha_r)_{LR} \cdot (1/b, \beta_r/b^2, \beta_l/b^2)_{LR}$$

$$A : B = (a, \alpha_l, \alpha_r)_{LR} : (b, \beta_l, \beta_r)_{LR} = \left(\frac{a}{b}, \frac{a\,\beta_r + b\,\alpha_l}{b^2}, \frac{a\,\beta_l + b\,\alpha_r}{b^2}\right)_{LR}$$

(for a>0, b>0)

$$A : B = (a, \alpha_l, \alpha_r)_{LR} : (b, \beta_l, \beta_r)_{LR} = \left(\frac{a}{b}, \frac{b \cdot \alpha_l - a \cdot \beta_l}{b^2}, \frac{b \cdot \alpha_r - a \cdot \beta_r}{b^2}\right)_{LR}$$

(for a<0, b>0)

$$A : B = (a, \alpha_l, \alpha_r)_{LR} : (b, \beta_l, \beta_r)_{LR} = \left(\frac{a}{b}, \frac{-b \cdot \alpha_r - a \cdot \beta_l}{b^2}, \frac{-b \cdot \alpha_l - a \cdot \beta_r}{b^2}\right)_{LR}$$

(for a<0, b<0)

Fuzzy analysis

Definition. Fuzzy equation
Equations in which the coefficients and unknowns are fuzzy numbers are called fuzzy equations.
In this section we discuss two types of equations

$$A + X = B \tag{2.75}$$

$$A \cdot X = B \tag{2.76}$$

where, A,B are given fuzzy numbers, X is unknown fuzzy numbers.
Let's consider equation (2.75). As it is proved in [52] a necessary and sufficient condition for the existence of a fuzzy numbers X satisfying (2.75) is $\alpha < \alpha^1$, it implies

$$(b_1^{\alpha^1} - a_1^{\alpha^1}) \geq (b_1^{\alpha} - a_1^{\alpha}) \tag{2.77}$$

$$(b_2^{\alpha^1} - a_2^{\alpha^1}) \leq (b_2^{\alpha} - a_2^{\alpha}) \tag{2.78}$$

Then the solution of (2.75) is

$$\forall\alpha \in [0,1] \quad X^{\alpha} = [b_1^{\alpha} - a_1^{\alpha}, b_2^{\alpha} - a_2^{\alpha}] \tag{2.79}$$

The solution X of the equation (2.75) is defined as

$$X = \bigcup_{\alpha\in(0,1]} X^{\alpha} \tag{2.80}$$

Now we consider the equation (2.76)
In [51,52] it is proved that necessary and sufficient condition for the existence of a fuzzy numbers X is

$$\forall\alpha, \alpha' \in [0,1]$$

$\alpha < \alpha'$ implies

$$1)\ (b_1^{\alpha'} / a_1^{\alpha'}) \geq (b_1^{\alpha} / a_1^{\alpha}) \tag{2.81}$$

$$2)\ (b_2^{\alpha'} / a_2^{\alpha'}) \leq (b_2^{\alpha} / a_2^{\alpha}) \tag{2.82}$$

The solution is given by

$$\forall\alpha \in [0,1] \quad X^{\alpha} = [b_1^{\alpha} / a_1^{\alpha}, b_2^{\alpha} / a_2^{\alpha}] \tag{2.83}$$

The solution of the equation (2.76) is defined as

$$X = \bigcup_{\alpha\in(0,1]} X^{\alpha} \tag{2.84}$$

Fuzzy functions

A fuzzy function can be interpreted according to where fuzziness is introduced. Here we consider mainly two types of fuzzy functions:

1. Fuzzy extension of classical functions by the extension principle;
2. Fuzzy function on nonfuzzy argument.

In first case classical function carries the fuzziness of its argument without introducing any fuzziness themselves. In accordance with extension principle for given a classical function $f: X \rightarrow Y$ and a fuzzy argument A in X, the fuzzy image B can be defined as

$$\propto_B(y) = \sup_{x\in f^{-1}(y)} \propto_A(x) \tag{2.85}$$

Fuzzy derivatives

Chang and Zadeh first have defined the derivative of a function $\tilde{f}(x,\tilde{a})$ with a fuzzy parameter $\tilde{a}$ [29]

$$\propto_{\tilde{f}'(x,\tilde{a})}(y) = \sup_{a,y=\tilde{f}(x,\tilde{a})} \propto_{\tilde{a}}(a) \tag{2.86}$$

In [110] the derivative of a fuzzy function of the crisp variable t $\tilde{f}(t)$, where $-\infty < t < \infty$, is defined as follows.

Definition. Derivative of a fuzzy function

If for an arbitrary fuzzy function $\tilde{f}(t)$ a fuzzy number $\tilde{f}'(t_0)$ such that

$$\lim_{h\to 0} d_p\left(\frac{\tilde{f}(t_0+h)-\tilde{f}(t_0)}{h}, \tilde{f}'(t_0)\right) = 0 \tag{2.87}$$

exists, $\tilde{f}'(t_0)$ is called the derivative of $\tilde{f}(t)$ at t_0.

Here d_p defines a metric in the Banach space $L_p[0,1]$.

Fuzzy maximum of a fuzzifying function

Let $\tilde{f}(x)$ be a fuzzifying function from $[a,b] \subseteq R$ to R. $\forall\alpha \in [0,1]$ the equations

$$\propto_{\tilde{f}(x)}(y) = \alpha \tag{2.88}$$

with x and α as parameters are assumed to have exactly two solutions in the case $\alpha \neq 1$

$$y = f_\alpha^+(x),\ y = f_\alpha^-(x);\ \alpha \neq 1$$

and

$$y = f(x). \tag{2.89}$$

for $\alpha = 1$, $f_\alpha^+(x), f_\alpha^-(x), f(x)$ are α-level curves of $\tilde{f}(x)$. f_α^+ and f_α^- are defined in such a way that

$\forall x \in X,\ \forall \alpha' \leq \alpha:$

$$f_{\alpha'}^+(x) \geq f_\alpha^+(x) \geq f(x) \geq f_\alpha^-(x) \geq f_{\alpha'}^-(x)$$

(see Fig. 2.19)

In order to define fuzzy maximum of $\tilde{f}(x)$ we maximize each f_α over D and define the maximum of $\tilde{f}(x)$ by the fuzzy set B which contains the maximums of $f_\alpha(x)$

$$\propto_B(y) = \left\{\left(\sup_{x\in D} f_\alpha^-(x), \alpha^-\right), \left(\sup_{x\in D} f(x), 1\right)\left(\sup_{x\in D} f_\alpha^+(x), \alpha^+\right)\right\} \tag{2.90}$$

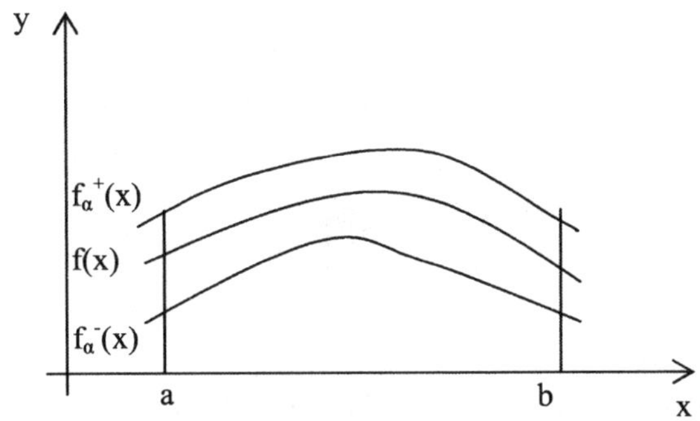

Fig. 2.19. α -level curves of $\tilde{f}(x)$

Integration of fuzzy function

At first we consider the integral of a fuzzifying function over a crisp interval [a,b]. An intuitive way of defining the integral $\tilde{I}(a,b)$ of $\tilde{f}(x)$ over [a,b] is to assign the membership value α to the integral of any α-level curve

$$\tilde{I}(a,b) = \int_{\alpha\in(0,1]} \alpha \left| \int_a^b f_\alpha^-(x)dx + \int_{\alpha\in(0,1]} \alpha \right| \int_a^b f_\alpha^+(x)dx \tag{2.91}$$

If $\tilde{f}(x)$ is L-R type fuzzifying function, i.e. if

$$\forall x \in [a,b] \quad \tilde{f}(x) = [f_1(x), l(x), r(x)]_{LR} \tag{2.92}$$

where $f_1(x)$, $l(x)$ and $r(x)$ are assumed to be integrable functions on [a,b], then [36]

$$\tilde{I}(a,b) = \left(\int_a^b f_1(x)dx, \int_a^b l(x)dx, \int_a^b r(x)dx, \right)_{LR} \tag{2.93}$$

It is necessary to mention that

$$\int_a^b \left[\tilde{f}(x) \oplus \tilde{g}(x) \right] dx = \int_a^b \tilde{f}(x)dx \oplus \int_a^b \tilde{g}(x)dx \tag{2.94}$$

Now we consider integral of crisp function f(x) over the fuzzy interval [A,B], where A and B are two fuzzy sets on R. The extension principle allows expressing the integral $\tilde{I}(A,B)$ of a nonfuzzy function over a fuzzy interval as

$$\propto_{\tilde{I}(A,B)}(z) = \sup_{x,y:z=\int_x^y f(u)du} \min\left[\propto_A(x), \propto_B(x)\right]. \tag{2.95}$$

There is another formula for calculating $\tilde{I}(A,B)$. If we denote an antiderivative of f as F, then we can write

$$\propto_{\tilde{I}(A,B)}(z) = \sup_{x\in R}\min\left\{\propto_A(x), \sup_{\substack{y \\ z=F(y)-F(x)}} \propto_B(y)\right\}$$

By using α-cuts it is easy to see that

$$\int_A^B [f(x)+g(x)]dx = \int_A^B f(x)dx + \int_A^B g(x)dx\,. \qquad (2.96)$$

Fuzzy relations, linguistic variables

In modeling systems the internal structure of a system must be described first. An internal structure is characterized by connections (associations) among the elements of system. As a rule these connections or associations are represented by means of relation. We will consider two forms of relations: crisp and fuzzy relations. Each of them characterizes a special property of the relation given on a set. A crisp relation shows only the fact of existence of the connection among the elements of sets and doesn't give the representation about a measure of degree of this connection. On the other hand, fuzzy relation gives us the representation about degree or strength of this connection.

There are several definitions of fuzzy relation [50,110,115]. Each of them depends on various factors and expresses different aspects of modeling systems.

Definition. Fuzzy relation

Let $X_1, X_2, ..., X_n$ be nonempty crisp sets. Then, a $R(X_1, X_2, ..., X_n)$ is called a fuzzy relation of sets $X_1, X_2, ..., X_n$, if $R(X_1, X_2, ..., X_n)$ is the fuzzy subset given on Cartesian product X_1 x X_2 x...x X_n.

If n=2, then fuzzy relation is called binary fuzzy relation, and is denoted as $R(X,Y)$. For three, four, or n sets the fuzzy relation is called ternary, quaternary, or n-ary, respectively.

In particular, if $X_1 = X_2 = ... = X_n = X$ we say that fuzzy relation R is given on set X among elements $x_1, x_2, ..., x_n \in X$.

Let fuzzy relation R be given by the membership function [101]

$$\propto_R(x_1, x_2) = 1/[1+100(x_1 - 5x_2)^2],$$

which approximately represents crisp relation $x_1 = 5x_2$.

It is easy to see that for all elements satisfying conditions $x_1 = 5x_2$ the membership function has value of one. On the other hand, the values of membership will decrease when the elements move further from straight line $x_1 = 5x_2$. This tendency can be seen in Fig. 2.20.

Notice, that fuzzy relation can be defined in another way. Namely, by two ordered fuzzy sets.

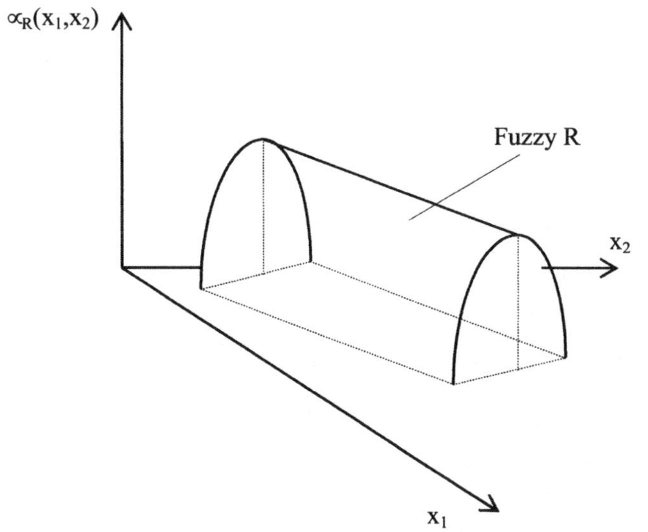

Fig. 2.20. Membership function of R

Assume, two fuzzy sets $\propto_A(x)$ and $\propto_B(y)$ are given on crisp sets X and Y, respectively. Then, it is said, that fuzzy relation $R_{AB}(X,Y)$ is given on sets X and Y, if it is defined in the following way

$$\propto_{R_{AB}}(x,y) = \min_{x,y}[\propto_A(x), \propto_B(y)]$$

for all pairs (x,y), where $x \in X$ and $y \in Y$. As above, fuzzy relation R_{AB} is defined on Cartesian product.

For example, let $X = \{1,2,3\}$ and $Y = \{Q,L,M\}$. Let membership functions of fuzzy sets A of set X and B of set Y, respectively, have forms

$$\propto_A = \{0.3/1, 0.8/2, 0.5/3\}, \propto_B = \{0.2/Q, 0.9/L, 0.1/M\}$$

Then

$$\begin{aligned} R_{AB} = \{&0.2/(1,Q), 0.3/(1,\ L), 0.1(1,\ M),\\ &0.2/(2,\ Q), 0.8/(2,\ L), 0.1(2,\ M),\\ &0.2/(3,\ Q), 0.5/(3,\ L), 0.1\,(3,\ M)\}\,. \end{aligned}$$

The frequently used way of representing finite fuzzy relations is matrices.

Let X and Y be crisp finite sets and Cartesian product $X \cdot Y$ is given. Then, we will say that fuzzy relation R on $X \cdot Y$ is given by

$$R = \begin{bmatrix} r_{11}\ r_{12} \cdots r_{1m} \\ r_{21}\ r_{22} \cdots r_{2m} \\ \cdots\cdots\cdots\cdots \\ r_{n1}\ r_{n2} \cdots r_{nm} \end{bmatrix}$$

Here

$$r_{ij} = \propto_R (x_i, y_j),\ i = \overline{1,n}\ ,\ j = \overline{1,m}$$

represents membership degree of pair (x_i, y_j).

Let A be a fuzzy subset of a set X. In this case, any fuzzy relation R given on A can be defined as a fuzzy graph [28,33,88], or weighted graph, where the edge $(x,y) \in X \cdot X$ has weight $R(x, y) \in [0,1]$.

For more simple and visual representation of fuzzy relation, we can use directed graph and mapping.

In general case a fuzzy graph can be defined the following way: a fuzzy graph G=(X, A, R) is a nonempty set X with a pair of functions $A: X \rightarrow [0,1]$ and $R: X \cdot X \rightarrow [0,1]$ such that for all $x, y \in X, R(x, y) \le A(x) \wedge A(y)$. A and R are called the fuzzy nodes set and the fuzzy edge set of G, respectively. Here, R is a fuzzy relation on A. For simplicity, we will use notation G=(A, R), omitting X. A set X is assumed to be finite.

In the most general case, both nodes and edges have membership values. But in special case, when A(x)=1, for all $x \in X$, edges alone have fuzzy membership. Then fuzzy graph is be denoted G=(X, R).

In case of representation of fuzzy relations by directed graphs we use membership degree of the respective ordered pair.

Let's consider operations on fuzzy relations some of which are analogous to the operations on fuzzy sets and others are specific only for fuzzy relations.

Intersection of fuzzy relations T and Q is a fuzzy relation $T \cap Q$ which is defined by the membership function:

$$\propto_{T \cap Q}(x, y) = \propto_T (x, y) \wedge \propto_Q (x, y) = \min\{\propto_T (x, y), \propto_Q (x, y)\}, \forall x, y \in X \cdot X$$

A fuzzy relation $T \cup Q$, defined by the membership function

$$\propto_{T \cup Q}(x, y) = \propto_T (x, y) \vee \propto_Q (x, y) = \max\{\propto_T (x, y), \propto_Q (x, y)\}, \forall x, y \in X \cdot X$$

is called a union of fuzzy relations T and Q on $X \cdot X$.

A relation $\overline{R}$ with membership function

$$\propto_R (x, y) = 1 - \propto_R (x, y), \forall x, y \in X \cdot X.$$

is called a complement of fuzzy relation $R \subseteq X \cdot X$

A relation R^{-1} with the membership function

$$\propto_{R^{-1}} (x, y) = \propto_R (y, x), \forall x, y \in X \cdot X$$

is called a relation reverse to R. Obviously, the matrix R^{-1} is the transposed R matrix.

A very important operation used in application problems is a composition of fuzzy relations. Like ordinary relation the composition can be defined in many ways. Let's list some of the most commonly used definitions.

For two relations T and Q on $X \cdot X$, a relation with membership function

$$\propto_{T \circ Q}(x,z) = \sup_{y \in X} \min\{\propto_T(x,y), \propto_Q(y,z)\}$$

is named a maximin composition.

For two relations T and Q on $X \cdot X$, a relation with membership function

$$\propto_{T \circ Q}(x,z) = \inf_{y \in X} \max\{\propto_T(x,y), \propto_Q(y,z)\}$$

is named a minimax composition.

For two relations T and Q on $X \cdot X$, a relation with membership function

$$\propto_{T \circ Q}(x,z) = \sup_{y \in X}\{\propto_T(x,y) \cdot \propto_Q(y,z)\}$$

is named a maxi-prod composition.

For two relations T and Q on $X \cdot X$, a relation with membership function

$$\propto_{T \circ Q}(x,z) = \sup_{y \in X}\{\propto_T(x,y) \odot \propto_Q(y,z)\}$$

is named a maxi $\odot$ composition of T and Q.

Now consider fuzzy equivalence relations and compatibility relations. The concept of equivalence relation is of significant value in set, algebraic and topology theories.

It is a fundamental concept and has wide application in many fields of mathematics and systems theory. The main specialty of an equivalence relation consists in following: if an equivalence relation is given on a set, then this set can be represented as separated into non-intersecting subsets. Each of these subsets consists of elements non-distinguishable from the viewpoint of this equivalence relation.

Let fuzzy binary relation on set X is given. Consider the following three properties of relation R:

1. Fuzzy relation R is reflexive, if

$$\propto_R(x,x) = 1$$

for all $x \in X$. If there exist $x \in X$ such that this condition is violated, then relation R is irreflexive, and if $R(x,x)=0$ for all $x \in X$, the relation R is antireflective;

2. A fuzzy relation R is symmetric if it is satisfied following condition:

$$\propto_R(x,y) = \propto_R(y,x)$$

for all $x, y \in X$. If from $R(x,y) > 0$ and $R(y,x) > 0$ follows $x = y$ for all $x, y \in X$ relation R is called antisymmetric;

3. A fuzzy relation R is transitive (or, more specifically, max-min transitive) if

$$\propto_R(x,z) \geq \max_{y \in Y} \min(\propto_R(x,y), \propto_R(y,z))$$

is satisfied for all pairs $(x,z) \in X$.

Definition. Equivalence (similarity) relation
If fuzzy relation R is reflexive, symmetric and transitive then relation R is an equivalence relation or similarity relation.

A fuzzy relation R is a fuzzy compatibility relation if it is reflexive and symmetric. This relation is cut worthy. Compatibility classes are defined by means of α -cut. In fact, using α -cut a class of compatibility relation is represented by means of crisp subset.

Therefore a compatibility relation can also be represented by reflexive undirected graph.

Now consider fuzzy partial ordering.

Let X be nonempty set. It is well known, that to order a set it is necessary to give an order relation on this set. But sometimes our knowledge and estimates of the elements of a set are not accurate and complete. Thus, to order such set the fuzzy order on set must be defined.

Definition. Fuzzy partial ordering relation
Let R be binary fuzzy relation on X. Then fuzzy relation R is called fuzzy partial ordering, if it satisfies the following conditions:

1. Fuzzy relation R is reflexive;
2. Fuzzy relation R is antisymmetric;
3. Fuzzy relation R is fuzzy transitive.

If fuzzy partial order is given on set X then we will say that set X is fuzzy partially ordered.

Next we consider projections and cylindrical extension.

Let R be n-dimensional fuzzy relation on Cartesian product $X = X_1 \cdot X_2 \cdot \ldots \cdot X_n$ of nonempty sets $X_1, X_2, \ldots, X_n$ and $(i_1, i_2, \ldots, i_k)$ be a subsequence of $(1,2,\ldots,n)$.

Definition. Projection of fuzzy relation
It is said, that fuzzy relation R_p on $X_{i_1} \cdot X_{i_2} \cdot \ldots \cdot X_{i_k}$ is the projection of fuzzy relation R on, if R_p is defined by the following membership function

$$\propto_{R_p}(x_{i_1}, x_{i_2}, \ldots, x_{i_k}) = \max_{x_{j1} \in X_{j1}, \ldots, x_{j(n-k)} \in X_{j(n-k)}} \propto_R (x_1, \ldots, x_n)$$

where $\{x_{j1},\ldots,x_{j(n-k)}\}$ is the complement of $\{x_{i1},\ldots,x_{ik}\}$ with respect to $\{x_1,\ldots,x_n\}$

In particular if R is a fuzzy binary relation on $X \cdot Y$, then the projection of R on X , denoted R_1, is a fuzzy set in X defined by

$$\propto_{R_1}(x) = \max_{y \in Y} \propto_R (x,y)$$

Definition. Cylindrical extension of relation
The cylindrical extension of R_p to $X_1 \cdot X_2 \cdot \ldots \cdot X_n$, is a fuzzy relation R_{PE} in $X_1 \cdot X_2 \cdot \ldots \cdot X_n$ defined by

$$\propto_{R_{PE}}(x_1,...,x_n) = \propto_{R_p}(x_{i_1}, x_{i_2},..., x_{i_k}).$$

Let A be a fuzzy subset of X and $X \cdot Y$ is Cartesian product. Then the fuzzy relation R in $X \cdot Y$ is the cylindrical extension of A to $X \cdot Y$, if R is defined by $\propto_R(x,y) = \propto_A(x)$ on set X.

Three important concepts of fuzzy logic are: if-then rule; linguistic variable whose values are words rather than numbers; fuzzy graph provides a machinery for manipulation of imprecisely defined relations.

Let us consider IF-THEN fuzzy relation. Let A and B be fuzzy subsets on the universes of discourse X and Y.

To relate the fuzzy subsets A and B of disparate universes of discourse X and Y, the concept of a fuzzy conditional statement (linguistic implication) is introduced, that is

A==>B or " if A then B"

A==>B or " if A then B"

Then implied relation R is expressed in terms of the Cartesian product of the subsets A and B denoted by $R = A \cdot B$ and its membership function is defined by

$$\propto_R(x,y) = \propto_{AxB}(x,y) = \min[\propto_A(x), \propto_B(y)], x \in X, y \in Y.$$

It is possible to encounter the following fuzzy relation. In such cases, the fuzzy conditional statement is nested, that is of the form

if A and B *then* C.

Then the fuzzy relation R is written as

$$R = A \cdot (B \cdot C) = A \cdot B \cdot C.$$

Fuzzy implication may be composed of two implications. These simple implications are connected by using the connectives 'or' ('else'), 'and', and others.

This can be extended to the case of more than two fuzzy implications.

Linguistic variable

The research have shown lately that traditional analysis methods for systems analysis and computer modeling, based on precise processing of numerical data, are not capable of dealing with huge complexity of real-world problems. This leads to the fact that when solving those real-world problems we need to move away from traditional requirements for measurement accuracy which are necessary for mathematical analysis of precisely defined mechanical systems.

The necessity to sacrifice the precision and determinary is dictated also by the appearance of some classes of real-world problems, which are connected with decision-making by operator in the "man-computer" interface. Implementation of the dialog in such interface is impossible without application of languages close to natural ones and capable of describing fuzzy categories near to human notions and imaginations. In this connection, it is valuable to use the notion of linguistic variable first introduced by L.Zadeh [114]. Such linguistic variables allow an adequate reflection of approximate in-word descriptions of objects and phenomena in the case if there is no any precise deterministic description. It should be noted as well that many fuzzy categories described linguistically even appear to be more informative than precise descriptions.

For example, the phrase "Jone is 25 years old" can be expressed approximately by "Jone is young". In this sense, the word "young" can be considered as a linguistic value for the variable age". Here we take into account that the linguistic value plays the same role as the numerical value 25 does. The same can be said about the linguistic values such as "very young", "a bit older than", "almost of middle age" and so forth, if we match them with numerical values 17, 24, and 31.

The set of values of a linguistic variable has made the term-set of this variable. This set in general may have unlimited number of elements though in practice it is finite. For example, a term-set of the linguistic variable "age" can be written as follows:

T(age)="very young"+"almost young"+"young"+...+"almost of middle age"+" of average age"+...+"old"+"old enough", "very old".

Here the sign " +" stands for union. Note, that in case of the linguistic variable "age" a numerical variable "age" , for example, taking the values of 17, 24, 31, 38,..., 94, is so-called basis variable of the linguistic variable of age. In this case such a linguistic value as "old" can be interpreted as a name for some fuzzy restriction on the values of the basis variable. This restriction essentially is the sense of the linguistic value "old".

The fuzzy restriction on the values of the basis variable is characterized by a compatibility function which maps each value of the basis variable into some number from the interval [0,1], characterizing the relation of this value with the fuzzy restriction. Note that mathematically and conceptually the compatibility function is identical to the membership function introduced below.

If u is the name for a fuzzy variable, then the restriction associated with that name can be interpreted as the meaning of the fuzzy variable u. For example, if the restriction associated with the value "old" of the fuzzy variable "age" represents a fuzzy subset of the set X=[17,94] of the type:

$$R(\text{"old"})= \int \{1-[1-\frac{x-17}{77}]\}/x, \ x \in X,$$

then this fuzzy set can be considered as the meaning for the variable "old".

Another important aspect of the notion of linguistic variable is that a linguistic variable is associated with the two rules: the syntactic rule, which can be set as a grammar, generating names for the variable; and the semantic rule, which determines an algorithmic procedure for calculating the meaning of each value. Thus these rules make the essential part of the description of the structure of linguistic variable.

Definition. Linguistic variable

A linguistic variable is characterized by the set (u,T,X,G, M), where u is the name of variable; T denotes the term-set of u that refer to as base variable whose values range over a universe X; G is a syntactic rule (usually in form of a grammar) generating linguistic terms; M is a semantic rule that assigns to each linguistic term its meaning, which is a fuzzy set on X.

A certain $t \in T$ generated by the syntactic rule G is called a term. A term consisting of one or more words, the words being always used together, is named an atomary term. A term consisting of several atomary terms is named a composite term. The concatenation of some components of a composite term (i.e. the result of linking the chains of components of the composite term) is called a sub term. Here t_1, t_2, ... are terms in the following expression

$$T = t_1 + t_2 + \ldots$$

The meaning of M(t) of the term t is defined as a restriction R(t; x) on the basis variable x conditioned by the fuzzy variable X:

$$M(t) \equiv R(t; x),$$

it is assumed here that R(t; x) and, consequently, M(t) can be considered as a fuzzy subset of the set X named as t.

The assignment equation in case of linguistic variable takes the form in which t-terms in T are names generated by the grammar G, where the meaning assigned to the term t is expressed by the equality

$$M(t) = R(\text{term in } T)$$

In other words the meaning of the term t is found by the application of the semantic rule M to the value of term t assigned according to the right part of equation. Moreover, it follows that M(t) is identical to the restriction associated with the term t.

It should be noted that the number of elements in T can be unlimited and then for both generating elements of the set T and for calculating their meaning, the application of the algorithm, not simply the procedure for watching term-set, is necessary.

We will say that a linguistic variable u is structured if its term-set T and the function M, which maps each element from the term-set into its meaning, can be given by means of algorithm. Then both syntactic and semantic rules connected with the structured linguistic variable can be considered algorithmic procedures for generating elements of the set T and calculating the meaning of each term in T, respectively.

However in practice we often encounter term-sets consisting of a small number of terms. This makes it easier to list the elements of term-set T and establishes a direct mapping from each element to its meaning.

Fuzzy relational equations

Assume that the relations Q, R and S constrain each other the following way [5,76-78,96]

$$Q * R = S, \tag{2.97}$$

where Q is a fuzzy relation on $X \cdot Y$, R is a fuzzy relation on $Y \cdot Z$, and S a fuzzy relation on $X \cdot Z$.

Definition. Fuzzy relational equation

When two of the relations in the relationship (2.97) are given and one is unknown, (2.97) is referred to as fuzzy relational equation.

Having knowing Q and R in (2.97), it is easy to find S. The converse problem, i.e. find Q (resp. R) knowing S and R (resp. Q) is interesting one but may seem more difficult [5].

Now we consider the problem: find Q knowing S and R (or find R knowing S and Q).

Two fundamental results that give the greatest solution of this problem consist of following.

1. Let Q be a fuzzy relation on $X \cdot Y$, S a fuzzy relation on $X \cdot Z$, and $\Re$ the set of fuzzy relations R on $Y \cdot Z$ such that $Q \circ R = S$, then either $\Re = \varnothing$ or $Q^{-1} @ S \in \Re$. If $\Re \neq \varnothing$, $Q^{-1} @ S$ is the greatest element in R;
2. Let R be a fuzzy relation on $Y \cdot Z$ and S a fuzzy relation on $X \cdot Z$ and Q the set of fuzzy relations on such that $Q \circ R = S$, then either $Q = \varnothing$ or $(R @ S^{-1})^{-1} \in Q$. If $Q = \varnothing$, $(R @ S^{-1})^{-1}$ is the greatest element in Q.

Now we consider two particular cases of this problem. The first problem is formulated as follows: find such R that

$$A \circ R = B \tag{2.98}$$

where A is a fuzzy set on X, B a fuzzy set on Y, and R is an unknown fuzzy relation in $X \cdot Y$, with value in [0,1]. X and Y are assumed to be finite.

The fuzzy relation $A @ B$ in X⊗Y has membership function

$$\propto_{A\sigma B}(x, y) = \propto_A(x) @ \propto_B(y)$$

Let $\Re = \{R, A \circ R = B\}$, if $\Re \neq \varnothing$, then $A @ B \in \Re$

The second problem consists in the following: find A such that $A \circ R = B$ where A is an unknown fuzzy set on X (finite), B a fuzzy set on Y (finite), and R a fuzzy relation in $X \cdot Y$.

$$\propto_A(x) = \sup_{y \in Y} \min(\propto_B(y), \propto_R(x, y))$$

This problem has no solutions since

$$\forall x, \exists y, \propto_R(x, y) < \propto_B(y)$$

When the solution $\propto_A$ exists, the fuzzy set A satisfies the following conditions: A is a solution iff

$$1)\ \forall y, \quad (\exists x, \ \propto_A(x) \geq \propto_R(x, y) = \propto_B(y))$$

$$\text{or} \qquad \exists x, \ \propto_A(x) = \propto_B(y) < \propto_R(x, y)),$$

$$2)\ \forall x \ \ \forall y, \quad \propto_A(x) \leq \propto_B(y) \text{ if } \ \propto_R(x, y) > \propto_B(y).$$

Now we consider more generalized version of the first problem. It is known that [10] there exists three types of the equation (2.97), i.e.

$$B = A \circ R \tag{2.99}$$

$$B = A * R \tag{2.100}$$

$$B = A \,\square\, R \tag{2.101}$$

where A and B are treated as fuzzy sets defined on the space universes of discourse X and Y, R denotes a fuzzy relation expressed on the Cartesian product X and Y. $\circ$,* and $\square$ are sup-min, sup-prod, and inf-max composition operators, respectively. Equations (2.99)-(2.101) can be rewritten in the following form:

$$\propto_B(y) = \sup_{x \in A} \min(\propto_A(x), \propto_R(x, y)) \tag{2.102}$$

$$\propto_B(y) = \sup_{x \in X}(\propto_A(x) \cdot \propto_R(x, y)) \tag{2.103}$$

$$\propto_B(y) = \inf_{x \in X} \max(\propto_A(x), \propto_R(x, y))$$

$y \in B$.

If fuzzy sets A and B are given, and they satisfy the sup-min fuzzy relational equation (2.99), the fuzzy relation R, which is a solution of this problem, is defined as follows.

If fuzzy sets A,B satisfy the equation $B = A \circ R$, then the greatest fuzzy relation $\hat{R}$, such that $B = A \circ \hat{R}$ holds, is equal to

$$\hat{R} = A \circledcirc B$$

where α stands for the α -composition of A and B.

If A and B satisfy A $\square$ R=B, then the least fuzzy relation $\tilde{R}$, such that A $\square$ $\tilde{R}$ =B is satisfied, is equal to

$$\tilde{R} = A \,\textcircled{\scriptsize ε}\, B$$

where ε - stands for ε -composition of A and B.

If X and B satisfy the sup-prod fuzzy relational equation A*R=B then the greatest fuzzy relation R', such that $A * R' = B$ is satisfied, is expressed as

$$R' = A \;\phi\; B$$

In the above equations we consider only statical fuzzy relational equations.

Fuzzy relational equation may be dynamical, i.e. they are time-dependent. Dynamical fuzzy relational equations has a form:

$$A_{t+1} = U_t \circ A_t \circ R \tag{2.104}$$

where A_t, A_{t+1} denote the states of the process at the t-th and (t+1)-th discrete time instants, respectively, U_t is the control parameter at the t-th time instant and R stands for the time-dependent relation describing the system. Both states and control are treated as fuzzy sets defined in finite spaces of state X and control U.

Methods for solving equation (2.104) are analogous to those for solving of statical fuzzy relational equations.

Below we give a family of composition operations:

max-min composition

$$u \circ x = \vee_i (u_i \wedge x_i)$$

α *-composition*

$$u[\alpha]x = \begin{cases} 1, & \text{if } u \le x \\ x, & \text{if } u > x \end{cases}$$

β *-composition:*

$$u[\beta]x = \begin{cases} 1, & \text{if } u < x \\ x, & \text{if } u \ge x \end{cases}$$

ϕ *-composition:*

$$u\,[\phi]\,x = \min\{1,\ x/u \mid u,x \in [0,1]\}$$

ε *-composition:*

$$u[\varepsilon]x = \begin{cases} x, & \text{if } u < x \\ 0, & \text{if } u \ge x. \end{cases}$$

2.1.3 Fuzzy Logic and Approximate Reasoning

We will consider the logics with multi-valued and continuous values (fuzzy logic). Let's define the semantic truth function of this logic. Let P be statement and T(P) its truth value, where $T(P) \in [0,1]$.

Negation values of the statement P are defined as:

$$T(\leftarrow P) = 1 - T(P).$$

Hence

$$T(\leftarrow\leftarrow P) = T(P).$$

The implication connective is always defined as follows:

$$T(P \rightarrow Q) = T(\leftarrow P \vee Q),$$

and the equivalence as

$$T(P \leftrightarrow Q) = T[(P \rightarrow Q) \wedge (Q \rightarrow P)].$$

It should be noted that exclusive disjunction ex, disjunction of negations (Shiffer's connective) |, conjunction of negations ↓ and ~→ (has no common name) are defined as negation of equivalence ↔, conjunction ∧, disjunction ∨, and implication →, respectively.

The tautology and contradiction will be, respectively:

$$T(\overset{\square}{P})=T(P \vee \leftarrow P);\ T(\overset{\circ}{P})=T(P \wedge \leftarrow P).$$

More generally

$$T(\overset{\square}{PQ})=T((P \vee \leftarrow P) \vee (Q \vee \leftarrow Q));$$

$$T(\overset{\square}{PQ})=T((P \wedge \leftarrow P)\ \wedge (Q \wedge \leftarrow Q)).$$

Let us define the basic connectives of fuzzy logic in the following two fuzzy set theories.

Logic based on ($\tilde{P}$ (x), ∩, ∪, –)

In this case disjunction and conjunction are defined as:

$$T(P \vee Q)=\max(T(P),T(Q));\ T(P \wedge Q)=\min(T(P),T(Q)).$$

It is clear, that ∨ and ∧ are commutative, associative, idempotent and distributive and do not satisfy the law of excluded-middle, i.e. $T(P \vee \leftarrow P)\neq 1$ and $T(P \wedge \leftarrow P)\neq 0$, but satisfy absorption law

$$T(P \vee (P \wedge Q))=T(P);\ T(P \wedge (P \vee Q))=T(P),$$

and also De Morgan's laws:

$$T(\leftarrow (P \wedge Q))=T(\leftarrow P \vee \leftarrow Q);$$

$$T(\leftarrow (P \vee Q))=T(\leftarrow P \wedge \leftarrow Q).$$

Equivalence is defined as

$$T[(\leftarrow P \vee Q)\ \wedge\ (P \vee \leftarrow Q)]=T[(P \wedge Q)\ \vee\ (\leftarrow P \wedge \leftarrow Q)].$$

Law of excluded disjunction:

$$T[(\leftarrow P \wedge Q)\ \vee\ (P \wedge \leftarrow Q)]=T[(P \vee Q)\ \wedge\ (\leftarrow P \vee \leftarrow Q)].$$

The expressions for 16 connectives are presented in Table 2.3. It is assumed here that T(P)=p and T(Q)=q.

The quantifiers in the statements will be:

$$T(\exists xP(x))=\sup_x (T(P(x)));\qquad T(\forall xP(x))=\inf_x (T(P(x))),$$

where x denotes an element of the universe of discourse.

Multi-valued logic based on ($\tilde{P}$ (x),∪,∩,–) usually is called K- standard sequence logic. In this logic the connectives satisfy the following properties:
implication $T[P\rightarrow(Q\rightarrow R)]=T[(P \wedge Q)\rightarrow R]$;

tautology and contradiction

$$T(P\to P)=T(\overset{\square}{P});\ T(\overset{\circ}{P}\to P)=T(P);\ T(P\to \overset{\circ}{P})=T(\overset{\square}{P});$$

$$T(P\leftrightarrow P)=T(\overset{\square}{P});T(\overset{\circ}{P}\to P)=T(\overset{\square}{P});\ T(P\to \overset{\circ}{P})=T(\leftarrow P);\ T(P\leftrightarrow \leftarrow P)=T(\overset{\circ}{P});$$

the Shiffer's and Pierce's connectives

Table 2.3. Expressions for connectives

$$T(\leftarrow P)=T(P|P);\ T(P\to Q)=T(P|(Q/Q));\ T(\overset{\circ}{P})=T(P|(P|P)).$$

PQ	$\overset{\square}{PQ}$	$P\vee Q$	$Q\to P$	P
pq	max(p,1-p,q, 1-q)	max(p, q)	max(p, 1-q)	p
PQ	$P\to Q$	Q	$P\leftrightarrow Q$	$P\wedge Q$
pq	max(1-p, q)	q	min(max(1-p,q), max(p, 1-q))	min(p, q)
PQ	P | Q	P ex Q	$\leftarrow Q$	$Q \sim\to P$
pq	max(1-p,1-q)	max(min(1-p, q), pqmin(p,1-q))	1-q	min(p, 1-q)
PQ	$\leftarrow P$	$P\sim\to Q$	$P\downarrow Q$	$\overset{\circ}{PQ}$
pq	1-p	min(1-p, q)	min(1-p, 1-q)	min(p,1-p,q,1-q)

It is shown in [36,52] that the multi-valued logic is fuzzification of the traditional propositional calculus (in sense of extension principle). In this logic each proposition P is assigned a normalized fuzzy set in [0,1], i.e. the pair $\{\propto_P(0), \propto_P(1)\}$ is interpreted as the degree of false or truth, respectively. Because, the logical connectives of standard propositional calculus are functionals of truthness, that is they are represented as functions, they can be fuzzified.

Logic based on $(\tilde{P}(x), \underset{\square}{\cap}, \overset{\square}{\cup}, -)$

In this case disjunction and conjunction are defined as

$$T(P \dot{\vee} Q)=\min(1,T(P)+T(Q));\qquad T(P \underset{\cdot}{\wedge} Q)=\max(0,T(P)+T(Q)-1).$$

It is clear that $\overset{\square}{\vee}$ and $\underset{\square}{\wedge}$ are commutative, associative, not idempotent and not distributive and they satisfy De Morgan's laws

$$T(\leftarrow(P \underset{\cdot}{\wedge} Q))=T(\leftarrow P \dot{\vee} \leftarrow Q);$$

$$T(\leftarrow(P \dot{\vee} Q))=T(\leftarrow P \underset{\cdot}{\wedge} \leftarrow Q);$$

and the law of excluded-middle

$$T(P \vee \leftarrow P)=1,\ T(P \wedge \leftarrow P)=0.$$

The 16 connectives are given in the Table 2.4. Here $\vee$, →, ↔, $\wedge$, |, ex, ~→, ↓ are denoted by $\overset{\square}{\vee}$, ⇒, ⇔, $\underset{\square}{\wedge}$, ||, ex, ≈⇒, ↓↓, respectively. Tautology and contradiction satisfy the following properties:

$$T(P\Rightarrow P)=T(\overset{\square}{P}),T(\overset{\square}{P}\Rightarrow P)=T(P),\ T(P\Rightarrow \overset{\square}{P})=T(\overset{\square}{P}),\ T(P\Leftrightarrow P)=T(\overset{\square}{P}),$$

$$T(\overset{\circ}{P}\Rightarrow P)=T(\overset{\square}{P}),T(P\Leftrightarrow \overset{\circ}{P})=T(\leftarrow P).$$

Table 2.4. Expressions for connectives

pQ	$\overset{\square}{PQ}$	$P \overset{\square}{\vee} Q$	Q⇒P	P
pq	1	min(1, p+q)	min(1, p+1-q)	p
PQ	P⇒Q	Q	P⇔Q	$P \underset{\square}{\wedge} Q$
pq	min(1,1-p+q)	Q	1-|p-q|	max(0,p+q-1)
PQ	P||Q	P ex Q	←Q	Q≈⇒P
pq	min(1,1-p+1-q)	|p-q|	1-q	max(0,p-q)
PQ	←P	P≈⇒Q	P↓↓Q	$\overset{\circ}{PQ}$
pq	1-p	Max(0,q-p)	max(0,1-p-q)	0

In Zadeh's notation the implication ⇒ corresponds to the usual inclusion for fuzzy sets, ex and ≈⇒ correspond to symmetric ∇ and bounded | - | to differences, respectively. This logic is known as Lukasiewicz logic (L-logic).

It should be noted that these two theories of fuzzy sets and logics constructed on the basis of these theories are not only known at the present time. In connection with this it is necessary to give semantic analysis of the major known multi-valued logics. For this purpose we will use power sets which are necessary for formalization of some operations on fuzzy sets.

Semantic Analysis of Different Fuzzy Logics

Let A and B be fuzzy sets of the subsets of non-fuzzy universe U; in fuzzy set theory it is known that A is a subset of B iff

$$\propto_A \le \propto_B,\ \text{i.e. } \forall x \in U,\ \propto_A(x) \le \propto_B(x).$$

Definition. Power fuzzy set

For given fuzzy implication → and fuzzy set B from the universe U, the power fuzzy set PB from B is given by membership function $\propto_{PB}$ [23]:

$$\propto_{PB} A = \bigwedge_{x \in U} \left(\propto_A (x) \rightarrow \propto_B (x)\right).$$

Then the degree to which A is subset of B, is

$$\pi(A \subseteq B) = \propto_{PB} A.$$

Definition
If fuzzy implication operator [23] is given on the closed unit interval [0,1] then

$$A \leftarrow b = b \rightarrow a$$

$$a \leftrightarrow b = (a \rightarrow b) \wedge (a \leftarrow b) = (a \rightarrow b) \wedge (b \rightarrow a).$$

Definition
Under the conditions of the definition PB the degree to which fuzzy sets A and B are equivalent or the degree of "equivalency" is:

$$\pi(A \equiv B) = \pi(A \subseteq B) \wedge \pi(B \subseteq A);$$

or

$$\pi(A \equiv B) = \bigwedge_{x \in U} (\propto_A x \rightarrow \propto_B x).$$

For practical purposes [23] in most cases it is advisable to work with multi-valued logics in which logical variable takes values from the real interval I=[0,1] divided into 10 subintervals, i.e. by using set V_{11}=[0, 0.1, 0.2, ..., 1].

We denote the truth values of premises A and B through T(A)=a and T(B)=b. The implication operation in analyzed logics [9,86] has the following form:
1) min-logic

$$a \underset{\min}{\rightarrow} b = \begin{cases} a, & \text{if } a \le b \\ b, & \text{otherwise.} \end{cases}$$

2) $S^{\#}$ - logic

$$a \underset{S^{\#}}{\rightarrow} b = \begin{cases} 1, & \text{if } a \ne 1 \text{ or } b = 1, \\ 0, & \text{otherwise.} \end{cases}$$

3) S - logic ("Standard sequence")

$$a \underset{S}{\rightarrow} b = \begin{cases} 1, & \text{if } a \le b, \\ 0, & \text{otherwise.} \end{cases}$$

4) G - logic ("Gödelian sequence")

$$a \underset{G}{\rightarrow} b = \begin{cases} 1, & \text{if } a \le b, \\ b, & \text{otherwise.} \end{cases}$$

5) G43 - logic

$$a \underset{G43}{\rightarrow} b = \begin{cases} 1, & \text{if } a = 0, \\ \min(1, b/a), & \text{otherwise.} \end{cases}$$

6) L - logic(Lukasiewicz's logic)

$$a \underset{L}{\rightarrow} b = \min(1, 1 - a + b).$$

7) KD - logic

$$a \underset{KD}{\rightarrow} b = ((1\text{-}a) \vee b) = \max(1\text{-}a, b).$$

In turn, suggested by as ALI1-ALI3 - logics, which will be used in further chapters are characterized by the following implication operations[4,13]:

8) ALI1 - logic

$$a \underset{ALI1}{\rightarrow} b = \begin{cases} 1 - a, & \text{if } a < b, \\ 1, & \text{if } a = b, \\ b, & \text{if } a > b \end{cases}$$

9) ALI2 - logic

$$a \underset{ALI2}{\rightarrow} b = \begin{cases} 1, & \text{if } a \leq b, \\ (1 - a) \wedge b, & \text{if } a > b \end{cases}$$

10) ALI3 - logic

$$a \underset{ALI3}{\rightarrow} b = \begin{cases} 1, & \text{if } a \leq b, \\ b / [a + (1 - b)], & \text{otherwise.} \end{cases}$$

The comparative analysis of the first seven logics has been given in [23]. The analysis of these seven logics has shown that only S- and G- logics satisfy the classical principle of Modus Ponens and allow development of improved rule of fuzzy conditional inference. At the same time the value of truthness of the implication operation in G-logic is equal either to 0 or 1; and only the value of truthness of logical conclusion is used in the definition of the implication operation in S-logic. Thus the degree of "fuzziness" of implication is decreased, which is a considerable disadvantage and restricts the use of these logics in approximate reasoning.

Considering the first seven logics to be well-known, we give below the main logical connectives and properties of fuzzy logics ALI1-ALI3 and their comparative semantic analysis[9,14,17]. Logical connectives in logic ALI1 are given in

Thus the considered theory of fuzzy sets differs from the existing theories in new operation of union ($\overset{1}{\cup}$) and intersection ($\underset{1}{\cap}$). The value of negation is calculated as T(←A)=1 - T(A).

Table 2.5. Logical connectives in logic ALI1

Name of connective	Designation of connective	Value of connective
Tautology	$\dot{A}$	1
Contradiction	$\overset{\circ}{A}$	0

Disjunction	$A \overset{1}{\vee} B$	$a,\ a+b>1$ $1,\ a+b=1$ $b,\ a+b<1$
Negation	$\leftarrow\!\!\!\!-A$	(1-a)
Conjunction	$A \underset{1}{\wedge} B$	$a,\ a+b<1$ $0,\ a+b=1$ $b,\ a+b>1$
Equivalence	$A \underset{1}{\leftrightarrow} B$	$1-ka,\ a<b$ $1,\ a=b$ $1-kb,\ a>b$
Sheffer's connective	$A \overset{1}{\mid} B$	$1-a,\ a+b<1$ $1,\ a+b=1$ $1-b,\ a+b>1$
Pearce's connective	$A \overset{1}{\downarrow} B$	$1-a,\ a+b>1$ $0,\ a+b=1$ $1-b,\ a+b<1$

In its turn, the connectives $\overset{1}{\mid}$ and $\overset{1}{\downarrow}$ are expressed as negation of $\underset{1}{\wedge}$ and $\overset{1}{\vee}$, respectively, and contradiction and tautology as

$$\overset{0}{T}(A) = T(A \underset{1}{\wedge} \leftarrow\!\!\!\!-A)\ ;\ \overset{1}{T}(A) = T(A \overset{1}{\vee} \leftarrow\!\!\!\!-A)\,.$$

For this fuzzy logic the following properties are true:

1.Non-commutativity: $a \overset{1}{\vee} b \neq b \overset{1}{\vee} a$; $a \underset{1}{\wedge} b \neq b \underset{1}{\wedge} a$. This property is directly derived from the equations:

$$a \overset{1}{\vee} b = \begin{cases} a, & a+b>1\ ; \\ 1, & a+b=1\ ; \\ b, & a+b<1\ ; \end{cases} \quad b \overset{1}{\vee} a = \begin{cases} b, & a+b>1\ ; \\ 1, & a+b=1\ ; \\ a, & a+b<1\ ; \end{cases}$$

Note that this fuzzy logic can become commutative, if the following type of inversion is used:

$$\leftarrow\!\!\!\!-a = \begin{cases} 1, & a=0; \\ 0, & a \neq 0. \end{cases}$$

2.Absence of associativity of disjunction and conjunction

$$a \underset{1}{\wedge} (b \underset{1}{\wedge} c) \neq (a \underset{1}{\wedge} b) \underset{1}{\wedge} c; \quad a \overset{1}{\vee} (b \overset{1}{\vee} c) \neq (a \overset{1}{\vee} b) \overset{1}{\vee} c.$$

3.Partial idempotence

$$a \underset{1}{\wedge} a = \begin{cases} a, & a \neq 0,5; \\ 0, & a = 0,5; \end{cases} \qquad a \overset{1}{\vee} a = \begin{cases} 1, & a = 0,5; \\ a, & a \neq 0,5. \end{cases}$$

4.Absence of distrubutivity

$$(a \overset{1}{\vee} b) \underset{1}{\wedge} c \neq (a \underset{1}{\wedge} c) \overset{1}{\vee} (b \underset{1}{\wedge} c),$$

$$(a \underset{1}{\wedge} b) \overset{1}{\vee} c \neq (a \overset{1}{\vee} c) \underset{1}{\wedge} (b \overset{1}{\vee} c).$$

5.Satisfaction of De Morgan's law generalized for the case of M=[0,1]

$$\overline{a \underset{1}{\wedge} b} = \bar{a} \overset{1}{\vee} \bar{b}; \quad \overline{a \overset{1}{\vee} b} = \bar{a} \underset{1}{\wedge} \bar{b}.$$

Note that the following three properties take place:

$$1)\ a \underset{1}{\wedge} 0 = \begin{cases} a, & a \neq 1, \\ 0, & a = 1; \end{cases} \quad 2)\ a \overset{1}{\vee} 1 = \begin{cases} 1, & a \leq 0, \\ a, & a > 0; \end{cases} \quad 3)\ (a) = a$$

ALI2 logic is a commutative fuzzy logic, in which the degree of truthness of the implication operation is determined considering the degrees of truthness of the premise and conclusion. This logic allows implementation of the fuzzy conditional inference rule, which satisfies criteria of both M.Mizumoto (see section 2.1.3) and Modus Ponens. The logical connectives used in ALI2 logic are shown in Table 2.6.

It is evident that the considered fuzzy sets theory differs from the existing ones with new operations of union ($\overset{2}{\cup}$) and ($\underset{2}{\cap}$) intersection. The values of negation, connectives $\overset{2}{|}$ and $\overset{2}{\downarrow}$, contradiction and tautology are defined as in ALI1 logic.

Note that the following properties are true for this fuzzy logic.

1.Commutativity

$$a \underset{2}{\wedge} b = b \underset{2}{\wedge} a; \qquad a \overset{2}{\vee} b = b \overset{2}{\vee} a.$$

2.Associativity

$$a \underset{2}{\wedge} (b \underset{2}{\wedge} c) = (a \underset{2}{\wedge} b) \underset{2}{\wedge} c; \qquad a \overset{2}{\vee} (b \overset{2}{\vee} c) = (a \overset{2}{\vee} b) \overset{2}{\vee} c.$$

3.Partial idempotence

$$a \underset{2}{\wedge} a = \begin{cases} 0, & a \leq 0,5; \\ a, & a > 0,5; \end{cases} \qquad a \overset{2}{\vee} a = \begin{cases} 1, & a \geq 0,5; \\ a, & a < 0,5. \end{cases}$$

Table 2.6. Logical connectives in logic ALI2

Name of connective	Designation of connective	Value of connective
Tautology	$\overset{\bullet}{A}$	1
Contradiction	$\overset{\circ}{A}$	0
Negation	$\leftarrow A$	1-a
Disjunction	$A \overset{2}{\vee} B$	$1, \quad a+b \geq 1$ $\min(a,b), \quad a+b<1$
Conjunction	$A \underset{2}{\wedge} B$	$0, \quad a+b \leq 1$ $\max(a,b), \quad a+b>1$
Implication	$A \underset{2}{\rightarrow} B$	$1, \quad a \leq b$ $\min[(1-a), b], \quad a>b$
Equivalence	$A \underset{2}{\leftrightarrow} B$	$\min\,[a, 1-b)], \quad a<b$ $1, \quad a=b,$ $\min\,[\,(1-a), b], \quad a>b$
Sheffer's connective	$A \overset{2}{\mid} B$	$1, \quad a+b \leq 1$ $\min[(1-a),(1-b)],$ $a+b>1$
Pearce's connective	$A \overset{2}{\downarrow} B$	$0, \quad a+b \geq 1$ $\max[(1-a),(1-b)],$ $a+b<1$

4.Distributivity

$$(a \overset{2}{\vee} b) \underset{2}{\wedge} c = (a \underset{2}{\wedge} c) \overset{2}{\vee} (b \underset{2}{\wedge} c); \qquad a \overset{2}{\vee} (b \underset{2}{\wedge} c) = (a \overset{2}{\vee} b) \underset{2}{\wedge} (a \overset{2}{\vee} c).$$

5.Satisfaction of De Morgan's laws generalized for the case M=[0,1]

$$\begin{cases} \overline{a \underset{2}{\wedge} b} = \bar{a} \overset{2}{\vee} \bar{b}; \\ \overline{a \overset{2}{\vee} b} = \bar{a} \underset{2}{\wedge} \bar{b}; \end{cases} \qquad \begin{cases} a \overset{2}{\vee} (a \underset{2}{\wedge} b) = a; \\ a \underset{2}{\wedge} (a \overset{2}{\vee} b) = a. \end{cases}$$

Note that the following three properties also take place

$$a \underset{2}{\wedge} 0 = 0; \qquad a \overset{2}{\vee} 1 = 1; \qquad \overset{=}{(a)} = a.$$

Main logical connectives of ALI3 logic are represented in Table 2.7.

Table 2.7. Logical connectives in logic ALI3

Name of Connective	Designation of Connective	Value of connective
Tautology	$\dot{A}$	1
Contradiction	$\overset{\circ}{A}$	0
Negation	$\leftarrow A$	$1-a$
Disjunction	$A\overset{3}{\vee}B$	$1,\quad a+b\geq 1$ $b/[2-(a+b)],\quad a+b<1$
Conjunction	$A\underset{3}{\wedge}B$	$0,\quad a+b\leq 1$ $1-[(1-a)/(a+b)],\quad a+b>1$
Implication	$A\underset{3}{\rightarrow}B$	$1,\quad a\leq b$ $b/[a+(1-b)],\quad a>b$
Equivalency	$A\underset{3}{\leftrightarrow}B$	$a/[b+(1-a)],\quad a<b$ $1,\quad a=b$ $b/[a+(1-b)],\quad a>b$
Sheffer's connective	$A\overset{3}{\mid}B$	$1,\quad a+b\leq 1$ $(1-b)/(a+b),\quad a+b>1$
Pearce's connective	$A\overset{3}{\downarrow}B$	$0,\quad a+b\geq 1$ $1-b/[2-(a+b)],\quad a+b<1$

ALI3 fuzzy logic is identical to ALI1 fuzzy logic for its properties. Hence we will illustrate only the most important property of them, the property of non-commutativity.

$$a\overset{3}{\vee}b=\begin{cases}1, & a+b\geq 1;\\ \dfrac{b}{2-(a+b)}, & a+b<1;\end{cases}\qquad b\overset{3}{\vee}a=\begin{cases}1, & a+b\geq 1;\\ \dfrac{a}{2-(a+b)}, & a+b<1.\end{cases}$$

Hence, it is abvious

$$a\overset{3}{\vee}b\neq b\overset{3}{\vee}a,\quad a+b<1;$$
$$a\underset{3}{\wedge}b=b\underset{3}{\wedge}a,\quad a+b\geq 1.$$

Implication operations are conveniently represented in form of the implicative transition tables for eleven-valued logic. Such table for the ALI1-logic has the following form (Table 2.8):

Table 2.8. Implicative transition table for logic ALI1

$\underset{ALI1}{\rightarrow}$											
b / a	0	0.1	0.2	0.3	0.4	0.5	0.6	0.7	0.8	0.9	1
0	1	1	1	1	1	1	1	1	1	1	1
0.1	0	1	0.9	0.9	0.9	0.9	0.9	0.9	0.9	0.9	0.9
0.2	0	0.1	1	0.8	0.8	0.8	0.8	0.8	0.8	0.8	0.8
0.3	0	0.1	0.2	1	0.7	0.7	0.7	0.7	0.7	0.7	0.7
0.4	0	0.1	0.2	0.3	1	0.6	0.6	0.6	0.6	0.6	0.6
0.5	0	0.1	0.2	0.3	0.4	1	0.5	0.5	0.5	0.5	0.5
0.6	0	0.1	0.2	0.3	0.4	0.5	1	0.4	0.4	0.4	0.4
0.7	0	0.1	0.2	0.3	0.4	0.5	0.6	1	0.3	0.3	0.3
0.8	0	0.1	0.2	0.3	0.4	0.5	0.6	0.7	1	0.2	0.2
0.9	0	0.1	0.2	0.3	0.4	0.5	0.6	0.7	0.8	1	0.1
1	0	0.1	0.2	0.3	0.4	0.5	0.6	0.7	0.8	0.9	1

Such table for ALI2 logic looks as follows (Table 2.9)

Table 2.9. Implicative transition table for logic ALI2

$\underset{ALI2}{\rightarrow}$											
b / a	0	0.1	0.2	0.3	0.4	0.5	0.6	0.7	0.8	0.9	1
0	1	1	1	1	1	1	1	1	1	1	1
0.1	0	1	1	1	1	1	1	1	1	1	1
0.2	0	0.1	1	1	1	1	1	1	1	1	1
0.3	0	0.1	0.2	1	1	1	1	1	1	1	1
0.4	0	0.1	0.2	0.3	1	1	1	1	1	1	1
0.5	0	0.1	0.2	0.3	0.4	1	1	1	1	1	1
0.6	0	0.1	0.2	0.3	0.4	0.4	1	1	1	1	1
0.7	0	0.1	0.2	0.3	0.3	0.3	0.3	1	1	1	1
0.8	0	0.1	0.2	0.2	0.2	0.2	0.2	0.2	1	1	1
0.9	0	0.1	0.1	0.1	0.1	C.1	0.1	0.1	0.1	1	1
1	0	0	0	0	0	0	0	0	0	0	1

Implicative transition table for ALI3 fuzzy logic is given in Table 2.10:

Let us consider the analytical expression for defining double implication operations for proposed fuzzy logic ALI1-ALI3 [9]:

$$a \underset{ALI1}{\leftrightarrow} b = \begin{cases} 1 - ka, & \text{if } a < b, \\ 1, & \text{if } a = b, \\ 1 - kb, & \text{if } a > b; \end{cases}$$

$$a \underset{ALI2}{\leftrightarrow} b = \begin{cases} 1, & \text{if} \quad a = b, \\ 0, & \text{if} \quad (ka = 1) \vee (kb = 1), \\ \min\left[(\leftarrow a \wedge b), (\leftarrow b \wedge a)\right], & \text{otherwise;} \end{cases}$$

$$a \underset{ALI3}{\leftrightarrow} b = \begin{cases} 1, & \text{if} \quad a = b, \\ 0, & \text{if} \quad (a = 0) \vee (b = 0), \\ \min\left[\dfrac{a}{b + \leftarrow a}, \dfrac{b}{a + \leftarrow b}\right], & \text{otherwise,} \end{cases}$$

where ka=max(a,1-a) .

Table 2.10. Implicative transition table for logic ALI3

					$\underset{ALI3}{\rightarrow}$						
b a	0	0.1	0.2	0.3	0.4	0.5	0.6	0.7	0.8	0.9	1
0	1	1	1	1	1	1	1	1	1	1	1
0.1	0	1	1	1	1	1	1	1	1	1	1
0.2	0	1/11	1	1	1	1	1	1	1	1	1
0.3	0	1/12	2/11	1	1	1	1	1	1	1	1
0.4	0	1/13	1/6	3/11	1	1	1	1	1	1	1
0.5	0	1/14	2/13	1/4	4/11	1	1	1	1	1	1
0.6	0	1/15	1/7	3/13	1/3	5/11	1	1	1	1	1
0.7	0	1/16	2/15	3/14	4/13	5/12	6/11	1	1	1	1
0.8	0	1/17	1/8	1/5	2/7	5/13	1/2	7/11	1	1	1
0.9	0	1/18	2/17	3/16	4/15	5/14	6/13	7/12	8/11	1	1
1	0	1/19	1/9	3/17	1/4	1/3	3/7	7/13	2/3	9/11	1

Definition. Top of a fuzzy set
The top of fuzzy set B is

$$HB = \underset{U}{\vee} \propto_B x.$$

Definition. Bottom of a fuzzy set
The bottom of fuzzy set B is

$$pB = \underset{U}{\wedge} \propto_B x.$$

Definition. Nonfuzziness
Nonfuzziness $a \in U$ is $ka = a \vee (1-a)$.

Then nonfuzziness of fuzzy set B is defined as:

$$kB = \underset{U}{\wedge} k\propto_B x$$

Let us give a brief semantic analysis of the proposed fuzzy logics by using the terminology accepted in the theory of power fuzzy sets. For this purpose we formulate the following.

Proposal 2.1
Possibility degree of the inclusion of set $\pi(A \subseteq B)$ in fuzzy logic ALI1-ALI3 is determined as:

$$\pi_1(A \subseteq B) = \begin{cases} 1-\propto_A x, & \text{if} \quad \propto_A x < \propto_B x, \\ 1, & \text{if} \quad \propto_A x = \propto_B x, \\ \propto_B x, & \text{if} \quad \propto_A x > \propto_B x; \end{cases}$$

$$\pi_2(A \subseteq B) = \begin{cases} 1, & \text{if } \propto_A x \leq \propto_B x, \\ (1-\propto_A x) \wedge \propto_B x, & \text{if } \propto_A x > \propto_B x; \end{cases}$$

$$\pi_3(A \subseteq B) = \begin{cases} 1, & \text{if} \quad \propto_A x \leq \propto_B x, \\ \dfrac{\propto_B x}{\propto_A x + (1-\propto_B x)}, & \text{if} \quad \propto_A x > \propto_B x. \end{cases}$$

We note, that if $\propto_A x = 0$ or $A \neq \varnothing$, then the crisp inclusion is possible for fuzzy logic ALI1. Below we consider the equivalence of fuzzy sets.

Proposal 2.2
Possibility degree of the equivalence of the sets $\pi(A \equiv B)$ is determined as:

$$\pi_1(A \equiv B) = \begin{cases} 1-\left[(1-\propto_A x) \vee \propto_B x\right], & \text{if} \quad \propto_A x < \propto_B x, \\ 1, & \text{if} \quad A = B, \\ 1-\left[(1-\propto_B x) \vee \propto_A x\right], & \text{if} \quad \propto_A x > \propto_B x, \end{cases}$$

$$\pi_2(A \equiv B) = \begin{cases} 1, & \text{if} \quad A = B, \\ \underset{T}{\wedge}\left\{\left[(1-\propto_A x) \wedge \propto_B x\right], \left[(1-\propto_B x) \wedge \propto_A x\right]\right\} & \text{if} \quad A \neq B, \\ 0, \quad \text{if } \exists x \,|||\, \propto_A x = 0, \propto_B x \neq 0 & (\text{or vice versa}), \\ \quad \text{and also } \exists x \,|||\, \propto_A x = 1, \propto_B x \neq 1 & (\text{or vice versa}), \end{cases}$$

$$\pi_3(A \subseteq B) = \begin{cases} 1, & \text{if } \propto_A x \leq \propto_B x, \\ \dfrac{\propto_B x}{\propto_A x + (1-\propto_B x)}, & \text{if } \propto_A x > \propto_B x. \end{cases}$$

Here the set $T = \{x \in U | \propto_A x \neq \propto_B x\}$ and A=B means that $\forall x$

$\propto_A(x) = \propto_B(x)$ or in other words, $T = \varnothing$.

The symbol ||| means "such as ".

From the expression $\pi_i(A \equiv B)$, $i = \overline{1,3}$, it follows that for ALI1 fuzzy logic the equivalency $\pi_1(A \equiv B) = 1$ takes place only when A=B.

It is obvious that the equivalence possibility is equal to 0 only in those cases when one of the statements is crisp, i.e. either true or false, while the other is fuzzy.

Proposal 2.3
Degree to which fuzzy set B is empty $\pi(B \equiv \varnothing)$ is determined as

$$\pi_1(B \equiv \varnothing) = \begin{cases} 1, & \text{if } B = \varnothing, \\ 0, & \text{otherwise;} \end{cases}$$

$$\pi_2(B \equiv \varnothing) = \begin{cases} 1, & \text{if } hB < 1 \text{ or } B = \varnothing, \\ 0, & \text{otherwise;} \end{cases}$$

$$\pi_3(B \equiv \varnothing) = \begin{cases} 1, & \text{if } B = \varnothing, \\ 0, & \text{otherwise.} \end{cases}$$

Here $B = \varnothing$ means that for $\forall x \propto_B x = 0$, or equivalently hB=0.

We introduce the concept of disjointness of fuzzy sets. There are two kinds of the disjointness. For a set A the first kind is defined by degree to which set A is a subset of the complement of B^c. The second kind is the degree to which the intersection of sets is empty. Therefore, we formulate the following.

Proposal 2.4
Degree of disjointness of sets A and B is degree to which A and B are disjoint

$$\pi(A \text{ disj}_1 \ B) = \pi(A \subseteq B^C) \wedge \pi(B \subseteq A^C),$$

$$\pi(A \text{ disj}_2 \ B) = \pi((A \cap B) = \varnothing) .$$

Proposal 2.5
Disjointness grade of sets A and B is determined as

$$1. \quad \pi_1(A \text{ disj}_1 \ B) = \begin{cases} 1, & \text{if } \exists x \,|||\, \propto_A x = 1 - \propto_B x, \\ (1 - \propto_B x) \wedge (1 - \propto_B x), & \text{otherwise,} \\ 0, & \text{never;} \end{cases}$$

$$\pi_2(A \text{ disj}_1 \ B) = \begin{cases} 1, & \text{if } \propto_A x \le 1 - \propto_B x, \\ 0, & \text{if } \exists x \,|||\, \propto_A x = 1, \text{ but } \propto_B x \ne 0, \\ & \text{or } \propto_B x = 1, \text{ but } \propto_A x \ne 0, \\ \underset{T}{\wedge}\left[(1 - \propto_A x), (1 - \propto_B x)\right], & \text{otherwise;} \end{cases}$$

$$3.\ \pi_3\left(A \text{ disj}_1 B\right)=\begin{cases}1, & \text{if } \propto_A x=\propto_B x \text{ or } \propto_B x=0,\\ \bigwedge\limits_T\left[\dfrac{1-\propto_B x}{\propto_A x+(1-\propto_B x)},\dfrac{1-\propto_A x}{\propto_B x+(1-\propto_A x)}\right], & \text{otherwise},\\ 0, & \text{never}.\end{cases}$$

Here

$$T=\{x \,|\, \propto_A x > 1-\propto_B x\}$$

We note that, the disjointness degree of the set is equal to 0 only for fuzzy logic ALI2, when under the condition that one of the considered fuzzy sets is normal, the other is subnormal.

Proposal 2.6
Degree to which set is a subset of its complement for the considered fuzzy logics $\pi_i\left(A \subseteq B^C\right)$ takes the following form

$$1.\ \pi_1\left(A\subseteq A^C\right)=\begin{cases}1, & \text{if } hA=0,\\ 0, & \text{if } hA=1,\\ 1-hA, & \text{otherwise};\end{cases}$$

$$2.\ \pi_2\left(A\subseteq A^C\right)=\begin{cases}1, & \text{if } hA\le 0.5,\\ 0, & \text{if } hA=1,\\ 1-hA, & \text{otherwise};\end{cases}$$

$$3.\ \pi_3\left(A\subseteq A^C\right)=\begin{cases}1, & \text{if } hA\le 0.5,\\ 0, & \text{if } hA=1,\\ (1-hA)/(2hA), & \text{otherwise}.\end{cases}$$

It is obvious that for the fuzzy logic ALI1 the degree to which a set is the subset of its complement is equal to the degree to which this set is empty. It should also be mentioned that the semantic analysis given in [4,13,15] as well as the analysis given above show a significant analogy between features of fuzzy logics ALI1 and KD. However, as it will be shown below, the fuzzy logic ALI1, unlike the KD logic, has a number of advantages. For example, ALI1 logic satisfies the condition $\propto_A x \wedge (\propto_A x \rightarrow \propto_B x) \le \propto_B x$ necessary for development of fuzzy conditional inference rules. ALI2 and ALI3 logics satisfy this inequality as well. This allows them to be used for the formalization of improved rules of fuzzy conditional inference and for the modeling of various systems, which is considered in next chapter.

Approximate Reasoning

Lately in fuzzy sets research the great attention is paid to the development of Fuzzy Conditional Inference Rules (CIR) [4,15,20,23,40,52,58,65,86,108]. This is connected with the feature of the natural language to contain a certain number of fuzzy concepts (F-concepts), therefore we have to make logical inference in which the preconditions and conclusions contain such F-concepts. The practice shows that there is a huge variety of ways in which the formalization of rules for such kind of inferences can be made. However, such inferences cannot be satisfactorily formalized using the classical Boolean Logic, i.e. here we need to use multi-valued logical systems. The development of the conditional logic rules embraces mainly three types of fuzzy propositions:

P_1=IF x is A THEN y is B

P_2=IF x is A THEN y is B

OTHERWISE C

P_3=IF x_1 is A_1 AND x_2 is A_2 AND ... AND x_n is A_n

THEN y is B.

The conceptual principle in the formalization of fuzzy rules is the Modus Ponens inference (separation) rule that states:

IF $(\alpha \rightarrow \beta)$ is true and α is true THEN β is true.

The methodological base for this formalization is the compositional rule suggested by L.Zadeh [111,113]. Using this rule, he formulated some inference rules in which both the logical preconditions and consequences are conditional propositions including F-concepts. Later E.Mamdani [58] suggested inference rule, which like Zadeh's rule was developed for the logical proposition of type P1. In other words the following type F-conditional inference is considered:

Proposition 1: IF x is A THEN y is B

Proposition 2: x is A' (2.105)

Conclusion: y is B',

where A and A' are F-concepts represented as F-sets in the universe U; B is F-conceptions or F-set in the universe V. It follows that B' is the consequence represented as a F-set in V.

To obtain a logical conclusion based on the CIR, the Propositions 1 and 2 must be transformed accordingly to the form of binary F-relation $R(A_1(x),A_2(y))$ and unary F-relation $R(A_1(x))$. Here $A_1(x)$ and $A_2(y)$ are defined by the attributes x and y which take values from the universes U and V, respectively. Then

$$R(A_1(x))=A' \qquad (2.106)$$

According to Zadeh-Mamdani's inference rule $R(A_1(x), A_2(y))$ is defined as follows.

The maximin conditional inference rule

$$R_m(A_1(x), A_2(y))=(A\cdot B)\cup(\leftarrow A\cdot V) \tag{2.107}$$

The arithmetic conditional inference rule

$$R_a(A_1(x), A_2(y))=(\leftarrow A\cdot V)\oplus(U\cdot B) \tag{2.108}$$

The mini-functional conditional inference rule

$$R_c(A_1(x), A_2(y))= A\cdot B \tag{2.109}$$

where $\cdot$, $\cup$ and $\leftarrow$ is the Cartesian product, union, and complement operations, respectively; $\oplus$ is the limited summation.

Thus, in accordance with [58,111,113] the logical consequence $R(A_2(y))$, (B' in (2.105)) can be derived as follows:

$$R(A_2(y))=A'\circ [(A\cdot B)]\cup[\leftarrow A\cdot U)]$$

$$R(A_2(y))=A'\circ [(\leftarrow A\cdot V)]\oplus[U\cdot B)]$$

or

$$R(A_2(y))=A'\circ (A\cdot B)$$

where $\circ$ - is the F-set maximin composition operator.

On the base of these rules the conditional inference rules for type P_2 were suggested in [20]:

$$R_4(A_1(x), A_2(y))=[(\leftarrow A\cdot V)\oplus(U\cdot B)]\cap[(A\cdot V)\oplus(U\cdot C)] \tag{2.110}$$

$$R_5(A_1(x), A_2(y))=[(\leftarrow A\cdot V)\cup(U\cdot B)]\cap[(A\cdot V)\cup(U\cdot C)] \tag{2.111}$$

$$R_6(A_1(x), A_2(y))=[(A\cdot B)\cup(\leftarrow A\cdot C)] \tag{2.112}$$

Note that in [20] also the fuzzy conditional inference rules for type P_3 were suggested:

$$R_7(A_1(x), A_2(y)) = \left[\bigcap_{i=1,n}(\leftarrow A_i\cdot V)\right]\oplus[(U\cdot B)] \tag{2.113}$$

$$R_8(A_1(x), A_2(y)) = \left[\bigcap_{i=1,n}(\leftarrow A_i\cdot V)\right]\cup[(U\cdot B)] \tag{2.114}$$

$$R_9(A_1(x), A_2(y)) = \left[\bigcap_{i=1,n}A_i\right]\cdot B \tag{2.115}$$

In order to analyze the effectiveness of rules (2.105)-(2.115) we use some criteria for F-conditional logical inference suggested in [40].The idea of these criteria is to compare the degree of compatibility of some fuzzy conditional inference rules with the human intuition when making approximate reasoning. These criteria are the following:

Criterion I

Precondition 1: IF x is A THEN y is B
Precondition 2: x is A

Conclusion: y is B

Criterion II-1

Precondition 1: IF x is A THEN y is B
Precondition 2: x is very A

Conclusion: y is very B

Criterion II-2

Precondition 1: IF x is A THEN y is B
Precondition 2: x is very A

Conclusion: y is B

Criterion III

Precondition 1: IF x is A THEN y is B
Precondition 2: x is more or less A

Conclusion: y is more or less B

Criterion IV-1

Precondition 1: IF x is A THEN y is B
Precondition 2: x is not A

Conclusion: y is unknown

Criterion IV-2

Precondition 1: IF x is A THEN y is B
Precondition 2: x is not A

Conclusion : y is not B

In [40] it was shown that in Zadeh-Mamdani's rules the relations R_m, R_a and R_c do not always satisfy the above criteria.

For the case of mini-operational rule R_c it has been found that criteria I and II-2 are satisfied while criteria II-1 and III are not.

In [40] an important generalization was made that allows some improvement to the mentioned F-conditional logical inference rules. It was shown there that for the conditional proposition arithmetical rule defined by Zadeh

$$P_1 = \text{IF x is A THEN y is B}$$

the following takes place

$$R_a(A_1(x), A_2(y)) = (\leftarrow A \cdot V) \oplus (U \cdot B) = \int_{U \cdot V} 1 \wedge (1 - \propto_A(u) + \propto_B(v))/(u,v)$$

The membership function for this F-relation is

$$1 \wedge (1 - \propto_A(u) + \propto_B(v))$$

that obviously meets the implication operation or the P*ly*-operator for the multi-valued logic L (by Lukasiewicz), i.e.

$$T(P \underset{L}{\rightarrow} Q) = 1 - (1 - T(P) + T(Q)) \tag{2.116}$$

where $T(P \underset{L}{\rightarrow} Q), T(P)$ and $T(Q))$ - are the truth values for the logical propositions $P \underset{L}{\rightarrow} Q$, P and Q respectively.

In other words, these expressions can be considered as adaptations of implication in the L-logical system to a conditional proposition.

Having considered this fact, the following expression was derived:

$$R_a(A_1(x), A_2(y)) = (\leftarrow A \cdot V) \oplus (U \cdot B) =$$

$$= \int_{U \cdot V} 1 \wedge (1 - \propto_A(u) + \propto_B(v))/(u,v) = \int_{U \in V} (\propto_A(u) \underset{L}{\rightarrow} \propto_B(v))/(u,v) = \tag{2.117}$$

$$= (A \cdot V) \underset{L}{\rightarrow} (U \cdot B)$$

In [40] an opinion was expressed that the implication operation or the Ply-operator in the expression (2.117) may belong to any multi-valued logical system. The following are guidelines for deciding which logical system to select for developing F-conditional logical inference rules [40].

Let F-sets A from U and B from V are given in the form:

$$A = \int_V \propto_A(u)/u, \; B = \int_V \propto_B(v)/v$$

Then, as mentioned above, the conditional logical proposition P1 can be transformed to the F-relation $R(A_1(x), A_2(y))$ by adaptation of the Ply-operator in multi-valued logical system, i.e.

$$R(A_1(x), A_2(y)) = A \cdot V \rightarrow U \cdot B = \tag{2.118}$$

$$= \int_{U \cdot V} (\propto_A(u) \rightarrow \propto_B(v))/(u, v)$$

where the values $\propto_A(u) \rightarrow \propto_B(v)$ are depending on the selected logical system.

Assuming $R(A_1(x))=A$ we can conclude a logical consequence $R(A_2(y))$, then using the CIR for $R(A_1(x))$ and $R(A_1(x), A_2(y))$, then

$$R(A_2(y)) = A \circ R(A_1(x), A_2(y)) =$$

$$= \int_U \propto_A(u)/u \circ \int_{U \cdot V} \propto_A(u) \to \propto_B(v))/(u,v) = \qquad (2.119)$$

$$= \int_V \bigvee_{u \in V} [\propto_A(u) \wedge (\propto_A(u) \to \propto_B(v))]$$

For the criterion I to be satisfied, one of the following equalities must hold true

$$R(A_2(y))=B$$

$$\bigvee_{u \in V} [\propto_A(u) \wedge (\propto_A(u) \to \propto_B(v))] = \propto_B(v)$$

or

$$\propto_A(u) \wedge (\propto_A(u) \to \propto_B(v)) \le \propto_B(v) \qquad (2.120)$$

the latter takes place for any $u \in U$ and $v \in V$ or in terms of truth values:

$$T(P \wedge (P \to Q)) \le T(Q) \qquad (2.121)$$

The following two conditions are necessary for formalization of F-conditional logical inference rules:

1. The conditional logical inference rules (CIR) must meet the criteria I-IV;
2. The conditional logical inference rules (CIR) satisfy the inequality (2.121).

Now we consider formalization of the fuzzy conditional inference for a different type of conditional propositions. As was shown above, the logical inference for conditional propositions of type P_1 is of the following form:

Proposition 1: IF x is A THEN y is B

Proposition 2: x is A'

Conclusion: y is B` (2.122)

where A, B, and A' are F-concepts represented as F-sets in U, V, and V, respectively, which should satisfy the criteria I, II-1, III, and IV-1.

For this inference if the Proposition 2 is transformed to an unary F-relation in the form $R(A_1(x))$=A' and the Proposition 1 is transformed to an F-relation $R(A_1(x), A_2(y))$ defined below, then the conclusion $R(A_2(y)$ is derived by using the corresponding F-conditional logical inference rule, i.e.

$$R(A_2(y))= R(A_1(x))^{\circ}R(A_1(x), A_2(y)) \qquad (2.123)$$

where $R(A_2(y))$ is equivalent to B' in (2.122).

Fuzzy Conditional Inference Rule 1
Theorem 2.3
If the F-sets A from U and B from V are given in the traditional form:

$$A = \int_U \propto_A(u)/u; \quad B = \int_V \propto_B(v)/v \tag{2.124}$$

and the relation for the multi-valued logical system ALI1

$$R_1(A_1(x), A_2(y)) = A \cdot V \underset{ALI1}{\rightarrow} U \cdot B = \int_{U \cdot V} \propto_A(u)/(u,v) \underset{ALI1}{\rightarrow} \int_{U \cdot V} \propto_B(v)/(u,v) = \\ = \int_{U \cdot V} \propto_A(u) \underset{ALI1}{\rightarrow} \propto_B(v)/(u,v) \tag{2.125}$$

where

$$\propto_A(u) \underset{ALI1}{\rightarrow} \propto_B(v) = \begin{cases} 1 - \propto_A(u), & \propto_A(u) < \propto_B(v), \\ 1, & \propto_A(u) = \propto_B(v), \\ \propto_B(v), & \propto_A(u) > \propto_B(v) \end{cases}$$

then the criteria I-IV are satisfied.

Fuzzy Conditional Inference Rule 2
Theorem 2.4
If the F-sets A from U, and B from V are the same as in (2.124) and the binary relation $R_2(A_1(x),A_2(y))$ for logical multi-valued system ALI2 is defined as follows:

$$R_2(A_1(x), A_2(y)) = A \cdot V \underset{ALI2}{\rightarrow} U \cdot B = \int_{X \cdot V} \propto_A(u) \underset{ALI2}{\rightarrow} \propto_B(v)/(u,v), \tag{2.126}$$

where

$$\propto_A(u) \underset{ALI2}{\rightarrow} \propto_B(v) = \begin{cases} 1, \ \propto_A(u) \le \propto_B(v); \\ (1 - \propto_A(u)) \wedge \propto_B(v); \\ \propto_A(u)) > \propto_B(v); \end{cases}$$

then the criteria I-IV are satisfied.

Fuzzy Conditional Inference Rule 3
Theorem 2.5
If F-sets A from U and B from V are the same as in (2.124) and the binary relation $R_3(A_1(x), A_2(y))$ for logical system ALI3 takes the form:

$$R_3(A_1(x), A_2(y)) = A \cdot V \underset{ALI3}{\rightarrow} U \cdot B = \\ = \int_{U \cdot V} \propto_A(u) \underset{ALI3}{\rightarrow} \propto_B(v)/(u, v) \tag{2.127}$$

where

$$\propto_A(u) \underset{ALI3}{\rightarrow} \propto_B(v) = \begin{cases} 1, & \propto_A(u) \le \propto_B(v) \\ \dfrac{\propto_B(v)}{\propto_A(u) + (1 - \propto_B(v))}, & \propto_A(u) > \propto_B(v) \end{cases}$$

then the criteria I-IV are satisfied.

The following example for the logical multi-valued system ALI1 can be shown as an illustration of the suggested inference rules.
Below we consider experimental selecting appropriate fuzzy implications. Let the fuzzy system can be appropriately described by a relational fuzzy equation:

$$B = A \circ R \qquad (2.128)$$

where $A = \{u, \propto(u)\}$ - is the input fuzzy set on U and $B = \{v, \propto(v)\}$ - is the output fuzzy set on V; u and v are elements of fuzzy sets A and B respectively; $\propto(u)$ and $\propto(v)$ are corresponding membership functions; R is a fuzzy relation on UxV in the form of rule with the structure IF...THEN...ELSE. The fuzzy relation R in UxV has membership function

$$\propto(u,v) = \int_{U \cdot V} [\propto_A(u) \to \propto_B(v)] \wedge [(1-\propto_A(u)) \to (1-\propto_B(v))]/(u,v) \qquad (2.129)$$

where $\to$ is implicat
ion; $\wedge$ is logical minimum.
In accordance with (2.102) solution of the equation in known R and A can be shown as;

$$\propto_B(v) = \max_u \min\{\propto(u), \propto_R(u,v)\} \qquad (2.130)$$

Let's represent the measuring system as equations [15]:

$$X(t) = u(t) + w_x,\ Y(t) = v(t) + w_y, \qquad (2.131)$$

where $u(t) \in U$ and $v(t) \in V$ is the input and output system parameters, w_x and w_y are the corresponding measurement errors.
Let's define operators for the formation of fuzzy sets A, B $Q_x : X \to A$, $Q_y : Y \to B$, where the parameters $X \subset [X_{min}, X_{max}]$ and $Y \subset [Y_{min}, Y_{max}]$ are considered as source data for the formation of linguistic variables and for calculating appropriate values of membership functions $\propto(u)$ and $\propto(v)$. The index of efficiency is considered as:

$$J(R) = \sum_{i=1}^{n} (\propto_{B_i}(v) - \propto_{B_i^\wedge}(v))^2, \qquad (2.132)$$

where $\propto_{B^\wedge}$ is the estimate of membership function of $B^\wedge$, calculated on the basis of the solution to the equation (2.130) with the estimation of the matrix $R^\wedge$. R is determined from (2.132) as:

$$R = \arg\min_{R \subset \{R\}}(J) \qquad (2.133)$$

where {R} is the finite space of fuzzy relational matrixes.

Thus a class of models of the system in form of fuzzy relational equations (2.128), (2.130), a measurement model (2.131), operators for creating fuzzy sets of system's input and output and membership functions, an index (2.132) and criterion (2.133) are given. It is required to determine a fuzzy relation R in form of a fuzzy relational matrix satisfying the given criterion.

In [83] it is shown in order for the system described by a fuzzy relational equation to be correct-identifiable it is necessary the associated relational matrix to be non-singular. Hence the domain of permissible fuzzy relations is determined as $\{R_D\}=\{R|\det R\neq 0\}$.

For experimental test and determination of model, which would provide the existence of a correct solution in [83] an imitating model was developed. It was used simple model like Y=100-X or IF X is X_{min} THEN Y is Y_{max} . To construct the matrix R the rule in form IF X is X_{min} THEN Y is Y_{max} ELSE Y is Y_{min} is used.

The following are the changing parameters of the model. Dimension of the system, i.e. the cardinality of fuzzy sets of input and output N=3, 5, 7, 9, 11. The type of membership function (MF) – three types:

a) stepped MF

$$\propto(u_i)=1-[1/(\text{card }U_x-1)]\ |u_i-\text{ent}[(\text{card }U_x-1)][(X-X_{min})/(X_{max}-X_{min})]|$$

where $u_i=\text{ent}\{(\text{card }U_x-1)[(X-X_{min})/(X_{max}-X_{min})]\}$, card U_x is cardinality of the set of fuzzy variables ; X is the measured value of the parameter ;

b) exponential MF

$$\propto(u_i)=\exp\{-\gamma|K_i-X|\},$$

where $\gamma=2\ln 0.5/\sigma$, $\sigma=a-b$, $\propto(a)=\propto(b)=0.5$ is crossover points; K_i is support value of the parameter;

c) triangle MF

$$\propto(u_i)=1-\gamma|K_i-X|,\ \gamma=1/\sigma.$$

All types of implications – 10 types (see section 2.1.3).

The following parameters have been watched: the count of events D during a series of tests, for which the identified fuzzy relation matrix R appears to be non-specific; and the count of correct solutions P.

A solution is considered to be correct if it allows the same solution to fuzzy equation (2.129) with the same input set A either for real or identified matrix $R^\wedge$; number of coincidences of events: identified matrix is non-singular, solution is correct N_c; number of coincidences of events. identified matrix is singular, solution is correct N_{nonc}. Relations of the above mentioned parameters with different combinations of changing parameters in series of 50 measurements with 39 test measurement have been researched under supposal of absence of errors in measurements (i.e. identification done with step 2, test done with step 2,564). Note that in all experiments the condition of necessity has been confirmed, N_c=P, N_{nonc}=0. Thus only parameters P and D have been used to create the table of results (Table 2.11). The table contains only those versions of the structures (combinations of system dimensions, MF types, and implication types) for which the parameters P and D are not equal to zero.

The analysis of the results has allowed the following conclusions. The necessary conditions are satisfied for all versions of structures. In case when the fuzzy sets of input and output are described by stepped MF (type a) the necessary conditions is also sufficient condition as the Zadeh composition minmax operator for this structure satisfies the condition of linearity. The versions of the structures are not equivalent, therefore, one can choose the best based on specific conditions. To make a choice of a type of implication one can use additional criteria, including human intuition criteria [4,15]. A priory definition of the domain $\{R_D\}$ allows the synthesis of methods for regularization of the identification problem to get correct solutions almost in all changing range of input and output values.

Assume that structure of model was chosen R_k, which provided the existence of a non-empty set of solutions.

Table 2.11. The analysis of the identification of a fuzzy system with different versions of structures. Count of test in a series is 50

Dimension of the system N	MFs of Input and output	Degree of satisfaction of the necessity condition of identification P/D with different types of implication								
		1	3	4	5	6	7	8	9	10
3	a-a		25/25	25/25	25/25	25/25	/25	25/25	25/25	25/25
	a-b	-	-	- /25	12/25	20/25	/25	- /25	-	- /25
	a-c	-	-	- /24	6/24	12/24	/24	- /24	- /24	-
	b-a	-	-	-	-	- /33	-	-	-	-
	b-b	/ 50	41/41	2/50	- /50	- /50	/50	4 /50	41/50	17/50
	b-c	-	-	-	-	- /47	-	-	-	-
	c-a	-	-	- /24	24/25	- /26	/24	- /24	- /24	- /24
	c-b	-	-	- /36	3/36	- /47	/48	- /36	-	9/48
	c-c	-	30/30	2/48	3/48	3/48	/48	32/48	2/48	14/48
5	a-a	-	13/13	13/13	13/13	13/13	-	13/13	13/13	13/13
	a-b	-	-	-	- /8	- /13	-	-	-	- /13
	b-a	-	-	-	-	- /11	-	-	- /2	- /5
	b-b	/ 23	-	- /12	- /47	- /47	/27	- /12	- /29	- /48
9	a-a	-	7/7	7/7	7/7	7/7	-	7/7	7/7	7/7
	b-b	/ 17	49/49	- /39	- /50	- /50	/15	- /39	49/49	- /50
11	a-a	-	5/5	5/5	5/5	5/5	-	5/5	5/5	5/5
	b-b	/ 20	41/41	- /42	- /50	- /50	/20	- /42	41/41	- /50

In other words, we succeed in solving the problem of structural identification. The solution is found in form of rules:

$$k{=}1 - \text{IF X is } X_{min} \text{ THEN Y is } Y_{max} \text{ ELSE Y is } Y_{min},$$

$$k{=}2 - \text{IF X is } X_{min} \text{ THEN Y is } Y_{min} \text{ ELSE Y is } Y_{max}.$$

The efficiency index can be represented as:

$$J(R) = \sum_{i=1}^{n} [\propto_i(v) - \hat{\propto}_i(v)]^2 + \lambda\Phi\|\Delta R\| \quad (2.134)$$

where $\Delta R = R_{Dk} - R^{\sim}$ is a matrix of mismatch between the constructed fuzzy relational matrix $R^{\sim}$ and the permissible area {Rd}; $\|\Delta R\|$ is the norm of the mismatch matrix; Φ is the operator to check the mismatch; λ is the parameter of regularization to define the weight of a priory data ($0<\lambda<1$). For Cartesian norm and operator $\Phi = \min_k$ the index will be changed to

$$J(R) = \sum_{i=1}^{n} [\propto_i(v) - \propto_i^{\wedge}(v)]^2 + \lambda \min_k \left(\sum_i \sum_j \left| \propto^{lj}(u_i, v_i) - \propto^{lj}_{Dk}(u, v) \right|^2 \right)^{1/2}.$$

The estimate of matrix $R^{\wedge}$ for the chosen version of the structure is

$$R^{\wedge}(u_i, v_i) = R^{\sim}(u_i, v_i) + \lambda \min_k \Delta R_k(u_i, v_i),$$

$$R^{\sim}(u_i, v_i) = (X_i \cdot V \to U \cdot Y_i) \cap (\neg X_i \cdot V \to U \cdot \neg Y_i),$$

where x is Cartesian product; $\neg$ is logical complement; $\to$ is implication. In terms of membership functions we have

$$\propto^{\wedge}(u_i, v_i) = \propto^{\sim}(u_i, v_i) + \lambda\Delta\propto(u_i, v_i),$$

$$\tilde{R}(u_i, v_i) = \int_{U \cdot V} [\propto(u_i) \to \propto(v_i)] \wedge [(1 - \propto(u_i)) \to (1 - \propto(v_i))]/(u_i, v_i), \quad (2.135)$$

$$\Delta\propto(u_i, v_i) = \min_k \{\propto(u_i, v_i) - \propto_{Dk}(u_i, v_i)\}.$$

where $\Delta\propto(u_i, v_i)$ is the difference of corresponding elements of the identified and k-th permissible fuzzy relational matrix; $\propto_{Dk}(u_i, v_i)$ is element of k-th matrix $R_k \subset \{R_D\}$.

The investigations worked out on the imitating model have shown an efficiency of the regularization of the fuzzy identification problem on the basis of (2.135). (Table 2.12). The table shows the results for cardinality of the term-set N=3. The elements of permissible solutions for the chosen structure $\{R_D\}$ have been found provided that X=1 and Y=99. The analysis of the results allows concluding that the application of the method of regularization extends the scopes of the local identification providing for some structures even the global identification ability.

Mentioned experiments have proved advantages of logics ALI1, ALI2, and ALI3 in comparison with other described logics. It is well conformed to evidences analyzed in the section 2.1.3 of the book. Further investigations of author [83] are based only on these three logics. A number of results of author [82] are shown in Table 2.12

Table 2.12. The influence of the regularization parameter on the identification ability of fuzzy system. Count of tests in a series is 50

Membership functions of input and Output	Value of the egularization parameter λ	Count of correct solutions P in the series with different implications							
		3	4	5	6	7	8	9	10
a-a	0	25	25	25	25	-	25	25	25
	0,25	36	36	42	38	12	40	37	40
	0,50	37	34	39	41	12	39	37	39
	0,75	38	39	36	35	10	34	38	37
	1	36	36	41	32	-	35	37	35
a-b	0	-	-	12	19	-	-	-	-
	0,25	-	-	26	30	-	-	-	-
	0,50	-	12	26	34	13	8	-	8
	0,75	-	-	-	16	34	-	-	-
	1	-	-	-	40	-	-	-	-
a-c	0	-	-	8	12	-	-	-	-
	0,25	-	-	5	22	-	-	-	-
	0,50	-	-	-	22	-	-	-	-
	0,75	-	-	-	-	-	-	-	-
	1	-	-	-	-	-	-	-	-
b-b	0	49	2	-	-	-	2	49	20
	0,25	49	15	6	-	-	15	49	34
	0,50	50	26	22	-	-	26	49	39
	0,75	49	-	-	-	-	-	49	30
	1	49	-	-	-	-	-	50	-
c-a	0	-	-	24	2	-	-	-	-
	0,25	-	12	31	3	2	11	-	13
	0,50	16	7	36	4	13	16	6	14
	0,75	-	13	41	-	-	15	-	15
	1	-	-	34	-	-	-	-	-
c-b	0	-	-	-	-	-	-	-	5
	0,25	-	4	4	1	6	4	-	19
	0,50	-	2	2	2	25	2	-	19
	0,75	-	-	-	-	-	-	-	-
	1	-	-	-	-	-	-	-	-
c-c	0	26	2	2	2	-	2	26	14
	0,25	26	11	11	7	8	11	34	21
	0,50	49	26	26	16	25	27	50	28
	0,75	49	-	-	-	-	-	49	-
	1	50	-	-	-	-	-	49	-

In Table 2.13 exp is an exponential MF and ent is a stepped MF.

Table 2.13. Results of analyses of some fuzzy logics

Dimension N of the system	Membership function of input and output	Types of implication ALI1	ALI2	ALI3
3	exp	*	*	*
	ent	***	**	***
5	exp	-	***	-
	ent	***	*	***
7	exp	-	*	-
	ent	***	***	***
9	exp	-	-	-
	ent	***	*	***
11	exp	-	*	-
	ent	***		***

Symbols "-, *, **, ***" are degrees of performing of necessary condition of correct identification: "***" marks practically global identification on the whole definitional domain of parameters and, consequently, absence of limitations on measurement procedure, "**" shows that in the center of definitional domain of parameters an area of incorrect solution exists, "*" - an increasing of its size and or "-" area of incorrect solution practically covers a definitional domain of parameters of input and output a system. Assigning known parameters (system dimensionality and type of membership functions of term sets) we can choose types of implication, ensuring existence of correct decision. For instance, the logic ALI2 is acceptable for N = 5 and exponential membership functions of exp type. In addition, the final choice should be generated upon conditions of solution problem, physical nature of prototyping object, as well as additional criterions, for instance, criterions of human intuitions [4,15].

2.1.4 Probability and fuzziness [14]

Randomness and fuzziness represent two different types of uncertainty. First, consider the case of randomness. We regard the sample space as the set of all possible values, which a random variable may take. And let's consider the points of this space as the outcomes of the system. Our purpose is to predict the probability of event, which can occur. This event is represented in the sample space in a well-defined region. Thus, we consider the uncertainty as generated by function from sample space in a well-defined region of that space. For more clear description of situation we must show that well-defined region has the crisp boundary, because we use a concept of characteristic function. If the well-defined region is given by characteristic function, then the sample space is divided into two subspaces: if the point belongs to region, the characteristic function value is one and the event is true, otherwise this point falls outside the region, the value of characteristic function is zero and event is false. That way, the probability value shows the frequency at which the point will occur inside the region.

Case of fuzziness is essentially different from randomness. Here the region is represented by a fuzzy subset. Thus fuzzy uncertainty is represented by partial membership of a point from the universe of discourse in an imprecisely defined region of space. This membership function gets its values in interval [0,1], while a characteristic function of well-defined region takes values in set {0,1}, that is, only two values: 0 or 1. Thus, the membership function will not divide the universe of discourse into two subspaces, as it was above. Membership function describes the degree to which the element of universe space corresponds to the property with which fuzzy set is defined. In this case we use fuzzy measure.

Definition. Fuzzy measure

A function g, defined on nonempty family B of subsets of universal set X, is called fuzzy measure if it satisfies the following conditions:

1. $g(\varnothing)=0$ and $g(X)=1$
2. If $A \subseteq B$, then $g(A) \leq g(B)$
3. For $A_1 \subset A_2 \subset \ldots$ in B, if $\bigcup_{n=1}^{\infty} A_n \in B$, then (2.136)

$$\lim_{n\to\infty} g(A_n) = g\left(\bigcup_{n=1}^{\infty} A_n\right)$$

Table 2.14 shows the comparison of Zadeh's Fuzzy Sets Theory and Probability Theory.

Table 2.14. Comparison of fuzzy set theory with probability theory

	Probability Theory	Fuzzy sets Theory
Domain	σ-Algebra	Universe of Discourse X
Range of membership degree	[0,1]	[0,1]
Uncertainty measure	Probability measure	Fuzzy measure, Possibility measure
Constraints	$\sum_{i}^{n} P(A_i) = 1$	
Union	Σ	Max
Intersection	Π	Min
Inference	Conditioning	Generalized Modus-ponens

2.1.5 Fuzzy sets and possibility theory

L.Zadeh in [109,112] proposed an idea that a membership function in some cases can be interpreted as a possibility distribution. That is, an arbitrary set can be considered as a restriction of the values of a particular variable.

Combination of theoretic-set and logical ways of representation of imprecise information with methods of qualitative description of imprecision is the main problem of the possibility theory. The possibility theory has been widely used in real-world problems of modeling, control and decision making for which the measurements do not allow precise information or the information is obtained from a human expert. This theory allows both modeling the imprecision and qualitatively characterizing the uncertainty. It is necessary to distinguish between the notions of imprecision and uncertainty of data and knowledge. Imprecision belongs to the content of information while uncertainty belongs to its truth level at which it corresponds to reality. The possibility theory is based on fuzzy sets.

Definition. Possibility measure

Assume that $P(X)$ is a power fuzzy set of the universe X. Then the mapping $\Pi : P(X) \to [0,1]$ with the following properties:

$\Pi(\varnothing) = 0, \Pi(X) = 1$;

$A \subseteq B \to \Pi(A) \leq \Pi(B)$;

$\Pi(\bigcup_{i \in I} A_i) = \sup_{i \in I} \Pi(A_i)$,

is the possibility measure. Here I is index set.

In possibility theory every possibility measure is uniquely determined by the possibility distribution function.

Definition. Possibility distribution function

Given a possibility measure Π on power set $P(X)$ of X, function $r : X \to [0,1]$ such that

$$r(x) = \Pi(\{x\}) \tag{2.137}$$

for all $x \in X$ called a possibility distribution function of Π.

We can write that

$$\Pi(A) = \max_{x \in A} r(x) \tag{2.138}$$

for each $A \in P(X)$.

For

$$\forall A, B \subseteq P(X),\ \Pi(A \cup B) = \max(\Pi(A), \Pi(B)) \tag{2.139}$$

When X is not finite (2.138) must be written as

$$\Pi(A) = \sup_{x \in A} r(x) \tag{2.140}$$

The probability value of an event is completely defined by the probability value of the opposite event. The possibility value of an event and the possibility value of the opposite one are weakly connected.

Possibility distribution function can be interpreted as a membership function of the fuzzy set; i.e. $r(x)$ is defined to be $\propto_A$.

Possibility distribution is an important technique to represent imprecise information in the problem of approximate reasoning.

2.2 Foundations of Neurocomputing

2.2.1 Basic Types and Architectures of Neural Networks [14]

Feed-Forward Neural Networks

As shown above the non-recurrent (feed-forward) network has no feedback connections. In this network neurons of the s-th layer receive signals from environment (when s=1) or the neurons of previous the (s-1)-th layer (when s>1) and pass their outputs to neurons of the next (s+1)-th layer or to the environment (when s is the last layer).

The hierarchical non-recurrent network can be single-layer or multi-layer. A non-recurrent network containing one input and output layer usually is called a single-layer network. The input layer serves to distribute signals out of each inputs a neuron to all neurons of the output layer. Neurons of the output layer are a computing unit i.e. they calculate their outputs as a function applied to the weighted sum of the input signals. That function may be linear and non-linear. For the linear activation function, the output of the network is calculated in the following manner

$$Y=WX+\theta,$$

where W is weight vector of the network, X and Y are input and output vectors correspondingly.

The use of the nonlinear activation function allows an increase in the computational power of the network. For the sigmoid activation function, the network outputs are calculated in the following manner

$$Y=\frac{1}{1-e^{-XW+\theta}}.$$

Multilayer neural networks consist of input, output and hidden layers. Single layer networks, which do not include hidden layers, cannot solve complicated problems. The use of hidden layers allows an increase in the computational power of the network. The outputs of the s-th layer are functions of (s-i) -th (here

$i = \overline{1, s-1}$) layers outputs. By choosing the optimal structure of a network, an increase in reliability and computational power, and a decreased processing time are provided. Such an increase in the number of layers with the non-linear activation function allows an increased computational power of the network. Also, the processing time decreases. Increased number of neurons in hidden layers allows increased reliability of the network.

Recurrent Neural Networks

The recurrent neural network consists of both feedforward and feedback connections between the layers and neurons forming a complicated dynamic system. Obviously, the ability of a recurrent neural network to approximate a continuous or discrete non-linear dynamic system by neural dynamics defined by a system of non-linear differential or difference equations offers the opportunities for applications to adaptive control problems.

From a computational point of view, a dynamic neural structure which contains a state feedback may provide a more computational advantage than that of a purely feedforward neural structure. For similar problems, a small feedback system is equivalent to a large and possibly infinite feedforward system. A non-linear dynamic recurrent neural network structure is particularly appropriate for identification, control and filtering applications. Multilayered recurrent neural network architecture (MRNN) is a hybrid feedforward and feedback neural network, with the feedback represented by the recurrent connections and cross-talk, appropriate for approximating a non-linear dynamic system. MRNN is composed of an input layer, a series of hidden layers, and an output layer. It allows for feedforward and feedback among the neurons of neighbouring layers, and cross-talk and recurrence in the hidden layers [48].

A basic structure of multilayered recurrent neural networks with feedforward and feedback connections is shown in Fig. 2.21 [48].

Let M be equal the total number of hidden layers of the MRNN. The i-th neuron in the s-th hidden layer as denoted by neuron (s,i), L_S is the total number of neurons in the s-th hidden layer, x_i is the i-th input of the MRNN, $z_{s,i}(t)$ is the state of the neuron (s,i), y_i is the i-th output of the MRNN, $w_{s,j}^{s,i}$ is the intralayer link-weight coefficient from the neuron (s,j) to the neuron (s,i), $w_{s-1,j}^{s,i}$ is the feedforward link-weight coefficient from the neuron (s-1,j) to the neuron (s,i), $w_{s+1,j}^{s,i}$ is the feedback link-weight coefficient from the neuron (s+1,j) to the neuron (s,i), and $\theta^{s,i}$ is the threshold of the neuron (s,i). Mathematically, the operation of the neuron (s,i) is defined by the following dynamic equations [48].

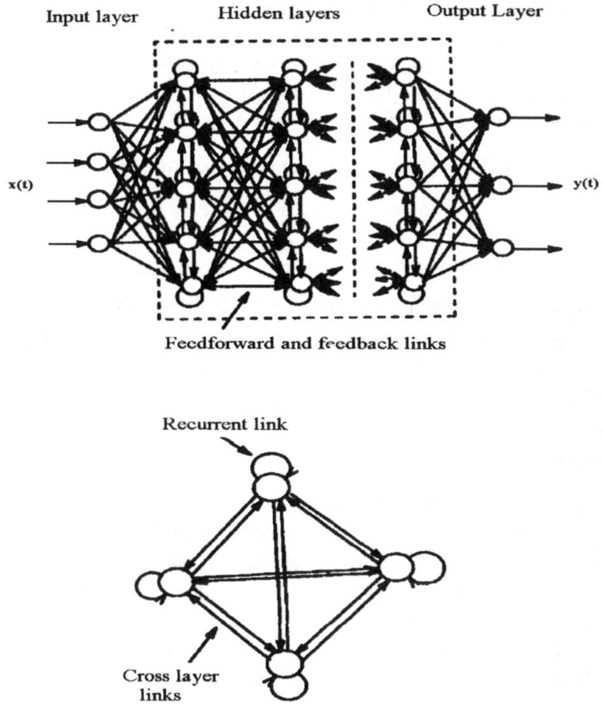

Fig. 2.21. Structure of MRNN

When s=1, i.e. for the first hidden layer:

$$z_{s,i}(t+1)=f[\sum_{j=1}^{L_{s+1}} w^{s,i}_{s+1,j}\cdot z_{s+1,j}(t)+\sum_{j=1}^{L_s} w^{s,i}_{s,j}\cdot z_{s,j}(t)+\sum_{j=1}^{L_{s-1}} w^{s,i}_{s-1,j}\cdot z_{s-1,j}(t)+\theta^{s,i}]=$$

$$=f[\sum_{r=-1}^{1}\sum_{j=1}^{L_{s+r}} w^{s,i}_{s+rj}\cdot z_{s+rj}(t)+\theta^{s,i}],\ s=2,3,...,M;\ i=1,2,...,L_s \tag{2.141}$$

$$z_{1,i}(t+1)=f\left[\sum_{j=1}^{L_2} w^{1,i}_{2,j}\cdot z_{2,j}(t)+\sum_{j=1}^{L_1} w^{1,i}_{1,j}\cdot z_{1,j}(t)+\sum_{j=1}^{L_0} w^{1,i}_{0,j}\cdot z_{0j}(t)+\theta^{1,i}\right],$$

$i = 1, 2, ..., L_1$.

For learning MRNN dynamic backpropagation is used, which will be described later.

Time-Delay Neural Networks

Typical structures of time-delay (tapped delay) neural networks (TDNNs), which consist of multilayered feedforward neural networks (MFNN) and some time delay operators with feedback are shown in Fig. 2.22 [48].

For time series analysis, the one-step and q-step prediction equations of the TDNNs can be given as

$$y(k+1)=F(W,y(k),y(k-1),...,y(k-t),u(k)) \tag{2.142}$$

and

$$y(k+q)=F(W,y(k),y(k-q),...,y(k-t),u(k)) \tag{2.143}$$

respectively, where $F(\cdot)$ is a continuous and differentiable function which may be obtained from the operation equations of the MFNN given in the last section, and the input components of the neural networks are the time delayed versions of the outputs of the networks or the time series systems. In this case, (2.142) and (2.143) represent a one-step ahead non-linear predictor and a q-step ahead non-linear predictor, respectively.

Radial Basis Function Neural Networks

Side by side with known neural-net constructions, having the effective arrangements for information processing, Radial Basis Function Neural Networks (RBFNN) are successfully applied to work out algorithms for solving difficult problems of artificial intelligence and control [62,63,81,84].

The models of RBFNN or similar ones were independently offered by many authors [26,56,67,68,72]. The most popular among them is RBFNN with hidden layer of RBF units, shown in Fig. 2.23. This network, consisting of n input and m output neurons, performs the mapping of the n-dimensional input vector X to the m-dimensional output vector Y ($R^n \to R^m$). The node elements φ_k (k=1,2,…,L) of the hidden layer represent RBFNN transformation of the input signals $x_i\ (i=1,2,...,n)$. The connections between the input layer and the hidden layer have no corresponding weights. Selection of an appropriate RBF depends on the type of problem we solve using the RBFNN. As RBF $\varphi_k=\varphi(\| X-a_k \|)= \varphi_k(r)$ the following radial function can be chosen :

$\varphi(r) = r$, a linear radial function;

$\varphi(r) = r^2$, a quadratic function;

$\varphi(r) = \exp(-r^2 / b^2)$, a gaussian function;

$\varphi(r) = r^2 \cdot \log(r)$, a thin-plate-spline-function;

$\varphi(r) = \sqrt{r^2 + b^2}$, a multiquadratic function.

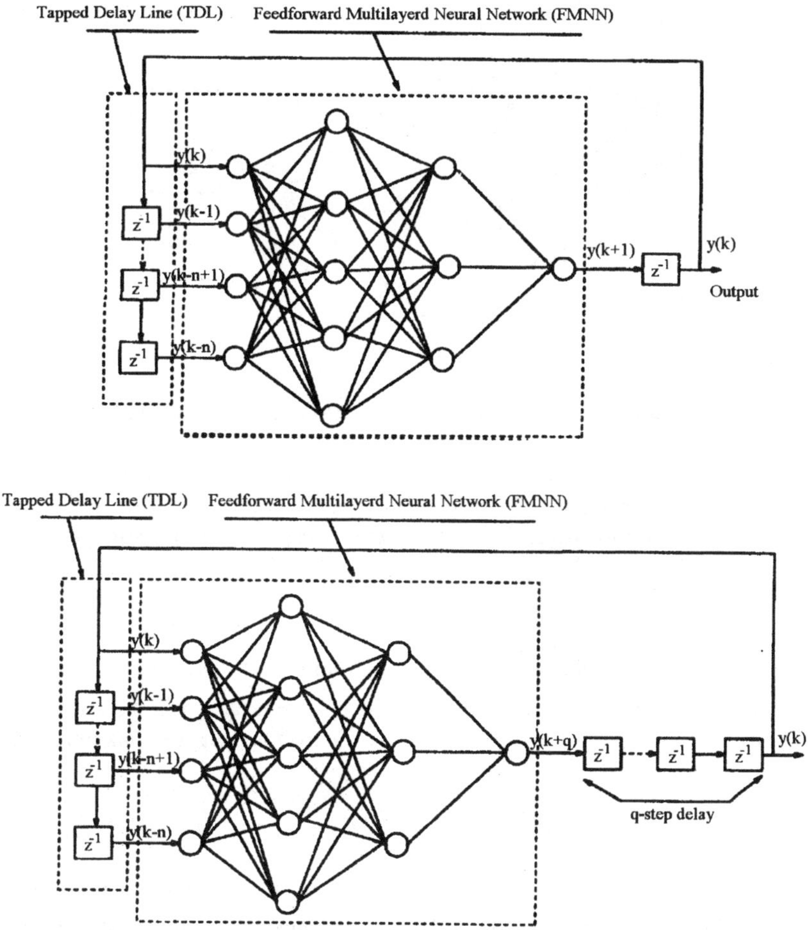

Fig. 2.22. Time-delay neural networks

In Fig. 2.23 the response characteristics of the k-th hidden unit are given by

$$\varphi_k(X) = \varphi\left(\frac{\|X - a_k\|}{b_k^2}\right) \tag{2.144}$$

where $\varphi(\cdot)$ is a strictly positive radially symmetric function with a unique maximum at k-th "center" a_k and which drops off rapidly to zero away from the center. The parameter b_k is the "width" of the receptive field in the input space for unit k. In other words, functions φ_k are defined in areas of the corresponding points a_k, which cause their "sensitive receptive" field. Parameter b_k defines geometrical size of k-th receptive field in the input space for unit k.

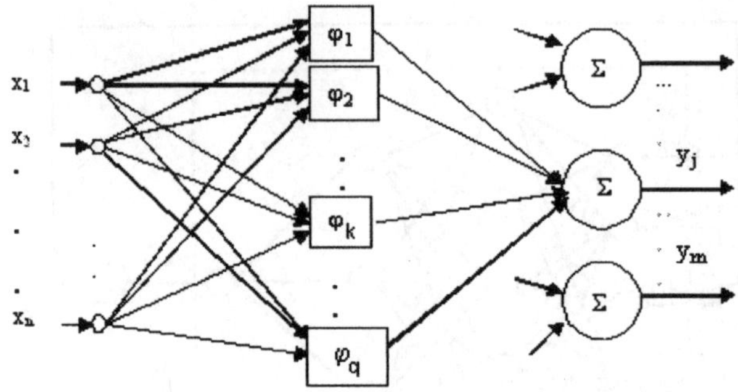

Fig. 2.23 Multi-input and multi-output structure of RBFNN with an only hidden layer

So, RBFNN are best suited for piecewise continuous real-valued mappings $R^n \to R^m$. Given an input n-dimensional vector $X=(x_1,x_2,\ldots,x_n)$, the output of the RBFNN is the m-dimensional activity vector $Y=(y_1,y_2,\ldots,y_m)$ whose jth component is given by

$$y_j(X) = \sum_{k=1}^{n} w_{kj}\varphi_k(X) \tag{2.145}$$

where w_{kj} is the weight of the connecting, uniting k-th RBF-unit of the hidden layer with j-th neuron of the output layer. If j=1, then the given structure will be with one output neuron. This network realizes the approximation function $F:R^n \to R$, which is described mathematically as

$$y(X) = \sum_{k=1}^{n} w_k\varphi_k(X) \tag{2.146}$$

The degree of accuracy of approximation $\varepsilon(I, a_i, b_i)$ depends on three parameters: number of RBF-units in the hidden layer, centers a_k and widths b_k of the hidden RBF-units.

Instead of the hidden j-th RBF-unit it is better to employ a Gaussian function given by

$$\varphi_k(X) = e^{\frac{\|X-a_k\|^2}{2b_k^2}} \tag{2.147}$$

where a_k and b_k (k= 1,2,…,n) are the standard mean deviation of k-th unit receptive field and dispersion, respectively, and the norm $\|X - a_k\|$ is the Euclidean norm in the form

$$\|X - a_k\| = \sqrt{\sum_{i=1}^{n}(x_k - a_{ki})^2}\ . \tag{2.148}$$

Unlike the traditional "feedforward" three-layer neural networks, the hidden layer of the RBFNN does not have corresponding weights and thresholds. It is formulated by other method. The hidden RBF-unit outputs are not calculated using the weighted-sum/sigmoidal activation mechanism. Here, each hidden RBF-unit output φ_k $(k=\overline{1,n})$ is obtained by calculating the input signal X to an n-dimensional parameter vector a_i.

2.2.2 Learning Algorithm of Neural Networks

We will at first consider supervised learning of Neural Networks.

Perception Learning Algorithm
In 1943 McCulloch and Pitts published the first systematic study of artificial neural networks. In later their work involved the simple neuron model shown in Fig. 2.24. Here the Σ multiples each input x_i by a weight w_i,$(i=\overline{1,n})$, and sums the weighted inputs. If this sum is greater than a neuron's threshold then the output is one; otherwise it is zero. The systems constructed on the basis of such neurons are called perceptrons. In general, they consist of artificial neurons connected by weights to a set of inputs. Perceptron systems may be single layer or multi layer.

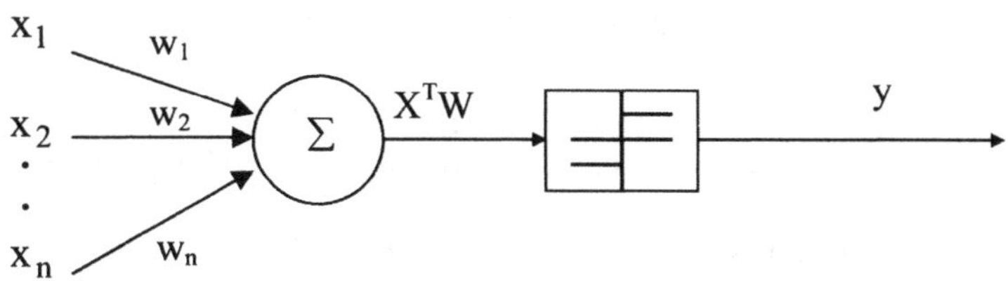

Fig. 2.24. Simple neuron model

A perceptron is trained by cyclic presenting a set of input patterns to its input and adjusting the connection weights until the desired output occurs for each of them.

Suppose that we wish to develop a perceptron system for classifying the figures from 0 to 9, shown on a rectangular card, into 2 classes: odd and even number. The card is divided into equal square cells by a set of horizontal and vertical lines. If the image line of a figure crosses a cell, then the perceptron's input, corresponding to this cell, receives 1 as its input signal, otherwise, it receives 0. The system has single output that should return 1 for every input representing an odd number and 0 for an even number.

Each input pattern can be represented as vector $X(x_1,x_2,\ldots,x_n)$, elements of which are zeros or ones. Output Y of the system is determined by comparing to the weighted sum of inputs signals with the threshold value θ: if the weighted sum of the elements of input vector exceeds θ, then the output of the system is 1, otherwise it is zero:

$$y = \begin{cases} 1, & \sum x_i w_i \geq \theta \\ 0, & \text{otherwise} \end{cases} \tag{2.149}$$

Learning is accomplished in the following manner. A pattern X is applied to the input of perceptron, and output Y is calculated. If the output is correct (i.e. corresponds to the desired one), the weights are not changed. If not, the weights, corresponding to input connections that cause this incorrect result, are modified to reduce the error.

Note, that all the weights are not changed so that to exclude any error for particular pattern at once. This can cause the system to "forget" the other, already learned patterns. Training must be global, the network must learn over the entire set of input patterns, applying them sequentially or randomly.

For our system, the learning process can be described as follows. Assume that the card bearing number 3 is the input to the system and the output is one: Y=1. Since that is the correct response, no weights are changed. If, however, the input is '4' and the output is 1, the weights of the connections that has 1 at their input must be decreased, because they are tending to produce an incorrect result. Analogously, if the card bearing '4' produces the output 0, then the weights corresponding to inputs that are 1 must be increased in order to correct erroneous result.

This training method may be generalized:

STEP 1. Accept the regular input pattern X and calculate output Y for it;

STEP 2.
a) If output Y is correct go to STEP 3;
b) If output Y is incorrect and equal to 0, then add each input to its corresponding weight;
c) If output Y is incorrect and equal to 1, subtract each input from its corresponding weight;

STEP 3. Repeat steps 1-3 until the learning error will be at acceptable level.

STEP 2 of the above-mentioned algorithm may be further generalized by introducing the term e that is the difference between the desired (target) output value y* and real (or actual) one y, and by the "learning rate" coefficient γ to regulate the average size of weight changes. Thus, we get STEP 2 in the following form:

$$\Delta_i = \gamma e x_i$$

$$w_i(t+1) = w_i(t) + \Delta_i,$$

where Δ_i -is the correction associated with input x_i, $w_i(t+1)$ -is the value of weight associated with i-th input after modification, $w_i(t)$ - is the value of weight associated with i-th input before adjustment.

This advanced algorithm, so-called "delta rule", allows the use of both binary values for input and output, and also gives a continuous representation (as a real value) for them.

Note that, the above-mentioned algorithm leads the system to correct the functioning in a finite number of steps (lessons). However, we cannot precisely evaluate this number. Furthermore, in some cases simply trying all possible adjustments of the weights may be superior. In addition to this, it should be mentioned that the representational ability of a single-layered network (perceptron in our case) is limited by condition of linear separability and there is no way to determine this condition if the dimension of input vectors is large.

Widrow-Hoff learning rule
Widrow and Hoff [106] suggested algorithm to train the linear unit, also known as the adaptive linear combiner element (Adeline), shown in Fig. 2.25. Their first model, the Adeline, had a single-output neuron; later model, Madeline, was extended to the multi-output case. In the figure the output of the linear unit in response to the input X is simply $Y=X^TW$. This rule is example of an error-correcting rule with a quadratic criterion function.

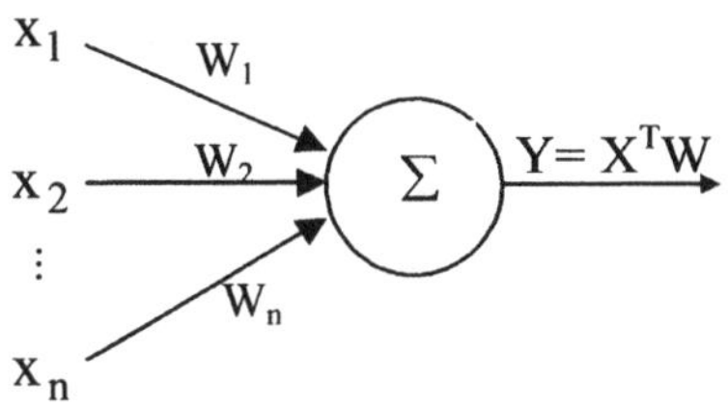

Fig. 2.25. Structure of linear neuron (adaptive linear combiner element)

Later, it was discovered that this rule converges in the mean square to the solution W* that corresponds to the least-mean-square (LMS) output error if all input patterns are of the same length (i.e., $||x^k||$ is the same for all k).

Widrow - Hoff training algorithm is as follows

1) Randomize all network weights

2)Apply an input vector X and calculate the network outputs by using following formula

$$y_j=\sum_i x_i w_{ij}$$

3)Compute the error for each neuron by subtracting the actual output from the target output.

$$e_i=y^*_i-y_i$$

4)Modify each weight as follows:

$$w_{ij}(t+1)=w_{ij}(t)+\gamma x_i e_i$$

5)Repeat steps 2 through 5 until the error is sufficiently low.

Delta rule

The delta rule is the learning algorithm for perceptrons having continuous activation function.

If activation function f is non-linear and differentiable then using formula for learning error one can get the delta rule in which the correction of weight coefficients is carried out in following manner.

$$\Delta w_i = \gamma\, (y^*_i - y_i)\, f'(I)\, x_i \qquad (2.150)$$

$$w_i(t+1) = w_i(t) + \Delta w_i \qquad (2.151)$$

where Δw_i is the correction associated to i-th input; $w_i(t)$ is the value of weight before adjustment; $w_i(t+1)$ is value of weight after adjustment.

The delta rule consists of following iterations:

a) Apply input pattern and calculate Y;

b)If the output and target signals are same, go to step a);

c)If the output and target signals aren't same, then change the weight coefficients using (2.150) and (2.151);

d)Go to step a).

Error backpropagation algorithm

In the previous section the procedure of training simple networks without hidden neurons was described. This procedure performs search for the network with minimal of error. However, unfortunately, such simple networks cannot solve complex problems, because of a lack of computational power.

For training multi–layer neural networks the least squares procedure must be generalized in order to provide adequate adjustment of the weight coefficients of connections, which come to hidden units. The error backpropagation algorithm [94,95] is a generalization of the least squares procedure for networks with hidden layers.

When such a generalization is built, the following question occurs: how to determine the measure of error for neurons of hidden layers? This problem is solved by estimating the measure of error through the errors of units of the subsequent layer. On every step of learning for each training pair of input/output set a first forward pass is performed. This means that the input of a neural network is given by the input vector and as a result, the activation flow passes through the network in the direction from the input layers towards the output.

After this process, states of all neurons of the network will already have been determined. Output neurons generate the actual output vector, which is compared with the desired vector, and the learning error is calculated. Then, this error is propagated backwards along the network in the direction of the input layer, modifying the values of weight coefficients.

Thus, the learning process is the consequence of interchanging forward and backward passes and during the forward pass, the states of network units are determined, while during the backward pass, the error is propagated and the values of the weights of the connections are updated. That is why this procedure is called the error backpropagation algorithm.

As we mentioned above, increasing the number of layers leads to enhancing the computational power of the network and, ultimately, to the possibility of providing much more complex mappings. It can be shown that the three-layer network can extract in input space convex regions. Adding a fourth layer would allow the extracting of the non-convex regions too [105]. Thus, by the use of four-layer neural network, practically any mapping can be provided. However, sometimes using more layers is effective.

On existence of hidden units the problem of their optimal use arises. The error backpropagation algorithm builds internal representations for the neurons of hidden layers. Note, that the automatic search of hidden units requires a significant expenditure of computer time and, therefore, increases to total amount of time needed for learning.

As one can see from the equation (2.150), for defining the step of modification of weight w_{ji}, calculating the value of derivative $\partial E / \partial w_{ji}$ is needed. This derivative, in turn, is determined through $\partial E / \partial y_j$. To define the latter derivative for hidden neurons, the following equation is used:

$$\frac{\partial E}{\partial y_j^{(s)}} = \sum_k \frac{\partial E}{\partial y_k^{(s+1)}} \frac{dy_k^{(s+1)}}{dI_k^{(s+1)}} \frac{dI_k^{(s+1)}}{dy_j^{(s)}} = \sum_k \frac{\partial E}{\partial y_k^{(s+1)}} \frac{dy_k^{(s+1)}}{dI_k^{(s+1)}} w_{kj}$$

where $y_j^{(s)}$ – is output of j-th neuron of the n-th hidden layer, $y_k^{(s+1)}$ is output of k-th neuron of (s+1)-st layer, $I_k^{(s+1)}$ is the total weighted sum of k-th neuron of the (s+1)-st layer. Thus, if the network includes M layers, then the derivative $\partial E / \partial y_j^{(M)}$, is calculated for output units and then values of $\partial E / \partial y_j^{(M-1)}$, $\partial E / \partial y_j^{(M-2)}, \ldots, \partial E / \partial y_j^{(1)}$ are defined consequently.

The activation function f must be differentiable everywhere.

The sigmoid function is used typically as an activation function. As we mentioned in the previous section, this function has the following derivative:

$$\frac{dy}{dx} = y(1-y)$$

Before the learning process one should assign small random values to all weight coefficients. It is very important that initial values of weights not be equal to each other.

Because the error backpropagation algorithm is mostly widespread from all methods of neural networks learning, let's consider it in detail.

Fig. 2.26 shows schematically the neuron used in this algorithm.

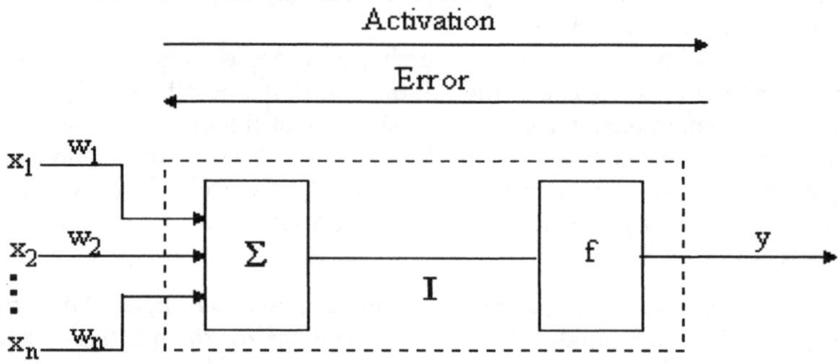

$$I = \sum_{i=1}^{n} x_i w_i$$

$$y = f(I)$$

Fig.2.26. Neuron used in the backpropagation algorithm

The given above formula for adjusting weight coefficients, is explicitly derived from the gradient descent method

$$\Delta w_i = -\gamma \frac{\partial E}{\partial w_i}$$

where E is the squared error that cumulatively measures the error through all cases given by the training set. In this case, all the input vectors are applied to the network consequently and therefore the measure of error is evaluated. According to this error, corrections are made. This procedure is called the batch version of the error backpropagation algorithm.
There is another approach for adapting weight coefficients. In the case where the single (current) input vector is applied, current output is generated and, in conformity with the error of the current case, the single step of weights correction is performed. Then the following input vector is selected and the process is continued. The latter procedure is called the real time version of error backpropagation algorithm.

Below we will consider the second type of algorithm (the real time version). We shall also discuss some modification of the basic procedure, intended for accelerating the learning process and avoiding the problems of stepping over the narrow minima.

Fig. 2.27 displays graphically the scheme of the error backpropagation algorithm. Let's now describe the steps of this algorithm [18].

Step 1. Start. Weight coefficients initialization.

Step 2. Repeat steps 3 - 6 for all vector pairs from the training set and then move to step 7.

Step 3. Apply the next input vector from the training set to the input neurons of network.

Step 4. Forward pass. Defining the states of all neurons of network layer by layer.

Step 5. Calculating the deviation (learning error) of actual output vector from desirable one.

Step 6. Backward pass. Propagating error back to the input layer. Modifying weight coefficients.

Step 7. If total learning error is not small enough, then return to the step 2.

Step 8. End of learning. Stop.

As mentioned above, at the starting point of procedure, weight coefficients are initialized by random numbers, distributed near zero. The initial settings often are crucial for success of learning. If initialization is not good, the network could fail in its attempt to learn. In this case, the learning procedure has to begin anew with another initial set of weight coefficients.

Step 4 is similar to the network functioning in recognition phase. At this step input vector X is connected to input nodes and for neurons of next layer total weighted inputs are defined:

$$I_j = \sum_i w_{ji} x_i$$

or in vector notation:

$$I = WX$$

After this is done, neurons of considered layer determine their output signals in accordance with the activation function.

$$y_j = f(\sum_i w_{ji} x_i)$$

or

$$Y = f(W X)$$

Outputs of this layer are inputs for the layer that follows. States of neurons are defined consequently: layer by layer until the output layer will be achieved and the output of the neural network will be achieved and the output of neural will generate actual output vector.

At Step 5 the deviation of the actual value of the output vector from the desired one from the current training pair is calculated. In accordance with the calculated value of deviation (error), at the 6-th step, modification of the weight coefficients of connections is performed consequently, layer by layer, but this time in reverse direction: from output layer – backwards to input one.

This weight adjustment process stops when the layer, which is most close to input, is achieved, after updating all weight coefficients of the network.

On the 7-th step the estimation of the magnitude of error is performed.If this magnitude is higher than the acceptable level then the learning process is continued. If this error is acceptable, then the process is stopped and the values of all weight coefficients are saved for using them in recognition phase.

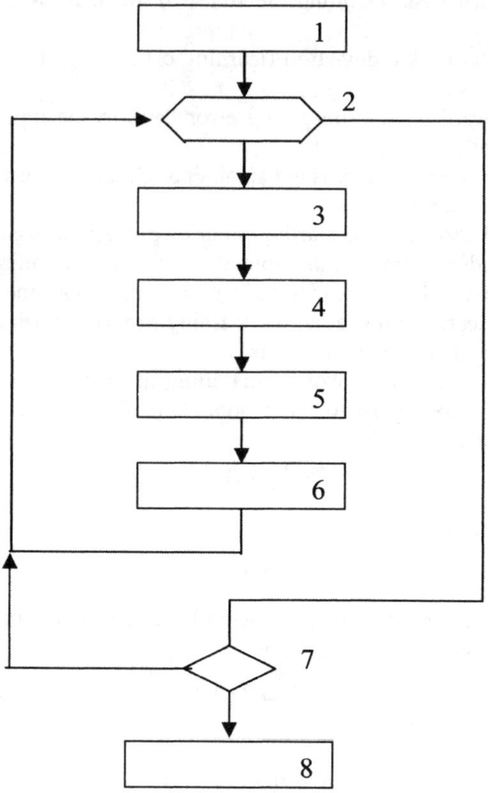

Fig. 2.27. Graphical scheme of backpropagation

Since the desirable output vector is immediately given by the training set, determining the value of error for every output neuron j is easily done by the formula:

$$e_j = y_j^* - y_j \tag{2.152}$$

where e_j – is the magnitude of error of j-th output neuron; y_j^* – value of j-th component of the actually present output vector (current output of j-th output neuron). Thus error is multiplied by the first derivative of the activation function $f'(I_j)$:

$$\delta_j = (y_j^* - y_j) \cdot f'(I_j)$$

If the activation function f is sigmoid, then the above formula will be transformed as follows:

$$\delta_j = (y_j^* - y_j) \cdot y(1 - y_j) \tag{2.153}$$

By means of a chosen estimate (2.153), which includes the error of the j-th output unit and derivative of activation function on the total weighted input in the point, given by the current states of neurons of foregoing layer, one can determine value of the modifications of the weight coefficients of connections between the last and previous layers or, the updated weights of the output neurons. The magnitude of the modification is determined by the following formula:

$$\Delta w_{ji} = \gamma \delta_j x_i \tag{2.154}$$

where Δw_{ji} is modification of weight coefficient of the output neuron j on i-th input; x_i is input signal of j-th output unit, which is the output signal of i-th neuron of the previous layer; γ is constant, the learning rate, which is selected within the boundaries [0.01; 1.0]. The updated values of the weights are calculated the following way:

$$w_{ji}(t+1) = w_{ji}(t) + \Delta w_{ji} \tag{2.155}$$

where $w_{ji}(t+1)$ – value of the weight after the (t+1)-th learning step and $w_{ji}(t)$ – value of that weight coefficient before the (t+1)-th step.

Thus, we have derived the formulas (2.152)-(2.155) for learning the last layer of multi-layer neural network. When one attempts to apply the above described algorithm for training the weights of the hidden layer, the problem is determining the error for the hidden neurons. In fact, the desirable output values y_j^* are given beforehand by the training set, but the objective values of the outputs of the hidden neurons cannot be known before the learning process. For a long time, this fact prevented the development of learning algorithms for multi-layer networks until Rummelhart et al proposed the idea of the error backpropagation algorithm.

During the forward pass, each neuron of each layer influences the global error. In order to determine the error put by the hidden neurons into the final error, it is necessary to establish the error of the output layer. Thus, the magnitude of error for neurons of the layer prior to last layer is determined through the error of output layer. Then error for all the neurons of layer s can be determined through the error of the neurons layer (s+1). Thus, errors, as it were, are propagated backward through the network.

Consider the s-th and (s+1)-st layers of the multi-layer neural network. Assume that the error of neurons of the (s+1)-st layer is known already. Then we can define δ for all neurons of the (s+1)-st layer.

$$\delta_j^{(s+1)} = e_j^{(s+1)} \cdot y_j(1 - y_j^{(s+1)}) \tag{2.156}$$

where $\delta_j^{(s+1)}$ is error of j-th neuron on layer (s+1); $y_j^{(s+1)}$ – state of j-th neuron of layer (s+1). In order to estimate the value of the error for j-th neuron of s-th layer propagate the value (2.156) through weighted connections from the neurons of the layer (s+1) to neuron i (Fig. 2.26). Consequently, the value of the error for neuron I on layer s:

$$e_i^{(s)} = \sum_j \delta_j^{(s+1)} w_{ji}^{(s+1)} = \sum_j e_j^{(s+1)} y_j^{(s+1)} (1 - y_j^{(s+1)}) w_{ji}^{(s+1)}$$

$$\delta_j^{(s)} = y_j^{(s)} (1 - y_j^{(s)}) \cdot \sum_i \delta_j^{(s+1)} w_{ji}^{(s+1)} \qquad (2.157)$$

Here $y_j^{(s)}$ is output of i-th neuron of s-th layer, $w_{ji}^{(s+1)}$ is weight of the neuron j of layer (s+1) on the input from neuron i of layer s.

Therefore, like in the case for the output layer neurons, the value of the weight modifications of neurons of k are defined as:

$$\Delta w_{il}^{(s)} = \gamma \delta_i^{(s)} x_l^{(s)} \qquad (2.158)$$

and finally,

$$w_{il}^{(s)}(t+1) = w_{il}^{(s)}(t) + \Delta w_{il}^{(s)} \qquad (2.159)$$

where $x_l^{(s)}$ – k-th input of neuron of s-th layer, $w_{ik}^{(s)}(t+1)$ – weight of neuron i of s-th layer line k, which is the output of neuron k of the previous layer after the (t+1)-st learning step, $w_{il}^{(s)}(t)$ – the same weight after t-th step (Fig 2.28).

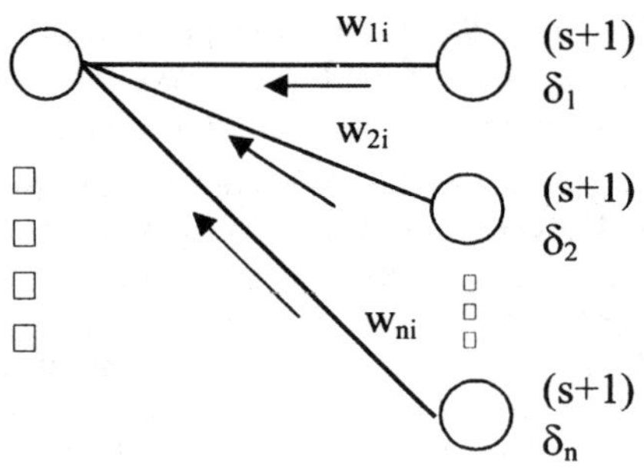

Fig 2.28. Error backpropagation

For solving many problems, the use of some bias in the formula of common weighted input of neuron would be desirable (sometimes this is necessary):

$$I_j = \sum_i w_{ji} x_i + \theta_i \, ,$$

which is close to the notion of threshold. This ensures the trainability of bias, which would have presented as weight of input, on which have always presented the signal "1" as we have accomplished that previously. To train the bias $\theta_i^{(s)}$ of neuron I on the layer s the following formulas are used:

$$\Delta\theta_i^{(s)} = \gamma \delta_i^{(s)}$$

$$\theta_i^{(s)}(t+1) = \theta_i^{(s)}(t) + \Delta\theta_i^{(s)} \qquad (2.160)$$

If the network consists of m layers, then the error of the neuron of output layer will be

$$e_q^{(M)} = y_q^{*(M)} - y_q^{(M)},\ q=1,2 \tag{2.161}$$

So, the formulas (2.156)-(2.161) represent the error backpropagation algorithm. Note that magnitude of γ stays constant during the learning process for all layers of the network.

Thus, for every training pair at every step the value Δw_{ji} is calculated only one time. The case in which for one pair of vectors the whole learning process is accomplished to the end and then next pair is selected is erroneous. In this case, when the next pair is selected, the weight, derived for the first pair is deteriorated. For example, if one would like to train a neural network to recognize a letter of the alphabet and begin to train it to letter "a", then, when we in a similar manner have trained it to recognize the letter "b", the network already would forget "a', because the corresponding weight coefficients would be lost.

One of the deficiencies of the error backpropagation algorithm is that when the addition of new mapping input/output to an already trained network is necessary, the whole learning process has to begin anew with the initial training set, added with a training pair (or pairs, if there are several pairs). On the basis of the error backpropagation algorithm a number of its modifications have been developed for increasing the efficiency of learning process.

In [95], so called the "momentum" method has been suggested which increases the stability of learning and decreases time expenses when backpropagation is used. Consider it in detail.

In the geometric interpretation, the error surface has multiple hills and valleys, when the hidden layers and non-layers activation function are used. In the beginning, the learning network finds itself in one of the points of this surface. During learning process the motion (descent) toward the closest valley (local minimum) is performed. The step of this motion depends on the selected learning rate. If the learning rate is too high then the network can stride over nearest valley, if it is sufficiently narrow. In the described method the previous weight correction is used for the determination of the current ones that would compel the network to descend to the bottom of these valleys. The corresponding equations for weight correction are written in the following form:

$$\Delta w_{ji}(t+1) = \gamma \cdot \delta_j x_i + \varepsilon \cdot \Delta w_{ji}(t) \tag{2.162}$$

$$w_{ji}(t+1) = w_{ji}(t) + \Delta w_{ji}(t+1) \tag{2.163}$$

Here ε is momentum, which typically is set to a value about 0.9.

The application of exponential smoothing in the learning process leads to the following equations

$$\Delta w_{ji}(t+1) = \varepsilon \Delta w_{ji}(t) + (1-\varepsilon)\delta_j x_i \tag{2.164}$$

$$w_{ji}(t+1) = w_{ji}(t) + \gamma\Delta w_{ji}(t+1) \quad (2.165)$$

The formula (2.165) was suggested by Sejnowsky and Rosenberg [99]. Here ε is a smoothing coefficient. As one can see from the above equations, when ε=0, the smoothing is not performed and formulas (2.164)- (2.165) represent the basic error backpropagation algorithm. When ε=1

$$\Delta w_{ji}(t+1) = \Delta w_{ji}(t)$$

That is, the previous weight modifications are repeated. When ε is chosen between 0 and 1, smoothing proportional to ε is performed.

Using the second derivative of the activation function in order to speed up learning was suggested in [74]. The corresponding method is called the second order error backpropagation algorithm. Parker in his work has shown that using derivatives of higher than the second order does not lead to improving the training time, but requires additional expenses for calculating the second derivatives.

The main drawback of the error backpropagation algorithm is a possibility of the network falling into the local minimum, i.e. finding such a local decision, which could be non-satisfactory. In geometric interpretation it means that the closest valley on the error surface is not sufficiently deep, having fallen into such a valley from which the network could not escape.

One way to solve this problem is to use additional hidden units, which could allow escaping the local minima.

The low speed of the considered learning method is another drawback. If we have the network, which contains Q_1 neurons and we wish to train it to provide Q_2 mappings and use Q_3 learning steps, then the total time of learning will be $Q_2 \cdot Q_1 \cdot Q_3$. Assuming, that approximately $Q_1 = Q_2 = Q_3$, in a very rough estimation the time of learning will be proportional to $O(Q_1^3)$. Using parallel computations decreases this estimation to $O(Q_1^2)$, since neurons work in parallel [45,116].

The error back-propagation algorithm is also very useful for training radial based function neural networks (RBFNN).

A series of supervised algorithms are employed for training RBFNN that was described in section 5.5. These algorithms are based on a training set of N pairs (X^l, Y^l) (l=1,2,…,N) that represent associations of a given mapping or samples of a continuous multivariate function. RBFNN have a differentiable nature, providing a differentiation approach to all the parameters from the point of view of network training. Therefore, it is possible to use a gradient method to train the parameters of RBFNN. This method consists in the following iteration procedures:

$$a_i(t+1) = a_i(t) + \Delta a_i,$$
$$b_i(t+1) = b_i(t) + \Delta b_i,$$
$$w_{ij}(t+1) = w_{ij}(t) + \Delta w_{ij},$$

here

$$\Delta a_i = -\gamma_a \frac{\partial E}{\partial a_i}, \quad \Delta b_i = -\gamma_b \frac{\partial E}{\partial b_i}$$

and

$$\Delta w_{ij} = -\gamma_w \frac{\partial E}{\partial w_{ij}},$$

where γ_a, γ_b, and γ_w are small positive constants; E is a mean-square error between outputs of a modelled object (or a function) y_{jr}^* and corresponding outputs of RBFNN y_{jr}

$$E = \frac{1}{2}\sum_{r=1}^{S}\sum_{j=1}^{M}(y_{jr}^* - y_{jr})^2$$

Recurrent backpropagation
In [80] and [19] the extension of the backpropagation algorithm, usually called recurrent backpropagation, for the training fully recurrent networks, was proposed. The neural network is trained on the basis of training pairs (x_l^*, y_l^*). According to these pairs, the algorithm on stable equilibria of the network is applied.

In Fig. 2.29 a simple recurrent network consisting of two units is shown. Here w_{ij} are weight connections, and x_j ,y_i , y_i^* , are the values of the inputs, outputs and desired signals respectively. The input units receive the value of element x_{li} of the input pattern. Output units are designated as units with prespecified desired outputs y_{lk}^* . In general, a unit may belong to the set of input units and the set of output units simultaneously, or it may be hidden unit that does not have a connection with input and output signals.

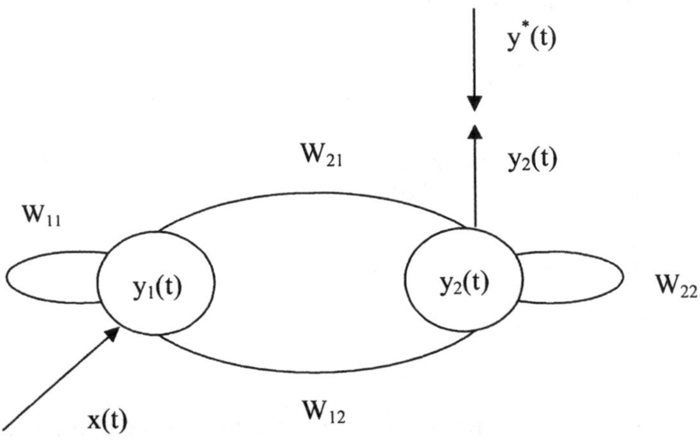

Fig. 2.29. Simple recurrent network

The state evolution of i-th unit [66] is given by

$$\frac{dy_i}{dt} = -y_i + f(\sum_j w_{ij} y_j + x_i) = -y_i + f(I_i), i=1,2,...,m, \tag{2.166}$$

here $f(I_i)$ is activation function, I_i is the total input activity of i-th unit. By setting $\frac{dy_i}{dt} = 0$, one could find equilibrium points y^* of the network. In this case

$$y_i^* = f(I_i^*) = f(\sum_j w_{ij} y_i^* + x_i) \tag{2.167}$$

The following is the derivation of the learning rule for the system/network in equation (2.166) that assumes the existence and asymptotic stability of at least one equilibrium point $y^+ = [y_1^+, y_2^+, ..., y_M^+]^T$ in equation (2.167). This equilibrium point represents the steady-state response of the network. Suppose that in the output of the network the equilibrium state y_i^+ in response to an input X are received. Then this output is compared with the desired response y_i^*, resulting in an error signal E_i. The goal of learning the network is to adjust the weight signal in such a way that the values of error signal E_i for all training pairs ultimately become equal to zero. In other words, our goal is to minimize the error function

$$E = \frac{1}{2}\sum_{i=1}^{M}(y_i^* - y_i^+)^2 = \frac{1}{2}\sum_{i=1}^{N}(E_i^+)^2 \tag{2.168}$$

with E_i=0 if unit is not an output unit. Using the error function on the basis of a gradient-descent search, one can write the weight update rule [66]

$$\Delta w_{pq} = -\gamma \frac{\partial E}{\partial w_{pq}} = \gamma \sum_{i=1}^{m} E_i^+ \frac{\partial y_i^+}{\partial w_{pq}} \tag{2.169}$$

by differentiating equation (2.167) we can obtain $\frac{\partial y_i^+}{\partial w_{pq}}$

$$\frac{\partial y_i^+}{\partial w_{pq}} = f'(I_i^+)[\delta_{ip} y_q^+ + \sum_j w_{ij} \frac{\partial y_j^+}{\partial w_{pq}}] \tag{2.170}$$

where δ_{ip} is the Kronecker delta function (δ_{ip}=1 if i=p and zero otherwise). Also one can write equation (2.170) by following formula

$$\sum_j \Psi_{ij} \frac{\partial y_i^+}{\partial w_{pq}} = \delta_{ip} f'(I_i^+) y_q^+ \tag{2.171}$$

where

$$\Psi_{ij} = \delta_{ij} - f'(I_i^+) w_{ij} \tag{2.172}$$

Now, one may solve for $\frac{\partial y_i^+}{\partial w_{pq}}$ by inverting the set of linear equations represented by equation (2.171) and get

$$\frac{\partial y_i^+}{\partial w_{pq}} = (\Psi^{-1})_{ip} f'(I_p^+) y_q^+ \tag{2.173}$$

where $(L^{-1})_{ip}$ is the ip-th element of the inverse matrix L^{-1}. Hence applying equation (2.173) in the equation (2.169) gives the desired learning rule:

$$\Delta w_{pq} = \gamma f'(I_p^+) \sum_i E_i^+ (\Psi^{-1})_{ip} y_q^+ \tag{2.174}$$

For a fully connected recurrent network, the matrix Ψ is $Q_l \cdot Q_l$ and then its inversion requires $O(Q_l^3)$ operations using standard matrix inversion methods. Pineda and Almedia independently showed that a more economical local implementation, utilizing a modified recurrent neural network of the same size as the original network, is possible. This implementation has $O(Q_l^2)$ computational complexity and is usually called recurrent backpropagation.

Now we consider unsupervised learning algorithms

Hebbian learning
Based on the achievements of physiology and psychology, Hebb proposed a simple idea for training neural networks [44]. It is based on strengthening the synapses between two neurons, which have activated simultaneously. Recent research in biology ensured this Hebbian idea [98]. The learning algorithm, based on this idea is the starting point in neural networks learning methods development.

To describe the idea of the algorithm, consider the network, which contains source neurons and destination neurons. Source neurons are immediately connected to destination neurons. Consider the vector X, which consist of activation levels of source neurons, and the vector Y, which is generated by output signals of destination units. Each source neurons is connected to each destination neuron through the synapse with the corresponding weight coefficient w_{ij}. The totality of weight coefficients forms the matrix W. Thus, learning process can be considered as a modification of each element of the matrix W. The magnitude of the modification of each element of the matrix is determined by the following formula:

$$\Delta W = \gamma\, Y \cdot X^T \tag{2.175}$$

where X^T is transposed vector X (since any vector can be considered as a matrix); ΔW - matrix, which consists of the magnitudes of modifications of coefficients, i.e. w_{ij}.

For updating weight coefficients at the (t+1)-st learning step, the following formula is used:

$$W(t+1)=W(t)+\Delta W \tag{2.176}$$

The formulas (2.175)-(2.176) represent the Hebbian learning algorithm.

Consider the case in which providing the single associations X_l Y_l is needed, and the learning rate γ=1. When this associations is considered, assuming that initial values of elements of matrix X are all equal to zero, their updated values are calculated as follows:

$$\Delta W = Y_1 X_1 \tag{2.177}$$

$$W = W^0 + \Delta W \tag{2.178}$$

where W^0 is W before adjustment. As one can see from the formulas (2.177), (2.178), in this case elements of the matrix W will be:

$$W = Y_1 X_1^T$$

Now when vector X_1 is connected to the input of network, the output units will generate the corresponding vector Y_1. However, in general, storing multiple associations $X_1 Y_1$ in the single matrix W is required. To do this, the Hebbian learning algorithm is applied repeatedly to the network for all associations. When the network has been trained, each element of the matrix W (i.e. weight coefficient) participates in providing all associations. Therefore, one can say that the matrix W (i.e. totality of the weights of network) is the model of distributed associative memory.

However, the Hebbian learning algorithm has a serious disadvantage. That is, in order to accomplish all associations successfully, orthogonality of all input vectors is needed, i.e. the following constraint must be satisfied:

$$X_1\ X_2\ X_3 \ldots\ X_n = 0 \tag{2.179}$$

where in the left side the scalar product of the all input is given. On the other hand, there always will be some level of error in the network functioning, caused by not satisfying the restriction (2.179). For example, if the input vectors $X_1\ X_2\ X_3 \ldots\ X_n$ are not orthogonal to each other, and the output vectors $Y_1\ Y_2\ Y_3 \ldots\ Y_n$ correspond to them, then, if some particular vector X_i is connected to the input of network, its output will generate vector Y, such as

$$Y = a_1 Y_1 + a_2 Y_2 + \ldots + a_n Y_n = \sum_{j=1}^{n} a_j Y_j$$

i.e. Y is some linear combination of particular vectors $Y_1\ Y_2\ Y_3 \ldots\ Y_n$.

Taking into account that in M - dimensional space at most M orthogonal vectors can be drawn, it becomes clear, that the network in this case could support correctly not more than M associations.

Hebbian learning algorithm plays an important role in the development of neural networks learning algorithms. The Hebbian learning algorithm is classified as an unsupervised learning method because in it, during the learning process, only the information about local interaction of two neurons is used. In addition, there are no instructions, given by the supervisor (teacher).

Competitive learning

As we mentioned above, in unsupervised algorithms there is no interaction with the external global supervisor. The training set consists of a totality of input vectors and the network is not informed about the desirable states of input neurons. The task of competitive learning, which is the most wide-spread algorithm of unsupervised learning techniques for neural networks, is the classification of coming output patterns into several clusters. Thus, the competitive learning method is closely related with the cluster analysis. In cluster analysis, several clusters are considered and the classification of a given pattern into of the clusters is performed. The neural network, which uses competitive learning, classifies two vectors, which are sufficiently close to each other in the same cluster. Otherwise, i.e. if vectors are not close to each other, they are classified into different clusters.

One form of competitive learning was investigated in [93]. In this work, Rummelhart and Zipser considered the network without feedback connections, which could include one or more layers, intended for features recognition.

The neurons of recognizing layers attempt to reveal the characteristic features of the coming input vector, which are intrinsic to the corresponding cluster.

In their attempts they gain different activation levels, because the weight coefficients of recognizing neurons are different to a sufficient degree to distinguish one from others. The process, appropriately called "competition", takes place. During this process, the neuron with the highest level of activation becomes the "winner". The "winner" is the neuron whose corresponding cluster is most close to the actual input vector. After the competition has been performed, the winning neuron inhibits other neurons to pretend recognition of the connected input vector, i.e. it takes all the rights to recognize this vector. The above described process implements the strategy called "winner-takes-all'.

There is the possibility that one of the neurons will respond to each coming input vector (if its weight coefficients are too large). In order to escape this situation, some restriction has to be made. One can make such restrictions by assuming that for each neuron, the sum of weight coefficients of connections coming from neurons of the previous layer must be constant. In the considered work of Rummelhart and Zipser, they assumed that sum of the neuron's weight must be equal to one.

For the sake of simplicity, consider the two-layer network, shown in Fig. 2.30.

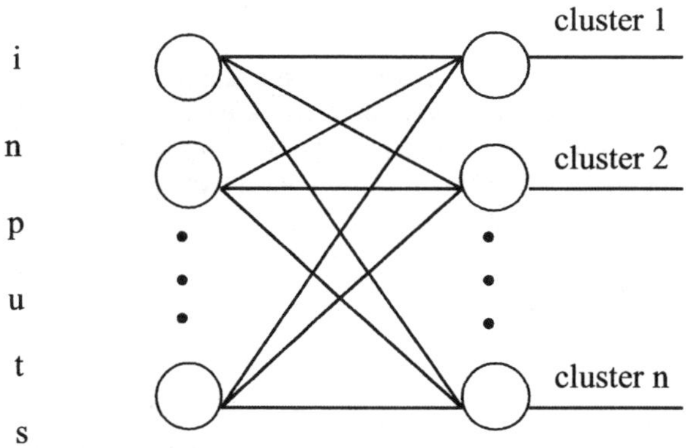

Fig. 2.30. Network for competitive learning

Every output neuron is associated with some cluster. When vector X is connected to the input layer of the network, the activation levels of output (recognizing) neurons is calculated in accordance with the formula of total weight sum:

$$y_j = \sum_i w_{ji} x_i$$

Here x_i is i-th input neuron; y_j - j-th output (recognizing) neuron, which correspond to the j-th cluster; w_{ij} - weight coefficient of connection from i-th input unit to j-th output one.

Before the learning process, it is desirable to normalize input vectors. This is made by enforcing the vectors to have the length equal to 1, using following formula:

$$x_i' = \frac{x_i}{\sqrt{\sum_j x_j^2}}$$

where x_i' is the value of the i-th component of input vector after the normalization has been done.

Rummelhart and Zipser in their work considered the case in which input vector X is binary. Before determining which neuron wins the competition, the activation levels of output neurons are distributed in the interval from zero to one. After this, the neuron with the highest magnitude of activation level becomes the winner, that is:

$$y_{win} = \max(y_1, y_2, \ldots, y_m)$$

where y_{win} is the neuron-winner. Activation level of the neuron-winner is set to one, while the other neurons are switched to zero state:

$$y_{win} = 1,$$
$$y_i = 0, \quad i \neq win$$

Thus, only one neuron wins the competition. All the other neurons, different from the winning one, are considered as lost neurons.

During the learning process, the current vector from the training set is connected to the input units of network and competition starts. After determining the winner, the connections, which come from the active input neurons to the winner, are strengthened. For calculating the weights of those connections, Rummelhart and Zipser proposed the following formulas:

$$\Delta w_{ji} = \gamma\left(\frac{x_{li}}{n_l} w_{ji}\right) \quad (2.180)$$

$$w_{ji}(t+1) = w_{ji}(t) + w_j \quad (2.181)$$

Here x_{li} is the i-th component of the l-th input vector from the training set; n_l- the number of active input neurons when the l-th vector is clamped to the input of the network; γ - the constant, learning rate.

As one can see from the equations (2.180)-(2.181), at every learning step the vector of weights of the neuron-winner moves closer to the connected input vector. After learning has been accomplished, when a new vector is applied to the network's input, the neuron, weight coefficients of which is closest to the actual input vector, will respond (fire). This can be clarified using the following reasoning.

Consider the vector X, which is connected to the input of network. For output neurons the magnitude of the weighted sum $\Sigma w_{ji}x_i$ is calculated. This is a scalar product of input vector X and vector W_j of weight coefficients of the neuron j. This means that for every output, the neuron scalar product of its weight vector and current input vector is defined, i.e. for the j-th neuron the activation level is calculated as follows:

$$y_i = \sum_j w_{ji} x_i = |W_i||X|\cos(W_i, X) \tag{2.182}$$

where W_j is the length of vector W_j, X is the length of input vector and (W_j, X) - angle between input and the j-th neuron's weight vector.
Thus, moving vectors W_j and X closer to each other maximizes the expression (2.182), because cosines of little angles are about one. Moving vectors W_j and apart minimizes the above level of activation.

One of the most important areas of the application of competitive learning is the training of ART neural networks, which have been considered in section 5.7. The idea of ART neural network learning is as follows. Each learning iteration includes a learning current vector X^1 and examines the weight vectors W_i that are sufficiently similar to X^1. If among all the weight vectors, W_i is the closest (in Euclidean norm sense) to the input vector X^1 then it is modified such as to better match X^1. Otherwise X^1 becomes the prototype for a new cluster.

The ART clustering procedure consists of the next concepts. Let $X \in \{0,1\}^n$ is an input binary vector (pattern). Instead of the j-th learned cluster, let's choose the weight binary vector $W_j \in \{0,1\}^n$ of the prototype unit j. Then the normalized output of the ART Neural Network will be

$$y_j = \frac{W_j^T \cdot X}{\|W_j\|^2}, \quad j=1,2,\dots,m$$

where $\|W_j\|^2$ is equal to the number of ones in W_j.

Let's realize the general idea ART training by submitting of next two steps for a given input X.

Step 1. Choose the *winner* prototype W_i being *close enough* to the input X, i.e. one must submit

$$y_i = \frac{W_i^T \cdot X}{\|W_i\|^2} > \frac{\|X\|^2}{n}.$$

This inequality guarantees that a sufficient fraction of the W_i and X bits is matched.

Step 2. For the given $\varepsilon \in (0,1)$ W_i must satisfy the following inequality

$$\frac{W_i^T \cdot X}{\|X\|^2} \geq \varepsilon$$

ε is called *vigilance parameter.* W_i is declared as «matched» to X, and that's why the ART nets are often called as *automatic scaling or self-scaling*.

As a result this procedure consists of steps X joins cluster i, and deals with its unit's prototype weight binary vector W_i is updated according to

$$W_i^{new} = W_i \wedge X$$

where "$\wedge$" is the logical AND operation applying to the corresponding components of vector W_i and X.

In [66] details of this training procedure are discussed.

Finally we consider reinforcement learning algorithms
The class of learning methods, in which the neural network adjusts its behaviour using one bit of information, which indicates approval or disapproval of current behavior of network, is called the reinforcement learning algorithms. The signal given to the network is called the reinforcement signal and is sent by a supervisor.

The child during its development can organize the behavior in an unfamiliar environment. If actions made by the child cause nondesirable effects (pain, disapproval of adults etc.), then the child will not repeat these actions in the future. For example if the child touches a hot surface he can burn his fingers and feel the pain. Or, if he finds throwing shoes out from the window, then he can be punished. Obviously, in future the child will try not to touch the hot surface anu throw articles from the window. If the actions of a child cause approval of environment (for example, praise of adults), then he will try to correct those actions.

Reinforcement learning algorithms recall the above mentioned aspects of child development. They are considered as biologically plausible. In these algorithms, as we mentioned, only a single reinforcement signal also called reward /penalty or award/ punishment signal, is used. When the network obtains reward signal (positive reinforcement) it tries to correct the current behavior to repeat in a similar situation. If a penalty signal (negative or zero reinforcement) arrives, then weight coefficients are modified in order to run away from the current behavior.

Using the reinforcement signal for learning is known from automata theory [70,116]. Consider an automaton, which selects one of the possible actions to perform on the base of corresponding probabilities. When some action has been selected, the environment responds to that action by sending the positive (reward) or negative (penalty) reinforcement signal. Based on this information, automaton can modify the corresponding probabilities in order to increase the expectation of positive reinforcement signal. This process is carried on until the frequency of deriving the positive reinforcement will be on a sufficiently high level.

Consider one of the algorithms of this class, so-called linear reward-penalty algorithm, L_{R-P} [70]. At first, consider the automaton, which has the set of selected actions $(a^1,a^2,\dots,a^n)$, which are selected on the basis of probabilities $(p^1,p^2,\dots,p^n)$. Because during learning, the probabilities of action selecting are modified, introduce the additional index, so that $(p_t^1, p_t^2,\dots, p_t^n)$ is the set of the current probabilities on the t-th iteration. On the t-th learning step some action ai $(i=\overline{1,n})$ is chosen and the environment responds to that choice by sending the signal b_t, which can have two values only: -1(0) or +1. No additional information is given besides this one bit signal. Then the automaton modifies the probabilities p_t^n in accordance with the following rules:

IF the selected actions $a_t = a^i$ AND $b_t = 1$, THEN

$$\begin{cases} p_{t+1}^i = p_t^i + \gamma_1(1-p_t^i) \\ p_{t+1}^j = (1-\gamma_1)p_t^j, \quad \text{for } j \neq i. \end{cases} \tag{2.183}$$

IF b_t=-1 THEN

$$\begin{cases} p_{t+1}^i = (1-\gamma_2)p_t^i \\ p_{t+1}^j = \dfrac{\gamma_2}{r-1} + (1-\gamma_2)p_t^j, \quad \text{for } j \neq i \end{cases} \tag{2.184}$$

Here γ_1 and γ_2 are the learning rates, r- number of automata. Thus, the probability of successful action is increased to the value, proportional to the difference between one and that probability before the learning step, while probabilities of other actions are decreased. Certainly, γ_1 and γ_2 must be within the interval from zero to one.

Let's move now to neural networks. Consider number of neurons, represented in Fig. 2.31. Instead of the i-th action a^i consider the activation of the i-th neuron. As in the case of the automaton, here learning is accomplished by modifying the probabilities of neuron firing.

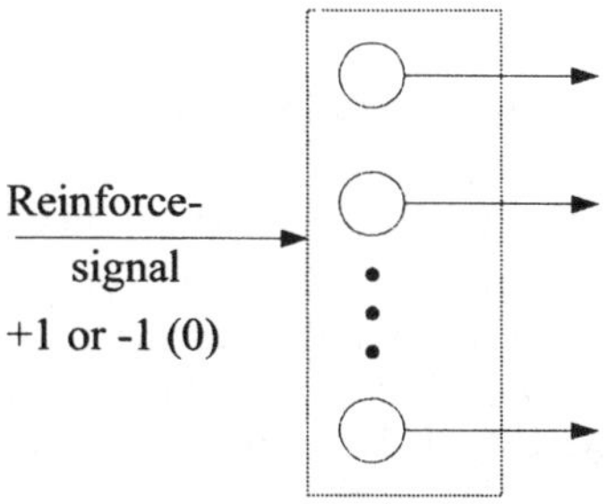

Fig. 2.31. Neurons for L_{R-P} algorithm

However, in this case, individual neurons don't compose a global network. They avoid the computational power, derived by using connections, and cannot accomplish associations and classification.

Taking into account the above-mentioned in [24] the enhanced L_{R-P} algorithm has been developed, which they called associative reinforcement learning. They considered the network, which contains both output and input neurons (Fig. 2.32). Input vectors are connected to these input units in order to accomplish classification. The network is trained to perform the correct classification with the use of a reinforcement signal, derived from the teacher. In order to indicate the correctness of network behavior, the teacher should have information on the relation between input vectors X_k and output ones Y_k. One way to store such information is to use the array $d(X_k,Y_k)$.

Assume that there are two neurons in the output layer of network. This means that input vectors correspond to two different classes. When a particular input vector X is connected to the input units, the classification error is minimized if the neuron y_i fires, under condition of $P(y_1/X)>P(y_2/X)$ or if the neuron y_2 activates and $P(y_2/X)>P(y_1/X)$. However, here we face the problem of determining the above given conditional probabilities.

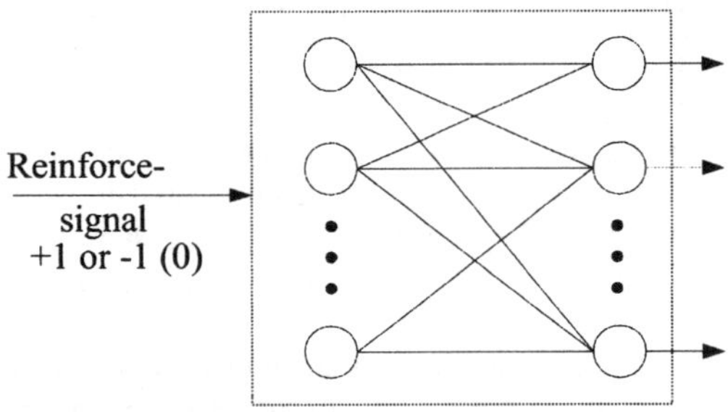

Fig. 2.32. Network for A_{R-P} algorithm

Barto and Anandan have suggested using the vector Θ, which approximate those probabilities:

$$\Theta X = P(y_1/X) - P(y_2/X)$$

Thus, if the condition $\Theta X > 0$ is satisfied then the output neuron y_1 activates, if, however $\Theta X < 0$ is true, then the second output neuron fires. The vector Θ is adjusted during learning process. Besides the above the label Z of class of input vector is introduced. This label is equal to one if X is in the class, which corresponds to y_1, otherwise Z=-1. It can be shown that minimizing the mathematical expectation of $(\Theta X\text{-}Z)^2$ causes minimizing the classification error.

Barto and Anandan used algorithm of Robinson-Monro [100]. Partial derivative of error E on Θ is defined as follows:

$$\frac{\partial E}{\partial \Theta} = 2(\Theta X\text{-}2)X \tag{2.185}$$

The equation (2.186) is used to adjust the vector Θ in learning process:

$$\Theta_{t+1} = \Theta_t - \gamma_t(\Theta_t - Z_t)X_t, \qquad (2.186)$$

where γ_t are constants, which have different values on different steps of learning. They decrease during learning process and influence the convergence.

Components of vector Θ can be considered as weights of connections, which connect input neurons with one of the output unit. These output neurons activate when their total input exceeds zero and does not activate otherwise.

For developing an associative reward penalty algorithm, $A_{R-P,}$ the randomness element had been used. Barto and Anandan assumed that each output neuron can be in two states: 1 and -1. The activation rule in their work had the following form:

$$y_t = \begin{cases} 1 & \text{if } X_t + \xi_t > 0, \\ -1 & \text{otherwise.} \end{cases}$$

Here ξ_t -is random variable with known distribution. When X_t and Θ_t are given, the mathematical expectation $E(y_t/\Theta_t,X_t)$ is known too. The formula for Θ updating stays like in the Robinson-Monro algorithm (2.185).

To distinguish the case of positive reinforcement (b=1) from negative one (b=-1) the coefficient λ is introduced. In the reward case, we have:

$$\Theta_{t+1} = \Theta_t - \gamma_t(E\{y_t/\Theta_t, X_t\} - b_t y_t)X_t$$

In case of penalty:

$$\Theta_{t+1} = \Theta_t - \lambda\gamma_t(E\{y_t/\Theta_t, X_t\} - b_t y_t)X_t$$

When $\lambda = 0$, the above algorithm is called the associative reward in action algorithm, A_{R-P}.

It can be shown that if:

1) input vectors are linearly independent;

2) the appearance of each input vector has finite probability;

3) distribution of random variable is continuous and monotonous;

4) the sequence of γ_k satisfies certain requirements, providing minimizing γ_k to zero as k increases, the weight vector will converge.

The main drawback of reinforcement learning algorithms is that they are not efficient in solving large problems. In large networks, it is difficult to adjust the behavior, based only on the single global signal.

Another drawback is that the ascent towards increasing the reinforcement signal expectation can lead to local optimum. When the network moves closer to that optimum, it derives less information about other possible solution.

2.3 Probabilistic Computing

There are various technologies that can be used for making inferences and decisions in uncertain environment. Usefulness of these approaches considerably depends on nature of uncertainty. Hence it is necessary to take into consideration the nature of uncertainty and capabilities of the reasoning techniques when one chooses suitable means for decision making and inferencing in an uncertain environment. In this chapter we will analyze three measures of uncertainty (additive probabilities, coherent lower or upper previsions and belief functions) and corresponding reasoning techniques [14,31,34,39,47,103]. Criteria for evaluating measure of uncertainty will be considered as well.

Human knowledge is basically uncertain. There are various sources of uncertainty. Uncertainty may be related to the domain, expert, knowledge engineer, user, and data.

We can compare measures of uncertainty according to the several criteria [103,104]. The common criteria for comparing measures of uncertainty are *Interpretation, Imprecision, Calculus, Consistency, Assessment, and Computation*.

The measure utilized in systems should have clear interpretation. Without clear interpretation the measure cannot be used to guide assessment and to understand the conclusions of the system. It also cannot support the rules for combining and updating measures.

The measure also should be able to deal with ignorance, conflicting information, information inadequacy and deficiency, and imprecision [2,7,14].

Inferencing is related to different computational procedures. Hence there should be rules for processing measures of uncertainty, updating them on the basis of the new information, and using them to make conclusions and decisions.

Special methods should support checking consistency of uncertainty assessments and assumptions used by the system. The rules of the calculus should be able to guarantee consistency of the conclusions with assessments. For example, in the Bayesian theory and theory of lower previsions this requirement is formulated and reflected in the principles of *coherence.*

The system should be able to handle and to combine qualitative judgments with quantitative assessments of uncertainty. It is necessary to provide the computational feasibility of inferences and conclusions from the assessments. In common cases it is desirable to satisfy all above mentioned criteria, but actually it is hardly achievable objective. Depending on system and problem being under consideration, various criteria play different roles. The fundamental interpretation criterion is essential in order to give meaning to the conclusions of the system. An imprecision criterion is necessary because low quality of information is available. If we want to derive conclusions from the uncertainty assessments a calculus is needed. A consistency criterion allows us to avoid erroneous conclusions. If we are going to implement in practice intelligent by system the assessment and computation criteria will be very useful.

2.3.1 Bayesian Approach

In the Bayesian theory uncertainty is measured by unconditional probabilities P(A) or by conditional probabilities P(A/B) [14,47]. The Bayesian theory satisfies by criteria of interpretation, calculus and consistency very well but it satisfies by criteria of imprecision and assessment relatively poorly.

The conditional probability $P(A_i|B)$ defined by the ratio

$$P(A_i|B)=P(A_i \cap B)/P(B) \tag{2.186}$$

describes the probability of concluding the hypothesis (A_i), given the evidence (B).

Updated probabilities after observing an event B are related to unconditional probabilities. We have the following equalities for the conditional probabilities :

$$P(A_i \cap B)=P(A_i \mid B)P(B)=P(B|A_i)P(A_i) \tag{2.187}$$

where $P(A_i \wedge B)$ is the probability that both A_i and B occur.

Finally, we get the Bayes formula:

$$P(A_i|B)=P(B|A_i)P(A_i)/P(B) \tag{2.188}$$

Bayesian approach has been used in several expert systems. Let us assume that the knowledge base has a number of rules that have B in their antecedent, with different consequents $A_1,A_2,...,A_n$. In this case, the probability of the event B is the sum of the intersections of B with all the possible hypotheses, as:

$$P(B)=\sum_{j=1}^{n} P(B \cap A_j)= \sum_{j=1}^{n} P(B \mid A_j)P(A_j) \tag{2.189}$$

In Bayesian theory of uncertainty probability $P(A_i)$ is called the prior probability, and $P(A_i|B)$ is called the posterior probability. The $P(A_i)$ is the expert's belief in the truth of the hypothesis A_i and the $P(A_i|B)$ is the corrected belief of the expert, after observing the evidence of B as an actual fact.

Let us demonstrate the rules and the inference process in graphical form. Assume that there are two hypotheses: A_1= *to approve* the applicant and A_2=*do not approve* the applicant. The evidence that can be used in both of the above hypotheses is B = the applicant has a previous work experience. The belief net for this example is depicted in Fig 2.33. The expert should specify the prior probabilities as:

$P(A_1)=0.8$
$P(A_2)=0.2$

and the conditional probabilities as:

$P(B/A_1)=0.85$
$P(B/A_2)=0.10$

The conditional probability $P(B/A_1)$ reflects the percentage of the applicants who had an experience of those applicants whose background were professionally appropriate to be approved.

$$P(A_1 \mid B) = \frac{P(B \mid A_1)P(A_1)}{P(B \mid A_1)P(A_1) + P(B \mid A_2)P(A_2)}$$

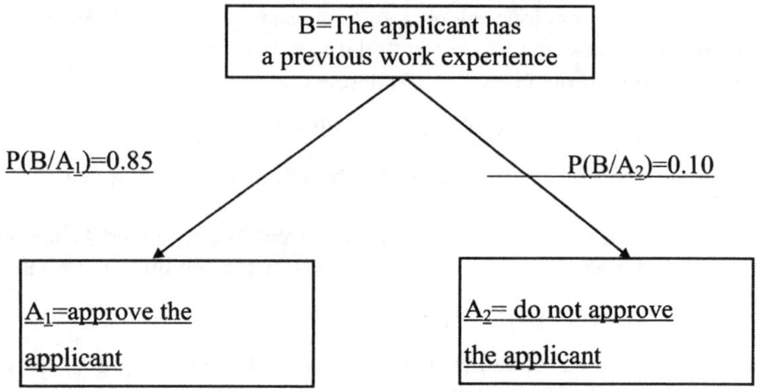

Fig. 2.33. Graphical interpretation of the Bayesian approach

$$P(A_1 \mid B) = \frac{(0.85)(0.8)}{(0.85)(0.8)+(0.10)(0.2)} = \frac{0.68}{0.68+0.02} = \frac{0.68}{0.70} = 0.971$$

$$P(A_2 \mid B) = \frac{0.02}{0.68+0.02} = \frac{0.02}{0.70} = 0.029$$

If there are more than one B in the rule, we have

$$P(A_i \mid B_1 \cap B_2 \cap \ldots \cap B_k) = \frac{P(B_1 \cap B_2 \cap \ldots \cap B_k \mid A_i)}{\sum_{j=1}^{n} P(B_1 \cap B_2 \cap \ldots \cap B_k \mid A_j)P(A_j)} \tag{2.190}$$

To compute (2.190), we need the following:

$$P(B_1 \cap B_2 \cap \ldots \cap B_k \mid A_i) =$$
$$= P(B_1 \mid B_2 \cap \ldots \cap B_k \cap A_i)P(B_2 \mid B_3 \cap \ldots \cap B_k \cap A_i) \;\ldots\; P(B_k \mid A_i)$$

where $P(B_1 \mid B_2 \cap B_k \cap A_i)$ is the probability that the evidence B_1 is true, given that other chunks of evidence $B_2 \ldots B_k$ and the hypothesis A_i are true.

When evidence B is uncertain , the probability of the hypothesis is

$$P(A_i \mid \widetilde{B}) = P(A_i \mid B)P(B \mid \widetilde{B}) + P(A_i \mid \leftarrow B)P(\leftarrow B \mid \widetilde{B}) \tag{2.191}$$

where $\leftarrow B$ means that the evidence B is not true. In case of uncertain conclusion the value of $P(A_i|B)$ would be multiplied by the probability of the conclusion, assuming conditional independence.

The Bayesian approach is useful and can be applied in various inferencing and decision making problems involving uncertainty. The problems arise when we have to deal with partial information and ignorance.

2.3.2 Dempster-Shafer Theory of Belief

Let's first give a brief overview of the evidence theory.

There are other measures of uncertainty with finite universe, such as belief measure, plausibility measure etc. Evidence theory is based on these two measures.

Belief and plausibility measures can be represented by mass distributions of a unit of belief, called basic probability assignment (BPA). This concept is used as basic for Probabilistic Reasoning based on Dempster-Shafer's theory of belief.

Belief functions were first introduced by Dempster and Shafer for the purpose of computing the degree of belief of statements made by different sources of witnesses from a subjective probability of the sources reliability [14,31,100]. Many other applications of belief functions have been known [34,47].

In all interpretations a belief function *Bel* is a real-valued function, defined on all subsets of a possibility space.

The mass function that satisfies the following three conditions:

$$m(\varnothing) = 0,$$

where $\varnothing$ is the empty set

$$0 < m(A) < 1$$

$$\sum_{A \subseteq P(X)} m(A) = 1$$

is called a basic probability assignment.

The mass distribution of units of belief across the propositions in P(X) represent the sets of partial beliefs. In accordance with these formulas, the total certainty over the space is 1 and non-zero value of basic probability assignment reflects the degree of ignorance.

Definition. Belief measure

A function $Bel: P(X) \to [0,1]$, where P(X) denotes power set of X is called a belief measure if it satisfies the following conditions [54]:

$$1.\ Bel(\varnothing)=0, \qquad Bel(X)=1; \tag{2.192}$$

$$2.\ Bel(A_1 \cup A_2 \cup \ldots \cup A_n) \geq \sum_{j}^{n} Bel(A_j) - \sum_{j<k}^{n} Bel(A_j \cap A_k) + \ldots + \\ + (-1)^{n+1} Bel(A_1 \cap A_2 \cap \ldots \cap A_n), \tag{2.193}$$

for all families of subsets of X.

Given $A \in P(X)$, Bel(A) on the basis of available evidence determines degree of belief to which a given element of X belongs to A.

As shown in [54] probability measures are special cases of belief measures, when the equality in (2.193) is always satisfied.

It should be mentioned that the basic formulas of probability are the same as in the frequentist and subjective cases, but collection of data, use of the formulas, and evaluation of the results are different.

The following is the fundamental property of belief measure

$$\mathrm{Bel}(A) + \mathrm{Bel}(\overline{A}) \le 1 \tag{2.194}$$

Belief function can be represented in the form

$$\mathrm{Bel}(A) = \sum_{B \subseteq A} m(B)$$

where *m* is a probability mass function on subsets of P(X).

Definition. Plausibility measure

Plausibility measure is defined by the equation

$$\mathrm{Pl}(A) = 1 - \mathrm{Bel}(\overline{A}) \tag{2.195}$$

Similar to (2.194) for plausibility measure the following inequality is satisfied

$$\mathrm{Pl}(A) + \mathrm{Pl}(\overline{A}) \ge 1 \tag{2.196}$$

A possibility space or environment is defined as a set X, which includes all mutually exclusive hypotheses. In the set of mutually exclusive hypotheses no two hypotheses can be true at the same time. In general, environment X can include n elements. On the base of these elements it is possible to form 2^n subsets of n.

Let's consider as an example the expected product demand. In this case one can represent an environment as

$$P(X) = \{d, r, i, s\},$$

where d, r, i and s stand for decline, remain the same, increase, and increase significantly. The demand can take only one of these elements as an expected value, so we consider them as mutually exclusive elements.

The set θ could have various subsets, including such as

$$X_1 = \{d, r, i, s\}$$
$$X_2 = \{d\}$$
$$X_3 = \{r, i\}$$
$$X_4 = \varnothing = \{\ \}$$

Different pieces of evidence produce various m values. Let's assume that one piece of evidence has arrived in the form of a market survey and the following m values are formed by the experts:

$$m_1(\{i, s\}) = 0.6$$
$$m_1(\{d, r\}) = 0.3$$
$$m_1(\{X\}) = 0.1$$

where $m_1(\{X\})$ is the part for which no belief is formed. The decision maker assigns weights (m values) to the focal elements and the environment. Sum of weights is equal to 1.

The result of the market survey produces the following m values:

$$m_2(\{s\}) = 0.7$$
$$m_2(\{X\}) = 0.3$$

The combination of the two sets representing different sources of evidence is shown in Table 2.15.

Table 2.15. Different sources of evidence

	$m_2(\{g\}) = 0.7$	$m_2(\{X\}) = 0.3$
$m_1(\{i, s\}) = 0.6$	$\{s\}$ 0.42	$\{i, s\}$ 0.18
$m_1(\{d, r\}) = 0.3$	$\{\varnothing\}$ 0.21	$\{d, r\}$ 0.09
$m_1(\{X\}) = 0.1$	$\{s\}$ 0.07	$\{X\}$ 0.03

We can define the combined belief values as

$$m_3(\{s\}) = m_1 \oplus m_2(\{s\}) = 0.42 + 0.07 = 0.49$$

$$m_3(\{\varnothing\}) = m_1 \oplus m_2(\{\varnothing\}) = 0.21$$

$$m_3(\{i, s\}) = m_1 \oplus m_2(\{i, s\}) = 0.18$$

$$m_3(\{d, r\}) = m_1 \oplus m_2(\{d, r\}) = 0.09$$

$$m_3(\{X\}) = m_1 \oplus m_2(\{X\}) = 0.03$$

where $\oplus$ represents the operation of combining the belief values.

In this example the empty set $\{\varnothing\}$ takes the value 0.21. But in accordance with Dempster-Shafer theory m-value of the empty set should be zero. In order to achieve this, m-values of the focal elements should be normalized and m-value of the empty set made equal to zero. The normalization factor is $1 - \sum m(\varnothing)$

In accordance with procedures described above

$$m_3(\{s\}) = 0.49 \frac{1}{1 - 0.21} = \frac{0.49}{0.79}$$

$$m_3(\{i, s\}) = 0.18 \frac{1}{1 - 0.21} = \frac{0.18}{0.79}$$

$$m_3(\{d, r\}) = 0.09 \frac{1}{1 - 0.21} = \frac{0.09}{0.79} = 0.11$$

$$m_3(\{X\}) = 0.03 \frac{1}{1 - 0.21} = \frac{0.03}{0.79}$$

The belief value of the demand set {i,s} is:

$$\text{Bel}(i, s) = \frac{0.49}{0.79} + \frac{0.18}{0.79} = \frac{0.67}{0.79} = 0.848$$

The plausibility of the set is

$$Pls(\{i,s\})=1-Bel(\leftarrow\{i, s\})=1-0.11=0.89$$

The evidential interval of the set {i,s} is the interval [0.848,0.890].

Belief functions satisfy various criteria for evaluating measures of uncertainty with different degrees of certainty.

There exists a variety of exact and approximate methods that perform inferences using belief functions.

Below we consider upper and lower prevision.

The simplest model for imprecise probabilities is a lower prevision $\underline{P}$. Let's assume that Ω is a set of possible states of worlds, L is a set of all gambles, and gamble is a bounded mapping from the possibility space of interest Ω, to the real numbers, that describe an uncertain reward for the gamble [31,102,103,107]. In this case we can consider lower prevision and upper prevision correspondingly as buying and a selling price for a gamble. $\underline{P}$ (X) is a supremum buying price for the gamble X.

It means that we are inclined to pay any price less than $\underline{P}$ (X). The interpretation of $\overline{P}$ (X) is that we incline to sell gamble X for any price greater than $\overline{P}$.

$\underline{P}(X)$ and $\overline{P}$ (X) can be considered as marginal buying and selling prices. The quantity of $\underline{P}(X)$ and $\overline{P}$ (X) depends on information available.

It is necessary to take into account that the quantities $\underline{P}(X)$ and $\overline{P}$ (X) might be considered only as lower and upper bounds for quantities that could be specified more precisely. Upper and lower previsions can be considered as useful tools in modeling of incompleteness or contradiction in data.

The theory of coherent lower prevision satisfies criteria of interpretation, imprecision, calculus and consistency.

Coherence of the lower prevision $\underline{P}$ is characterized by three axioms:

(P1) $\underline{P}(X) \geq \inf\{X(\omega) : \omega \in \Omega\}$ for all $X \in L$,

(P2) $\underline{P}(\lambda X) = \lambda \underline{P}(X)$ for all $X \in L, \lambda > 0$,

(P3) $\underline{P}(X+Y) \geq \underline{P}(X) + \underline{P}(Y)$ for all $X, Y \in L$.

Coherence implies the inequalities (for all gambles X and Y)

$$\underline{P}(X)+\underline{P}(Y) \leq \underline{P}(X+Y) \leq \underline{P}(X)+\overline{P}(Y) \leq \overline{P}(X+Y) \leq \overline{P}(X)+\overline{P}(Y)$$

A readiness to pay any price less than infimum value of X for the gamble X is confirmed by the axiom (P1). In accordance with axiom (P2) a gamble Z is acceptable iff λZ is acceptable for any positive λ.

Axioms (P2) and (P3) are reasonable only if gambles are defined in a linear utility scale, otherwise the combinations of the transactions might be unacceptable.

The degree of imprecision in previsions depends on the amount of information on which they are based and the degree of collision between different pieces of information.

Let's assume that a number of experts assess precise previsions $P_1(X),...,P_n(X)$. In this case we can define $\underline{P}(X)=\min\{P_i(X):1\le i\le n\}$, $\overline{P}(X)=\max\{P_i(X):1\le i\le n\}$.

Based on this information we can consider lower prevision $\underline{P}$ as a model acceptable for all experts, and a range $\overline{P}(X)-\underline{P}(X)$ as the measure of disagreement of the opinions among the experts.

There are several ways of definition of the coherent lower prevision. One can represent coherent lower prevision as the lower envelope of a closed convex set M of linear previsions $\underline{P}(X)=\min\{P(X):P\in M\}$, the lower envelope of the set of extreme points extM

$$\underline{P}(X)=\min\{P(X):P\in \mathrm{extM}\},$$

or by specifying the probability mass function or probability density function of each linear prevision in extM.

It is known that the general method for making inferences and decisions is natural extension. Let's consider the general procedure of natural extension. Assume that a user makes number of judgments. These judgments can be transformed into conditional lower previsions, $\underline{P}(X_i\mid B_i)\ge \propto_i$ for $1\le i\le k$, or in the case of precise judgments $P(X\mid B)=\propto$ by taking $\underline{P}(X\mid B)\ge\propto$, and $\underline{P}(-X\mid B)\ge -\propto$). The natural extension to any conditional lower prevision $\underline{P}(X\mid B\ )$ can be found as a solution to the problem

$$\underline{P}(X\mid B\ )=\sup\left\{\propto: B(X-\propto)\ge\sum_{i=1}^{k}\lambda_i B_i(X_i-\propto_i)\quad \text{for some}\quad \lambda_i\ge 0\right\},$$

where $Y\ge Z$ denotes $Y(\omega)\ge Z(\omega)$ for all $\omega\in\Omega$,B and B_i stand for indicator functions, and $\propto$ and $\propto_i$ denote constant gambles.

All the inferences, that can be derived from the primary judgments by means of the rules of coherence, are embraced by the natural extension.

The rules that follow from the principles of coherence and natural extension, can be justified via the behavioral interpretation of lower previsions.

The decisions by means of the upper and lower previsions can be derived in the following way. Assume we have to choose an operation from a finite set of possible operations $\{a_1,a_2,...,a_k\}$. The utility $U(a,\omega)$ of operation a depends on the unknown element ω. In order to compare two operations a_i and a_j we have to compute the upper and lower previsions on information available.

If $\underline{P}(X_i-X_j)>0$, then operation a_i is preferable to operation a_j, otherwise in case of $\overline{P}(X_i-X_j)<0$, operation a_j is preferable to operation a_i, and if neither above given conditions satisfy, preference cannot be defined on the basis of information available . In case of imprecise probability judgments, there may be several actions available.

As it was shown above the natural extension is the general method for inference and decision making. The natural extension can be brought to a single linear programming problem, or to a finite sequence of linear program problems.

2.4 Evaluation Computing

2.4.1 Evolution programming and Genetic Algorithms

During the last three decades there has been a growing interest in algorithms, which rely on analogies to natural processes [41,64]. The emergence of massively parallel computers made these algorithms of practical interest. The known algorithms in this class include evolutionary programming, genetic algorithms, evolution strategies, simulated annealing etc.

Evolution programs can be perceived as a generalization of genetic algorithms. Classical genetic algorithms operate on fixed-length binary strings, which need not be the case for evolution programs. Also, evolution programs usually incorporate a variety of "genetic" operators, whereas classical genetic algorithms use binary crossover and mutation.

The beginnings of genetic algorithms can be traced back to the early 1950s when several biologists used computers for simulations of biological systems [64]. However, the works done in late 1960s and early 1970s at the University of Michigan under the direction of John Holland led to genetic algorithms, as they are known today.

Genetic algorithms (GA) are global optimization algorithms based on the mechanics of natural selection and natural genetics.

GA have a number of specific peculiarities by which they differ from the other methods of optimization. These are the following ones:

1. GA employ only the objective function, not the derivative one or some other information on the object. It is very convenient in case that the function is neither differentiable nor discrete.
2. GA employ a parallel multipoint search strategy by maintaining a population of potential solutions, which provides wide information on the function behavior and exclude the possibility of sticking in local extremum of the function, while the traditional search methods, such as gradient etc. can not cope with this problem.
3. GA use probability-transitive rules instead deterministic ones. Besides, GA are very simple for computer implementation.

Genetic algorithms use a vocabulary borrowed from natural genetics. A candidate solution is called an individual. Quite often this individual is called also string or chromosome. This might be a little bit misleading: each cell of every organism of a given species carries a certain number of chromosomes; however, we talk about one-chromosome individuals only. Chromosomes are made of units - genes arranged in linear succession; every gene controls the inheritance of one or several characteristics.

Each gene can assume a finite number of values, called alleles (feature values).

In binary representation, chromosome is a vector, consisting of the bits succession, i.e. the succession of zeroes and ones.

A set of chromosomes makes a population. A number of chromosomes in population define a population size.

The genetic algorithms evaluate a population and generate a new one iteratively, with each successive population referred to as a generation.

The population undergoes a simulated evolution: at each generation the relatively "good" solutions reproduce while the relatively "bad" solutions die. To distinguish between different solutions we use an objective (evaluation) function, which plays the role of an environment. Quite often the objective function is called also fitness function.

The structure of a simple genetic algorithm is the same as the structure of any evolution program. During iteration t, a genetic algorithm maintains a population of potential solutions (chromosomes, vectors), G(t)= $\left\{x_1^t,...,x_n^t\right\}$. Each solution x_i^t is evaluated to give some measure of its "fitness". Then, a new population (iteration t+1) is formed by selecting the more fit individuals. Some members of this new population undergo reproduction by means of crossover and mutation, to form new solutions. Crossover combines the features of two parent chromosomes to form two similar offspring by swapping corresponding segments of the parents. For example, if the parents are represented by five-dimensional vectors (a_1, b_1, c_1, d_1, e_1) and (a_2, b_2, c_2, d_2, e_2), then crossing the chromosomes after the second gene would produce the offspring (a_1, b_1, c_2 , d_2, e_2) and (a_2, b_2, c_1, d_1, e_1).

Mutation arbitrarily alters one or more genes of a selected chromosome, by a random change with a probability equal to the mutation rate.

For a particular problem, GA can be presented by the chart (see Fig.2.34).

We discuss the actions of a genetic algorithm for a simple parameter optimization problem.

Now suppose we wish to maximize a function of k variables $f(x_1,...,x_k)$: $R^k \rightarrow R$.

If the optimization problem is to minimize a function f, this is equivalent to maximizing the function g, where g=-f, i.e.,

$$\min\{f(x)\}=\max\{g(x)\}=\max\{-f(x)\}.$$

Suppose further that each variable x_i can take values from a domain D_i =[a_i, b_i]$\subseteq$R and $f(x_1, ..., x_k)>0$ for all $x_i \in D_i$. We wish to optimize the function f with some required precision: suppose six decimal places for the variables' values are desirable.

It is clear that to achieve such precision each domain D_i should be cut into $(b_i - a_i) \cdot 10^6$ equal size ranges. Let us denote by m_i the smallest integer such as

$(b_i - a_i) \cdot 10^6 \le 2^{m_i} - 1$. Then, a representation having each variable x_i coded as a binary string of length m_i clearly satisfies the precision requirement. Additionally, the following formula interprets each such string:
where decimal ($string_2$) represents the decimal value of that binary string.

$$x_i = a_i + \text{decimal}(1001...001_2) \cdot \frac{b_i - a_i}{2^{m_i} - 1},$$

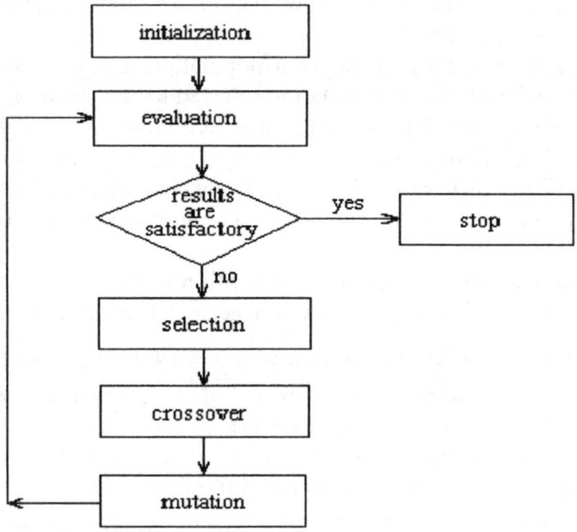

Fig. 2.34. The structure of a simple GA

Now, each chromosome (as a potential solution) is represented by a binary string of length $m = \sum_{i=1}^{k} m_i$; the first m_1 bits map into a value from the range $[a_1, b_1]$, the next group of m_2 bits map into a value from the range $[a_2, b_2]$, and so on; the last group of m_k bits map into a value from the range $[a_k, b_k]$.

To initialize a population, we can simply set some ps number of chromosomes randomly in a bitwise fashion. However, if we have some knowledge about the distribution of potential optima, we may use such information in arranging the set of initial (potential) solutions.

The rest of the algorithm is straightforward: in each generation we evaluate each chromosome (using the function f on the decoded sequences of variables), select new population with respect to the probability distribution based on fitness values, and recombine the chromosomes in the new population by mutation and crossover operators. After some number of generations, when no further improvement is observed, the best chromosome represents an (possibly the global) optimal solution. Often we stop the algorithm after a fixed number of iterations depending on speed and resource criteria.

For the selection process (selection of a new population with respect to the probability distribution based on fitness values), we must implement the following actions at first:

- Calculate the fitness value eval (v_i) for each chromosome v_i $(i = \overline{1, ps})$.
- Find the total fitness of the population

$$F = \sum_{i=1}^{ps} \text{eval}(v_i)$$

- Calculate the probability of a selection p_s^i for each chromosome v_i $(i = \overline{1, ps})$:

$$p_s^i = \text{eval}(v_i)/F$$

- Calculate a cumulative probability p_{cum}^i for each chromosome v_i $(i = \overline{1, ps})$:

$$p_{cum}^i = \sum_{j=1}^{i} p_s^j$$

The selection process is implemented ps times; each time we select a single chromosome for a new population in the following way:

- Generate a random (float) number r from the range [0,1].
- If $r < p_{cum}^1$ then select the first chromosome (v_1); otherwise select the i-th chromosome v_i $(2 \le i \le ps)$ such that

$$p_{cum}^{i-1} < r < p_{cum}^i$$

Obviously, some chromosomes would be selected more than once; the best chromosomes get more copies; the average ones remain, and the worst die off.

Now we are ready to apply the first recombination operator, crossover, to the individuals in the new population. One of the parameters of a genetic system is probability of crossover p_c. This probability gives us the expected number $p_c \cdot ps$ of chromosomes, which undergo the crossover operation. We proceed in the following way:

For each chromosome in the (new) population:

- Generate a random (float) number r from the range [0,1];
- If $r<p_c$, select given chromosome for crossover.

Now we mate selected chromosomes randomly: for each pair of coupled chromosomes we generate random integer number pos from the range [1, m-1] (m is the total length - number of bits in a chromosome). The number pos indicates the position of the crossing point. Two chromosomes

$$(b_1\, b_2 \dots b_{pos}\, b_{pos+1} \dots b_m)$$

and

$$(c_1\, c_2 \dots c_{pos}\, c_{pos+1} \dots c_m)$$

are replaced by a pair of their offspring:

$$(b_1\, b_2 \dots b_{pos}\, c_{pos+1} \dots c_m)$$

and

$$(c_1\, c_2 \dots c_{pos}\, b_{pos+1} \dots b_m)$$

The intuition behind the applicability of the crossover operator is information exchange between different potential solutions.

The next recombination operator, mutation, is performed on a bit-by-bit basis. Another parameter of the genetic system, probability of mutation p_m, gives us the expected number of mutated bits $p_m \cdot m \cdot ps$. Every bit (in all chromosomes in the whole population) has an equal chance to undergo mutation, i.e., change from 0 to 1 or vice versa. So we proceed in the following way.

For each chromosome in the current (i.e., after crossover) population and for each bit within the chromosome:

- Generate a random (float) number r from the range [0,1];
- If $r < p_m$, mutate the bit.

The intuition behind the mutation operator is the introduction of some extra variability into the population.

Following selection, crossover, and mutation, the new population is ready for its next evaluation. This evaluation is used to build the probability distribution (for the next selection process). The rest of evolution is just cyclic repetition of the above steps.

However, as it frequently occurs, in earlier generations the fitness values of some chromosomes are better than the value of the best chromosome after a finite number of generations.

It is relatively easy to keep track of the best individual in the evolution process. It is customary (in genetic algorithms implementations) to store "the best ever" individual at a separate location; in that way, the algorithm would report the best value found during the whole process (as opposed to the best value in the final population).

It is necessary to note, that classical GA may employ the roulette wheel method for selection, which is a stochastic version of the survival-of-the-fittest mechanism. In this method of selection, candidate strings from the current generation G(t) are selected to survive to the next generation G(t+1) by designing a roulette wheel where each string in the population is represented on the wheel in proportion to its fitness value. Thus those strings, which have a high fitness, are given a large share of the wheel, while those strings with low fitness are given a relatively small portion of the roulette wheel. Finally, selections are made by spinning the roulette wheel ps times and accepting as candidates those strings, which are indicated at the completion of the spin.

Example
As an example, suppose ps =5, and consider the following initial population of strings: G(0)={(10110), (11000), (11110), (01001), (00110)}. For each string v_i in the population, the fitness may be evaluated: eval(v_i). The appropriate share of the

roulette wheel to allot the i-th string is obtained by dividing the fitness of the i-th string by the sum of the fitnesses of the entire population:

In Table 2.16 a listing of the population with associated fitness values is given. Fig. 2.35 describes the corresponding roulette wheel

$$\frac{eval(v_i)}{\sum_{i=1}^{ps} eval(v_i).}$$

The next population is computed by the spinning the roulette wheel five times [66]. In order to put these strings into the new population they must undergo crossover and mutation.

Table 2.16. Listing of the population

	String v_i	Fitness $eval(v_i)$	Relative Fitness
V_1	10110	2.23	0.14
V_2	11000	7.27	0.47
V_3	11110	1.05	0.07
V_4	01001	3.35	0.21
V_5	00110	1.69	0.11

Pairs of the ps (assume ps even) candidate strings, which have survived selection, are next chosen for crossover. The probability that the crossover operator is applied will be denoted by p_c. Pairs of strings are selected randomly from G(t), without replacement, for crossover. A random integer k called the crossing site, is chosen from {1, 2,..., m-1}, and then the bits from the two chosen strings are swapped after the k-th bit with a probability p_c. This process is repeated until G(t) is empty. Fig. 2.36 illustrates a crossover for two 6-bit strings.

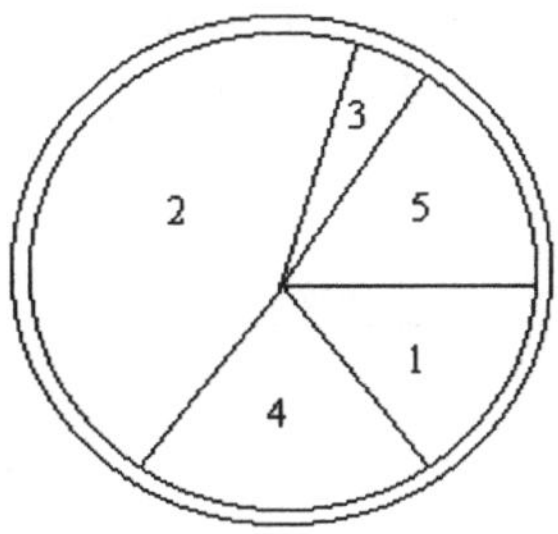

Fig. 2.35. Roulette wheel for string selection

The integers shown on the roulette wheel correspond to string labels

(a)	(b)	(c)
110101	1101 | 01	110100
100100	1001 | 00	100101

Fig. 2.36. An example of a crossover for two 6-bit strings

After crossover, mutation is applied to the candidate strings. The mutation operator is a stochastic bit-wise complementation applied with uniform probability p_m. That is, for each single bit in the population, the value of the bit is flipped from 0 to 1 or from 1 to 0 with probability p_m (Fig.2.37).

Before mutation: 11100
After mutation: 11000

Fig. 2.37 An example of a mutation

Finally the candidate strings are copied into the new population of strings G(t+1).

2.4.2 Computation with Genetic Algorithms

Let's consider genetic algorithm based optimization.
The problem of optimization of function f(x):

$$f(x) \rightarrow \min$$

subject to $x \in L \subset R$ with the use of a GA can be represented as given in Fig. 2.38 [64].

In Fig. 2.38 L denotes search space for x, Ω denotes space for binary strings, f(s) denotes fitness function.

Let's consider a simple problem of maximization of the function $F(x) = x^2$, where $x \in [0,31]$ (see Fig.2.39).

In order to use GA [1], we should first code variables into bit strings. As any integer number between 0 and 31 may be represented as binary number of 5 symbols (between (00000) = 0 and (11111)=31), the length of chromosomes will be five.

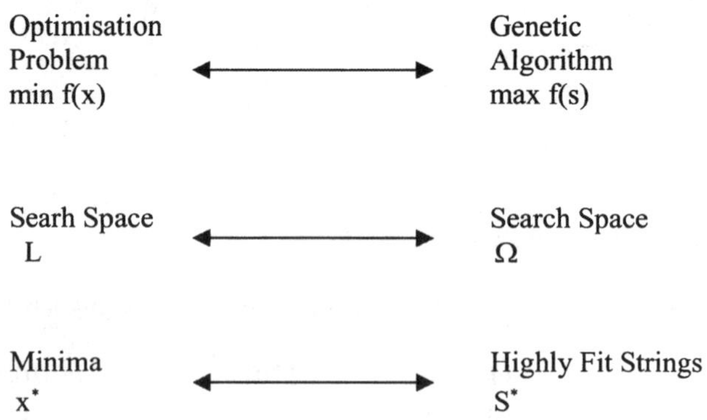

Fig. 2.38. A schematic representation of the optimization process by GA

Let's take four chromosomes with random set of 5 genes as an initial population. For example,

01101
11000
01000

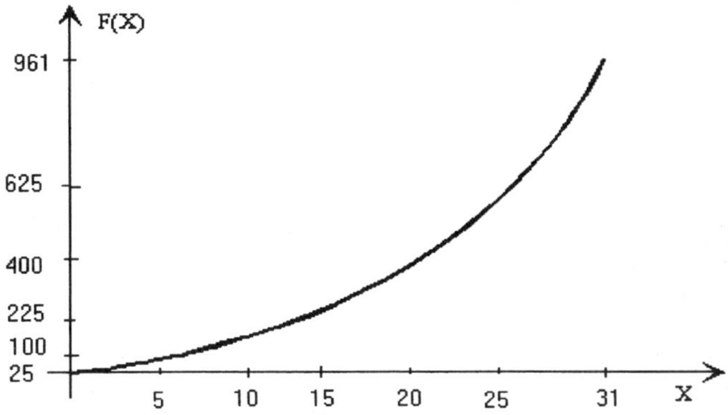

10011

Fig. 2.39. To the problem of the F(x) maximization

Then define their fitness, inserting appropriate real values into the function. For example,

$$F(01000)_2 = f(8) = 64$$

Thus the fitness of all individuals is calculated. Then in accordance with the formula

$$p_s^i = f_i / \sum_{j=1}^{4} f_i \ , \quad i = \overline{1,4}$$

survival probability for each individual is calculated, whereas cumulative probabilities are calculated according to the formula

$$p_{cum}^i = \sum_{j=1}^{i} p_s^i \qquad i = \overline{1,4} \,.$$

Let's enter all the calculated values in Table 2.17.

For the selection process we generate 4 random numbers from the range [0,1]. Suppose, that we generated 0.1;0.25;0.5 and 0.8.

Comparing these values with cumulative probabilities, we obtain the following

$$0.1 < p_{cum}^1$$

$$p_{cum}^1 < 0.25 < p_{cum}^2$$

$$p^1_{cum} < 0.5 < p^2_{cum}$$

$$p^3_{cum} < 0.8 < p^4_{cum}$$

Table 2.17. On application of GA

Initial population	Their Integer value	$F(x)=x^2$	p_s	p_{cum}	Number after selection
01101	13	169	0.14	0.14	1
11000	24	576	0.49	0.63	2
01000	8	64	0.06	0.69	0
10011	19	361	0.31	1	1
Average		293	0.25		1
Maximum		576	0.49		2
Sum		1170	1.00		4

Looking at the right sides of these inequalities, one can easily see, that the first and the fourth chromosomes have passed the selection, each fourth taking a place in the new generation, the second chromosome - the most highly fitted has got 2 copies, while the third one did not survive at all. These indices are written in the right column of Table 2.17.

Then crossover operation is applied. If the probability of crossover $P_c = 1$ is given, it means that, $4 \cdot 1 = 4$ chromosomes will participate in crossover process.

Let us choose them at random. Suppose, that the first and the second strings mate from crossover point 4 and the third with the forth-from crossover point 2.The strings obtained may be checked by Table 2.18.

$$0110|1$$

$$1100|0$$

$$11|000$$

$$00|011$$

Table 2.18. On application of GA

Population After selection	Crossover point	New Population	value of x	$F(x)=x^2$
01101	4	01100	12	144
11000	4	11001	25	625
01000	2	11011	27	769
10011	2	10000	16	256
Average				439
Maximum				729
Sum				1754

Having compared both of the tables, we see, for ourselves, that the population fitness is improved and we have come close to the solution.

The next recombination operator, mutation, is performed on a bit-by-bit basis. If P_m =0.05 is given, it means that only one of twenty bits in population will be changed: $20 \cdot 0.05 = 1$. Suppose, that the third bit of the fourth string undergoes mutation, i.e. $x_4' = 10100$.

Repeating these operations in a finite number of generations we will get the chromosome (11111) corresponding to optimal solution.

It is necessary to mention, that GA is especially effective for multiextreme problems in the global solution search process.

For example, if the function is of the type shown in the Fig. 2.40, it is rather difficult to find its global maximum by means of traditional methods.

Suppose that the function is defined as [64]

$$F(x) = x \cdot \sin(10\pi \cdot x) + 1.0$$

and is drawn in Fig. 2.40. The problem is to find x from the range [-1,2], which maximizes the function f, i.e. to find x_0 such that

$$F(x_0) \geq f(x), \text{ for all } x \in [-1,2]$$

It is relatively easy to analyze the function f. The zeros of the first derivative $f'(x)$ should be determined

$$f'(x) = \sin(10\pi \cdot x) + 10\pi \cdot x \cdot \cos(10\pi \cdot x) = 0 .$$

This formula is equivalent to

$$tg(10\pi \cdot x) = -10\pi \cdot x .$$

It is clear that the above equation has an infinite number of solutions,

$$x_i = \frac{2i-1}{20} + \xi_i , \quad \text{for } i=1, 2, ...,$$

$$x_0 = 0 ,$$

$$x_i = \frac{2i-1}{20} + \xi_i , \quad \text{for } i=-1, -2, ...,$$

where terms ξ_i represent decreasing sequences of real numbers (for i=1, 2, ..., and i=-1, -2, ...) converging to zero.

Note also that the function f reaches its local maxima for x_i if i is an odd integer, and its local minima for x_i if i is an even integer (see Fig. 2.40).

Since the domain of the problem is $x \in [-1,2]$, the function reaches its maximum for

$$x_{19} = \frac{37}{20} + \xi_{19} = 1.85 + \xi_{19} ,$$

where $f(x_{19})$ is slightly larger than

$$f(1.85) = 1.85 \cdot \sin\left(18\pi + \frac{\pi}{2}\right) + 1.0 = 2.85 .$$

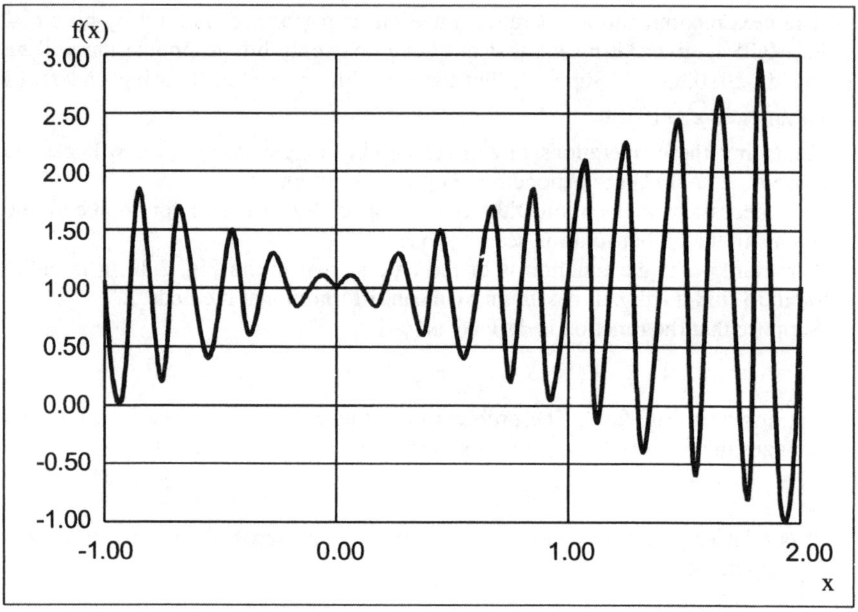

Fig. 2.40. Graph of a multiextreme function

Assume that we wish to construct a genetic algorithm to solve the above problem, i.e., to maximize the function f. Now let us discuss the major components of such a genetic algorithm in turn.

Representation we use a binary vector as a chromosome to represent real values of the variable x. The length of the vector depends on the required precision, which in this example is six places after the decimal point.

The domain of the variable x has length 3; the precision requirement implies that the range [-1,2] should be divided into at least $3 \cdot 1000000$ equal size ranges. This means that 22 bits are required as a binary vector (chromosome):

$$2097152=2^{21}<3000000\leq 2^{22}=4194304$$

The mapping from a binary ($b_{21}\, b_{20} \ldots b_0$) string into a real number x from the range [-1,2] is straightforward and is completed in two steps:

- convert the binary string ($b_{21}\, b_{20} \ldots b_0$) from the base 2 to base 10:

$$(< b_{21} b_{20} \ldots b_0 >)_2 = \left(\sum_{i=0}^{21} b_i \cdot 2^i \right)_{10} = x',$$

- find a corresponding real number x:

$$x = -1.0 + x' \, \frac{3}{2^{22}-1},$$

where -1.0 is the left boundary of the domain and 3 is the length of the domain.
For example, a chromosome

$$(1000101110110101000111)$$

represents the number 0.637197, since

$$x' = (1000101110110101000111)_2 = 2288967$$

and

$$x = -1.0 + 2288967 \cdot \frac{3}{4194303} = 0.637197$$

Of course, the chromosomes

$$(0000000000000000000000) \text{ and } (1111111111111111111111)$$

represent boundaries of the domain -1.0 and 2.0, respectively.

Initial population the initialization process is very simple: we create a population of chromosomes, where each chromosome is a binary vector of 22 bits. All 22 bits for each chromosome are initialized randomly.

Evaluation function evaluation function eval for binary vectors v is equivalent to the function f:

$$eval(v) = f(x),$$

where the chromosome v represents the real value x.
As noted earlier, the evaluation function plays the role of the environment, rating potential solutions in terms of their fitness. For example, three chromosomes:

$$v_1 = (1000101110110101000111)$$

$$v_2 = (0000001110000000010000)$$

$$v_3 = (1110000000111111000101)$$

correspond to values x =0.637197, x =-0.958973 and x_3 =1.627888, respectively. Consequently, the evaluation function would rate them as follows:

$$eval(v_1) = f(x_1) = 1.586345$$

$$eval(v_2) = f(x_2) = 0.078878$$

$$eval(v_3) = f(x_3) = 2.250650$$

Clearly, the chromosome v_3 is the best of the three chromosomes, since its evaluation returns the highest value. Genetic operators during the reproduction phase of the genetic algorithm we would use two classical genetic operators: mutation and crossover.

As mentioned earlier, mutation alters one or more genes (positions in a chromosome) with a probability equal to the mutation rate. Assume that the fifth gene from the v_3 chromosome was selected for a mutation. Since the fifth gene in this chromosome is 0, it would be flipped into 1. So the chromosome v_3 after this mutation would be

$$v_3^{/}=(1110100000111111000101)$$

This chromosome represents the value $x_3^{'}=1.721638$ and $f(x_3^{'})=-0.082257$. This means that this particular mutation resulted in a significant decrease of the value of the chromosome v_3 .On the other hand, if the 10th gene was selected for mutation in the chromosome v_3, then

$$v_3^{//}=(1110000001111111000101).$$

The corresponding value $x_3^{//}=1.630818$ and $f(x_3^{//})=2.343555$, the improvement over the original value of $f(x_3)=2.250650$.

Let us illustrate the crossover operator on chromosomes v_2 and v_3. Assume that the crossover point was (randomly) selected after the 5th gene:

$$v_2=(00000|01110000000010000),$$

$$v_3=(11100|00000111111000101).$$

The two resulting offspring are

$$v_2^{/}=(00000|00000111111000101),$$

$$v_3^{/}=(11100(01110000000010000).$$

These offspring evaluate to

$$f(v_2^{'})=f(-0.998113)=0.940865,$$

$$f(v_3^{'})=f(1.666028)=2.459245.$$

Note that the second offspring has a better evaluation than both of its parents.

Parameters for this particular problem we have used the following parameters: population size ps =50, probability of crossover p_c =0.25, probability of mutation p_m =0.01. The following section presents some experimental results for such a genetic algorithm.

Experimental results in Table 2.19 we provide the generation number for which we noted an improvement in the evaluation function, together with the value of the function. The best chromosome after 150 generations was

$$v_{max}=(1111001101000100000101),$$

which corresponds to a value $x_{max}=1.850773$.

As expected $x_{max}=1.85+\xi$ and $f(x_{max})$ is slightly larger than 2.85.

Consider the following 2-dimensional function [8]:

$$f(x_1,x_2) = 21.5+ x_1\sin(4\pi x_1)+x_2 \sin(20\pi x_2) \tag{2.197}$$

Table 2.19. Experimental results of optimization by GA

Generation number	Evaluation function
1	1.441942
6	2.250003
8	2.250283
9	2.250284
10	2.250363
12	2.328077
39	2.344251
40	2.345087
51	2.738930
99	2.849246
137	2.850217
145	2.850227

The given function (2.197) is multiextremal and possesses high frequency. Let's investigate, if it is reasonable to attempt to find a global extremum of this function with the help of classical methods of optimum finding. For example let's choose gradient method that is enough successfully applied for a certain class of problems. The considered function has a very little period, and, consequently, to determine the length of step that doesn't step over the extremum, one needs to select its value several times less than the length of period. But, even under such successful choice of step, we can achieve just that local extremum which is the closest to the chosen initial approximation. The probability that this exstremum is the global one is rather small. And if we don't know the period of function beforehand, finding the global extremum seems the difficult problem. The results of solving this problem by genetic algorithm are given in Table 2.20.

Table 2.20. Simulation results of optimization by GA

gener.number 10 total fit 571.919 max of pop f(7.6,5.1)=32.901
gener.number 20 total fit 536.692 max of pop f(8.1,5.4)=32.564
gener.number 30 total fit 574.263 max of pop f(6.1,5.0)=32.624
gener.number 40 total fit 525.889 max of pop f(8.1,5.0)=33.149
gener.number 50 total fit 574.603 max of pop f(7.1,5.5)=33.825
gener.number 60 total fit 612.590 max of pop f(7.1,5.5)=34.086
gener.number 70 total fit 642.664 max of pop f(7.1,5.5)=34.114
gener.number 80 total fit 625.976 max of pop f(7.1,5.5)=33.829
gener.number 90 total fit 624.194 max of pop f(7.1,5.7)=34.328
gener.number 100 total fit 591.221 max of pop f(7.1,5.5)=34.113
gener.number 110 total fit 570.381 max of pop f(7.1,5.7)=33.320
gener.number 120 total fit 616.179 max of pop f(7.1,5.7)=33.562
gener.number 130 total fit 621.786 max of pop f(7.1,5.7)=32.978
gener.number 140 total fit 597.910 max of pop f(7.1,5.5)=33.291
gener.number 150 total fit 630.848 max of pop f(7.1,5.7)=33.730
gener.number 160 total fit 636.193 max of pop f(7.1,5.7)=33.832

gener.number 170 total fit 609.329 max of pop f(8.1,5.5)=33.538
gener.number 180 total fit 628.802 max of pop f(7.1,5.6)=34.167
gener.number 190 total fit 643.545 max of pop f(8.1,5.6)=33.930
gener.number 200 total fit 654.736 max of pop f(8.1,5.6)=34.089
gener.number 210 total fit 609.340 max of pop f(8.1,5.6)=33.708
gener.number 220 total fit 630.769 max of pop f(8.1,5.6)=33.680
gener.number 230 total fit 636.263 max of pop f(7.1,5.6)=34.016
gener.number 240 total fit 622.702 max of pop f(7.6,5.6)=34.372
gener.number 250 total fit 610.381 max of pop f(7.6,5.6)=34.516
gener.number 260 total fit 539.640 max of pop f(7.1,5.4)=33.962
gener.number 270 total fit 636.173 max of pop f(7.1,5.5)=34.023
gener.number 280 total fit 597.196 max of pop f(7.1,5.4)=33.941
gener.number 290 total fit 556.904 max of pop f(7.1,5.7)=34.225
gener.number 300 total fit 622.570 max of pop f(7.1,5.5)=33.302
gener.number 310 total fit 601.822 max of pop f(7.1,5.5)=33.930
gener.number 320 total fit 639.573 max of pop f(7.1,5.5)=34.047
gener.number 330 total fit 578.657 max of pop f(7.1,5.5)=34.038
gener.number 340 total fit 602.191 max of pop f(7.1,5.5)=34.052
gener.number 350 total fit 469.975 max of pop f(11.1,5.5)=38.030
gener.number 360 total fit 565.759 max of pop f(7.1,5.5)=34.052
gener.number 370 total fit 613.330 max of pop f(7.1,5.5)=34.052
gener.number 380 total fit 607.128 max of pop f(7.1,5.5)=34.150
gener.number 390 total fit 632.374 max of pop f(7.1,5.5)=33.726
gener.number 400 total fit 576.375 max of pop f(7.1,5.4)=33.675
gener.number 410 total fit 622.438 max of pop f(7.1,5.5)=34.151
gener.number 420 total fit 553.456 max of pop f(7.1,5.4)=32.598
gener.number 430 total fit 593.215 max of pop f(7.1,5.4)=32.604
gener.number 440 total fit 620.224 max of pop f(7.1,5.5)=32.629
gener.number 450 total fit 592.319 max of pop f(6.1,5.5)=32.797
gener.number 460 total fit 595.806 max of pop f(6.1,5.5)=32.963
gener.number 470 total fit 606.655 max of pop f(6.1,5.5)=33.112
gener.number 480 total fit 627.283 max of pop f(6.6,5.5)=33.223
gener.number 490 total fit 616.025 max of pop f(6.6,5.5)=33.275
gener.number 500 total fit 578.026 max of pop f(6.7,5.5)=33.092
gener.number 510 total fit 575.476 max of pop f(6.6,5.6)=32.840
gener.number 520 total fit 643.179 max of pop f(6.6,5.6)=33.101
gener.number 530 total fit 622.002 max of pop f(6.6,5.6)=33.282
gener.number 540 total fit 605.838 max of pop f(6.6,5.7)=33.458
gener.number 550 total fit 553.371 max of pop f(6.6,5.6)=33.281
gener.number 560 total fit 576.080 max of pop f(6.6,5.6)=33.616
gener.number 570 total fit 602.460 max of pop f(6.6,5.6)=33.244
gener.number 580 total fit 552.331 max of pop f(6.6,5.6)=33.428
gener.number 590 total fit 548.172 max of pop f(5.6,5.6)=32.411
gener.number 600 total fit 600.481 max of pop f(5.7,5.6)=31.965
MAXIMUM f(11.13488, 5.52456)=38.0297 after 24000 eval

Thus, this demonstrates the case when to find extremum by classical methods is practically impossible, while this problem has been solved with the help of GA.

Now we consider genetic algorithm with group principle. There are a number of modifications of GA's [12,25,41,46] proposed by different authors at the different stages of GA development. They differ from each other basically in the way of coding or in the forms of crossover, mutation, selection etc.

Here we consider a new approach to genetic algorithms constructing. It is a group principle genetic algorithm, which combines an individual chromosome evolution with group chromosome evolution and the directional search with possibility theory. Such an approach allows observing and estimating the behavior of different groups of individuals, where the groups fight with each other for surviving. The groups with a higher fitness get more possibility to survive and to reproduce, whereas the lower fitness groups have less possibility for surviving. On the other hand, the individuals inside the groups do also fighting for life, except the lower fitness individuals from the groups and supplementing them with high fitness individuals, improving an average fitness of the groups.

Thus we consider a two level genetic algorithm, called Group Principle Genetic Algorithm (GPGA), where the top level provides the group search while the bottom level provides the individual search [12]. Principles of grouping the individuals and tuning the size of groups (the number of individuals within the groups) are based on the possibility theory.

In GPGA at the first phase the set of points (individuals) is scattered on the domain of variables and gives us an approximate information about the function behavior on the whole domain. Then we get rid of the «bad» individuals (the individuals which have a less average fitness), and retain the high fitness individuals. After the first phase the groups of individuals are formed by use of acquired information about the whole population. The nearest points in the parameter space organize single groups of individuals. Further we concentrate most attention on these groups, potentially containing the global solution on the problem. For this aim we use an estimation of their fitness.

As we have not complete information about groups, we may give only approximate estimation for every group. That is why we should use fuzzy logic. Having information about some members of groups and using the possibility theory one can give a quantitative evaluation of fitness for every group. This fitness of groups means the possibility that just given group may contain the global solution of the problem. It allows us to regulate the size of groups where the fittest groups get possibility to have more members (individuals). Then genetic operators, such as crossover, mutation and reproduction are conducted over the individuals yet inside of groups. This is the second level of algorithm where the evolution occurs inside of each group. In Fig.2.41 the algorithm's chart is demonstrated.

Reduction of the size of population takes place only once- on the first generation after defining the possibilities. Further the algorithm works with a fixed number of individuals. The groups of individuals are actually subpopulations inside of which the ordinary genetic operations are conducted. Due to these operations, after some number of generations these groups become, in genetic terms, families. On the top level these families fight with each other to get more possibility (chances) for reproduction. Due to this fight an average fitness value of whole population and subpopulations (groups) rises from generation to generation.

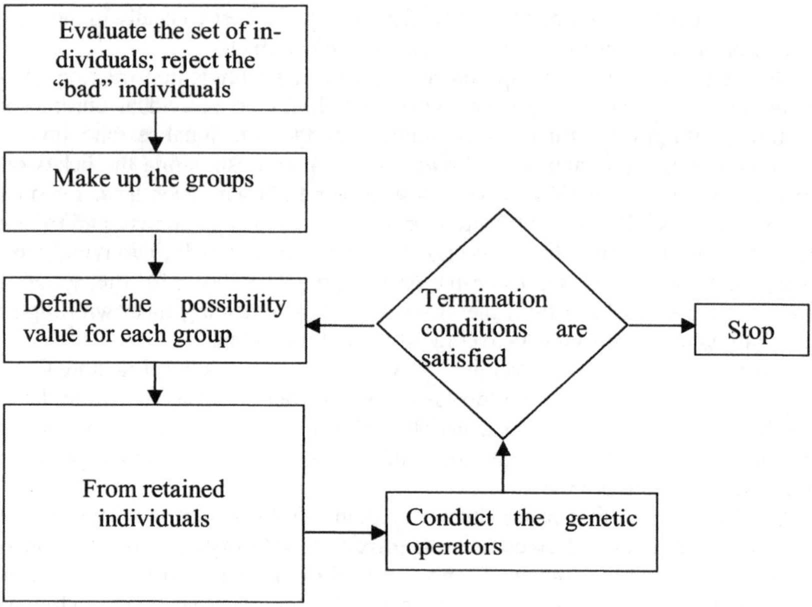

Fig. 2.41. GA with group principle

More detailed description of GPGA is as follows:

1. Selecting the set of M random individuals and their evaluating. $eval_i = F(x_i)$, where F(x) is a fitness function and x_i-i-th individual.
2. Calculating the average fitness of population

$$evpop = \frac{\sum_{1}^{M} eval_i}{M} \tag{2.198}$$

□ and discarding low average fitness individuals.

3. Organizing groups according to the principle of space proximity of individuals

$$\|x_i - x_j\| < \delta$$

4. Calculating the group average fitness by using

$$evgr_j = \frac{\sum_{1}^{M_j} x_{ji}}{M_j} \tag{2.199}$$

where M_j is the size of j-th group.

5. Calculating the possibility that just given group contains the global solution by using

$$P_i = \frac{evgr_i}{\sum_j evgr_j} \tag{2.200}$$

6. Reducing the population size and retaining only necessary number of individuals. Distributing them among the groups according to their possibility values.
7. Applying the crossover, mutation and selection operations by an ordinary way inside of each group.
8. If termination condition isn't satisfied, go to the step 4 and stop elsewise.

Genetic algorithms are effective, first of all, for solving the multiextremal function optimization problems. Experiments have shown that classical genetic algorithms (CGA) are irreplaceable for such functions, behavior of which is not specified logically (Fig 2.42). As an example one can adduce the Brown chaotic behavior. In such a problem there is no ary group behavior, hence a GPGA may be less effective than CGA. But in the most of function optimization problems, it is ob served group behavior logic, where advantage of GPGA is considerable.

In Fig. 2.43 a multiextremal function's graph is demonstrated. In this case we have some groups, which are defined after the first phase and each of them has a possibility of containing the global maxima. As shown in Fig. 2.43 the rightist group will have the highest possibility. Using this information algorithm directs the search towards the groups having higher possibility. It gives us the possibility to avoid unnecessary iterations and move towards the global solution.

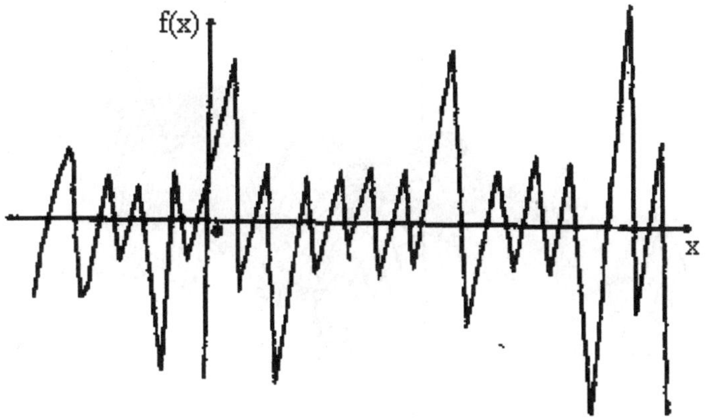

Fig. 2.42. Example of the chaotic function

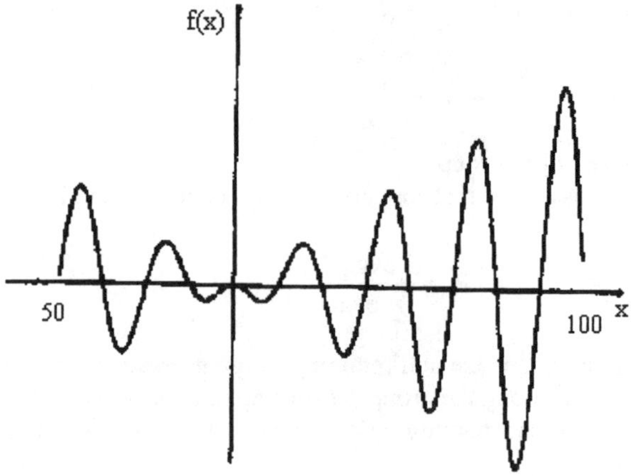

Fig. 2.43. The graph of the function $f(x)=\frac{x}{3}\sin\left(\frac{x}{4}\right)$

On this simple example one can see that there are some classes of functions for which the proposed algorithm has better performance characteristics than classical genetic algorithms.

For comparison of two methods we show experimental results of optimizing the function

$$f(x,y)=x\sin(\pi x)+y\sin(\pi x), \tag{2.201}$$

for x from [-2; 3.1] and y from [-0.8; 1]. The graph of this function is demonstrated in Fig. 2.44

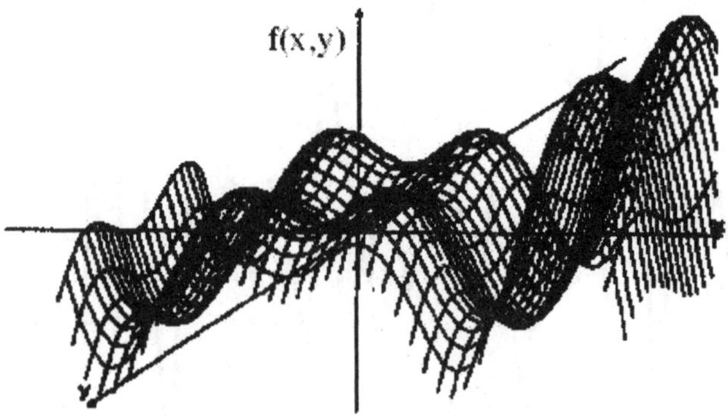

Fig. 2.44. The graph of the function $f(x,y)=\sin(\pi x)+y\sin(\pi x)$

As shown in the Table 7.6 the number of generations in CGA is almost twice as much as in GPGA.

Table 2.21. Comparison of two GA-based methods

Technique	Number of evaluations	Number of generations	Reached solution
CGA	880	40	3.5024
GPGA	570	21	3.5029

The experiments have shown that the GPGA is acceptable also for such problems as vector optimization, group interest games, business games etc.
The new type of group principle genetic algorithms [6,11], which use directed mutation is considered in this section. Below the main goal of this method is to make the search more effective and the results more accurate. Localization of mutation makes the search less global but faster and accurate. It allows searching global and local peaks of certain class of multimodal functions. Algorithm description is given below.
On the first stage the large size population is taken in order to embrace all areas of parameter space. Then higher-than-average individuals survive and form the groups (subpopulations). Further search is performed inside the different groups.

The detailed description of group genetic algorithm with directed mutation (GGA-DM) is following:

1. Generate an initial population with N random individuals (N vectors in the parameter space)

$$x_i, i = \overline{1,n}$$

2. Evaluate all individuals and calculate the average fitness of population.

$$f = [\sum f(x_i)]/N$$

3. At the selection stage the individuals satisfying

$$f(x_j)>f \quad (j=\overline{1,N})$$

are selected.

4. Make up the groups according to the principle of space neighborhood of individuals [12]

$$\|x_i - x_j\| < \delta$$

where δ is a neighborhood parameter: $\delta = 1{,}5\ L^k/N$, and L is the length of range of k-th component of vectors x_i and x_j.
As shown in Fig. 2.45 the 'bad' individuals are rejected, and each formed group searches for one peak.

5. Conduct selection: retain the best individual in each group.

6. Conduct mutation over all retained individuals by following way:

$$x_i^{t+1} = \begin{cases} x_i^t + r, \text{if} \quad f(x_i^t + r) > f(x_i^t), \\ x_i^t - r, \text{ if } f(x_i^t - r) > f(x_i^t), \\ x_i^t, \text{ otherwise.} \end{cases}$$

Here r-is random value from range $[0, \delta (1-t/T)]$, where t is the generation number, T is the maximal number of generations. This operation is conducted until all individuals become unimproved. If after this step the desired result isn't reached, set t=0 and return to step 1.

Probability estimation is performed as follows. Let L be the length of the range [a,b] in Fig. 2.45, and l is the length of range containg the global maximum. If we generate N random points in the range [a,b], then probability of falling of at least one of the points into the range [c,b] is described by the formula:

$$P=1- [(L-l)/L]^N$$

Therefore, for P to be greater than some fixed P_0, the inequality

$$N>1/ \log_{1-P_0}[(L-1)/L]$$

must be satisfied.

In another approach for the choice of the initial population size N, an irregular grid with a random step h is set in the range [a,b]

$$0{,}5\ L / N<h<1{,}5\ L / N$$

Such an approach is more convenient for creating the groups, and probability of falling into the range [c,b] (Fig. 2.45) will be equal to 1, if the inequality

$$N>1{,}5 \cdot L / l$$

is satisfied.

However, the value l can be only predicted by use of some knowledge about described physical process or another prior information if available.

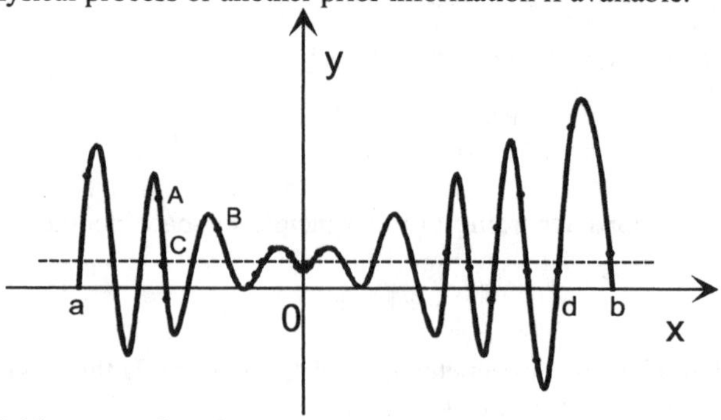

Fig. 2.45. Multiextreme function

In order to test an effectiveness of proposed algorithm, we have carried out experiments with following functions (Fig. 2.46) [6].

$$f_1(x,y) = \begin{cases} |x+y|/2, & \text{if trunc(x/2) and trunc(y/3.5) are odd,} \\ 1, & \text{otherwise.} \end{cases}$$

$$f_2(x,y) = 0.022 \cdot e^{-2\sqrt{x^2+y^2+3(\cos 5x+4\sin 4y)}}$$

$$f_3(x,y) = 0.5x \cdot \sin x + 0.6\sin(y/2) + 1.5\sin(9(x-y))$$

$$f_4(x,y) = |x-y|(4e^{-2\sqrt{x^2+y^2}} + e^{\cos x+\sin y})$$

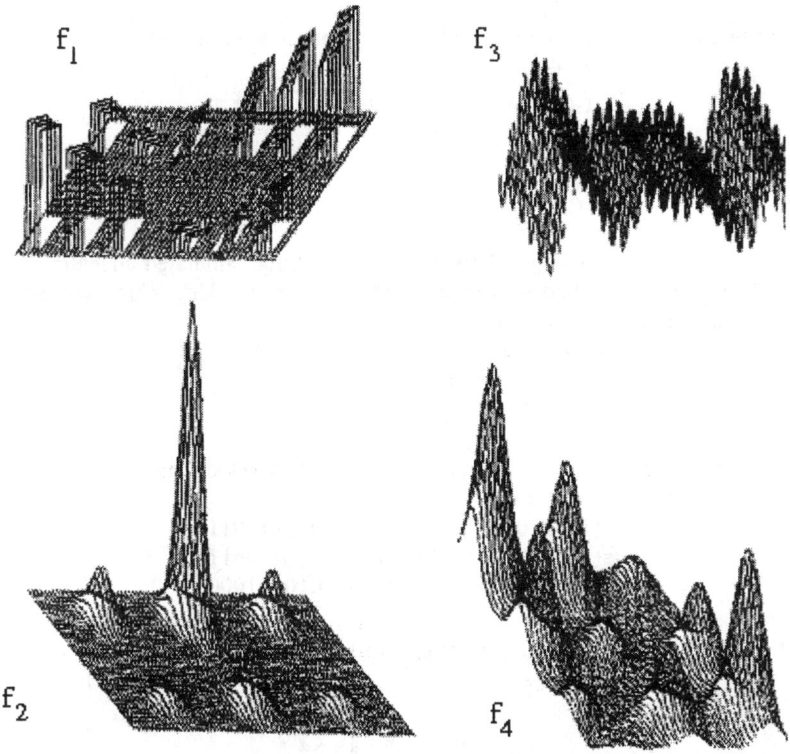

Fig. 2.46. Test functions

Here *trunc* (x) means the whole part of x. For comparison an iterated genetic hill climbing (IGH) [6], group principle genetic algorithm (GPGA), and simple genetic algorithm (SGA) [1] have been used. Each experiment has been conducted 10 times and the average results are shown in Table 2.22.

As shown in Table 2.22 for 3 functions from 4 the best result has been reached by GGA-DM. But GPGA gives better result than others for $\square_2$ because that function has not only group behavior but also plenty of local peaks in each group. Due to the global search inside of groups, GPGA becomes better for $\square_2$.

The numbers in the Table 2.22 describe the time of performance. Thus, one can note, those group types of genetic algorithms aren't such universal as SGA, but for functions with group behavior they are more effective due to directed evolution. The search of not only global but also local extremes is another advantage of the group approach.

Note that GAGP is also effective for multicriterial optimization problems and for some problems of the game theory.

Below we compare computational abilities of conventional and evolutionary computing.

Table 2.22. Comparison of four GA-based optimization methods

Method	f_1	f_2	f_3	f_4
GGADM	11	91	6	7
GAGP	20	69	10	12
SGA	31	107	12	25
IGH	24	97	10	19

Let's discuss three algorithms, i.e., hill climbing, simulated annealing, and genetic algorithm, applied to a simple optimization problem. This example underlines the uniqueness of the GA approach.

The search space is a set of binary strings v of the length 30. The objective function f to be maximized is given as

$$f(v) = |11 \cdot one(v) - 150|,$$

where the function one(v) returns the number of 1s in the string v.

For example, the following three strings

$$v_1=(110110101110101111111011011011),$$
$$v_2=(111000100100110111001010100011),$$
$$v_3=(000010000011001000000010001000),$$

would evaluate to

$$f(v_1)=|11 \cdot 22-150|=92,$$

$$f(v_2)=|11 \cdot 15-150|=15,$$

$$f(v_3)=|11 \cdot 6-150|=84,$$

($one(v_1)=22$, $one(v_2)=15$, and $one(v_3)=6$).

The function f is linear and does not provide any challenge as an optimization task. We use it only to illustrate these three algorithms. However, the interesting characteristic of the function f is that it has one global maximum for

$$v_g=(111111111111111111111111111111),$$

$f(v_g)=|11 \cdot 30-150|=180$, and one local maximum for

$$v_l=(000000000000000000000000000000),$$

$$f(v_l)=|11\cdot 0-150|=150 .$$

There are several versions of hill climbing algorithms. They differ in the way a new string is selected for comparison with the current string. One version of a simple (iterated) hill climbing algorithm is given in Fig. 2.47 (steepest ascent hill climbing).

Initially, all 30 neighbors are considered, and one v_n which returns the largest value f (v_n) is selected to complete with the current string v_c. If $f(v_c)<f(v_n)$, then the new string becomes the current string. Otherwise, no local improvement is possible: the algorithm has reached the optimum. In this case, the next iteration ($t\leftarrow t+1$) of the algorithm is executed with another string selected at random.

It is interesting to note that the starting string (randomly selected) determines the success or failure of the single iteration of the above hill climbing algorithm (i.e., return of the global or local optimum). It is clear that if the starting string has thirteen 1s or less, the algorithm will always terminate in the local optimum (failure). The reason is that a string with thirteen 1s returns a value 7 of the objective function, and any single-step improvement towards the global optimum, i.e., increase the number of 1s to fourteen, decreases the value of the objective function to 4. On the other hand, any decrease in the number of 1s would increase the value of the function: a string with twelve 1s yields a value of 18, a string with eleven 1s yields a value of 29, etc. This would push the search in the «wrong» direction, towards the local maximum.

For problem with many local optima, the chances of hitting the global optimum (in a single iteration) are weak.

procedure iterated hill climber
begin
$t\leftarrow 0$
repeat
local←FALSE
select a current string v_c at random
evaluate v_c
repeat
select 30 new strings in the neighborhood of v_c by flipping
single bits of v_c select the string v_n from the set of new strings
with the largest value of objective function f
if $f(v_c)<f(v_n)$
then $v_c\leftarrow v_n$
else local←TRUE
until local
$t\leftarrow t+1$
until t=MAX
end

Fig. 2.47. A simple (iterated) hill climber

The structure of the simulated annealing procedure is given in Fig. 2.48
.p = exp{(f(vn)-f(vc))/T} = exp{(7-18)/T}
The function random [0, 1] returns a random number from the range [0, 1].

```
procedure simulated annealing
begin
t←0
initialize temperature T
select a current string vc at random
evaluate vc
repeat
repeat
select a new string vn
in the neighborhood of vc
by flipping a single bit of vc
if  f(vc) < f(vn)
then vc ← vn
else if  random [0,1) < exp{(f(vn )-f(vc))/T}
then vc←vn
 until (termination condition)
T ← g(T,t)
t ← t+1
until (stop-criterion)
end
```

Fig. 2.48. Simulated annealing

The termination-condition checks whether «thermal equilibrium» has reached, i.e., whether the probability distribution of the selected new strings approaches the distribution [1,55]. However, in some implementations [64], this loop is executed just k times (k is an additional parameter of the Boltzmann method).
The temperature T is lowered in steps (g(T, t)<T for all t). The algorithm terminates for some small value of T: the stop-criterion checks whether the system is «frozen», i.e., virtually no changes are accepted anymore.
As mentioned earlier, the simulated annealing algorithm can escape local optimums.
Let us consider a string

$$v_4=(111000000100110111001010100000\,),$$

with twelve 1s, which evaluates to $f(v_4)= |11 \cdot 12 - 150| = 18$. For v_4 as the starting string, the hillclimbing algorithm (as discussed earlier) would approach the local maximum

$$v_l=(000000000000000000000000000000),$$

since any string with thirteen 1s (i.e., «towards» the global optimum) evaluates to 7 (less than 18). On the other hand, the simulated annealing algorithm would accept a string with thirteen 1s as a new current string with probability which, for some temperature, say, T=20, gives

$$p = e^{-\frac{11}{20}} = 0.57695,$$

i.e., the chances for acceptance are better than 50 %.

Genetic algorithms maintain a population of strings. Two relatively poor strings

v_5=(111110000000110111001110100000)

v_6=(000000000000110111001010111111)

each of which evaluates to 16 can produce much better offspring (if the crossover point falls anywhere between the 5th and the 12th position):

v_7=(111110000000110111001010111111)

The new offspring v_7 evaluates to

$$f(v_7) = |11 \cdot 19 - 150| = 59.$$

For a detailed discussion on these and other algorithms (various versions of hillclimbers, genetic search, and simulated annealing) tested on several functions with different characteristics, the reader is referred to [12].

2.5 Chaotic Computing

2.5.1 Elements of Chaotic Computing

What is chaos ?

It might seem curious, but there is no general definition of chaos [3,79].

The order inside a disorder, the regularity in randomness - these obvious paradoxes are in the kernel of chaotic systems. More strictly, chaos is a random behavior in a deterministic nonlinear system. This definition is symbolic enough– we will further consider also the behavior of fuzzy dynamic systems.

For a long time scientists and engineers have been unaware of the chaotic behavior of systems as they were approximating nonlinear systems by linear ones.

In spite of the presence of considerable nonlinearities in physical systems, in some cases their linear approximation appear to be useful. The behavior of linear systems can be easily forecasted in contrast to the nonlinear ones. The behavior of linear systems can be described by a system of differential equations with solutions in a closed form; rare nonlinear systems might have such an advantage.

A linear system has only three types of attractors:

- fixed point attractor;
- periodic attractor;
- quazy-static attractor.

Possible movements in nonlinear systems are much different and complex. A nonlinear system, as opposed to a linear system, is capable of acting in chaotic modes (possessing chaotic attractors) [57,75, 79].

Getting back to the question stated, we formulate the following definition [79].

Definition. Chaos

Chaos is the trajectory of a certain nonlinear dynamic system. Below we consider dissipative systems described by the equations [14, 57]

$$\dot{X} = V(X), X(0) = X_0 \tag{2.202}$$

where

$$X = \{x_1, x_2, ..., x_n\}$$

is a n-dimensional vector,

$$V = \{v_1, v_2, ..., v_n\}$$

is a vector function.

The solution of (2.202)

$$\dot{x}_i(t) = F_i(t, x_0), \quad i = \overline{1, n}$$

represents a phase trajectory in the phase space of system (2.202) created by the components

$$\{x_1, x_2, ..., x_n\}$$

Note, that if div V=0 then the system will be conservative and its phase volume will be the same value. If div V<0 then the system is dissipative and as the time progresses its phase volume decreases and all decisions converge at $t \to \infty$ to fall inside some subset B, called an attractor of the phase space M [57].

All of the attractors considered here are simple attractors and belong to stable stationary points, or stable limit cycles, or invariant torus. Such attractors are sub-varieties of the phase space of dynamical systems. For example, a limit cycle and two-dimensional invariant torus are one-dimensional and two-dimensional sub-varieties, respectively.

In dissipative dynamical systems with $n \geq 3$, some limited attracting sets can exist which are attractors and are not sub-varieties in the same time. Such attractors are named strange attractors [43, 57]. In systems with strange attractors, the phase curves tend to converge to these attractors and once get inside the field of these attractors never move out. The movement at the attractor, however, is unstable: any two trajectories of the system diverge exponentially fast still residing on the strange attractor. These strange attractors are examples of chaotic movement in dissipative dynamical systems.

2.5.2 Non-linear dynamics and chaotic analysis.

It has been established that a simple attractor has regular dynamics while strange atrractors have chaotic dynamics. So it is necessary to determine some criteria to distinguish regular movements from chaotic ones or in other words to distinguish one type of attractors from the others. Let's consider system (2.202).

Because chaotic behaviour is consequence of instability in phase trajectories, it is reasonable to apply the value of run-out of the phase curves of the dynamical system as the distinction criteria, i.e.

$$d(t) \approx d(0)e^{kt} \tag{2.203}$$

where d(t) is the distance between two nearest phase points, $d(t) = |x_2(t) - x_1(t)|$, k is the average speed of exponential run-out of the trajectory.

k is defined from (2.203) as

$$k \approx t^{-1} \ln[d(t)/d(0)] \tag{2.204}$$

If in formula (2.204) $d(0) \to 0$ and $t \to \infty$ then we get

$$h = \lim_{\substack{d(0)\to 0 \\ t\to\infty}} t^{-1} \ln[d(t)/d(0)] \tag{2.205}$$

which is the enthropy by Kolmogorov and Sinay, i.e. the KS-enthropy [57]. In case of emerging chaotic movement the value h>0. In case when the movement corressponds to a stable point then h<0. When h is equal to zero this means either presence of a periodic or a quazi-periodic movement. Numerical methods for determining criteria of chaotic behaviour have been considered in [75].

In [57] the following function:

$$\Lambda(\vec{\xi}(0) = \lim_{t\to\infty} t^{-1} \ln\left(|\vec{\xi}(t)|/|\vec{\xi}(0)|\right)$$

determined on initial bias vectors

$$\vec{\xi} = x_2(0) - x_1(0) = d_0,$$

such that $|\vec{\xi}(0)| = \varepsilon$, where $\varepsilon \to 0$ is composed for system (2.202). It has been stated in [57] that for different turns of vector $\vec{\xi}(0)$ the function Λ changes by steps and returns a finite range of values $\{\lambda_i\}$, $i = \overline{1,n}$, where λ_i is named a Lyapunov's index. In [57] it is also shown that h, the enthropy by Kolmogorov-Sinay, is actually a maximum characteristic index by Lyapunov. Based on the latter, in [43,57] criteria have been formulated that allow identification of the types of attractors in dynamical systems.

In two-dimensional systems attractors of two types may originate: stable stationary point and limited cycle. They can be identified by using the $\{\lambda_i\}$ this way. If $\{\lambda_1, \lambda_2\}$={-,-}, then the attractor is stable point, else if $\{\lambda_1, \lambda_2\}$={-,0}, then the attractor is limited cycle. If in system (2.202) n=3, all types of attractors can be identified as follows: $(\lambda_1,\lambda_2,\lambda_3)$=(-,-,-) is either stable focus or node; $(\lambda_1,\lambda_2,\lambda_3)$=(-,-,0) is stable limited cycle; $(\lambda_1,\lambda_2,\lambda_3)$=(-,0,0) is stable torus; $(\lambda_1,\lambda_2,\lambda_3)$=(-,0,+) is a strange attractor. Numerical methods to define the Lyapunov indices are considered in [75].

Another important criterion to distinguish between regular and chaotic movement of non-linear systems is the autocorrelation function. Obviously, either periodic or quazi-periodic is the behavior of system, the corresponding autocorrelation functions will also be periodic or quazi-periodic, respectively. As mentioned in [57], if the auto-correlation function converges to zero and there are no steady points in the system, then the system will be subject to chaotic movement mode.

The Fractal Dimensions of Strange Attractors

Assume that system (2.202) is characterized by n-dimensional phase space. Let's place n-dimensional cubes into the set A of the phase space M so that they contain inside all the points of the set A. Let N be the minimum number of such cubes covering A. In the limit

$$d(A) = \lim_{\varepsilon \to 0}[\ln N(\varepsilon) / \ln(1/\varepsilon)]$$

$d(A) \equiv dF$ is a metric dimension and is named a fractal dimension [57,60,61]. For regular sets the fractal dimension is the same of the ordinary dimension, for example, for a surface dF=2, for a line dF=1. So for a limited cycle dF=2 (both attractors are simple). For strange attractors the fractal dimension is fractional. Consequently, the fractal dimension can be used as a criterion to identify simple and strange attractors, i.e. to distinguish between regular chaotic movements. The problem now is how to calculate the fractal dimension. In [57] it is shown that the fractal dimension dF coincides with the Lyapunov's dimension dL calculated as:

$$dL = j + \sum_{i=1}^{j} \lambda_i / |\lambda_{j+1}| \qquad (2.206)$$

where $\lambda_1 \geq \lambda_2 \geq \ldots \geq \lambda_n$, and value of j is found from the requirements:

$$\sum_{i=1}^{j} \lambda_i \geq 0, \sum_{i=1}^{j+1} \lambda_i < 0 \qquad (2.207)$$

Assuming n=3 in (2.202) and that the movement takes place on a strange attractor, (2.206) takes the form

$$dL = 2 + \lambda_1 / |\lambda_3|$$

as for such systems $\lambda_1 \geq \lambda_2 \geq \lambda_3$ and $(\lambda_1,\lambda_2,\lambda_3)$=(+,0,-).

Numerical methods for calculating the fractal dimension are considered in [75]. In [57] the fractal dimension, dF of a strange attractor has been numerically calculated for the system given below:

$$\dot{x} = y(z - 1 + x^2) + \gamma x, \quad \dot{y} = x(3z + 1 - x^2) + \gamma y, \quad \dot{z} = -2z(\nu + xy)$$

$\nu = 1.1$; $\gamma = 0.87$

It has been established that the strange attractor in this system takes a medium position between the surface and the three-dimensional figure $dF = 2.318 \pm 0.002$. If the considered system depends on a parameter (the most of the researched here systems are suchlike), by changing the parameter, one can alternate the characteristics of the attractors and, consequently, the behaviour of the system. In particular, by causing a little change in the parameters of system (2.202) with n=2 the stable point may shift and the form of limited cycle and period may change. If we continue changing, the parameter soon comes to some critical value. The attractor may undergo some qualitative reconstruction in the topology that will cause the behavior of the system change considerably. The reconstruction in the topology of the steady movement modes is named a bifurcation. The continuous changings in the parameters of the system correspond to a series of bifurcations.

Let's consider a system depending on a control parameter $\propto$

$$\dot{x} = V(x, \propto) \tag{2.208}$$

Assume there is a solution $x_0 = x_0(\propto)$ with some value of $\propto$. Also let's assume that with the condition $\propto < \propto_0$ satisfied, the stationary point $x_0(\propto)$ of system (2.202) is stable while with $\propto > \propto_0$ it is unstable. The point $\propto_0$ is named the bifurcation point and the area of the solution with $\propto > \propto_0$ is the most interesting as concerned with chaos progressing.

The control of chaos progressing consists in revealing a series of bifurcations that provide chaotic movements by way of adjusting the parameter $\propto$.

For a specialist in automatic control, a well-known example of bifurcation of stationary state is the bifurcation by Andronov-Hofp, where the stationary decision x_0 losses its stability as a result of what a pair of the complexly-conjuncted eigenvectors $\lambda_{1,2} = \xi + 1\omega$ of the linearization matrix fall into the right semi-space. This leads to a bifurcation - the stable stationary point is transformed to a limited cycle. If the generated limited cycle appears to be stable, then a lower loss in stability takes place (Fig. 2.49).

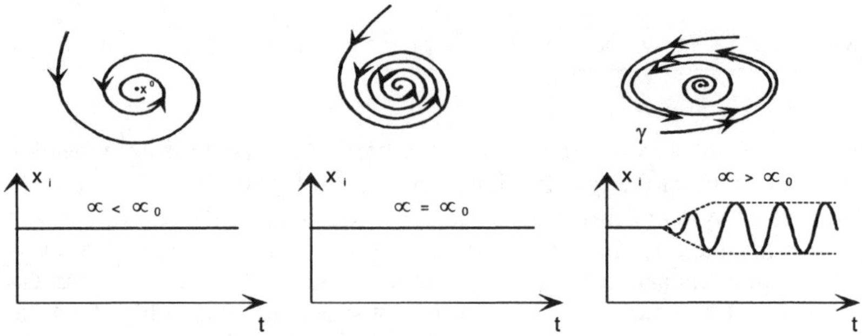

Fig. 2.49. Phase portrait of losing of stability

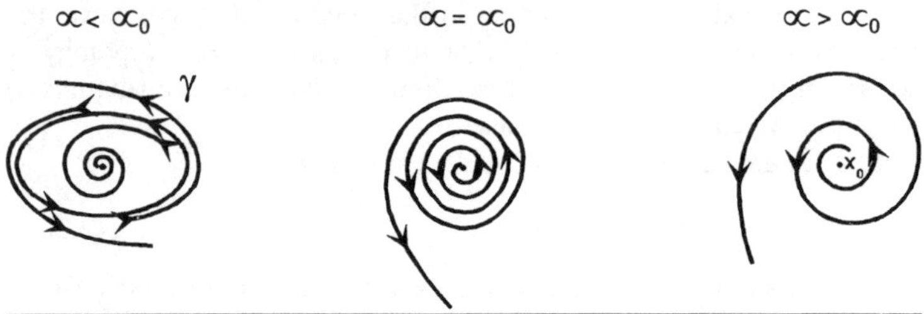

Fig. 2.50. Phase portrait of losing of stability

When the loss in stability is high (Fig. 2.50), the system passing through α_0 immediately switches to the other movement mode. Assume that a stable limited cycle has appeared as a result of bifurcation. Continuously changing the parameter α may lead to different bifurcations in the system. Let's focus on such of them that will contribute to chaos progressing. First let's assume that some decrease in the stability of the original cycle and change to an invariant two-dimensional torus (Fig 2.51) is caused.

The invariant two-dimensional torus may then lose its stability and change to a form of a three-dimensional toroidal multi-variety. The system's behavior then is characterized by three independent frequencies.

Further change of the control parameter α may lead to a series of bifurcations as a result of which invariant torus of still higher dimension is originated.

Finally this leads to a complex quazi-periodic movement with k uncommensurable frequencies, the movement being as more chaotic as large k is.

In [92] the authors first noted that continuously changing the parameter α after three bifurcations leads to a three-frequency quazi-periodic movement, that appears to be, as a rule, unstable and is easily destructed to form a strange attractor in place of three-dimensional torus.

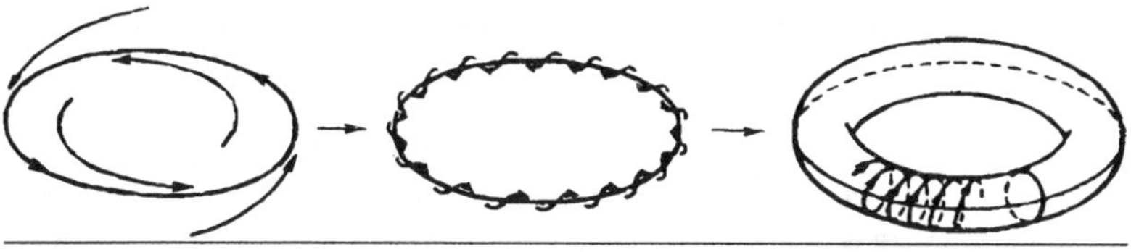

Fig. 2.51. Transition of limit cycle in two-dimensional torus

Secondly, the bifurcation may lead to generating a limited torus with double period. Further increase of the control parameter may lead to another bifurcation, loss in stability, and generation of a cycle of four-time period and so on. Such consequence of bifurcations takes place within a certain value range of $\propto$ and moves the system from a stable periodic movement to a chaotic movement. In [57] this type of chaos progressing is illustrated by the following system of equations

$$\dot{x} = -(y + z), \dot{y} = x + y/5, \dot{z} = 1/5 + z \qquad (x - \propto),$$

that describes a chemical reaction.

The certain value of $\propto = \propto_1$, to be precise $\propto$=2.6, a stable point as a result of bifurcation changes to a limited cycle with the period τ_1 (Fig. 2.52a).

Another bifurcation $\propto = \propto_2$, (i.e. $\propto$=3.5) changes the cycle with the period τ_1 to an eight-shaped stable cycle with double period $\tau_2 = 2\tau_1$ (Fig. 2.52b). Further increase of $\propto$ ($\propto$=4.1, $\propto$=4.18) leads to stable periodic movements with periods $2^m \tau_1$, m=3, 4,... and so on taking place in the system (Fig. 2.52 c,d). At the limit point, i.e. when $\propto = \propto_\infty$, the limited cycle transforms to an unclosed attracting phase trajectory, that provided that $\propto > \propto_\infty$, i.e. $\propto$=4.23 (Fig. 2.52e) generates a strange attractor.

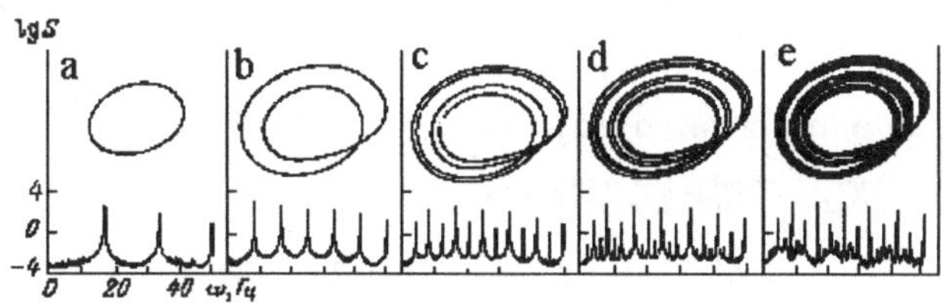

Fig 2.52. Transition to a strange attractor

2.5.3 Empirical Chaotic analysis

Sometimes it is useful to reconstruct the dynamics from observed data. For example, a specific type of a empirical chaos is time series. In this case we deal with evidence of chaos when no knowledge of the underlying structure is available. Dynamical invariants, which characterize chaos, are tied to the parameters of the underlying structural model, even if the functional form of the model must remain unspecified. Therefore it is possible to infer chaotic properties without any knowledge of the functional form of chaotic system.

A positive Lyapunov exponent is the operational definition of chaos and it is primary dynamical invariant used to characterize a chaotic process. Let's first give one of methods to estimate the largest Lyapunov exponent [73, 85]. Let chaotic system is characterized by

$$x_t = f(x_{t-L}, x_{t-2L}, \dots, x_{t-dL}) + e_t \tag{2.209}$$

where L is a lag, x is the only observed variable, d>2D+1, 1 is the embedding dimension (the number of lags in the observation vector) and e_t is noise. Suppose that

$$X_t^T = (x_t, \dots, x_{t-d+1})$$

and

$$E_t^T = (e_t, 0, \dots, 0)$$

Define the mapping F

$$X_t = F(X_{t-1}) + e_t \tag{2.210}$$

Choose two initial vectors that start close together:

$$X_0^A - X_0^B = U\delta, \tag{2.211}$$

where U is fixed vector of unit length and δ is small. After one iteration with identical random shocks the new distance between the state vectors is

$$\|X_1^A - X_1^B\| = \|F(X_0^A) - F(X_0^B)\| = \|DF(X_0^A)(X_0^A - X_0^B)\| + d(\delta) \tag{2.212}$$

Where

$$DF(X_t^A) = J_t \tag{2.213}$$

is the Jacobean of the map F. Further define $T_M = J_M J_{M-1} \dots J_1$. When M and $\delta \to 0$ and D^2F is uniformly bounded [85], then :

$$\|X_M^A - X_M^B\| = \|T_M U\| \delta + 0(\delta) \tag{2.214}$$

If the largest eigenvalue T_M of T_M^1 is denoted $v_1(M)$, then the largest Lyapunov exponent can be determined as

$$\lambda = \lim_{M \to \infty} \left\{ \frac{1}{2M} \log |v_1(M)| \right\} \tag{2.215}$$

The method to estimate this largest Lyapunov exponent (LE) is to find estimation for the Jacobean matrices and then, using this estimation, estimate Lyapunov exponent. There are several algorithms for estimation of LE in empirical chaos. The method given in [38] uses neural network regression to estimate the relation of the variable with its lagged values. This function is used to determine Jacobean matrices at every point, and the average maximum divergence, the largest Lyapunov exponent, is calculated from those matrices only.

Now we consider other important dynamical invariant characterizing chaos – empirical correlation dimension as measure of fractal dimension of a strange attractor [75,85].

The correlation dimension is derived from the correlation integral

$$C^{(d)}(r,n) = \sum_{i=1}^{N} C_i^{(d)}(r,n) = \frac{1}{N(N-1)} \sum_{i,j=1,\dots,N, i\neq j} I\left(\left\| y_j^{(d)} - y_i^{(d)} \right\|\right) \leq r \tag{2.216}$$

where d is the embedding dimension, r is a distance measure, N is the sample size, y is the observation vector, I is Heaviside function that equals one if distance between two vectors is less than r, and otherwise takes the value of zero. The theoretical correlation dimension is defined as [42]

$$P^{(d)} = \lim_{r\to 0} \lim_{N\to\infty} \ln C^{(d)}(r,n) / \ln r \tag{2.217}$$

Since the sample size N is finite in case of empirical chaos (2.217) must be approximated with real data. Empirically, correlation dimension can be determined as regression ln[C(r,n)] on ln[r]. The slope coefficient from the regression yields an estimation of the correlation dimension, and hence for the fractal dimension, of the strange attractor.

References

1. Aarts EH, Korst J (1989) Simulated Annealing and Boltzmann Machines. Chichester, UK: Wiley
2. Aliev R, Bonfig K, Aliew F (2000) Soft Computing. Berlin: Verlag Technic
3. Aliev RA, Aliev RR (1997) Soft Computing, volume II. Baku: ASOA Press
4. Aliev RA, Aliev RR (1997-1998) Soft Computing, vol I, II, III. Baku: ASOA Press, (in Russian).
5. Aliev RA, Mamedova GA (1993) Identification and optimal control of fuzzy systems. J News of Academy of Sciences of Russia, Tech.Cybernetics 6: 118-126
6. Aliev RA, Nasirov IM (1996) Group genetic algorithms with directed mutation. In: Second International Conference on Applications of Fuzzy Systems and Soft Computing, ICAFS'96. Siegen. Germany. pp 443-436
7. Aliev RA (1994) Fuzzy Expert Systems. In Aminzadeh F, Jamshidi M. (eds) SOFT COMPUTING: Fuzzy Logic. Neural Networks and Distributed Artificial Intelligence. pages 99-108. NJ: PTR Prentice Hall
8. Aliev RA (1997) Special representation: Advances in Fuzzy Logic, Neural Networks. Genetic Algorithms and Hybrid Systems. Atlanta. USA: GSU
9. Aliev RA (1995) Fuzzy knowledge based Intelligent Robots. Moscow: Radio i svyaz. (in Russian).
10. Aliev RA, Abdikiev NM, Shachnazarov MM (1990) Intelligent control systems.Moscow: Radio b svyaz (in Russian)
11. Aliev RA, Abilov YA, Nasirov IM (1996) Genetic Algorithms with group search and directed mutation. J Computing and System Sciences 5: 21-24

12. Aliev RA, Abilov YA, Nasirov IM (1994) Theory of possibility and group principle in genetic algorithms. In Aliev RA, Kanarangui (eds) Inter. Conf. on Application of Fuzzy Systems. Iran. Tabriz. pp 361-365
13. Aliev RA, Aliev FT, Babaev MD (1991) Fuzzy process control and knowledge engineering. Koln: Verlag TUV Rheinland
14. Aliev RA, Aliev RR (2001) Soft Computing and its Application. World Scientific. New Jersey. London.Singapure. Hong Kong
15. Aliev RA, Bonfig KW, Aliev FT (1993) Messen. Steuern und Regeln mit Fuzzy-Logik. München: Franzis-Verlag (in German)
16. Aliev RA, Mamedova GA and Aliev RR (1993) Fuzzy Sets Theory and its application. Tabriz: Tabriz University Press
17. Aliev RA, Mamedova GA, Tserkovny AE (1991) Fuzzy control systems. Moscow: Energoatomizdat (in Russian)
18. Aliev RA, Vahidov R, Aliev RR (1995) Artificial Neural Networks. Theory and practice. Tabriz: Tabriz University Press
19. Almedia LB (1987) A learning rule for asynchronous perseptrons with feedback in a combinatorial environment. In: Caudill M. and Butler C.(eds.) IEEE First Inter.Conf. on Neural Networks. vol II pp 609-611
20. Baldwin JF, Pilsworth BW (1979) A model of fuzzy reasoning through multi-valued logic and set theory. J Int. J. Man-Machines Studies 11:351-380
21. Bandemer H, Gottwald S (1995) Fuzzy sets, Fuzzy logic, Fuzzy methods with applications. England: John Wiley and Sons
22. Bandemer H, Nather W (1992) Fuzzy data analysis. Boston: Kluwer Academic Publishers
23. Bandler W, Kohout L (1980) Fuzzy power sets and fuzzy implications operators. J Fuzzy Sets and Systems 1:13-30
24. Barto AG, Anandan P (1985) Pattern Recognizing Stochastic Learning Automata. J IEEE Trans. Systems, Man and Cybernetics 15: 360-375
25. Bethke AD (1980) Genetic Algorithms As Function Optimizers. PhD Dissertation (CCS). Ann Arbor. MI: The Univercity of Michegan Press
26. Bromhead DS, Lowe D (1988) Multivariate functional interpolation and adaptive networks. J Complex Systems 2: 321-355
27. Buckley JJ, Eslami E. (1997) Fuzzy plane geometry I: Points and lines. J Fuzzy Sets and Systems 86: 179-187
28. Cerruti V (1982) Graphs and fuzzy graphs. J Fuzzy Information and Decision Processes. Amsterdam – NY:77-95
29. Chang SZ, Zadeh LA (1972) On fuzzy mapping and control. J IEEE Trans. Systems, Man, and Cybernetics 2: 30-34
30. Deboeck J (ed) (1994) Trading on the edge: neural, genetic, and fuzzy systems for chaotic and financial markets. John Wiley and Sons
31. Dempster AP (1967) Upper and lower probabilities induced by multivalued mapping. J Ann. Math. Stat 38: 325-339
32. Diederich J (1990) Artificial Neural Networks. Concept Learning. Los Alamitos CA: J IEEE Computer Society Press
33. Dodsow CTJ (1981) A new generalization of graph theory. J Fuzzy Sets and Systems 6: 293-308
34. Dubois D, Prade H (1992) Evidence, knowledge and believe functions. J Int. J.Approx. Reasoning 6: 295-319

35. Dubois D, Prade H (1979) Fuzzy real algebra: Some results. J FSS2: 327-348
36. Dubois D, Prade H (1980) Fuzzy Sets and Systems: Theory and Applications. NY: Academic Press
37. Dubois D, Prade H (1998) Possibility Theory: Approach to Computerized Processing of Uncertainty. NY: Plenum Press
38. Ellner S, Nychka DW, Gallant A.R.(eds) (1992) LENNS, a program to estimate the dominant Lyapunov exponent for noisy nonlinear systems from time series data. Ralegh, NC 27695-8203: North Carolina State University Press, 1992.
39. Farreny H, Prade H, Wyss E (1986) Approximate reasoning in rule-based expert system using possibility theory: a case study. In: Kugler H.J. (ed) Information. Nothern-Holland. Amsterdam. pp 407-413
40. Fukami S, Mizumoto M, Tanaka K (1980) Some consideration of fuzzy conditional inference. J Fuzzy Sets and Systems 4: 243-273
41. Goldberg DE (1989) Genetic Algorithms in Search, Optimization and Machine Learning. Reading, MA: Addison-Wesley
42. Grassberger P, Procaccia I (1983) Characterization of strange attractors. J Phys.Rev. Lett. 50: 346-349
43. Haken H (1988) Information and Self-organization. Berlin: Springer
44. Hebb DO (1949) The Organization of behaviour. NY: John Willey and Sons
45. Hinton GE (1989) Connectionist Learning Procedures. JArtificial Intelligence 40(1): 143-150
46. Holland JH(ed) (1975) Adaptation in Natural and Artificial Systems. In:Ann Arbor. MI: The Univercity of Michigan Press
47. Jaffray Y (1992) Bayesian updating and belief functions. J IEEE Trans. Systems, Man, and Cybernetics 22: 1144-1152
48. Jin L, Gupta MM, Nikiforuk PN (1994) Computational Neural architectures for control applications. In: Aminzadeh F, Jamshidi M.(eds) Soft Computing: Fuzzy logic, Neural Networks and Distributed Artificial Intelligence. pp 121-152
49. Kasyap RL, Blaydon CC, Fu KS (1970) Stochastic approximation. In: Fu KS, Mendel JM(eds) Adaptation, Learning and Pattern Recognition Systems: Theory and Applications. NY: Academic Press
50. Kaufman A.(1973) Introduction to theory of fuzzy sets, volume 1. Orlando: Academic Press
51. Kaufmann A, Gupta MM (1985) Introduction to Fuzzy arithmetic. Theory and Application. NY: Van Nostrand Reinhold
52. Klir GJ, Yuan B (1995) Fuzzy sets and fuzzy logic: Theory and Applications. NJ: PRT Prentice Hall
53. Klir GJ, Clair US, Yuan B (1997) Fuzzy Set Theory. Foundations and Applications. NJ: PTR Prentice Hall
54. Klir GY, Yuan B (1995) Fuzzy Sets and Fuzzy Logic. Theory and Applications. NY: PTR Prentice Hall
55. Laarhoven PJ, Van M, Aarts EH (1987) In: Reidel D.(ed) Simulated annealing: Theory and Applications. Dordrecht. Holland
56. Lee S, Kil R (1988) Multilayer feedforward potential function networks. In: The IEEE Second Inter. Conf. on Neural Networks. vol 1: pp 161-171
57. Loskutov AYu, Mikhailov AS (1990) Introduction to synergetic. Moscow: Nauka. (in Russian)

58. Mamdani EH (1977) Applications of fuzzy logic to approximate reasoning using linguistic systems. J IEEE Trans. Comput. C-26: 1182-1191
59. Mamedov RM, Balayev RS, Mammadova AT, Jabbarova K, Alieva KR (2000) Chaotical analysis of petroleum production time-series. In: Fourth Inter. Conf. on Application of Fuzzy Systems and Soft Computing, ICAFS'2000. Siegen. Germany. pp 402-404
60. Mandelbrat V (1982) The Fractal Geometry of Nature. San Francisco: W.H. Freeman Company
61. Mangelbrat BB (1977) Fractals: Form , Chance and Dimension. San Francisco: W.H. Freeman Company
62. Medgassy P (1961) Decomposition of Superposition of Distributed Functions. Budapest: Hungarian Academy of Sciences Press
63. Miccehelly CA (1986) Interpolation of Scattered data. Distance and conditionally positive definite functions. J Constructive Approximation 2: 11-22
64. Michalewicz Z (1992) Genetic Algorithms+Data Structures=Evolution Programs. Berlin: Springer-Verlag
65. Mizumoto M (1982) Fuzzy Conditional Inference under Max-Composition. J Information Sciences 27: 183-209
66. Mohamad H.Hassoun (1995) Fundamentals of Artificial Neural Networks. Cambridge: MIT Press
67. Moody J, Darken C (1989) Fast learning in networks of locally-tuned processing units. J Neural Computation 1(2): 281-294
68. Moody J, Darken C (1989) Learning with localized receptive fields. In: Touretzky D, Hinton G, TSejnowski (eds) Proc. of the 1988 Connectionist Models Summer School. San Mateo, Calif.: Morgan Kaufmann. pp 133-143
69. Mördeson JN, Premchand SN (1998) Fuzzy Mathematics. An introduction for Engineers and Scientists. Heidelberg: Physica-Verlag
70. Narenda KS, Thathachar MA (1974) Learning Automata-a Survay. J IEEE Trans. Systems, Man, and Cybernetics 4: 323-334
71. Nguyen HT, Walker EA (1996) A first Course in Fuzzy logic. Boca Raton: CRC Press,
72. Niranjan M, Fallside F (1988) Neural Networks and Radial Basis Functions in Classifying Static Speech Patterns. Technical Report. Cambridge: University Press
73. Nychka D, Ellner S, Gallant AR, Mc Caffrey D (1992) Finding chaos in noisy systems. J Royal Statistical Society 54(2): 399-426
74. Parker DB (1987) Optimal algorithms for adaptive networks: Second order backprop, second order direct propagation, and second order Hebbian learning. In: Caudill M, Butler C. (eds) IEEE First Inter. Conf. on Neural Networks. vol II: pp 593-600
75. Parker TS, Chua LO (1989) Practical Numerical Algorithms for Chaotic Systems. NY: Springer-Verlag
76. Pedrycz W (1987) Fuzzy models and relational equations. J Mathematical Modeling 9(6): 427-434
77. Pedrycz W 91983) Fuzzy relational equations with generalized connectives and their applications. J Fuzzy Sets and Systems 10(2): 185-201
78. Pedrycz W (1989) Fuzzy Control and Fuzzy Systems. NY:John Wiley and Sons
79. Philip D, Wasserman PD (1993) Advanced methods in neural computing. NY: Van Nostrand Reinhold
80. Pineda FJ (1987) Generalization of back-propagation to recurrent neural networks. J Physical Review Letters 59: 2229-2232

81. Poggio T, Girosi F (1989) A Theory of Networks for Approximation and Learning. A.I.Memo 1140. Massachusetts. Cambridge: MIT Press
82. Potiupkin AYu (1999) Estimation technical state of an objects described by fuzzy relational equations. J News of Academy of sciences of Russia 4: 111-119
83. Potiupkin AYu (1996) Solving of the problem of fuzzy systems identification. J News of Academy of sciences of Russia 4: 40-46
84. Powell MJD (1997) Radial basis functions for multivariate interpolation. In: Mason JC, Cox MG (eds) A review in Algorithms for the Approximation of Functions and Data. England.Oxford: Clarendon Press
85. Reidar Hagtvedt (1999) Chaos in Stock Returns. PhD Dissertation J.Mack Robinson College of Business Administration. Atlanta. Georgia: Georgia State University
86. Rescher (1969) Many-Valued Logic. NY: McGraw–Hill
87. Rosenfeld A (1985) Distances between fuzzy sets. J Pattern Recognition Letters 3: 229-233
88. Rosenfeld A (1975) Fuzzy graphs. In: Zadeh A, Fu KS, Shimura MA(eds) Fuzzy Sets and Their Applications. NY: Academic Press pp 77-95
89. Rosenfeld A (1994) Fuzzy plane geometry: triangles. J Pattern Recognition Letters 15: 1261-1264
90. Rosenfeld A (1990) Fuzzy rectangles. J Pattern Recognition Letters 11: 677-679
91. Rosenfeld A (1984) The diameter of a fuzzy set. J Fuzzy Sets and Systems 13: 241-246
92. Ruelle D (1980) Strange attractors. J The mathematical Intelligenser 2: 126-137
93. Rummelhart DE, Zipser D (1985) Competitive Learning. J Cognitive Science 9: 75-112
94. Rummelhart DE, Hilton GE, Williams RJ (1986) Learning internal Represebtations by Back Propagating Errors. J Nature 323: 533-536
95. Rummelhart DE, Hinton GE, Williams RJ (1986)Learning intenal representations by errors propagation. In: Rummelhart DE, McClelland JL, the PDP Research Group (eds) Parallel Distributed Processing: Explorations in the Microstructure of Cognition, vol 1 Foundations. Cambridge. MA: MIT Press
96. Sanchez E (1984) Solution of fuzzy equations with extended operations. J Fuzzy Sets and Systems 12(3): 237-248
97. Sankar K Pal, Ghosh A (1992) Fuzzy geometry in image analysis. J Fuzzy Sets and Systems 48: 23-40
98. Schwartz PB, Samalam VK, Solla SA, Denker JS (1990) Exhaustive learning. J Neural Computation 2(3): 374-385
99. Sejnovski T.J. and Rosenberg C.R. Parallel Networks that Learn to Pronounce English Text. *Complex Systems* 1: 145-161, 1987.
100. Shafer G (1976) A mathematical theory of evidence.NJ: Princeton University Press
101. Silva CW (1995) Intelligent Control. Florida: CRC Press
102. Walley P (1996) Inferences from multinomial data: Learning about a bag of marbles. J.Roy. Stat. Soc. Ser. B-58: 3-57
103. Walley P (1996) Measures of uncertainty in expert systems. J Artifical intelligence 83: 1-58
104. Walley P (1991) Measures of uncertainty in expert systems. Statistic Research Report. Department of Mathematics. Western Australia. Perth University Press
105. Wasserman PD (1989) Neural computing. Theory and practice. NY: Van Nostrand Reinhold

106. Widrow B, Hoff MEJr (1960) Adaptive switching circuits. J IREE Western Electric Show and Convention Record 4: 96-104
107. Wilson N, Moral S (1994) A logical view of probability. In: Cohn A (ed) Eleventh European Conf. On Artificial Intelligence. London: Wiley pp 386-390.
108. Yager RR, Filev DP (1994) Essentials of Fuzzy Modelling and Control. NY: John Wiley and Sons
109. Yager RR, Ovchinikov S, Tong RH, Nguyen HF (eds) (1987) Fuzzy sets and applications. Zadeh's selected papers. NY: John Wiley and Sons
110. Yeh RT, Bang SY (1975) Fuzzy relations, fuzzy graphs, and their applications to clustering analysis. In: Zadeh LA, Fu KS, Shimura MA (eds) Fuzzy Sets and Their Applications. pp 125-149
111. Zadeh LA (1971) Fuzzy orderings. J Information Sciences 3: 117-200
112. Zadeh LA (1978) Fuzzy sets as a basis for a theory of possibility. J Fuzzy Sets and Systems 1: 3-28
113. Zadeh LA (1965) Fuzzy Sets. J Information and Control 8: 338-353
114. Zadeh LA (1973) Outline of a new approach to the analysis of complex system and decision processes. J IEEE Trans. Systems, Man, and Cybernetics 3: 28-44
115. Zadeh LA (1971) Similarity relations and Fuzzy orderings. J Information Sciences 3: 177-200
116. Zeidenberg M (1990) Neural network in Artificial Intelligence. England:Ellis Horwood Limited
117. Zimmermann HJ (1993) Fuzzy Set Theory and its applications. Boston: Kluwer Academic Publishers. Second Revised Edition

3 Emerging Combined Soft Computing Technologies

3.1 Neuro-Fuzzy Technology

Fuzzy neural networks

Fuzzy logic [161] has been successfully applied in many areas, such as control of industrial processes, control of robotic manipulators, control of servo-motors, complex decision making, diagnostic systems and others [10,93,94,98]. When using fuzzy logic, input data in form of linguistic values are represented by membership functions which are used for defining fuzzy sets of crisp values and their corresponding membership degrees related to these sets. Many parameters of systems based on fuzzy logic should be defined with help of an expert. In the same time, however, the associated parameters construction processes are performed by the method of trial and error or some heuristic algorithms [143]. Moreover, a designer that knows the characteristics of the system also needs initial rules. Some researchers have suggested a number of mechanisms to generate fuzzy rules and developed methods of their modification on the basis of experience [131,134,140,141,142]. Among them we must distinguish the self-organizing fuzzy controller that is capable of forming and modifying fuzzy rules [134,142], the clustering algorithm for fuzzy partitioning the input data space [140], and the least square algorithm for defining a succession of parameters [140,141] that can be used to construct systems based on fuzzy logic.

Despite utilization of all these algorithms, they still remain to be heuristic in some sense, and the selection of membership functions in fuzzy rules is connected to the method of trial and error. In order to remove this disadvantage recently some interesting methods were developed that utilize the learning ability of neural networks [25,75,83,113]. Multi-layer neural networks with such learning methods as backpropagation, reinforcement, and unsupervised algorithms, or the combinations of the above, have been used to select fuzzy rules and precisely adjust membership functions used in those rules.

For fuzzy neural network the signals and/or the weights must be fuzzy set. There are three types of fuzzy neural networks (FNN)- FNN_1, FNN_2, FNN_3 [68].

The first type of fuzzy neural net (abbreviated FNN_1) has crisp signals (X_i) $(i=\overline{1,n})$ but fuzzy weights. The second type of FNN (FNN_2) has fuzzy signals and real number weights. The last FNN (FNN_3) has both fuzzy signals and fuzzy weights.

General learning algorithm for fuzzy neural network

Here we consider a general fuzzified learning rule for the third type FNN. The total output of the considered FNN $\tilde{y}_l$ is determined as follows

$$\tilde{y}_l = f\left[\sum_{j=1}^{m} \tilde{k}_j \cdot \tilde{z}_{lj}\right], \quad l=\overline{1,N} \tag{3.1}$$

where $\tilde{z}_{lj}$ - is the output of j-th neuron in the hidden layer when the vector $\tilde{X}_l = (\tilde{x}_{l1}, \tilde{x}_{l2}, ..., \tilde{x}_{ln})$ is the input to the network; $\tilde{k}_j$ - is the fuzzy weight coefficient of the connection between j-th neuron of the hidden layer and output layer; f is the activation function which is taken to be

$$f(I) = [1 + e^{-I}]^{-1}.$$

For FNN_3 $\tilde{z}_{lj}$ is defined as

$$\tilde{z}_{lj} = f\left[\sum_{i=1}^{n} \tilde{x}_{li} \cdot \tilde{w}_{ij}\right] \tag{3.2}$$

where $\tilde{w}_{ij}$ - is the weight coefficient of the connection between i-th neuron of the input layer and j-th neuron of the hidden layer.

The fuzzy error between the current output of the FNN_3 and the desired output is determined as

$$\tilde{E} = \frac{1}{2}\sum_{l=1}^{N} (\tilde{y}_l - \tilde{y}_l^*)^2 \tag{3.3}$$

where $\tilde{y}_l^*$ - is the desired fuzzy value of the output of the FNN_3. The problem is determination of an algorithm for learning of FNN_3 which would provide such values of $\tilde{w}_{ij}$ and $\tilde{k}_j$ $(i=\overline{1,n};\ j=\overline{1,m})$ that minimize (3.3) making $\tilde{E}$ possibly close to zero.

However, because of the fuzzy subtraction operation, used for calculation of $\tilde{E}$, it will not be equal to zero even if $\tilde{y}_l = \tilde{y}_l^*$ for all l. So, a termination condition should be defined for the FNN_3 learning procedure. Let's assume that the interval

$[\tilde{y}_{l1}, \tilde{y}_{l2}]$ ($l = \overline{1,N}$), is the support of the fuzzy set $\tilde{y}_l$. Then if $\tilde{y}_l = \tilde{y}_l^*$ then the interval $[-\omega, \omega]$ will be the support of $\tilde{E}$, where

$$\omega = \frac{1}{2}\sum_{l=1}^{N} (\tilde{y}_{l1}^* - \tilde{y}_{l2}^*)^2. \tag{3.4}$$

If $\varepsilon > 0$ is an admissible error of $\tilde{E}$ for $\tilde{y}_l = \tilde{y}_l^*$, then the termination condition can be formulated as follows [6,16,68]. The iterations to search $\tilde{w}_{ij}$ and $\tilde{k}_j$ $\left(i = \overline{1,n};\ j = \overline{1,m}\right)$ should be terminated when $\tilde{E}$ falls into set

$$\Omega = [-\omega - \varepsilon, \omega + \varepsilon] \cdot [0,1] \tag{3.5}$$

Now let's describe the learning algorithm itself. The learning rules for $\tilde{k}_j$ will be the following

$$\tilde{k}_j(q+1) = \tilde{k}_j(q) + \gamma\Delta\tilde{k}_j(q), \qquad (j=\overline{1,m}),\ q = 1,2,... \tag{3.6}$$

where

$$\Delta\tilde{k}_j = \frac{\partial\tilde{E}}{\partial\tilde{k}_j}, \tag{3.7}$$

$$\frac{\partial\tilde{E}}{\partial\tilde{k}_j} = \sum_{l=1}^{N} (\tilde{y}_l - \tilde{y}_l^*)(\tilde{y}_l)(1 - \tilde{y}_l)\tilde{z}_{lj} \tag{3.8}$$

γ is the learning rate.

The following rules can be used to determine the weight coefficients $\tilde{w}_{ij}$

$$\tilde{w}_{ij}(q+1) = \tilde{w}_{ij}(q) + \gamma\Delta\tilde{w}_{ij}(q) \tag{3.9}$$

where

$$\frac{\partial\tilde{E}}{\partial\tilde{w}_{ij}} = \sum_{l=1}^{N} (\tilde{y}_l - \tilde{y}_l^*)(\tilde{y}_l)(1 - \tilde{y}_l)\tilde{k}_j \frac{\partial\tilde{z}_{ij}}{\partial\tilde{w}_{ij}} \tag{3.10}$$

$$\frac{\partial z_{lj}}{\partial\tilde{w}_{ij}} = (\tilde{z}_{lj})(1 - \tilde{z}_{lj})\tilde{x}_{lj} \tag{3.11}$$

The algorithm given above is based on fuzzy arithmetic and fuzzy analysis [6,16,36]. Lee was the first who introduced fuzzy sets into a neural network [96,97]. In [36] a review of the works uniting neural networks and fuzzy sets is given. Yamakawa [156-159] considers in his works FNN_1 and its modifications. FNN_2 were considered in [125], particularly in works of Ishibushi and others [77,78], where beside the basic features of fuzzy neural networks, appropriate learning algorithms are considered.

[63,64,65,66,79] consider different types of fuzzy neural networks (i.e. FNN_1, FNN_2, FNN_3). [16,67,76] consider the fuzzification of the delta-rules to learn

FNN_3 and their various applications, in particular, for fuzzy regression, fuzzy control, fuzzy classification etc.

Learning of FNN by applying α-cuts

A 3-layer feed-forward neural network is considered with n input, n_h hidden and m output neurons. The input and desired signals and weight coefficients are fuzzy, e.g. trapezoidal numbers. The essence of the considered algorithm [78,79] is in solving the problem of learning of the fuzzy neural network by crisp learning rules, utilizing α -cuts of fuzzy inputs, desired outputs, and connection weights.

Let $Y_l^* = (y_{l1}^*, y_{l2}^*, ..., y_{lm}^*)$, $(l = \overline{1, N})$, be the desired vector corresponding to the input vector. Having considered [78,79], the mean square error for the fuzzy outputs is found as

$$[e_{lk}]^\alpha = [e_{lk}]_L^\alpha + [e_{lk}]_U^\alpha, \tag{3.12}$$

where $[e_{lk}]_L^\alpha, [e_{lk}]_U^\alpha$ are mean square error for lower and upper bounds of the α -level set, respectively.

$$[e_{lk}]_L^\alpha = \alpha\left([y_{lk}^*]_L^\alpha - [y_{lk}]_L^\alpha\right)^2 / 2, \tag{3.13}$$

$$[e_{lk}]_U^\alpha = \alpha\left([y_{lk}^*]_U^\alpha - [y_{lk}]_U^\alpha\right)^2 / 2. \tag{3.14}$$

The total mean square error function for each pair (x_l, y_l^*) is calculated as

$$e_l = \sum_\alpha \sum_{k=1}^{n} [e_{lk}]^\alpha \tag{3.15}$$

Like in [78,79] assume that the learned fuzzy weights are trapezoidal fuzzy numbers. Then for each weight coefficient we find the four parameters:

$$W_{jk} = (w_{jk}^{(1)}, w_{jk}^{(2)}, w_{jk}^{(3)}, w_{jk}^{(4)})$$

In accordance with the classical backpropagation algorithm for each parameter $w_{jk}^{(q)}$, $q = \overline{1,4}$ we have:

$$\Delta w_{jk}^{(q)}(t+1) = -\gamma \frac{\partial [e_l]^\alpha}{\partial w_{jk}^{(q)}} + \xi \cdot \Delta w_{jk}^{(q)}(t), \quad q = \overline{1,4} \tag{3.16}$$

where γ is the learning rate, ξ is momentum coefficient. The calculation of the derivatives has been shown in [78, 79] and is performed as follows:

$$\frac{\partial [e_l]^\alpha}{\partial w_{jk}^{(1)}} = (1-\alpha) \frac{\partial [e_{lk}]_L^\alpha}{\partial [w_{jk}]_L^\alpha} \tag{3.17}$$

$$\frac{\partial[e_l]^\alpha}{\partial w_{jk}^{(2)}} = \alpha \frac{\partial[e_{lk}]_L^\alpha}{\partial[w_{jk}]_L^\alpha} \tag{3.18}$$

$$\frac{\partial[e_l]^\alpha}{\partial w_{jk}^{(3)}} = \alpha \frac{\partial[e_{lk}]_U^\alpha}{\partial[w_{jk}]_U^\alpha} \tag{3.19}$$

$$\frac{\partial[e_l]^\alpha}{\partial w_{jk}^{(4)}} = (1-\alpha) \frac{\partial[e_{lk}]_U^c}{\partial[w_{jk}]_U^\alpha} \tag{3.20}$$

For N learning pairs (X_l, Y_l^*), $l = \overline{1,N}$ the learning is performed for r α-levels, i.e. $(\alpha_1, \alpha_2, ..., \alpha_r)$, and the learning algorithm can be structured as given below [78, 79]:

Step 1. Repeat steps 2-5 for α_i, $i = \overline{1,r}$;

Step 2. Repeat steps 3-5 for l, $l = \overline{1,N}$;

Step 3. Calculate α-level set of the fuzzy output vector Y_l corresponding to input vector X_l;

Step 4. Adjust the four parameters of the fuzzy weights (thresholds) by the value function;

Step 5. If the predefined terminating condition is not satisfied go to step1.

A maximum number of iterations of algorithm can be used as the terminating condition.

Here several recommendations should be given. First, if the condition

$$w_{jk}^{(1)} \le w_{jk}^{(2)} \le w_{jk}^{(3)} \le w_{jk}^{(4)}$$

is violated, then all w should be rearranged to satisfy the condition. Secondly, if the fuzzy weights are represented in the triangle form, they still can be considered as trapezoidal with the two middle parameters taking equal values and treated by the considered algorithm.

Below we will consider basic principles of neuro-fuzzy technology.

Architecture of Neuro-Fuzzy Systems

Fig. 3.1 shows the top level description of a neuro-fuzzy system to generate fuzzy rules and membership functions. The kernel of the system is a multilayered network-based structure. As it can be seen the system generates both fuzzy rules and membership functions. Source data for 0the generation are input-output vectors describing the desired system's behavior. The data can be obtained either by experimenting or through some other method.

A more detailed scheme of Neuro-Fuzzy system is represented in Fig.3.2. For simplicity the structure with two inputs X_1 and X_2 and one output Y is represented. As it can be seen in the figure the first layer of neurons simply distributes inputs of the system among neurons of the following layer. The second layer comprices several groups of neurons (depending on number of inputs, in this case 2).

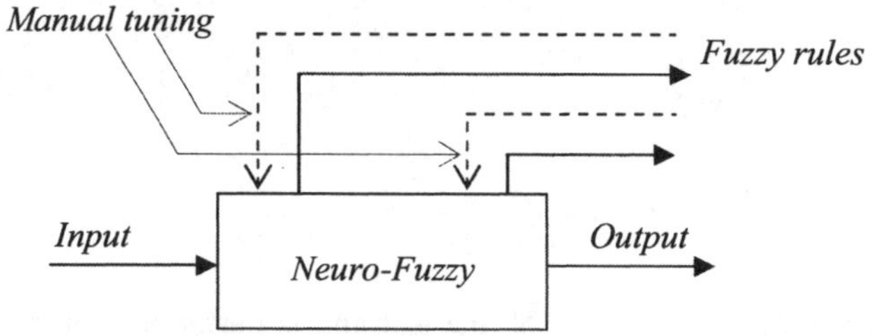

Fig. 3.1. Neuro-fuzzy system

Neurons in each group represent membership functions for fuzzy labels used as values for the input connected with this group. Output of every such neuron is value of membership of the input to the corresponding fuzzy label. This process is called "fuzzification" and these neurons are "fuzzifiers".

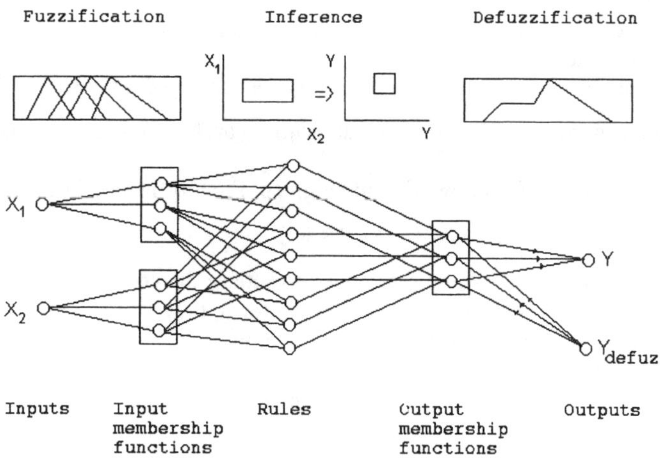

Fig. 3.2. Detailed scheme of Neuro-Fuzzy system

Neurons of the third layer represent fuzzy rules. Number of neurons in this layer can be the same as the number of rules in the logical system IF-THEN, but often in order to minimize the size of the network for fast learning (this, although can complicate extraction of the rules), the traditional neural networks are used.

Neurons of the fourth layer determine membership functions of fuzzy labels. Neurons of this layer perform the most complex operation. First the operation MAX (or some smooth approximation) is performed by each neuron over the set of input signals. The found value is used for modifying the corresponding membership function (MF) - membership values above this found value are truncated to be equal to it. After this, the MAX operation (or analogue) on all the MFs modified in this way is performed. Thus the output membership function is determined.

In the fifth layer the defuzzification procedure is performed. This means determination of crisp value of output based on the inferred fuzzy value.

Fig. 3.2 shows a detailed structure of a Neuro-Fuzzy system. As one can see the structure is similar to one for a backpropagation neural network, and hence allows investigation by the appropriate methods. But some distinctions should be noticed. In this case each neuron is characterized not only by a set of weights, a threshold and a universal activation function, but is a complex processing unit with an individual function and set of parameters. The neurons between consecutive layers are not fully-connected unlike in the case of traditional backpropagation neural network.

As we just said the set of adjustable parameters for a particular neuron depends on its type and location in the network. For example, if we used a traditional backpropagation neural network for fuzzifiers then the parameters of the corresponding "complicated" neuron are weights of that backpropagation network. The parameters are adjusted during learning. In case of triangle membership functions each neuron has 3 parameters, in case of trapezoid or Gaussian function - 4 or 2 parameters, respectively.

When using gradient learning methods we need the functions in the neurons to be smooth, unlike the functions MIN and MAX. These functions should be substituted, for example, by $y = x_1 x_2$ and $y = 1-(1-x_1)(1-x_2)$, respectively.

When using genetic algorithms to find the parameters' optimal values and optimal network structure, we do not need the used functions to be smooth and, hence, less calculation is required. However, we may need much time for learning (especially as we go farther from the initial point, and the current point is not near to a local minimum - this will be a result of less constrained direction of search).

Let's summarize the main advantages of Neuro-Fuzzy system (Fig. 3.2):

- Fast computation based inference;
- Automatic generation of rules and membership functions from input-output data;
- Possibility to switch between automatic and manual analysis and modification of the rules and membership functions.

Generation of Fuzzy Rules and Membership Functions

Unlike the traditional neural network, which allows construction of system's model from the input-output data, the model obtained on the basis of neural-fuzzy technology is not a "black box", though supporting the same way to produce the model. In addition to the correct implementation of the input/output mapping, it allows acquisition of heuristic knowledge and description of qualitative information on the system.

We say "generation of rules" to express the possibility of directly mapping the structure of the built and learned neural network into a set of fuzzy rules with required characteristics of fuzzy terms used in rules and defuzzification algorithm. The generation of rules implies as well the definition (generation) of a set of necessary membership functions describing the terms used in rules as values for input and output variables.

Depending on the characteristics of a problem to be solved, i.e. the number of inputs and outputs, requirements to membership function types (e.g. necessity for them to have smooth curves; inputs are fuzzy while outputs are crisp, the outputs are singletons etc.), and the types of used networks, that procedure may differ and be more or less difficult. Application of the special type of neurons, implementing operations intrinsic for fuzzy logic (e.g. min, max, union, etc.), simplifies the conversion of the rules into the neuro-fuzzy network and vice-versa extraction of the rules. It should be noted, however, that implementation of the necessary computation sometimes causes difficulties in learning because of application of neurons with complex non-linear signals transformations requiring large memory for processing.

Often, the traditional neural networks are used for all phases of fuzzy inference: fuzzification, inference, and defuzzification.

If it is necessary the appropriate neurons should implement the following membership function:

$$y = e^{-b(x-a)^2} \tag{3.21}$$

Such a neuron uses 2 parameters, defining the center a and width b of the area under the membership function. Triangle-shaped membership functions require neurons with three parameters, trapezoidal membership functions requires neurons with four parameters etc. If there are no restrictions concerning to the shape of the membership function, then a traditional backpropagation feed-forward neural network can be used as a membership function generator. In some cases to assure the convexity of generated membership functions (so that the associated fuzzy terms may express real notions or natural fuzzy values), the corresponding neural networks are learned by specific learning algorithms. To this end, for example, the back-propagation learning algorithm is supplemented by an additional error component evaluating the deviation from the convexity. This prevents from losses in the convexity though may slow down the total learning.

The block of a neuro-fuzzy system responsible for conducting fuzzy logical inference, represents a set of fuzzy rules and can be based on neurons implementing the MIN operation (or some smooth analog). The number of neurons necessary to realize the fuzzy inference in some cases can directly determine the number of rules taking part in fuzzy logical system IF-THEN. In some other cases on purpose to minimize the size of the network and to simplify the learning the traditional neural networks are used. In the first case the extraction of the rules from the neural network is easier; here the maximum (enough) number of neurons in the 3-rd layer can be calculated as the multiplication of the numbers of neurons in each group of the second layer. In the case shown in Fig. 3.2 we can have 9 (3·3) neurons as a maximum in the 3-rd layer, this is the same as the number of rules that can be formed by making possible combinations of fuzzy terms for each input variable taking values from the sets containing the three fuzzy terms. In accordance with the fuzzy inference rules, the neurons in layer three, perform the MIN operation (or multiplication) over all their inputs.

In the first case the automatic generation of the rules is performed by a special algorithm, which at first generates the maximum possible number of rules, and then eliminates some of them on the basis of the values of the associated weight coefficients at certain connections (these values indicate the importance of the connection in this case) and the value of the error at the output. The parameters of the neurons in the associated layers of the neural network found after learning, allow directly obtaining membership functions for all fuzzy terms used in rules.

In the second case, where the layer for rules is implemented as a traditional Feed-Forward Back-Propagation Neural Network (this function is realized by one or two consequent layers of neurons with sigmoidal activation function), the acquisition of rules is done by passing the 1.0-valued signal to the inputs (condition part) of the layer for rules. Thereby the appropriate right part of the rule is found - the "winner-neuron" will correspond to a suitable fuzzy term.

Fuzzification and Defuzzification in Neuro-Fuzzy Systems

The difference of the fuzzification and defuzzification used in neuro-fuzzy systems from those used traditionally is in that here the role of adjusting or learning is more important. A fuzzifier is a unit that has one input and several outputs; each of the outputs represents a certain fuzzy term and returns the membership value of a given input to that term. As a rule, fuzzifiers are adjustable and can be realized as neurons and neural networks.

By analogy, a unit that allows obtaining an appropriate crisp value based on a given fuzzy one is named defuzzifier (see section 2.1). Several methods of defuzzification are known; difference among them is in the way they calculate the average value based on the area below the membership function curve. There are no recommendations on which formula is the most universal and precise - each separate application may have its own reasons for selecting one out of these three. An attractive feature of the defuzzification in neuro-fuzzy systems is that the formula of defuzzification can be adjusted just like the membership function.

The method to create fuzzifier (and also defuzzifier) depends mostly on the requirements for the shapes of generated membership functions, that is on the formula that allows general appearance for membership function shapes necessary for this application. On the base of the obtained formula one can select the necessary type of neuron. For example, if all membership functions should be described by “Gaussian bells”, the appropriate fuzzifier is the neuron shown in Fig. 3.3 and performing at its output function (3.21). This neuron has two parameters - center (a), and width (b). This type of neurons is used very often since its output is a smooth function of its input, which allows easy application of gradient optimization methods.

If one needs triangle-shaped membership functions for application, then the appropriate fuzzifier is the neuron implementing the function

$$Y = \max(\min(\frac{x - a_2}{a_2 - a_1}, \frac{x - a_3}{a_2 - a_3}), 0) \tag{3.22}$$

This neuron has three parameters, directly defining the associated fuzzy number by the triple (a_1, a_2, a_3), i.e. left bound, middle point and right bound.

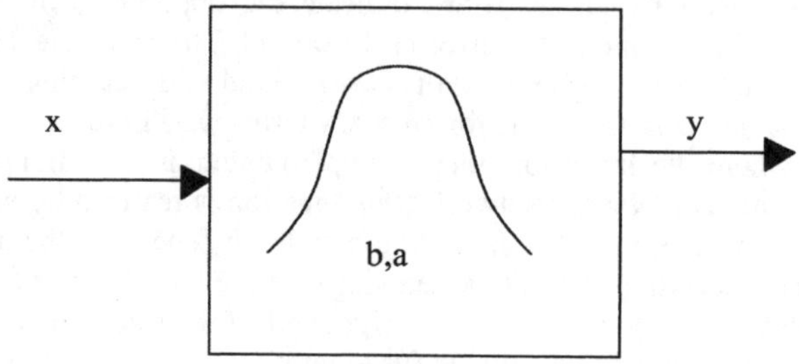

Fig. 3.3. Neuron for implementing a Gaussian fuzzifier with parameters a and b

If there are no recommendations or restrictions on the selection of one or another shape for membership functions, one can delegate the problem of the selection of the fuzzifier to the machine. The machine, having replaced the fuzzifier by a traditional neural network and learned it, will obtain so-called neural fuzzifier (Fig. 3.4).

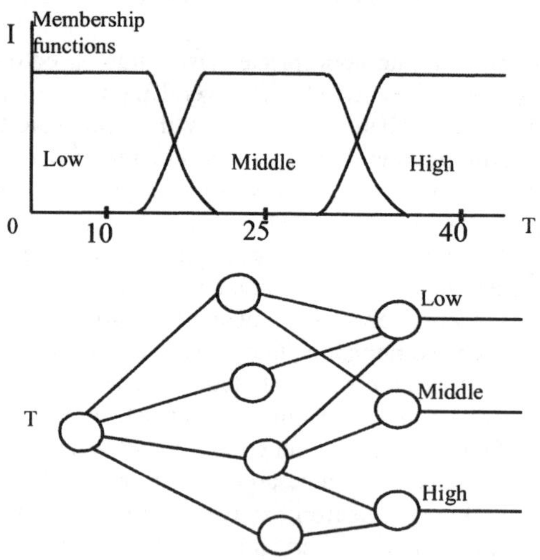

Fig. 3.4. Neural fuzzifier

The most commonly used defuzzification formula that consists in finding the line that divides the area below the membership function (describing the defuzzified number) into two equal-size sub-areas. The associated method is known by name "Center-of-gravity method".

$$y_{defuz} = \frac{\int y \cdot \alpha(y)dy}{\int \alpha(y)dy}. \tag{3.23}$$

Because the output membership functions may consist of a combination (union) of some parts of input membership functions, and this combination does not in general inherit the features of the original shapes ("Gaussian bell"), then the resulting defuzzifying function will hardly be of a compact mathematical form.

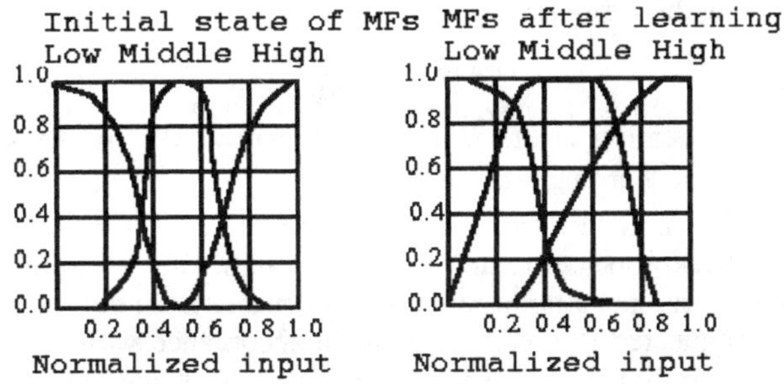

Fig. 3.5. Change of MFs after learning

However, in most cases, especially when concerned with the monolith (neuro- and fuzzy in a single chip) compact, simple and effective neuro-fuzzy systems for limited applications, the output membership functions are expressed as singletons. In this case the output membership functions are assumed to be singletons. They are completely defined by the parameters of the neuron (W_j) after learning. Here the division may be excluded by way of normalization of the coefficients (W_j), i.e. the defuzzification is performed by a neuron that just weighs the input signal forming its output.

In the latter case the generated rules can be represented as a table (multi-dimensional in the general case) like Table 3.1.

Table 3.1. Rules represented as a table

Pressure	Low	Average	High
Temperature			
Near zero	0.21	0.05	0.11
Low	0.19	0.25	0.15
Average	0.05	0.31	0.12
High	0.28	0.01	0.09

3.2 Neuro- Genetic Approach

As shown in [7,25], one of the main problems connected with the development of neural network based systems is application of an appropriate learning method to adjust the network parameters (weights, biases etc.) so that the system acts appropriately. Many approaches have been suggested to solve this problem. Some of the learning algorithms make structural modifications to the network; some others only change the weights. The most popular of them is the back-propagation algorithm.

The backpropagation algorithm uses the gradient to calculate the weights in accordance with the formula:

$$\Delta W = -\gamma \frac{\partial E}{\partial W},$$

where E is learning error, W is weight matrix, ΔW is the vector of weights change, γ is the learning rate coefficient.

As it is shown in section 2.2 unfortunately there are some difficulties when applying the backpropagation algorithm. The initial set of weights is chosen randomly and represents a point in the weight space. Each point (the weights combination) is connected to a certain error. In accordance with geometric view, each vector of the n-dimensional space corresponds to some height of learning error. All possible weight combinations make an error surface. The aim of learning algorithm is to find such a weight matrix that points to a possibly lower error value. The lower the error level, the more precise the solution.

The back-propagation algorithm, like other gradient-based algorithms does not avoid falling into local minima - unlucky initial weights combination may lead to getting stuck in a local minimum. Thus here the success considerably depends on how successful was the initially selected weight combination.

The other difficulty is selection of an effective value for the learning rate. If it appears to be very small, this leads to much time consumption, if else wise it is too big, the algorithm will not converge.

Also the back-propagation algorithm requires differentiability of the activation function, which does not hold for many networks used.

As it is shown in section 2.4 genetic algorithms (GA) is an optimization method successfully applicable to a wide range of problems. They are used when the traditional methods are not able to find required solution.

Different approaches to apply GA to NN learning have been suggested. Some of them change the weight coefficients, others change the structure of the network.

A commonly used approach consists in application of GA to the optimum selection of the weight coefficients of an NN with predefined structure [7,107,110,151].

GA can be used independently as well. For example, Montana and Davis [110] used them to recognize the depth of signals from sonars. In their work a GA with only mutation and crossover, after 10000 iterations, had managed to find a solution twice better than that the traditional backpropagation algorithm found. If there were an opportunity to use the gradient of the object function, then the hybrid

GA/gradient-based algorithm would be more effective. This time, the step of the gradient descent can be used as one of the operators of the GA [110]. In [28] it is shown that the feature of evolution in GA and the feature of learning in NN are intercomplementing and give better results when used in combination rather than separately. In particular, GA can provide an evolution model, and a supervised learning (or some other learning paradigms) can provide a powerful learning mechanism.

GA need fitness function to perform optimization. A NN is described by the learning error:

$$E = \sum_{i=1}^{N}\sum_{j=1}^{M}(y_{ij}^{*} - y_{ij})^2$$

where y_i^* and y_i are M-dimensional desired and current outputs from neurons, respectively.

The summation is done over all the output neurons and learning pairs. That estimation can be averaged like given:

$$\bar{E} = \frac{1}{M}\frac{1}{N}\sum_{i=1}^{N}\sum_{j=1}^{M}(y_{ij}^{*} - y_{ij})^2$$

These estimations of errors can be used to check the neural network for fitness. Some modification to the error formula should be made to turn the problem to a maximization problem. The common object function can be written as follows:

$$\text{fitness} = \text{eval}(E) \text{ or } \text{eval}(\bar{E}),$$

For example,

$$\text{fitr.ess} = 1 - (\bar{E})$$

Obviously, other ways are possible as well, e.g. those considering the maximum error of neurons. One can see that there is no need for the function to be differentiable unlike in the case of gradient algorithm.

The population of individuals is specially selected from a certain interval. Each individual represents a complete set, e.g. the vector

$$W = (w_1, \ldots, w_n),$$

consists of n weight coefficients from some set.

In classical GA, individuals are represented as bit-strings. However in this case the number of weights can be very large. In addition, there exist difficulties with the representation accuracy. A multitude of bits are required to get a high accuracy. To avoid it one need to use the real number representation. The regeneration of the population can be done by different ways. The most common way considers that the value of individuals should determine its inclusion to the next generation. In accordance with these probabilities of regeneration of individuals are determined.

In some algorithms a non-uniform mutation is used. At such a mutation the value of the change depends on the "maturity" of the population.

As the age of the population increases, the mutation is decreased to provide a fine-tuning, unlike the uniform mutation when each gene is selected randomly independently on the age. There exist a number of operators capable of improving solutions in case of specific problems.

Several approaches exist concerning how to use GA to NN learning. In one case, first GA is used and after the backpropagation algorithm is launched. In other case the set of local maxima found by the backpropagation algorithm are then used as the initial population for GA.

Let's consider some characteristics of GA. As compared to the gradient method, GA provide a lower percentage of falling to local minima. GA are free of fluctuations that are possible for the gradient method.

The fitness function does not have to be differentiable or even continuous. This allows learning of NN using the step (threshold) activation function.

However there still exist a possibility for GA to fall to a local minimum. One "super-individual", residing at a local minimum may be included into the population, generate a multitude of children, and fill up all the population. There exist a number of strategies to prevent GA from falling to local minima.

Another shortcoming of GA is weakness in fine tuning as the solution becomes close to the individuals. GA hardly find it with a required precision. GAs spend a lot of time for the multipoint search in large Neural Networks.

Now we consider use of genetic algorithms for neural network learning. Let's first consider the application of a genetic algorithm to an ordinary optimization problem given in the form:

$$y(x) \to \min,$$

$$x \in Q \subset R^n \tag{3.24}$$

To this end we need, first, to find a mapping of the area of real values of variables Q in some space of binary strings $\Omega : Q \to \Omega$. Secondly we should define a suitable fitness function $f(s)$, the maximization of which will result in the minimization of $y(x)$ in (3.24). As applied to learning of neural networks, the above can be interpreted as follows. If $E = E(W)$ is a total squared estimate of the error (criterion of sum of squared error) of type:

$$E(W) = \frac{1}{2}\sum_{k=1}^{N}\sum_{\ell=1}^{M}(y^*_{k\ell} - y_{k\ell})^2 ,$$

Where N, M are the numbers of learning pairs and output neurons, respectively, the fitness function can be considered as $-E$. As proposed in (3.24), the complete set of weight coefficients of the neural network is coded as binary strings s_i with the fitness function $f(s_i)$. For example, assume that the weight search space is $Q = (0.2, 0.8)$. We transform Q into $\Omega = \{0,1\}^6$. That is the binary search space consists of 6-bits strings. The transformation may be done in accordance with

$$V(S) = 0.2 + \frac{\nu(s)}{64}(0.8 - 0.2), \qquad (3.25)$$

where $\nu(s)$ is the ordinary decimal representation of the 6-bits string s. For example, the decimal representation of 000011 is 3, then $V(000011) \approx 0.229$. Obviously, $V(000000) = 0.2$, $V(111111) = 0.8$.

The fitness function can be written as

$$f(s_i) = -EV(s_i) \qquad (3.26)$$

where $V(s_i)$ is the decoded transformation.

The solution of the problem of definition of optimal values of neural networks weights with application of genetic algorithms can be described geometrically as given in Fig. 3.6. Before application of the standard genetic algorithm to find optimal weights for a neural network, the values of L, P_c and P_m should be set, where L is the number of binary strings in the population, P_c is the probability for the crossover, and P_m is the probability for mutation. The values of these parameters usually are defined empirically. To define L it is useful to consider the inequality $n \le L \le 2n$, where n is the number of bits in the strings [1,109,124]. On the basis of the empirical results, it is recommended using the values for P_c and P_m from the following intervals [61,130].

Optimization problem		Genetic algorithm
$E(W) \to \min$	$\leftarrow$ - - - - - - - $\rightarrow$	$f(s) \to \max$
$W \in Q$ $Q \in R^n$	$\leftarrow$ - - - - - - - $\rightarrow$	$\Omega = \{0,1\}^n$
Minimum W^*	$\leftarrow$ - - - - - - - $\rightarrow$	Strings with high fitness S^*

Fig. 3.6. Geometric interpretation of the search by genetic algorithms

$$0.6 \le P_c \le 0.99,$$
$$0.01 \le P_m \le 0.05$$

To start with the genetic search we need at first to form the initial population

$$S(0) = \{ S_1, S_2, \ldots, S_L \} \in \Omega,$$

each string in which has n bits. From this initial population, by using the three known genetic operators - selection, crossover, and mutation - we will get a succession of populations $S(1), S(2), ..., S(t), ...$ The operation of string selection can be done, for example, by way of roulette wheel [109]. The operation of selection of the maximum from the set of numbers can be used as well. The string found after the selection operation are only candidate members to the next generation. For them to be included to the new population they still need to pass through the operations of crossover and mutation.

The pairs of L candidate-members survived after the selection are subject to crossover. These strings are selected randomly from $S(t)$. A random k, named the crossover point, is chosen from $(1,2, ..., n-1)$, and then the two selected strings exchanges their bits starting from the k-th bit with the probability P_c. This process is continued until $S(t)$ becomes empty.

After the crossover, the strings undergo the mutation operation. This time, the value of each separate bit of population is altered 0 to1, or 1 to 0, with the probability P_m.

After the mutation the strings are included into the new population of strings $S(t+1)$ and all the process is repeated, i.e. calculation of fitness, string selection by roulette wheel, and application of operators of crossover and mutation.

Hybrid learning method

As mentioned above, in some cases, for example, when the gradient of the error function can be easily found, the combination of the genetic algorithm and the gradient descent (GD) method result in a more effective search. The hybrid GA/GD method includes two phases: genetic search in the hidden desired space; and the weights updating on the base of the gradient at each layer.

We consider a full-connected feed-forward neural network with one hidden layer of neurons with the linear threshold activation function [109].

The set of learning pairs is given as $\{x^k, d^k\}$, $k = 1, 2, ..., N$. The output neurons can have different activation functions. As a rule, we have no information in advance about the desired values of the neurons' outputs in the hidden layer $\{h^k\}$. The GA can be used to obtain the vector $\{h^k\}$, and further we can get the optimized values of the weight coefficients of the neural network by application of different learning methods including the GD method. As the desired outputs of the neurons in the hidden layer are binary values, their natural coding can be represented as a string $S = \{S^1 S^2 ... S^N\}$, where S^i is a string formed from the bits of the vector $h^i, h^j \in \{0,1\}^j$. Thus the search "point" can be represented as a $J * N$ matrix $H = [h^1 h^2 ... h^N]$. At first, N random binary strings $\{H_j\}, j = 1, 2, ..., L$ are generated as initial populations of the search "point".

The fitness function is defined as $f(H_i) = -E_j$. Beginning with generation of random values for weights and outputs of hidden neurons by application of the least square method, the weights of the hidden layer of the network are determined for each of the L networks, provided that the learning pairs $\{x^k, h^k\}, k = 1,2,...,N$ are specified. Similarly, the weights of the output layer of the neural network are determined if the appropriate learning data $\{h^k, d^k\}$ are specified.

After determination of the weights, passing the signal through tests, each network and calculations, thereby the fitness values are found. This time the output signals of the hidden layer become the input signals to the output layer.

Then the selection, crossover, and mutation operators of GA are used to produce next generation of the hidden layer neurons output values $\{H_j\}$. When reproducing, $L/2$ sets of the hidden layer neurons outputs with the highest values of fitness are doubled and included into a temporary buffer for further selection. The pair $\{H_i, H_j\}$ is selected randomly from the generated buffer. If the learning pair $\{x^k, d^k\}$ is not properly learned by the neural network i (the network j) during the previous learning phase, then the corresponding column h^k from H_i is substituted for the k-th column from H_j.

Then the standard mutation operator is applied. This time each bit of the array H_i, after the crossover, is modified with the probability P_m.

This cycle is repeated until the population $\{H_i\}$ converges to the dominant representation or until the last of the generated networks has the error less than some predefined value.

In [61] a simulation and comparative estimation of the effect of the GA/GD hybrid learning algorithm is given. A simulation of the problem of non-linear identification of a dynamic system is considered there. The comparative estimations for the following learning methods are given; incremental back-propagation; standard GA in the weights space; and the described above GA/GD hybrid method.

In all simulations, the GA/GD method led to successful convergence. Simulation results verify the effectiveness of the GA/GD method.

3.3 Fuzzy-Genetic Paradigm

At first we will consider GA-Based Method for Defining of Relational Matrix and Membership Functions. Let's briefly recall that the most of fuzzy systems are built in accordance with the following procedure. At first, linguistic term-sets for input X and output Y variables and the appropriate membership functions are defined. Then, the relational matrix R is built between the linguistic descriptions of X and Y on the base of available knowledge. The output vector Y is calculated by the compositional inference rule:

$$Y = X \circ R$$

where $\circ$ is the composition operator.

By using some or other defuzzification methods, the current crisp values of the output Y are found. It should be noted, that the effect of any fuzzy system created on the base of the given procedure depends mainly on the three factors: selected type of fuzzy implication, construction of fuzzy relational matrix, and definition of the membership functions. The last two problems, i.e. construction of fuzzy relational matrix and definition of the membership functions, are naturally solved subjectively and are supported by the experience and acquired knowledge. Earlier in the book it was pointed out that some difficulties arise when learning the membership functions for fuzzy systems. In [10,122] procedures are suggested to learn membership functions for a fuzzy logical controller. However, they are not universal and are based on subjective and heuristic factors and reasoning.

In this section a method is considered that is based on a genetic algorithm and free of the mentioned disadvantages [9,15,86,105,120]. For simplicity, let's assume that the model of a fuzzy system is described as follows:

$$\begin{aligned}&\text{IF x is } A_1, \text{THEN y is } B_1\\&\text{ALSO}\\&\text{IF x is } A_2 \ \text{THEN y is } B_2\\&\text{ALSO}\\&\dots\dots\dots\\&\text{IF x is } A_n, \text{THEN y is } B_n\\&\text{ALSO}\end{aligned} \tag{3.27}$$

(3.27) can be implemented by using a fuzzy relational matrix $R(X,Y), R(X,Y) = \text{ALSO}(R_1, R_2, ..., R_n)$

$$R_i = \int \propto_i(x) \otimes \propto_i(y)/(x,y), \qquad i = \overline{1,n},$$

where $\otimes$ is the operator of fuzzy implication.

The effect of the fuzzy model is described by the least square error e^2, evaluated as a mismatch between the measured values and those calculated by using the constructed matrix R

$$e^2 = \sum_{i=1}^{N} (y_i^m - y_i^R)^2 / \sum_{i=1}^{N} (y_i^m)^2 \tag{3.28}$$

where N is the number of sampling intervals of y.

One can minimize e^2 by appropriate selection of the type of fuzzy implication [5,15,90,120] and optimal membership function. Let's consider the construction of a $n \cdot m$ dimensional fuzzy relational matrix and membership functions by application of genetic algorithms.

Construction of fuzzy relational matrices

As mentioned in section 2.4 the first step necessary to apply a genetic algorithm is coding the problem by changing from the real decimal representation of numbers to the bit strings space.

Each string is represented as an array with the length equal to $n \cdot m$ (or a $n \cdot m$ matrix). A population is initialized that contains M strings. The fitness function that evaluates each string at the current operation may have different representations, for example, $(1-e^2)$. In [120] it is taken in the form

$$y = (1+e^2)^{-1} \tag{3.29}$$

Then, by applying the three known GA operators to manipulate the genetic composition of the population, a succession of new generations are formed.

The terminating condition of the algorithm is defined by setting a value for maximum number of generations, Q.

Optimization of membership functions

We consider a mechanism based on a genetic algorithm to optimize membership functions on criterion (3.28). The membership function for each term-set A_i, B_i can be created by using a basis function, with parameters, that allow definition of a variety of suitable shapes for those membership functions by way of varying the parameters. Such basis functions are considered in [16,120]

$$f(x; a_{i1}, a_{i2}, a_{i3}) = \begin{cases} 0, & \text{if } x \le a_{i1} \\ \dfrac{x - a_{i1}}{a_{i2} - a_{i1}}, & \text{if } a_{i1} \le x \le a_{i2} \\ \dfrac{a_{i3} - x}{a_{i3} - a_{i2}}, & \text{if } a_{i2} \le x \le a_{i3} \\ 0, \text{ if } x \ge a_{i3} & \end{cases} \tag{3.30}$$

where a_{i1}, a_{i2}, a_{i3} are tuning parameters for i -th fuzzy subset

$$a_{i1}^n = (a_{i1} + \delta_i) - \eta_i,$$

$$a_{i2}^n = (a_{i2} + \delta_i),$$

$$a_{i3}^n = (a_{i3} + \delta_i) + \eta_i,$$

where δ_i, η_i are some tuning coefficients.

The parameter δ_i shifts the membership function to the left or to the right. The parameter η_i allows changing the shape of the membership function (extends or shrinks the membership function).

Then GA is used to determine optimal values for δ_i and η_i Assume that δ_i and η_i are real numbers within the interval $[-d, d]$. If we need to create k membership functions, then the length of each string should be $2k$.

At first, a population consisting of M strings of length $2n$ is formed. Having selected (3.29) as the fitness function, a succession of generations is made, by applying the genetic operators, to ensure the membership functions are optimal by criterion (3.28).

Parallel optimization of relational matrices and membership functions

A string of length $n \cdot m + 2k$ is formed. First n·m genes correspond to the fuzzy relational matrix. 2k genes corresponds to k membership functions and each gene takes a value from the interval $[-d, d]$.

As each string consists of two different sub strings, they are considered independently when applying the operators of crossover and mutation. When applying the genetic operators, each sub string is considered a part of the whole string.

After formation of an initial population of M strings, a succession of generations is made, by applying the mentioned above operators, to provide an optimal relational matrix and membership functions. Function (3.29) is used as the fitness function.

Now we consider design of fuzzy knowledge base by using genetic algorithms. For designing a fuzzy logical system, in particular, a FLC, the main problem is the selection of an appropriate knowledge base, i.e. optimal number of rules with effective membership functions. Different methods to form fuzzy knowledge bases are known. These include obtaining the knowledge from experts, the method based on using mathematical models, neural-fuzzy approach and others. The basic criteria thereat are presence of redundancy, conflicts, and defects in knowledge. Designers of fuzzy logical systems usually try to use least number of fuzzy rules to get an acceptable solution. This is of special importance when considering the effect of a hardware or software implementation of the designed fuzzy system. One way to solve the problem of designing an effective set of fuzzy rules is the approach based on GAs [53]. The essence of this approach is in the following. Assume that we consider N fuzzy rules with M parameters. For example, if we consider a controller with two inputs and one output and each input is divided into 11 parts, then total number of rules will be 121. The designer wants to know what minimum number of rules (n from 121) is enough to solve the problem. Moreover, if it is required to set the parameters of 11 membership functions of the controller's output and each membership function is characterized by 4 parameters, then we have to determine 44 parameters of the membership functions in addition to the number of rules. Such a function (or a number of functions if the problem is multi-criterial) should be selected as the fitness function that would characterize the effect of the system. For example, if a FLC is created, then the integral square error, mismatch between the set point and actual value of the output of the system, can be set as the fitness function.

Each rule considered as a candidate to be member in the system's set of rules is assigned a value between 0 and $N-1$. Also the varied parameters of the rules that have real values should be coded. Each parameter is described as a coded m-element array of length m with elements taking values between 0 and 9. Each formed string consists of two parts: the first part includes n integer values between

0 and $N-1$ and the second one include $m-M$ integer values between 0 and 9. The length of each string is $n+m\cdot M$.

2p strings are selected with their elements initialized randomly. The operators of GA are applied to these strings, and by using the fitness function we conduct the search in the space of the rules and their parameters. After the selection of 2 strings by the selection operator, parts of these strings (one from each) are selected to apply the crossover operator. Then the parts are processed independently to make a new string. Then the strings undergo the mutation. After p repetitions of the above mentioned operators, new strings are generated. This completes one generation of genetic algorithm.

Fuzzy-Genetic Modeling

Performance of fuzzy systems may be improved if the fuzzy reasoning model is supplemented by a genetic-based learning mechanism. Whereas the selection of acceptable membership functions is generally a subjective decision, change in the membership functions may alter the performance of the fuzzy models. The genetic algorithm enables us to generate an optimal set of parameters for the fuzzy model based either on their initial subjective selection or on a random selection [129].

Throughout this section we adopt the following fuzzy conditional statements to describe a particular situation [17,23]:

$$\begin{aligned}
&\text{IF } x \text{ is } A_1 \text{ THEN } y \text{ is } B_1,\\
&\text{ALSO}\\
&\text{IF } x \text{ is } A_2\text{, THEN } y \text{ is } B_2,\\
&\text{ALSO}\\
&\dots\dots\dots\dots\\
&\text{ALSO}\\
&\text{IF } x \text{ is } A_n\text{, THEN } y \text{ is } B_n
\end{aligned} \tag{3.31}$$

where x and y are linguistic variables, and $A_1,\dots,A_n$ and $B_1,\dots,B_n$ are fuzzy sets on X and Y, respectively. The fuzzy conditional statements (3.31) can be formalized in the form of the fuzzy relation $R(X,Y)$

$$R(X,Y)=\text{ALSO}(R_1,R_2,\dots,R_i,\dots,R_n)$$

where ALSO represents a sentence connective which combines the R_i into the fuzzy relation $R(X,Y)$, and R_i denotes the fuzzy relation between X and Y determined by the i -th fuzzy conditional statement.

It is necessary to find fuzzy sets A_i, B_i $i=\overline{1,n}$, such that the mean square error e^2 between the fuzzy model output values and experimental output values would be the smallest. The mean square error is calculated by formula

$$e^2 = \frac{\sum_{i=1}^{L}(y_i^* - y_i)^2}{\sum_{i=1}^{L} y_i^{*2}} \tag{3.32}$$

where y_i^* is the experimental output value of the object for some current value i; y_i is the corresponding fuzzy model output value; L is number of experiments.

The calculated degrees of membership for each input value are the truth levels of linguistic rules preconditions. The truth level of i-th rule precondition defines the part of i-th rule in fuzzy output value.

In other words, because of the partial matching attribute of fuzzy rules and the fact that the preconditions of rules overlap, more than one fuzzy rule can fire simultaneously. The consequences of multiple rules can be combined by a conflict - resolution process that results in a fuzzy output.

The precision of fuzzy model obtained is determined by means of mean square error calculated by formula (3.32).

To see how the fuzzy logic inference can be applied let's, for example, take the demand function $y = x \cdot \sin(1/x)$ to be approximated in [0.15, 0.45]. Observing the relationship between x and y we may formulate the following linguistic descriptions to specify the characteristic of the function:

IF x=small THEN y=zero
ALSO
IF x=a bit larger than small THEN y=negative small
ALSO
IF x=larger than small THEN y=negative large
ALSO
IF x=smaller than medium THEN y=negative large
ALSO
IF x=a bit smaller than medium THEN y=negative medium
ALSO
IF x=medium THEN y=negative small (3.33)
ALSO
IF x=a bit larger than medium THEN y=zero
ALSO
IF x=larger than medium THEN y=positive small
ALSO
IF x=smaller than large THEN y=positive medium
ALSO
IF x=a bit smaller than large THEN y=larger than medium
ALSO
IF x=large THEN y=smaller than large

From Fig. 3.7 we see that the variable x has 11 linguistic values, whereas the variable y has 8. The membership functions of the fuzzy sets for x and y are shown in Fig. 3.7 and Fig. 3.8, respectively

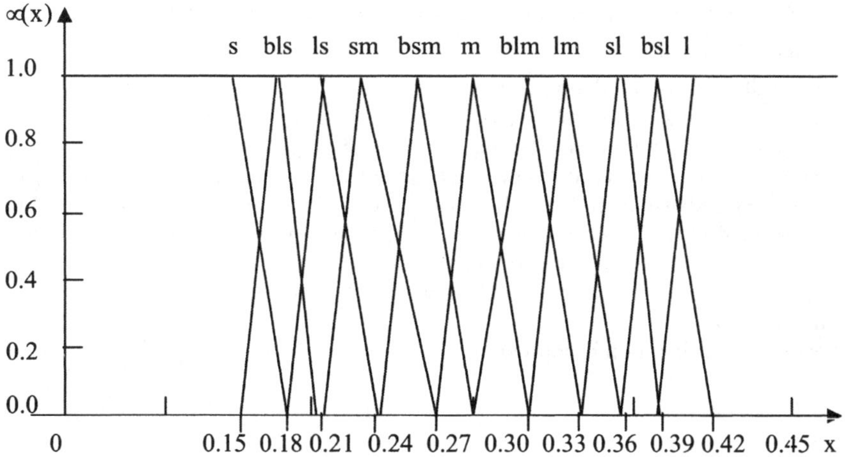

Fig. 3.7. Membership function of the fuzzy sets for x

Assume it is necessary to find the crisp output value corresponding to the input value $x = 0.26$ by successive applying the fuzzification, fuzzy logic inference mechanism and defuzzification.

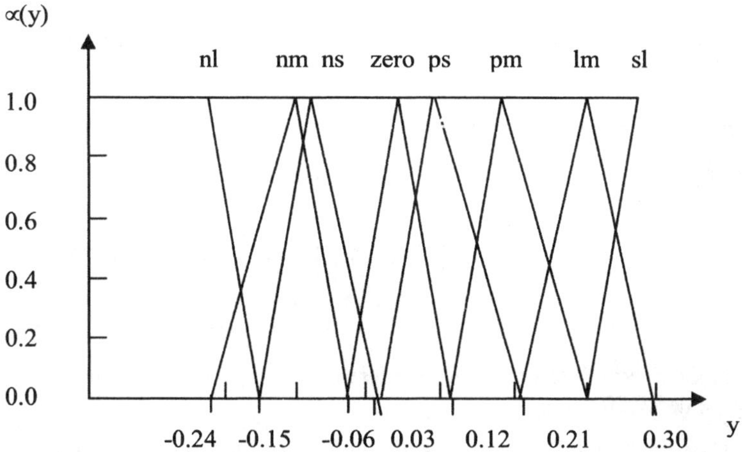

Fig. 3.8. Membership function of the fuzzy sets for y

Experimental output value, found by formula $y = x \cdot \sin(1/x)$ will be $y = 0.26 \cdot \sin(1/0.26) = -0.17$.

Let's determine membership degrees of the given input value to fuzzy sets $A_i, i=\overline{1,11}$, shown in Fig. 3.7.

The following is simulation results:

m[1]=0.00; m[2]=0.00; m[3]=0.00; m[4]=0.33; m[5]=0.67; m[6]=0.00;
m[7]=0.00; m[8]=0.00; m[9]=0.00; m[10]=0.00; m[11]=0.00

Now knowledge-based inference mechanism can be applied. Here we use rule base (3.33) consisting of fuzzy linguistic rules. Consequences of multiple (11) rules resulted in the fuzzy output set (see Fig. 3.9), constructed on universe V and bounded by the following membership function:

m[1]=0.33; m[2]=0.67; m[3]=0.00; m[4]=0.00; m[5]=0.00; m[6]=0.00
m[7]=0.00; m[8]=0.00

Then the defuzzification is applied.

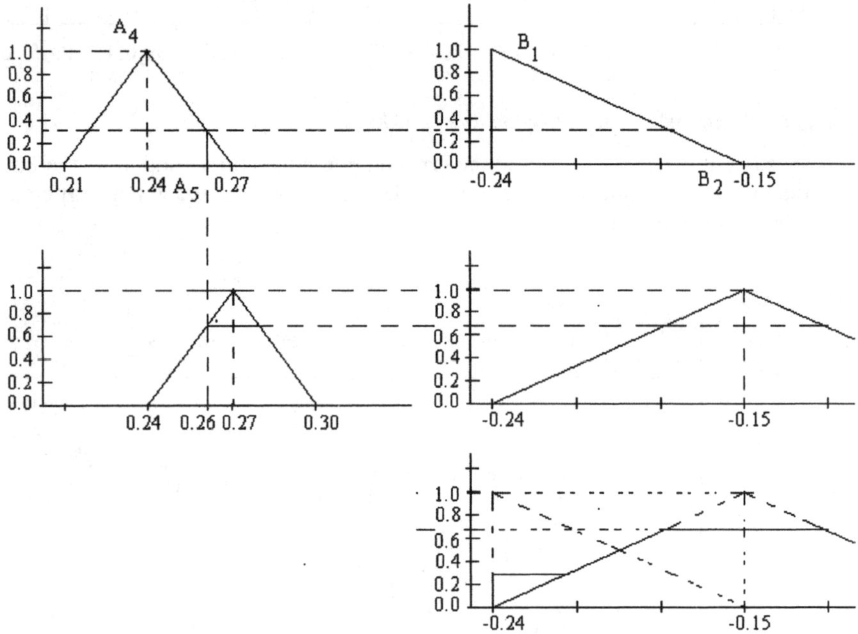

Fig. 3.9. Geometric interpretation of inference mechanism and center of gravity method of defuzzification

Let's use the "center" of gravity method of defuzzification.

$$y_c = \frac{\int_{-0.24}^{0.39} y \cdot \alpha_Y \, dy}{\int_{-0.24}^{0.39} \alpha_Y \, dy} = -0.155$$

Crisp value and the mean square error e are shown in the following listing:

$$bx[1] = 0.26$$

$$cro[1] = -0.155$$

$$e^2 = 7.7854 \cdot 10^{-3}$$

Output values for the given input values are calculated in the same way (see Table 3.2).

Fuzzy rules and membership functions have been generated heuristically. Obviously they can't provide the model precision required. To achieve the latter it is necessary to tune appropriately the rules, as well as the shape and the center of the membership functions. To this end we use GA.

Table 3.2. Comparison of models

N	Input values	experimental output values	Output values of the GA-generated fuzzy model	Output values of the fuzzy model
1	0.15	0.056	0.030	0.030
2	0.18	-0.120	-0.091	-0.060
3	0.21	-0.210	-0.209	-0.210
4	0.24	-0.205	-0.210	-0.210
5	0.27	-0.140	-0.160	-0.150
6	0.30	-0.057	-0.060	-0.060
7	0.33	0.037	0.027	0.030
8	0.36	0.128	0.120	0.120
9	0.39	0.213	0.196	0.210
10	0.42	0.290	0.300	0.300
11	0.45	0.358	0.360	0.360

11 fuzzy sets are used in linguistic rules preconditions. Consequently, we must encode $11 \cdot 3 - 2 = 31$ points. Each point u_i, $i = \overline{1,31}$, takes value from a domain $D = [a_i, b_i] \subseteq U$. Suppose that $u_1 = 0.15$ takes values from $[0.12, 0.16] \subseteq U$, $u_2 = 0.18$ from [0.17, 0.20] and etc.

Then the processes of encoding and decoding are applied. They have been described in section 2.4.

As more than one point is needed for tuning membership functions, then a finite number of chromosomes is initialized randomly in a bitwise fashion. These chromosomes represent an initial population. Given an initial population $G(t), t = 1$ the GA generates a new generation $G(t+1)$ based on the previous generation $G(t)$.

For the selection process (selection of a new population with respect to the probability distribution based on fitness values) the following actions are implemented first:

1. Calculate the fitness value J_i for each chromosome c_i, $i = \overline{1,ps}$ $J_i = \dfrac{1}{1+e_i^2}$.

2. Find the total fitness of the population $J = \sum_{i=1}^{ps} J_i$.
3. Calculate the probability of the selection p_s^i for each chromosome c_i, $i = \overline{1, ps}$, $p_s^i = \frac{J_i}{J}$.
4. Calculate the cumulative probability p_{cum}^i for each chromosome c_i, $i = \overline{1, ps}$, $p_{cum}^i = \sum_{j=1}^{i} p_s^j$

Then selection, crossover and mutation are applied successively. We stop the GA after a fixed number of iterations (Table 3.3).

Table 3.3. Parameter Setting for the GA

Population Size	30
Generation Number	100
Crossover Probability	0.9
Mutation Probability	0.1

To summarize, the above mentioned fuzzy model construction algorithm may be represented as follows:

1. Fuzzy sets A_i, $i = \overline{1, l}$ and $B_j, j = \overline{1, p}$ are constructed on universes U and V, respectively.
2. Linguistic (fuzzy) rules are determined.
3. The justification process is implemented. During the fuzzification the values of input variable are transformed by using stored membership functions to produce fuzzy input values.
4. Knowledge-based fuzzy logic inference mechanism is activated. Fuzzy output value is generated.
5. Defuzzification process is executed. It results in a crisp value of the output fuzzy value.
6. By formula (3.32), the mean square error e^2 for each input value is calculated.
7. If e is less than the given precision, go to step 17.
8. The GA starts work $t = 1$.
9. The initial population is created.
10. G(t) is evaluated. This step also consists of fuzzification, inference, defuzzification, which precede calculation of the mean square error for each chromosome c_i, $i = \overline{1, ps}$. Besides, minimum mean square error is stored in memory.
11. If some termination conditions are met, go to step 15.
12. Generate new generation $G(t+1)$ from $G(t)$. Then crossover and mutation are applied.
13. Evaluate $G(t+1)$.
14. Return to step 11.

15. The termination of GA's work.
16. Among all minimum errors stored in memory the smallest one is found. The fuzzy sets A_i, $i=\overline{1,n}$ and crisp output values, by which the smallest mean square error is obtained, are selected
17. End.

The resulted GA-generated optimal membership functions for inputs are shown in Fig. 3.10. The mean square error is $e^2 = 5.77550 \cdot 10^{-3}$.

GA-generated fuzzy model output values are shown in Table 3.2

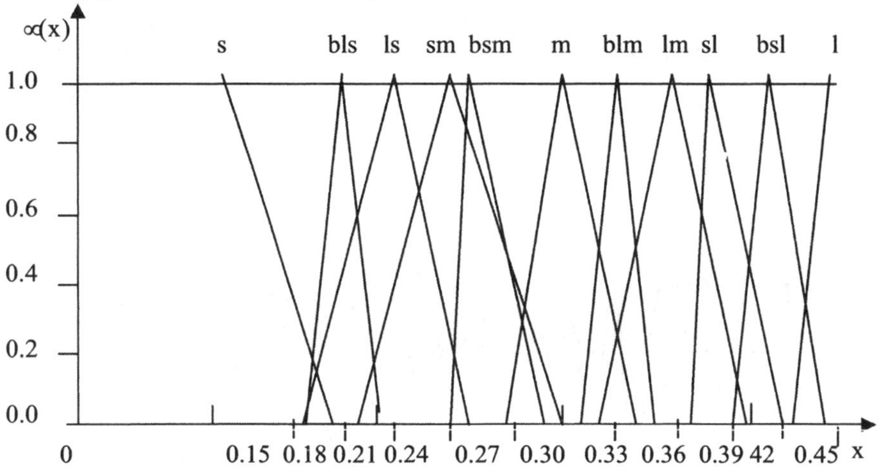

Fig. 3.10. GA-generated membership functions

3.4 Genetic Algorithms with Fuzzy Logic

Use of fuzzy approach in genetic algorithms for tuning the search directions significantly improves the behaviour of GA. Simple genetic algorithms manipulating various combinations of genes, improve fitness of individuals and, consequently, increase fitness of the whole population. Selection of individuals and their recombination occur randomly-evolutionary search is not managed by supervisor. But how the search should be directed to avoid large number of iterations and make the algorithm more effective? How to use the cumulative experience of the previous iterations and current information in order to define the directions of further search? How to use data about parents when crossover is processed to obtain a "good" offspring? Just these questions are aimed to be answered in this section.

As we have not, as a rule, enough information about object investigated, we cannot precisely define directions of the search. Random selection of search directions, as noted, does not give high efficiency adding some degree of chaos. As we have no possibility (time) to perform a chain of experiments, we also cannot use the probability logic. That is why we use the fuzzy logic having experience of the previous iterations only. Just by the use of the fuzzy logic we try to interfere to evolutionary process on each iteration and tune it.

At the macro level, we investigate the search space, and starting from received information, we form separated groups of individuals (subpopulations). These subpopulations are arranged in those areas of the space where possibility of finding the global solution is high [13]. For example, in case of optimizing a multimodal function each group searches for one peak (Fig. 3.11). Using the possibility theory we assign the weights to groups in accordance to their average fitness. By this we avoid unnecessary crossover operations between members of different groups.

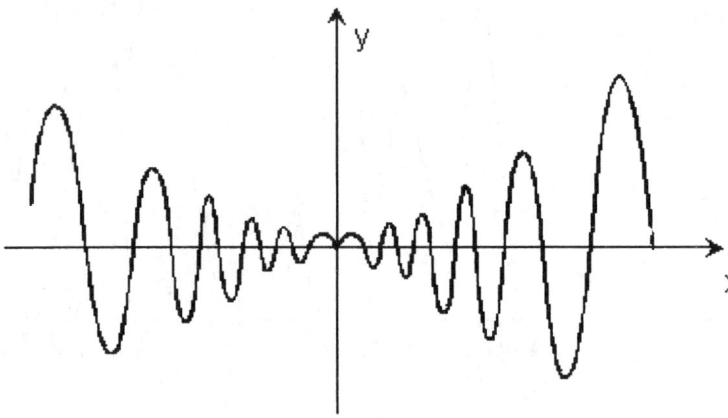

Fig. 3.11. A simple example of multimodal functions for which a group genetic algorithm is suitable

The random approach is used for crossover as usual. But for generating new offspring it should be better to use information about his parents. Thus, if the "father" has higher fitness than the "mother", then the child should probably be nearer to father or vice versa. If, for example, A and B are selected as parents and $C = dB + (1-d)A$ will be the offspring, we suggest a version where d is fuzzy variable. If A has larger fitness than B, d is set small for being C nearer to A in parametric space or vice versa (Fig. 3.12). When A and B are of approximately equal fitness, d takes the value "average". Experiments carried out by us, showed that better result is reached in case of combining random and fuzzy approaches for selecting parameter d, i.e. in the preliminary stages of evolution the more weight is paid to random selection, and as the number of generations increases the more weight takes fuzzy approach in crossover. In case of use of such crossover type for group genetic algorithms, it accelerates significantly the search in accordance to experiments performed.

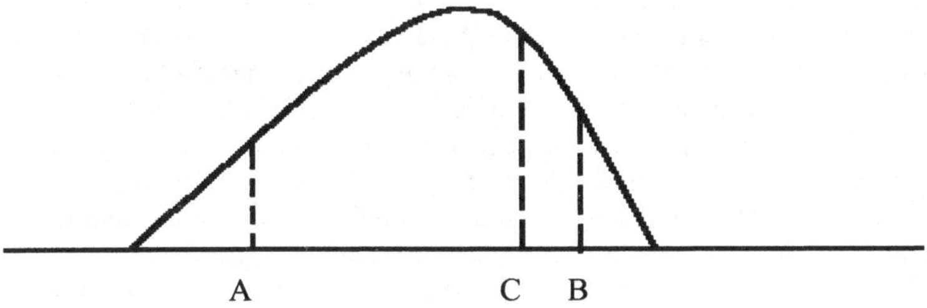

Fig. 3.12. Case of "big" d

Fuzzy logic based genetic algorithms

Fuzzy sets theory as well as systems based on fuzzy logic can be used to improve the behavior of genetic algorithms [12,74,95,128]. In [74] some fuzzy instrumental tools have been suggested to create effective fuzzy genetic algorithms. Fuzzy connectives are used there to design the crossover operator for a real-coding genetic algorithms (RGA) and systems for dynamically controlled of RGA parameters, based on the measure of distinction between chromosomes and alleles. In [128] a fuzzy crossover has been proposed on the base of templates and union operators.

In [7] an idea how to use the methods of fuzzy logic for control of genetic algorithms has been considered. In [12] the fuzzy approach is used in order to correct the search direction of genetic algorithms.

3.5 Neuro-Fuzzy-Genetic Paradigm

In this section we will consider use of genetic algorithms in learning of fuzzy neural networks. In spite of great importance of fuzzy feed-forward and recurrent neural networks (FNN) for solving wide range of real world problems, today there is no effective learning algorithm for FNN.

In [68] a learning algorithm for FNN with fuzzy signals and weights called the "fuzzified delta rule" is proposed. As shown in [62] this algorithm lacks theoretical basis and is derived only through use of real differentiation instead of fuzzy differentiation.

In [80] an architecture and learning algorithm of FNN with trapezoid fuzzy weights are proposed. Using Zadeh's extension principle, α-level sets and interval arithmetic, a crisp learning algorithm is derived for adjusting four parameters of each trapezoid fuzzy weight. Earlier in [79] similar crisp learning algorithm for FNN with triangular weights was proposed. In [42,62] the idea of using genetic algorithms for learning of FNN was expressed.

In this section we consider an effective genetic-based learning mechanism for FNN with fuzzy inputs, fuzzy weights expressed as LR-fuzzy numbers, and fuzzy outputs [2,18,19]. Hamming fuzzy distance, which is non-differentiable is used as a performance error index for learning of FNN. The main distinguishing features of the considered learning method from the existing approaches include the following. The proposed method does not require differentiability of error performance index of FNN. Fuzzy distance in general fuzzy number space is used for comparison of fuzzy actual output and fuzzy target of FNN, therefore the derived algorithm is not crisp. The considered method guarantees with higher degree the global minimum of error performance of FNN than existing algorithms.

Fig. 3.13 shows an example of fuzzy three-layered feed-forward network of type 3. We consider the three-layer network for simplicity sake, in general, any number of layers may be introduced. Input neurons receive a vector of fuzzy input signals

$$X = \{x_1, x_2, ..., x_n\}$$

from the environment. Input neurons do not perform operations on input signals, but simply transfer them to all fuzzy neurons of the consequent (hidden) layer. The connections between the layers may be represented as a matrix of fuzzy weights W^h, where an element w_{ji}^h represents a fuzzy weight of a connection between i-th neuron of the input layer, and j-th neuron of the hidden layer. The total fuzzy input of j-th neuron of the second layer is defined as:

$$I_j^h = \sum_{i=1}^{N} w_{ji}^h \cdot x_i + \theta_j^h \tag{3.34}$$

Here, I_j^h is the total fuzzy input of the j-th neuron of hidden layer, x_i is the i-th fuzzy input of that neuron, and θ_j^h is fuzzy bias of the j-th neuron. The operations on fuzzy numbers in (3.34) are defined by the rules of fuzzy arithmetic [52,87].

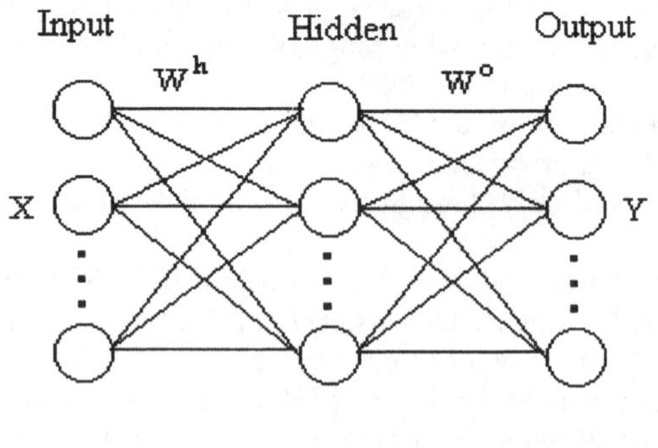

Fig. 3.13. A three-layered FNN

The fuzzy output of the j-th neuron is defined by the transfer function

$$y_j^h = f(I_j^h) \tag{3.35}$$

The output is calculated using (3.35) and the extension principle of Zadeh [162]. We use the following sigmoidal function:

$$f(x) = \frac{1}{1+e^{-x}}$$

The outputs are then passed to the neurons of the consequent layer, which perform similar operations. In three-layered neural nets the consequent layer will be the output layer which produces vector Y. In general multi-layer fuzzy neural networks, fuzzy signals propagate layer by layer until they reach the output layer.

The purpose of training fuzzy neural networks is in designing such an FNN that exhibits the desired behavior, i.e. provides an adequate mapping of fuzzy inputs to the desired fuzzy outputs. A set of pairs of given fuzzy inputs / desired fuzzy outputs represents a training set for FNN

$$(X_i; Y_i^*) \ , \quad i = \overline{1, N}$$

where N is the number of pairs in the training set. Learning algorithm should select such fuzzy values for the weights of interneuronal connections that minimize the fuzzy distance between the calculated and the desired outputs for all fuzzy input vectors in the training set. In other words, the objective is:

$$d = \sum_{i=1}^{N} \left| Y_i^* - Y_i \right| \rightarrow \min \tag{3.36}$$

Here Y_i^* , is the desired output for i -th training pair, and Y_i is the actual output of FNN. To achieve the objective (3.36) by traditional methods (e.g. gradient based techniques) one has to introduce α -level procedures, because (3.36) is not differentiable by itself. We employ an alternative approach, based on genetic algorithms to train fuzzy neural networks.

For training FNN we use a measure of distance (3.36) as a basis for a fitness function. We define the fitness function as follows:

$$F = \frac{1}{(1+d)} \tag{3.37}$$

Therefore, the larger the distance between the actual output of FNN and the desired output, the smaller is the fitness value, and vice versa.

Representation of candidate solutions is another important issue in GA-based optimization. Traditionally GA use binary encoding to build chromosomes. Therefore, an adequate mapping of problem variables to binary strings and vice versa is required. In our case a single chromosome carries the values of all fuzzy weights of FNN in a binary format. We use weights expressed as LR-type fuzzy numbers with n parameters. A fuzzy weight is represented by n binary strings of a specified length. Each of the n binary strings corresponds to a particular parameter of a fuzzy weight.

Consider, for example, the case of triangular representation of fuzzy weights. The following chromosome illustrates this example. Here, the precision level of five bits per parameter of a fuzzy weight is used:

$$\left(\ldots\underbrace{\underbrace{01100}_{w^h_{jiP}}\;\underbrace{11101}_{w^h_{jiL}}\;\underbrace{01000}_{w^h_{jiR}}}_{w^h_{ji}}\ldots\;\underbrace{\underbrace{10010}_{w^0_{mlP}}\;\underbrace{11101}_{w^0_{mlL}}\;\underbrace{01100}_{w^0_{mlR}}}_{w^0_{ml}}\ldots\right)$$

Here w^h_{ji} is a fuzzy weight between j-th hidden and i-th input neurons, $w^h_{jiP}, w^h_{jiL}, w^h_{jiR}$ are the peak, and the left and right spreads of a fuzzy weight. Similarly, w^0_{ml} is a fuzzy weight between m-th output and l-th hidden neurons.

The strings are translated into the real-valued numbers for parameters using the prespecified ranges for the weights. Therefore, for any combination of zeros and ones of a given chromosome length there exists an FNN with the corresponding weights, for which the distance can be calculated and, with the use of (3.37), converted into the value of fitness for a given chromosome.

Selection operator randomly chooses chromosomes from the population to produce a new population with the probabilities proportional to the values of fitness of chromosomes.

Two other genetic operators: crossover and mutation are applied to the new population. The crossover operator randomly picks two chromosomes and mates them producing two child chromosomes. To this end, first the split point in chromosomes is randomly chosen, then child chromosomes are generated, each of them inheriting the genes up to the split point from one parent, and from that point on from the other parent. Consider, for example, two chromosomes representing two different sets of candidate triangular weights for an FNN. The following illustrates an example of crossover operation:

Before crossover

$$\left(\ldots\underbrace{011001110101000}_{w^h_{ji}}\ldots\underbrace{11100101 \overset{SP}{|} 1111100}_{w^h_{sr}}\ldots\underbrace{100101110101100}_{w^0_{ml}}\ldots\right)$$

$$\left(\ldots\underbrace{101000011111010}_{w^{h'}_{ji}}\ldots\underbrace{01110001 \overset{SP}{|} 0100000}_{w^{h'}_{sr}}\ldots\underbrace{011110000110101}_{w^{0'}_{ml}}\ldots\right)$$

Here SP stands for “split point”. After crossover the child chromosomes become

$$\left(\ldots\underbrace{011001110101000}_{w^h_{ji}}\ldots\underbrace{11100101 \overset{SP}{|} 0100000}_{w^{h''}_{sr}}\ldots\underbrace{011110000110101}_{w^{0'}_{ml}}\ldots\right)$$

$$\left(\ldots\underbrace{1010000111111010}_{W^{h'}_{ji}}\ldots\underbrace{01110001\overset{SP}{|}1111100}_{W^{h'''}_{sr}}\ldots\underbrace{100101110101100}_{W^{0}_{ml}}\ldots\right)$$

Note, that one of the weights, most likely, will not belong to either parents, because the split point may fall within the sub string representing one of the fuzzy weights.

Mutation randomly picks one of the bits in a chromosome and flips it. The following example demonstrates the mutation operator:

Before mutation

$$\left(\ldots\underbrace{011001110101000}_{W^{h}_{ji}}\ldots\underbrace{111001\ \overset{Mt}{0}\ 11111100}_{W^{h}_{sr}}\ldots\underbrace{100101110101100}_{W^{0}_{ml}}\ldots\right)$$

After mutation

$$\left(\ldots\underbrace{011001110101000}_{W^{h}_{ji}}\ldots\underbrace{111001\ \overset{Mt}{1}\ 11111100}_{W^{h'}_{sr}}\ldots\underbrace{100101110101100}_{W^{0}_{ml}}\ldots\right)$$

Here M_t points at the mutated gene. In effect, mutation randomly changes one of the fuzzy weights of an FNN represented by the chromosome. Crossover enables information exchange between two directions of search, while mutation implements random search [106]. The probabilities of crossover and mutation control the exploration/exploitation level in the search process.

The algorithm proceeds as follows:

1) Initial population is randomly created (i.e. a population of fuzzy weights of FNN is randomly specified);
2) Each chromosome is evaluated to its fitness value;
3) Selection takes place to create new population;
4) Crossover operator is applied to the population;
5) Mutation operator is applied to the population;
6) If stop criteria are not met then go to step 2;
7) End.

The stop criterion may be, for example, maximum number of generations. This iterative process leads to the improved performance of candidate set of fuzzy weights.

We consider two examples for training FNN with the use of GA. For measuring the performance of FNNs fuzzy error (3.36) was used. The fuzzy error is compared with the case of perfect match of the FNN model outputs with the desired outputs. Because of fuzzy subtraction operation used, the right and left deviation of fuzzy error is not expected to be crisp zero [36]. Therefore, to assure learning is acceptable, it is necessary to define a set that would reflect the required accuracy

[16,36]. Let the interval [y^*_{iL}, y^*_{iR}] be the support of y^*_i. If y^*_i is represented as a triangle LR - fuzzy number and $y_i = y^*_i$, then the support of d is [0, λ], where

$$\lambda = \sum_{i=1}^{N} \left| y^*_{iR} - y^*_{iL} \right| \tag{3.38}$$

If the fuzzy error derived from (3.36) is within the boundaries specified by (3.38) plus some user-specified acceptance level ε then we accept the model.

Fuzzy Regression by FNN

We use the proposed FNN learning method for intermediate quality evaluation in the manufacturing process of aluminum heaters. The quality is determined by human operators, hence, it is expressed by fuzzy numbers defining estimations on a linguistic scale. Therefore, we are given the sets of fuzzy inputs (x_{1i}, x_{2i}) and fuzzy outputs y_i, $i = \overline{1,9}$. The problem is to construct a FNN for mapping the inputs to the outputs with an acceptable accuracy for predicting the quality of the semi-product of the aluminum $y = G(x_1, x_2)$ plates:

Fuzzy-valued data expressed as fuzzy numbers are given in Table 3.4.

Table 3.4. Fuzzy-valued data

N	Input 1	Input 2	Output
1	(0.1 0.1 0.1)	(0.1 0.1 0.1)	(0.037 0.120 0.129)
2	(0.6 0.2 0.2)	(0.5 0.2 0.1)	(0.912 1.470 1.372)
3	(0.9 0.1 0.3)	(0.5 0.3 0.4)	(0.729 2.394 3.339)
4	(0.7 0.1 0.1)	(0.95 0.05 0.05)	(1.325 1.750)
5	(0.5 0.1 0.1)	(0.6 0.4 0.3)	(1.360 2.305 2.068)
6	(0.1 0.1 0.1)	(0.5 0.2 0.1)	(0.861 0.883 0.542)
7	(0.7 0.1 0.1)	(0.6 0.4 0.3)	(1.321 2.568 2.504)
8	(0.5 0.1 0.1)	(0.5 0.3 0.4)	(0.937 1.586 2.194)
9	(0.1 0.1 0.1)	(0.8 0.1 0.1)	(2.149 0.967 0.924)

A fuzzy neural network for fitting the data contained two input, one output and four hidden fuzzy units. The accuracy level for the data using (3.38) and ε was set at [0, 30]. Table 3.5 shows the parameters of GA. After 2000 generations of GA the aggregate error of the FNN was (0, 0.612, 29.7). The fuzzy error is within the prespecified range, therefore, the performance of the FNN is acceptable.

Table 3.5. Parameters of GA

Population size	50
Number of generations	2000
Probability of crossover	0.7
Probability of mutation	0.005
Selection method	Elitist

This problem was also solved using α - cuts method. The fuzzy error, d was (0, 4.331, 34,362). Our results of the quality evaluation were compared with the results obtained by the α - cuts method using the criterion $E = \frac{1}{2}\sum_{i=1}^{k}(y_i - y_i^d)^2$. In our case this criterion was $E = (-46.419, 0.050, 37.020)$, while in the case of using α - cuts the error was $E = (-48.228, 2.115, 63.643)$. This shows the advantage of the GA-based learning method. Moreover, our experience, as we solved this problem, has shown that α - cuts method is associated with considerable computational complexity.

Estimation of Fuzzy Profit

We next consider a simplified example of a firm's profit model in an oligopolistic environment, where a number of firms compete in a market for a single product. In such industry the actions of one firm have an impact on the industry demand. The firm's management pursues profitability objectives. The variables that affect the profit include: firm's price and advertising expenditures for a product (controllable), as well as the average price and advertising expenditures of the competitors (uncontrollable).

The firm's management needs to estimate competitors' average prices and advertising expenditures before making decision on firm's price and advertising. However, the information about competitor future actions is not known precisely. Competitor actions may be estimated in fuzzy terms. For example: "the average competitor price will be about $315". To cope with this kind of data we employ an FNN for predicting the expected fuzzy profit.

Part of the inputs to the FNN is fuzzy (average price and advertising expenditures), and part is crisp (firm's price and advertising expenditures). The output of the FNN is fuzzy estimated profit. The obtained training pairs were used to genetically train a three layered neural network. The prespecified value of the aggregate fuzzy error is in the range (in million USD): [0; 3.5]. The calculated aggregate fuzzy error is: (0, 0.132, 3.37) . Table 3.6 shows the examples of calculated profits for different combinations of fuzzy inputs.

Table 3.6. Calculated profits

Avg.Price (in $)	Avg.Advertising	Firm Price	Firm Advertising	Profit
(295,4,4)	(60,5,5)	(300,4,4)	(60,5,5)	(383.5,116,115)
(320,4,4)	(90,5,5)	(310,4,4)	(70,5,5)	(384.6,82,88)
(330,4,4)	(90,5,5)	(300,4,4)	(80,5,5)	(459,100,80)
(330,4,4)	(100,5,5)	(330,4,4)	(60,5,5)	(332.1,49,51)
(310,4,4)	(70,5,5)	(330,4,4)	(90,5,5)	(419.4,116,113)

3.6 Multi-Agent Distributed Intelligent Systems Paradigm

Many economic, ecological, and business systems are often complex systems, and almost always, these systems are characterized by imprecision and uncertainty. It is known that in such systems soft computing based distributed multi-agent intelligence is the most effective approach for systems analysis, decision making, and control. The key problem in constructing such systems is the problem of granulation, i.e. decomposition of the monolith intelligence of the whole system into autonomous agents' intelligence. In this section we suggest a method for creation of Distributed Intelligent Systems which include intelligent agents capable of competing and cooperating with each other in order to propose optimal/near optimal local solution to the problem and to integrate local solutions into the final solution. The optimization includes determination of a rational number of autonomous agent and fuzzy rules, optimal scaling factors, shapes and centers of membership functions of fuzzy rules of agents' knowledge bases, and optimal inference engine by using genetic algorithms.

Complex systems such as industry, economy, finance, marketing, ecology are composed of a number of agents interacting in distributed mode. Advances in distributed artificial intelligence (DAI), intelligent agent theory and soft computing technology make it possible for these agents as components of complex system to interact, cooperate, contend and coordinate in order to form global behavior. Therefore the use of multi-agent distributed intelligence technology has been growing in planning, control, decision making and solving other problems of complex systems. Recently, there has been great interest in development of Intelligent Agents (IA) and Multi-agent systems [4,20,54,56,99,100,117,150,152,153]. The theory of Intelligent Agents and Multi-Agent Systems originates from the research in Distributed Artificial Intelligence [117, 153]. The field of Distributed Artificial Intelligence (DAI) has been rapidly growing over the last couple of decades [32]. In Distributed Artificial Intelligence the overall intelligence of the system is distributed among multiple nodes for cooperative problem solving. Some of the rationale for distribution of intelligence includes: better adaptability; cost efficiency; ease of development and management; greater efficiency/speed; autonomy; naturalness; reliability due to redundancy; "width" of expertise; openness; and possibility of having multiple conflicting goals [32,102]. Many publications are devoted to general problems of creating DAI [3,33,139,149]. [33] refers to over 500 works in the area of cooperation, coordination, distributed control, distributed knowledge, knowledge representation, multi-agent planning, etc.

Bond and Gasser point that there are two directions of research originating from DAI: Distributed Problem Solving, and Multi-Agent Systems [32]. Generally, agents are assumed to have their own goals that may be in conflict. The key question in Multi-Agent systems is: can cooperation among intelligent agents achieve better performance than each component separately [49, 144].

The important design-related issues with multi-agent systems include how agents coordinate, and how they solve problems. Gasser & Bond define coordination as "a property of interaction among some set of agents performing some collective activity" [32]. Cooperation can be viewed as a special case of coordination when the participating agents are not antagonistic [32, 126]. Competition is a form of coordination when antagonistic agents are involved [117]. Coordination may take place without communication among agents if they have each other's models in their possession [137]. In this case coordination may be achieved through organization [117]. Nwana et al. categorized different types of coordination schemes in multi-agent environments [117]. These include: organizational structuring; contracting; multi-agent planning; and negotiation. In organizational structuring each node of a distributed system has its own "role" in an overall organizational structure [32,117]. Durfee suggested that there is a priori defined long-term relationships among agents [51]. The coordination in this setup is mainly viewed in a client-server fashion. Blackboard architecture is often used for coordination of agents [43,44,70]. In blackboard architecture, agents communicate via writing messages to and reading from the system component called blackboard.

Contract-based coordination stems from the work by Smith and Davis in the late seventies/early eighties [47]. In a contract-based setup, the system uses contract-net protocol for coordinating agents. In this market-like structure, an agent-manager announces a job and a number of contractor agents submit their bids to the agent-manager, which evaluates the bids and awards the contract to the agent with the lowest bid. The contract-net protocol is a coordinating strategy that provides means of distributing tasks and self organizing a group of agents [137]. Examples of use of contract net protocol include: allocation of jobs among multiple CPUs [39]; transportation scheduling [55]. Another coordination strategy is to utilize multi-agent planning mechanisms which can be either centralized, or decentralized [32,117]. The necessity for coordination here is dictated by the fact that multiple agents may develop their own plans that could be inconsistent or conflicting with each other. One approach to detect the conflicts and inconsistencies would be to use central coordinating agent [58,59]. Another approach is to have agents exchange their individual plans to avoid conflicts and inconsistencies [43]. One example of implementing such an approach is through use of partial global planning and general partial global planning, where agents receive the commitments of other agents and plan their actions to avoid conflicts [48,51]. Negotiation is another important basis for coordination in a community of autonomous agents [117]. Negotiation is initiated by agents in order to reach certain agreement on some issue [108].

[112] presents multi-agent scenario-based methods involving human and artificial agents using different approaches to describe scenarios. The authors do not intend to propose general method for designing multi-agent systems. They illustrate the usefulness of the developed technique for supporting cooperative work between autonomous artificial and human agents.

An attempt to create the coordination theory in DAI is shown in [85]. The authors attempted to determine the concept of coordination on the basis of quantitative representation of components, structure and parameters, as well as performance measures. They considered both centralized and decentralized coordination structures of DAI. The proposed coordination theory and technique are illustrated in a real-world domain. The general problems and practical application of DAI for industrial plants are considered in [91]. It is shown that instead of using centralized artificial intelligence with a single agent carrying the global knowledge (which is usually implemented on computers with monolith programs), it is preferable to use the DAI with a proper distribution of functions, knowledge, and resources of the "center" among intelligent agents.

It should be noted that intelligent agents often deal with incomplete, contradictory, missing and inaccurate data and knowledge [3,22]. Furthermore, the agents have to make decision in uncertain situations, i.e. in the real world DAI functions within an environment of uncertainty and imprecision. Obviously, emerging technologies, especially fuzzy logic and soft computing are adequate approaches to deal with real application areas with such properties [3,81,161]. However, there are only few published works [14,115,147] devoted to fuzzy and soft computing based multi-agent distributed intelligent systems. Aliev et al. in [3,14] introduce the concept of fuzzy distributed multi-agent intelligent system (FDIS). They describe the problem of coordination of intelligent cooperative autonomous agents, namely, fuzzy planning, fuzzy scheduling, dispatching, and local online dynamic systems. They consider rules and conventions, protocol of interactions, unique internodes language of information representation among intelligent agents, models of transferring and processing distributed data and knowledge in FDIS. Experimental results of working of the FDIS on industry plant are given. In [147] fuzzy contract net responsible for planning and controlling the associated machines, service staff, and resources is presented. The cooperation and coordination of the fuzzy contract nets are considered. [115] presents fuzzy DAI system which consists of a network of processing nodes. These nodes cooperate to achieve global goals as a community. Nodes communicate by sending high level fuzzy-structured messages, allowing the representation of uncertainties throughout the system. It discusses applications of fuzzy traffic light control system.

Some application areas of Intelligent Agent based systems include: World Wide Web agents for finding useful information on WWW [46, 111]; recommending web pages [27]; electronic shopping [50]; advertising on the web [155]; electronic marketplace agents for buying and selling decisions [38]; trading on the stock market [114]; assistant agents for meeting scheduling [72,101]; document classification [41]; entertainment agents [60,69]; scheduling tasks in computer networks [39,104]; manufacturing and robotics [26,29,30,31,116]; different expert agents, for example for medical diagnosis [45]; business process management [118] and many others. The above list is a sample of applications of agent-based systems and is by no means complete.

As it is seen from the review of the most important works mentioned above and [32,57] it is reasonable to conclude that in conventional concept of multi-agent distributed intelligent systems the main idea is granulation of functions and powers from a central authority to local authorities. In these terms, DAI is composed of several agents, which can perform their own functions independently, and, therefore, have information, authority and power necessary to perform only their own function. These intelligent agents can communicate together to work, cooperate and be coordinated in order to reach a common goal.

This chapter is mainly based on [20] and introduces an alternative concept of a multi-agent distributed intelligent system (MADIS) with cooperation and competition among agents, distinguished from the conventional approach by the following: each intelligent agent acts fully autonomously; each intelligent agent proposes full solution of the problem (not only for own partial problem); each agent has access to full available input information; total solution of the problem is determined as proposal of one of the parallel functioning agents on the basis of a competition procedure (not by coordinating and integrating partial solutions of agents, often performed iteratively); agents' cooperation produces desired behavior of the system; cooperation and competition acts in the systems are performed simultaneously (not sequentially).

The effectiveness of the proposed concept is illustrated by a multi-agent distributed intelligent marketing DSS.

It should be noted that a similar idea of decomposition of the overall system into subsystems with cooperation and competition among subsystems for approximation of nonlinear functions is implemented in [84]. Distributed trajectory generation for multi-arm robots through cooperation and competition among subsystems is considered in [145, 146].

Zhang proposed a way to synthesize final solutions in systems where different agents use different inexact reasoning models to solve a problem [174]. In this approach a number of expert systems propose solutions to a given problem. These solutions are then synthesized using ego-altruistic approach [88]. In this approach the agents either cooperate (by "moving" their solutions closer to each other) or compete depending on the extent to which their proposals are different. The final solution generated using the mean and uniformity of solutions. Zhang & Zhang also reported the set of neural networks for the synthesis of solutions in a simulated medical diagnosis example [174]. A special case of synthesis is competitive synthesis, where the best proposed solution is selected [136].

Architecture of the Proposed MADIS

Fig. 3.14 shows the basic structure of the proposed MADIS. The MADIS is composed of N agents. All agents receive the same input information $x_1, x_2, ..., x_m$. Each agent performs inference and produces its own solution to the full problem:

$$U_j = [u_{j1}, u_{j2}, ..., u_{jl}]^T, \qquad j = \overline{1, N}$$

The estimator uses the agents' current solutions $U_1, U_2, ..., U_N$, input information $x_1, x_2, ..., x_m$, and any other necessary information to determine system outcomes for each agent's proposal (e.g. production volume in industry plant, expected profit in marketing mix, etc). These outcome values are the basis for competition among agents Ag1,...., AgN. The evaluator compares the outcome values and determines a "winner" (for example, ith) agent, with the best outcome value.

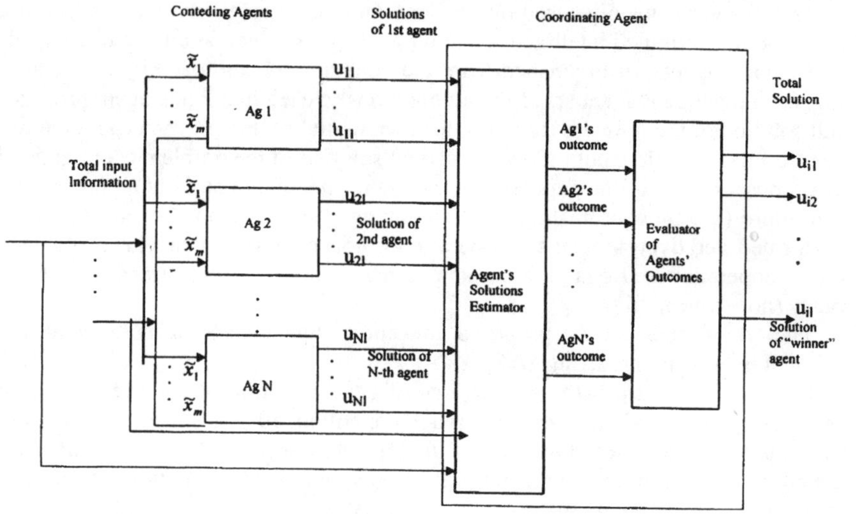

Fig. 3.14. The architecture of MADIS

The winner's solution is accepted as the total solution of the full system:

$$U_i = [u_{i1}, u_{i2}, ..., u_{iN}]^T .$$

Below we describe the components of the proposed MADIS.

Agent's Solution Estimator

This component estimates the expected values of outcome of the system on the basis of proposed solutions of each contending agent. Agents' solutions estimator is implemented as fuzzy neural network with crisp and fuzzy inputs and fuzzy weights ($\tilde{W}$ and $\tilde{V}$) represented as fuzzy numbers. Inputs to the fuzzy neural network consist of the current total input of the system and solution produced by agents. Fuzzy neural network transforms these data into the value of the outcome of the system. The number of input neurons is determined by the number of total inputs and agent's solution variables. The number of hidden layers and the number of neurons in hidden layers are determined by the required accuracy. The number of output neurons is equal to the number of system's outcomes. Learning of fuzzy weights, which are expressed by triangular membership functions is performed by genetic algorithm using experimental training set and fuzzy distance $\tilde{E}$ between actual and desired outputs of agent. The fuzzy-neural genetic estimator in testing mode produces set of fuzzy values of the system's outcomes $\tilde{O}_{j1}, \tilde{O}_{j2}, ..., \tilde{O}_{jq}$ for each agent proposal. The structure of the neuro-fuzzy genetic estimator is shown in Fig. 3.15.

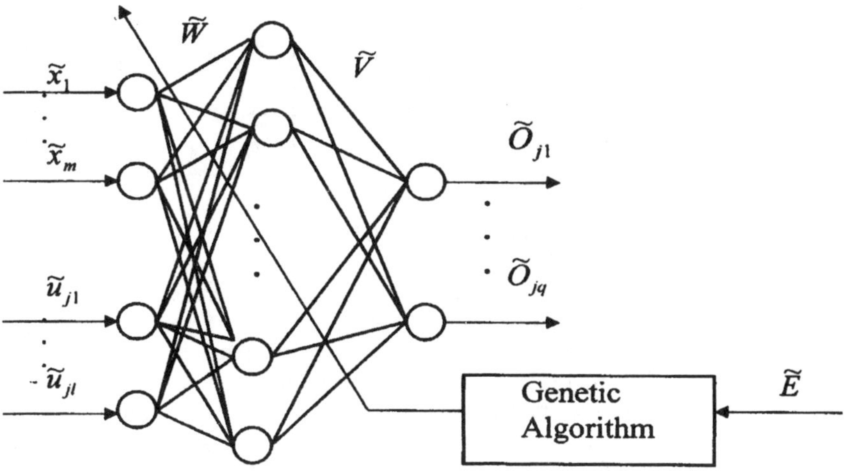

Fig. 3.15. Structure of estimator of agent's solution

Evaluation of Contending Agent's Solutions

The outputs of the estimator are fuzzy numbers $\tilde{O}_{j1}, \tilde{O}_{j2}, \ldots, \tilde{O}_{jq}$, $j=\overline{1,N}$ which describe system outcome values for each agent solution. The fuzzy solutions $\tilde{O}_{j1}, \tilde{O}_{j2}, \ldots, \tilde{O}_{jq}$ must be ranked to determine the best solution. The agent with the best solution will be the "winner" and its solution will be taken as the total solution of the system. The evaluator receives the fuzzy numbers, associated with the agent solutions and performs fuzzy ranking to obtain the ordering for the fuzzy numbers. Output of the evaluator, corresponding to the highest fuzzy value outcome determines the solution. The structure of the evaluator is shown in Fig. 3.16.

Architecture of Soft Computing Intelligent Agent

It is very difficult to precisely define a term like agent [56,100,117,153]. Very often the works on Multi-Agent Systems are criticized for inadequate expression by authors of what the agent is. That is why we give a detailed description of this matter below.

Authors in [138] define an agent as "persistent software entity dedicated to a specific purpose". Here, persistency and purposefulness are emphasized.

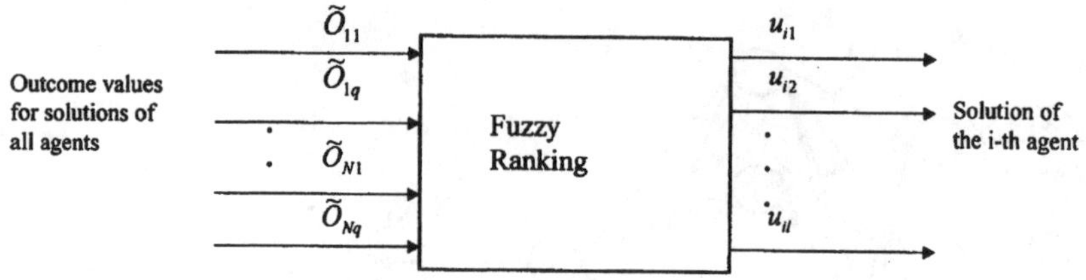

Fig. 3.16. Structure of the evaluator of contending agent

In [132] agents are defined as "computer programs that simulate a human relationship by doing something that another person could do for you". Implicitly, an idea of delegation is viewed as central to the concept of agency here.

In definition given in [135] an agent is any entity to which mental state can be ascribed. Shoham sees mental states (consisting of beliefs, capabilities, and commitments) as the key to defining agents. The Beliefs, Desires, and Intentions model of agents specify that agents have their mental attitudes in terms of what they believe is true, what their desires are, and what they intend to do [82,89,123].

In Russel and Norvig's definition an agent is anything that can be viewed as perceiving its environment through sensors and acting upon that environment through effectors [127]. Here the idea of agent actively involved in interaction with the environment is emphasized.

In [102,103] an agent is defined as a "system that tries to fulfill a set of goals in a complex, dynamic environment. Maes realizes that agents need to have goals to fulfill. Also, the agent environments are complex and dynamic.

Definition given in [35] is similar to the above: "Autonomous agents are systems capable of autonomous, purposeful action in the real world". The author emphasizes autonomy and real world environment for the agents.

In [71] a definition is proposed according to which intelligent agents continuously perform three functions: perception of dynamic conditions in the environment; actions to affect conditions in the environment; and reasoning to interpret perceptions, solve problems, draw inferences, and determine actions. Here the reasoning capability is introduced in order for an agent to be intelligent.

In [56] it is stated that "an autonomous agent is a system situated within and a part of an environment that senses that environment and acts on it, over time, in pursuit of its own agenda and so as to affect what it senses in the future". In this definition the authors point that affecting the environment in the future is an important part of notion of agency.

Along the dimensions of autonomy, learning and cooperation dimensions the autonomous agents that have learning capability are classified as interface agents; those that combine autonomy and cooperation are collaborative agents. The agents that possess all of these characteristics are smart agents [117,152].

A definition similar to [71,117,152] was suggested by us in 1986 [24] and we will use this definition in this work which embraces the following features: autonomy; interaction with an environment and other agents; perception capability;

learning; reasoning capability. In [24] a system with the mentioned characteristics was called a smart system.

The architecture of Soft Computing intelligent agent in accordance with this definition is given in Fig. 3.17. It is a prototype system of intelligent (smart) agents which will be used in the investigated MADIS.

Note that granulation fine, i.e. number of agents in MADIS is determined by the system designer (in our case using genetic algorithm).

The mathematical description of knowledge in the KB of agent is based on fuzzy interpretation of antecedents and consequents in production rules.

For the knowledge representation the antecedent of each rule contains a conjunction of logical connectives <name of object> $\begin{Bmatrix} = \\ \neq \end{Bmatrix}$ <linguistic value> named elementary antecedent.

The consequent of the rule is a list of imperatives, among which may be some operator-functions (i.e. input and output of objects' values, operations with segments of a knowledge base, etc). Each rule may be complemented with a confidence degree $cf \in [0,100]$ and with the author's comments on the rule. Each linguistic value has a corresponding membership function which is built using parametric LR-representation

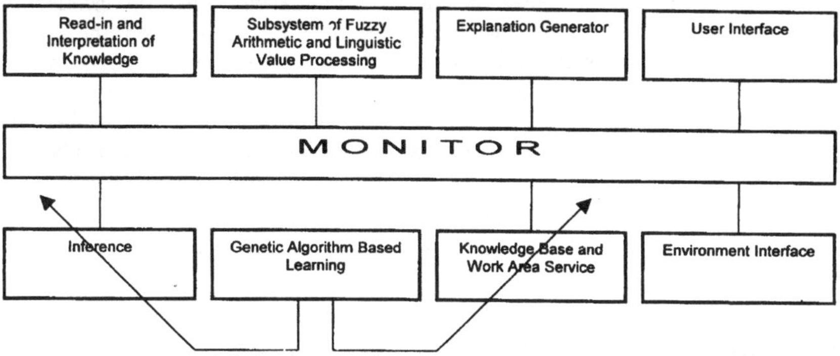

Fig. 3.17. Structure of intelligent agent

The subsystem of fuzzy arithmetic and linguistic values processing (see Fig. 3.17) provides automatic interpretation of linguistic values like "high", "low", "OK", "near...", "from ... to ..." and so on; i.e. for each linguistic value this subsystem automatically computes parameters of membership functions using universes of corresponding variable. The user of the system may define new linguistic values, modify built-in ones and explicitly prescribe a membership function in any place where linguistic values are useful.

Learning of agents is based on genetic algorithms which includes adjusting of agent's KB and choosing appropriate inference mechanism (see Fig. 3.17).

The agents knowledge-base includes certain number of fuzzy rules related through "ALSO":

$$R^k: \text{IF } x_1 \text{ is } A_{k1} \text{ AND } x_2 \text{ is } A_{k2} \text{ AND...AND } x_m \text{ is } A_{km} \text{ THEN}$$
$$u_{k1} \text{ is } B_{k1} \text{ AND } u_{k2} \text{ is } B_{k2} \text{ AND...AND } u_{kl} \text{ is } B_{kl}, k = \overline{1,K} \quad (3.39)$$

where $x_i, i = \overline{1,m}$ and $u_j, j = \overline{1,l}$ are total input and local output variables , A_{ki}, B_{kj} are fuzzy sets, and k is the number of rules. Note, that inputs $x_1, x_2, ..., x_m$ may be crisp or fuzzy variables. If input data are crisp, then fuzzifier will map these data into fuzzy sets. Decision of each agent is made by the composition rule, which is the basis of inference mechanism, as follows:

$$\tilde{U}_j = R_j \circ \tilde{X}, \qquad j = \overline{1,N} \quad (3.40)$$

where $\tilde{U}_j$ is fuzzy value of decision of j-th agent, R_j is the fuzzy relation corresponding to the fuzzy model (3.39), and $\tilde{X}$ is total input information after fuzzification.

Efficiency of inference engine considerably depends on the knowledge base internal organization and logic used for reasoning. Inference mechanism acts as follows. First, some objects take some values (initial data). Then, all production rules, containing each of these objects in antecedent, are chosen from the knowledge base. For these rules the truth degree is computed (in other words, the system estimates the truth degree of the fact that current values of objects correspond to values fixed in antecedents). If the truth degree exceeds some threshold then imperatives from consequent are executed. At that time the same objects as well as a new one take new values and the process continues till work area contains "active" objects ("active" object means untested one).

By using the defuzzifier fuzzy decisions $\tilde{U}_j$ are mapped into the crisp value of solutions.

Optimization of Agents Behavior

Optimality of the objective function of full MADIS is mainly related with solution produced by each agent. Effectiveness of these solutions is determined mainly by knowledge base and inference mechanism of each agent. Optimization of the agents knowledge base means finding of scaling factors, membership functions, and rule set. Scaling factors and membership functions taken together represent the semantics of the symbols used by agent and the rule set represents the syntactic mapping among the symbols [121].

As effectiveness of the solutions of an agent significantly depend on type of inference engine, it is important to use switching engines for approximate reasoning by agents.

It is important to note that not all rules are necessary for the optimal solution generation by agents. So it is necessary to determine which rules and how many should be in the agents' rule base, the centers and shapes of the membership functions of the rules, the value of the scaling factors, and the kind of inference engine for agent use.

To solve this problem we will use genetic algorithm, namely fuzzy-genetic approach. Assume that an agent has Q author pre-specified possible rules, each of which has M parameters (for example characteristic points of triangular membership functions).

At first step, the agents must be numbered. The inclusion of agent into multi-agent system should be represented by a 1 and rejection by a 0. Also, the rules in knowledge base of the agent must be numbered. The presence of rule should be represented by a 1 and the absence by a 0. Assume that number of switching logics for approximate reasoning is q and number of scaling factors is l. Suppose that each parameter needs a r coded array. Then a string of length $N+K+(M+q+l)r$ is formed. First N genes correspond to the size of decomposition (granulation) of total system intelligence (number of agents), K -fuzzy relational matrix. $(M+q+l)$ genes correspond to M parameters, q switching logics and l scaling factors.

After formation of an initial population of T strings each solution in the population is then evaluated to assign a fitness value. In this case, we have assigned a fitness value as

$$\widetilde{J} = [1+I]^{-1} \tag{3.41}$$

where I is fuzzy distance between desirable and actual outcomes of the system.

Since the objective of the considered multi-agent system is to minimize the fuzzy distance I, we use a GA to find a string that corresponds to the maximum fitness value (3.41). GA for learning rule bases is well described in [4,20].

After each solution in the population is evaluated and fitness is assigned, the population is modified by using three operators: selection, crossover and mutation [4]. After a new population of solutions is created, each of them is evaluated again to find a fitness value and all three operators are applied again. These generations proceed until pre-specified number of generations have elapsed.

Investigation on optimization of an agent's behavior by selecting appropriate fuzzy implications, showed that ALI-2 fuzzy logic is more appropriate then other logics described in section 2.1.

Decision Generation through Cooperation and Competition among Intelligent Agents

The objective of the proposed MADIS is to generate such solution which produces the outcome nearest to the desirable value of the system outcome. Therefore, we will consider the following objective function (with one outcome index):

$$I = \sum_{j=1}^{N} \alpha_j \left| \tilde{O}^d - \tilde{O}_j^c \right| \tag{3.42}$$

$$\sum_{j=1}^{N} \alpha_j = 1, \quad \alpha_j = \{0,1\}, \tag{3.43}$$

where $\tilde{O}^d$ is fuzzy desirable value of the outcome index, $\tilde{O}_j^c$ is the current fuzzy value of the j-th agent's outcome index, α_j, $j=\overline{1,N}$ is the relative weight values for the contending agents. In the case of very complex unstructured systems, a designer may not have information about the desirable value of the system outcome. To maximize the actual outcome index we may take a large fuzzy number (larger than expected value of $\tilde{O}_j^c$) as the value of $\tilde{O}^d$. Minimization of objective function (3.42) under conditions (3.43) is achieved through competition of agents with each other for the right to produce the total solution such that provides $\tilde{O}_j^c$ close to $\tilde{O}^d$.

For generation of solutions minimizing the objective function (3.42) each contending agent produces its own total solution as follows [11,119,160]:

1. Each rule in (3.39) is formalized in the form of fuzzy relation R_{kj}, $k=\overline{1,K}$, defined as

$$R_{kj} = \int \propto_{A_K}(x) * \propto_{B_K}(u_{kj}), \tag{3.44}$$

where $*$ is fuzzy implication operator

2. The set of fuzzy rules (3.39) is formalized using (3.44) in the form of composed fuzzy relation

$$R_j(X, U_j) = ALSO(R_{1j}, R_{2j}, \ldots, R_{Kj}) \tag{3.45}$$

with the membership function

$$\propto_{R_j}(u_{1j}, u_{2j}, \ldots, u_{lj}, x_1, x_2, \ldots, x_m) = \max\left[\propto_1(\cdot), \propto_2(\cdot), \ldots, \propto_K(\cdot)\right] \tag{3.46}$$

3. The generation of agent's solution is performed on the basis of resulting fuzzy relation R_j by applying the compositional rule of inference

$$\tilde{U}_{jr} = \bigcap_{i=1}^{m} \tilde{X}_i^c \circ R_j(X, U_{jr}) \qquad j=\overline{1,N}, \quad i=\overline{1,m}, \quad r=\overline{1,l} \tag{3.47}$$

where $\circ$ is a compositional operator and r is the index of output variable (solution).

To evaluate the agents' solutions $\tilde{U}_{jr}$, it is necessary to estimate the outcomes of the system in terms of their ability to minimize objective function (3.42) of the full system. Fuzzy neural networks with fuzzy and crisp inputs and fuzzy weights (Fig. 3.15) is used to estimate solutions. Fuzzy inputs and weights may be described by different shapes, such as triangular, trapezoidal and Gaussian membership functions traditionally used in fuzzy logic. We use triangular membership functions to describe fuzzy inputs and weights. For learning of fuzzy neural network, i.e. tuning of weights, we use fuzzy arithmetic and genetic algorithms. Also, we use fuzzy arithmetic in testing mode of neuro-fuzzy estimator to calculate system outcomes.

Learning of Fuzzy-Neural Network the fuzzy distance between actual output values $\tilde{O}_j$ and desired output values $\tilde{O}^d$ is calculated by the formula:

$$\tilde{E}_j = \sum_{s=1}^{s}\sum_{r=1}^{q} \left| \tilde{O}^d_{jrs} - \tilde{O}_{jrs} \right|, \qquad (3.48)$$

where q is the number of outcomes related with agents, s is the number of learning pairs. Note, that learning of FNN is performed independently of agents' solutions.

The fitness function that we use is given by (3.49):

$$\tilde{J} = \left[1 + \tilde{E}\right]^{-1} \qquad (3.49)$$

Each fuzzy weight is described by triangular membership function characterized with three parameters. Hence, if the number of adjusting weights is q, then we must encode $3q$ parameters. Each parameter takes value from a domain $D_i = [a_i, b_i]$.

The GA generates the optimal fuzzy membership function (in accordance with (10)) of weights as follows:

1. The GA starts its work, $t = 1$.
2. The initial population of candidate solutions is created.
3. The population of generation at time t, $G(t)$, is evaluated.
4. If some termination conditions are met, go to step 8.
5. Generate new generation $G(t+1)$ from $G(t)$. Then crossover and mutation are applied.
6. Evaluate $G(t+1)$.
7. Return to Step 4.
8. The termination of GA's work.
9. Among all the minimum errors stored in memory the smallest one is found. Membership functions by which the smallest error is obtained, are selected.
10. End.

Calculation of Fuzzy Value of System Outcome Indices the trained fuzzy neural network based estimator receives fuzzy and/or crisp input data and solution values and using fuzzy arithmetic [87] produces corresponding fuzzy outputs

$$\tilde{O}_{j1}, \tilde{O}_{j2}, \ldots, \tilde{O}_{jq}, \quad j = \overline{1, N} \quad .$$

Evaluation of Agents' Solutions and Selection of Total Problem Solution the evaluator (see Fig. 3.14) receives all the system outcome fuzzy values, corresponding to the solutions of each contending agent $\tilde{O}_1, \tilde{O}_2, \ldots, \tilde{O}_{N}$., and evaluates them according to the objective function (3.42). For this purpose we use fuzzy ranking procedure for N fuzzy numbers to give a total ordering based on compatibility measure of fuzzy sets [21,34,40,133]. Each fuzzy outcome value, corresponding to agents solution, is presented by fuzzy number described with triangular membership function. The determining of the best $\tilde{O}_j$ in terms of the objective function is performed by the following procedure.

1. Preparing of initial data by creating Table 3.7.
2. Calculation of the compatibility measure. As the compatibility measure we use the Jaccard index. The Jaccard compatibility measure can be presented by $N \cdot N$ relation $E_{\geq}$.
3. Calculation of the crisp ranking relation $R_{>}$ using operator">".

Table 3.7. Outcome fuzzy values

N	Set Name	Parameters
1	$\tilde{O}_1$	[a1, b1, c1]
2	$\tilde{O}_2$	[a2, b2, c2]
...	...	...
N	$\tilde{O}_N$	[aN, bN, cN]

4. Configuration of the order vector $O = [O_j]$ by summing the elements in each row of $R_{\succ}$. First component of vector O will be selected as nearest to the desirable value of system outcome in terms of (3.42). The solution that produced this outcome is accepted as system solution. For example, if j-th agent produced the nearest to the desirable solution, then this j-th agent receives the weight $\alpha_j = 1$, and rest $\alpha_i, i \neq j, i = \overline{1, N}$ receive weights $\alpha_i = 0.0$

3.7 Computing With Words Technology

The methodology of computing with words (CW) is an important event in the evolution of cognitive science, natural language processing, artificial intelligence, and different existing scientific theories. This is because CW can enrich the existing scientific theories and the above-mentioned science fields giving them the capability of using natural languages to operate on perception-based information, not only measurement-based information [148]. Indeed in many real-world problems in natural sciences as well as in industrial engineering, economics, and business, often there is a need to deal with both perception and measurement based information. In the case of perception based information, the available information is not precise enough to justify the use of numbers. Such information is usually described in natural languages rather than in strict (idealized) mathematical expressions. So a strong need has appeared for a new approach, theory and technology for the development of knowledge representation, computing, and automated reasoning tools that allow creation of systems with high MIQ. Such tools especially are of central importance for many of the postmodern information technologies [92]. A few of them that need special mention are intelligent information systems, knowledge discovery, intelligent control, and agent theory.

CW is a methodology in which words play the role of labels of perceptions expressed in natural language.

The origin of CW technology is related to the fundamental works by Prof. L.Zadeh: [160] on linguistic variable and Theory of Fuzzy Information Granulation (TFIG), [164-166] on fuzzy constraint and fuzzy constraint propagation, and [167, 168] on meaning representation and test-score semantics. However, CW as a distinct methodology was formulated in later works by Prof. L.Zadeh [73, 169-171]. The phases of evolution of CW into distinct methodology is schematically represented in Fig.3.18.

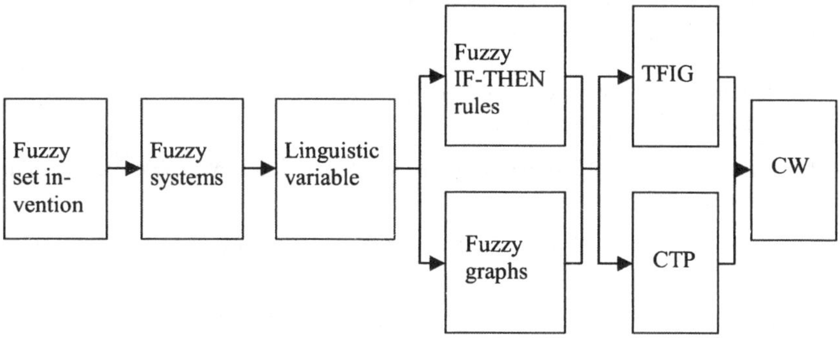

Fig. 3.18. The evolution of CW

At present, we can see the growing increase in the number of works devoted to CW technology both in scientific journals and conference proceedings. Book [148] is the first in the field of theory and technology of CW. This book encloses articles embracing a wide amount of works on the development, explanation, and prediction of the potential applications of the computing with words ranging from natural language processing and cognitive science to information systems and control. In [92] an approach to computing with words on an alternative interpretation of linguistic variables with using mass assignment theory is developed. A methodology for converting probabilistic inference into linguistic inference is suggested.

In the traditional sense, we understand "computing" as the manipulation with numbers and symbols. However, humans use primarily words in computing and reasoning. Realizing that the artificial intelligent systems created by human are still far to compare with human abilities in many spheres of intelligent activity, every step in science in this direction may get us close to creation of systems with an intelligent power similar to natural brain. In this direction, the CW theory and technology are the main way. As L.Zadeh stated "the role model for CW is the human mind" [163]. By L.Zadeh, computing with words is a methodology in which words play the role of labels of perceptions expressed in a natural language.

There are two main reasons for why we need CW. The first is when the available information is not precise enough to use numbers. The second is when there is a tolerance for imprecision, uncertainty, and partial truth. The main starting point for CW is the concept of granule. A word *w* is a label of a granule *g*. In CW a granule is viewed as a fuzzy constraint on a variable. In CW, the information is conveyed by constraining the values of variables. Natural words play the role of labels of fuzzy granules. So, we are given information a collection of propositions (viewed as an implicit fuzzy constraint on an implicit variable) expressed in a natural language. This is the initial data set (IDS) – input. On the base of perception input information, it is necessary to infer an answer to a query expressed in a natural language. Here we can evidently see the relationship between CW and Computational Theory of Perceptions (CTP). By CTP perceptions and queries are expressed as propositions in a natural language. The answer is identified as output (terminal data set – TDS) and also is a collection of propositions expressed in natural language. In order to derive the output from the input, the propositions and queries are processed by CW-based methods. These methods are based on the rule of inference in fuzzy logic, which is used for constraint propagation from premises to conclusion [169]. Thus, the solution of a given problem by CW technology boils down to the following [172]:

1. Propositions in natural language for verbal description of a problem domain are transformed into antecedent constraints.
2. By use of rules of constraints propagation, the antecedent constraints are transformed into consequent constraints.
3. Consequent constraints are transformed into a natural language by use of linguistic approximation.

Below we will briefly characterize each of these phases. Zadeh's approach to the representation of fuzzy constraints is based on test-score semantics. According to this semantics, a proposition of natural language is considered as a network of fuzzy constraints. The latter can be represented as X is R, where R is a constraining fuzzy relation and X is the constrained variable. As it is shown in [172], the expression in question is the canonical form of p. The canonical form is necessary for explicitation of the fuzzy constraints in evidence which is implicit in proposition. It is necessary to note that a constraint expressed as a canonical form may be conditional

$$\text{IF X is R THEN Y is S.}$$

The constraints in question are the basic constraints. For proposition meaning representation in general case the generalized constraints are used [173], which contain the basic constraints as a special case. A generalized constraint is represented as [172] X isr R. Here r is a variable copula and r is discrete variable, which defines the way in which R constraints X. Depending on the type of value of r in the practice of CW, the following constraints are possible [172]: equality,

$r = e$; possibilistic, $r =$ blank; veristic, $r = v$; probabilistic, $r = p$; probability value constraint, $r = \lambda$; random set type, $r = rs$; fuzzy graph constraint, $r = fg$.

Let's add a few words about fuzzy constraint propagation. It is necessary to note that fuzzy constraint propagation rules in principle are the same as fuzzy inference rules. With this in mind, the well-known compositional rule, generalized modus ponens, syllogistic rule and others usually are used.

To conclude the methodology of CW, it necessary to note the following. The processes of explicitation and constraint propagation play the key role in obtaining TDS from IDS. As can be evidently drawn from the above discussion, the explicitation and constraint propagation are considered as translation of propositions expressed in a natural language into generalized constraint language (GCL). Then, the rule of constraint propagation is applied to expressions in this language.

References

1. Alander JT (1992) On optimal population size of genetic algorithms. In: IEEE Computer Society press, NY, pp 65-70
2. Aliev RA (1997) Special representation: Basic principles of Soft Computing and its components. Atlanta, USA: GSU
3. Aliev RA, Aliev RR (1997) Fuzzy Distributed Intelligence System for Continuous production. Application of Fuzzy Logic. In: Jamshidi M, Titli M, Zadeh L, Bevrie S (eds.) Towards High machine Intelligence Quotient Systems. Prentice Hall PTR, Upper Sadle River, New Jersey, USA
4. Aliev RA, Aliev RR (2001) Soft Computing and its Application. World Scientific, New Jersey, London, Singapore, Hong Kong
5. Aliev RA, Aliev RR (1996) Soft Computing, vol I. ASOA Press, (in Russian) Baku
6. Aliev RA, Aliev RR (1997) Soft Computing, vol III. ASOA Press, (in Russian) Baku
7. Aliev RA, Aliev RR (1997) Soft Computing, vol II. ASOA Press, Baku
8. Aliev RA, Aliev RR (2001) Theory Intellectual Systems and its applications. Chashioglu, Baku
9. Aliev RA, Aliev RR (1996) Soft Computing based control system for catalytic cracking processes. In: Biennial Conference of the North American Fuzzy Information Processing Society-NAFIPS'96, Berkley, California, USA, pp 387-391
10. Aliev RA, Aliew FT, Babaev MD (1991) Fuzzy process control and knowledge engineering. Verlag TUV Rheinland, Koln
11. Aliev R, Aliew F, Babaev M (1991) Fuzzy Process Control And Knowledge Engineering In Petrochemical And Robotic Manufacturing. Verlag TUV Rheinland, Koln
12. Aliev RA, Abilov YA, Nasirov IM (1994) Theory of possibility and group principle in genetic algorithms. In: Aliev RA , Kenarangui R (eds) Inter. Conf. of Application of Fuzzy System. ICAFS-94, Tabriz
13. Aliev RA, Abilov YA, Nasirov IM (1995) Use of Fuzzy approach in Genetic Algorithms for tuning the Search Directions. In: Third European Congress on Intelligent Techniques and Soft Computing, EUFIT 95, Aachen, Germany, vol 1, pp 492-494

14. Aliev RA, Bonfig H, Aliev FT, Aliev RR (1993) The Distributed Intelligent Manufacturing Systems. In: First European Congress on Fuzzy and Intelligent technologies, EUFIT, Aachen, Part II, Germany, pp 229-235
15. Aliev R, Bonfig K, and Aliew F (2000) Soft Computing. Verlag Technik Berlin
16. Aliev RA, Bonfig KW, Aliev FT (1994) Messen, Steuern und Regeln mit Fuzzy Logik. Franzis-Verlag, (in German)
17. Aliev RA, Bonfig KW, Shirinova UK (1998) Genetic algorithm based fuzzy modeling. In:Third International Conference on Application Fuzzy Systems and Soft Coputing-ICAFS'98. Wiesbaden, Germany, pp 118-123
18. Aliev RA, Fazlollahi B,Vahidov R (1998) Learning of fuzzy neural networks by using genetic algorithms. In: World Automatoin Congress. Advances in Robotics, Manufacturing, Automation, Control, Soft Computing, Multimedia and Image Processing. Anchorage, Alaska, pp 139.1-139.8
19. Aliev RA, Fazlollahi B, Vahidov RM (2000) Genetic Algorithm Based Learning of Fuzzy Neural Networks, Part I, Feed-Forward Fuzzy Neural Networks. J *Fuzzy Sets and Systems*
20. Aliev RA, Fazlollahi B, Vahidov RM (2000) Soft computing based multi-agent marketing decision support system. J Intellegence and Fuzzy Systems 9: 1-9
21. Aliev RA , Mamedova GA, Aliev RR (1993) Fuzzy Sets Theory and its Application. Tabriz University, Tabriz
22. Aliev RA, Rashad RA, Aliev FT (1997) Fuzzy Distributed Multi-Agent Manufacturing System. In: NAFIPS 97, Syracuse, New-York, pp 311-316
23. Aliev RA, Shirinova UK (1997) Fuzzy objects modeling by genetic algorithms. Intelligent Control and Decision Making Systems. (Thematic Scientific Works Collection). Baku: ASOA, 1: pp90-105
24. Aliev RA, Tsercovny AE (1988) "Smart" manufacturing systems (in English and Russian). J News of Academy of Sciences of USSR,Tech. Cybernetics 6: 99-108
25. Aliev RA, Vahidov R, Aliev RR (1995) Artificial Neural Networks. Theory and practice. Tabriz University Press, Tabriz
26. Arlabosse F(1994) ARCHON and its Environment. In: Perram JW, Muller JP (eds.) Distributed Software Agents and Applications: MAAMA W'94. Berlin, Germany: Springer- Verlag, pp.11-18
27. Balabanowic M (1997) An Adaptive Web Page Recommendation Service. In: First International Conference on Autonomous Agents. New York: ACM, pp378-385
28. Belew R, Mcinerney J, Scraudolph NN (eds) (1980) Evolving Networks: Using the Genetic Algorithm with Connectionist Learning. (CSE Technical Report CS90-174).San Diego: University of California
29. Bonarini A, Basso F (1997) Learning to Compose Fuzzy Behaviors for Autonomous Agents. J International Journal of Approximate Reasoning 17: 409-432
30. Bonarini, A (1993) ELF: Learning Incomplete Fuzzy Rule Sets for an Autonomous Robot. In: European Conference on Fuzzy and Intelligent Technologies (EUFIT'93). Aachen, Germany: ELITE Foundation, pp 69- 75
31. Bonarini A (1994) Learning to Coordinate Fuzzy Behaviors for Autonomous Agents. In: European Conference on Fuzzy and Intelligent Technologies (EUFIT'94) Aachen, Germany: ELITE Foundation, pp 475-479
32. Bond AH, Gasser L (1988) Readings in Distributed Artificial Intelligence. Morgan Kaufmann, San Mateo, Ca

33. Bond AH, Gasser L (1992) A subject -indexed bibliography of distributed artificial intelligence (DAI). J IEEE Transactions on Systems, Man and Cybernetics 22(6): 1260-1281
34. Bortolan G, Degani R (1985) A review of some methods for ranking fuzzy subsets. J Fuzzy Sets and Systems 15: (1-19)
35. Brustoloni JC (1991) Autonomous Agents: Characterization and Requirements (Carnegie Mellon Technical Report CMU-CS-91-204). Carnegie Mellon University.
36. Buckley JJ, Hayashi Y(1994) Fuzzy Neural Networks. In:Yager R, Zadeh L(eds) Fuzzy Sets, Neural Networks, and Soft Computing. Van Nostrand Reinhold, pp 233-249
37. Caundel TP, Dolan CP (1989) Parametric connectivity: Training of constrained networks using genetic algorithms. In: Schaffer J.D.(ed) Third Inter. Conf. on Genetic Algorithms. San Mateo, Calif.· Morgan Kaufmann, pp 370-374
38. Chavez A, Dreilinger D, Guttman R, Maes P (1997) A Real-life Experiment in Creating an Agent Marketplace. In: Nwana HS, Azarmi N (eds.) Software Agents and Soft Computing. Berlin, Germany: Springer- Verlag, pp160-179
39. Chavez A, Moukas A, Maes P (1997) Challenger: A Multi-Agent System for Distributed Resource Allocation. In: First International Conference on Autonomous Agents. New York: ACM, pp 323-321
40. Chen SJ, Hwang CL (1991) Fuzzy Multiple Attribute Decision making: Methods and Applications. Springer-Verlag, Heidelberg
41. Clack Ch, Farringdon J, Lidwell P, Yu T (1997) Autonomous Document Classification for Business. In: First International Conference on Autonomous Agents. New York: ACM, pp 201-208
42. Cordon F, Herrera M (1997) On the combination of fuzzy logic and evolutionary computation. In: Pedrycz W (ed) Fuzzy evolutionary computation USA: Kluwer Academic Publishers, pp 33-56
43. Corkill DD, Gallagher KQ, Johnson PhM (1987) Achieving Flexibility, Efficiency, and Generality in Blackboard Architectures. In: Conference of the American Association for Artificial Intelligence, pp 18-23
44. Corkill DD, Gallagher KQ, Murray KE (1986) A Generic Blackboard Development System. In: Conference of the American Association for Artificial Intelligence, pp 1008-1014
45. Das SK, Fox J, Elsdon D, Hammond P (1997) Decision Making and Plan - Management by Autonomous Agents: Theory, Implementation and Applications. In: First International Conference on Autonomous Agents. New York: ACM, pp276-283
46. Davies NJ, Weeks R, Revett MC (1997) Information Agents for the World Wide Web. In: Nwana HS, Azarmi N (eds) Software Agents and Soft Computing. Berlin, Germany: Springer- Verlag, pp 81-99
47. Davis R, Smith RG (1983) Negotiation as a Metaphor for Distributed Problem Solving. J Artificial Intelligence 20(1): 63-109
48. Decker KS, Lesser VR (1995) Designing a Family of Coordination Algorithms. In: First International Conference on Multi-Agent Systems. San-Francisco, CA: AAAI Press/MIT Press, pp 173-180
49. Denzinger J (1995) Knowledge-Based Distributed Search Using Teamwork. In: First International Conference on Multi-Agent Systems. San-Francisco, CA: AAAI Press/MIT Press, pp 81-88

50. Doorenbos RB, Etzioni 0, Weld DS (1997) A Scalable Comparison-Shopping Agent for the World-Wide Web. In: First International Conference on Autonomous Agents. Marina del Rey. New York: ACM, pp 39-48
51. Durfee EH, Lesser VR (1987) Using Partial Global Plans to Coordinate Distributed Problem Solvers. In: International Joint Conference on Artificial Intelligence, pp 875-873
52. Dubois D, Prade H (1980) Theory and Applications. J Fuzzy Sets and Systems. NY: Academic Press
53. Fatehi A, Lucas C (1994) Optimization of the Rule Base and Parameters of a Fuzzy Logic Controller by Genetic Algorithms. In: Aliev RA, Kenarangui R (eds) Inter. Conf. of Application of Fuzzy System. ICAFS-94.Tabriz, pp 236-240
54. Fazlollahi B, Aliev RA, Vahidov RM (2000) Multi-agent distributed intelligent systems based on fuzzy decision-making. J International Journal of Intelligent Systems, vol 15: 849-858
55. Fischer K, Muller JP, Pischel M, Scier D (1995) A Model for cooperative Transportation Scheduling. In: First International Conference on Multi-Agent Systems. San-Francisco, CA: AAAI Press/MIT Press, pp109-116
56. Franklin S, Graesser A (1997) Is it an Agent, or Just a Program?: A Taxonomy for Autonomous Agents. In: Muller JP, Wooldridge MJ, Jennings NR (eds) Intelligent Agents III: Agent Theories, Architectures, and Languages. Berlin, Germany: Springer Verlag, pp 21-36
57. Gasser L, Huhns M (eds) (1989) Distributed Artificial Intelligence, vol II. Morgan Kaufmann, San Mateo, California, pp 259-290
58. Georgeff MP, Lansky AL (1987) Reactive Reasoning and Planning. In: 6th National Conference on Artificial Intelligence. pp 677-682
59. Georgeff MP, Ingrand FF (1989) Decision-making in an Embedded Reasoning System. In: 11th International Joint Conference on Artificial Intelligence. pp 972- 978
60. Grand S, Cliff D, Malhotra A (1997) Creatures: Artificial Life Autonomous --2]Software Agents for Home Entertainment. In: First International Conference on Autonomous Agents. New York, N.Y.: ACM, pp 22-29
61. Grefenstette JJ (1986) Optimization of control parameters for genetic algorithms. J IEEE Trans. Systems, Man, and Cybernetics, SMC – 16: 122-128
62. Gupta M, Ding H (1994) Foundations of Fuzzy Neural Computations. In: Aminzadeh F, Jamshidi M (eds) Soft Computing: Fuzzy Logic, Neural Networks, and Distributed Artificial Intelligence. NJ: PTR Prentice Hall, pp 165-195
63. Gupta MM, Knof JK (1990) Fuzzy neural network approach to control systems. In: First Inter. Symp. Uncertainty Model. Anal., Maryland, UO, pp 483-488
64. Gupta MM, Qi J (1991) Fusion of fuzzy logic and neural networks with application to decision and control problem. In: NAFIPS 91. Columbia, UO, pp 327-328
65. Gupta MM, Qi J (1991) On fuzzy neuron models. In: Inter. Joint Conf. Neural Networks (IJCNN 91).Seattle, pp 431-436
66. Gupta MM (1992) Fuzzy logic and neural network. In: Second Inter. Conf. Fuzzy Logic Neural Networks (IIZUKA 92). Iizuka, Japan, pp 157-160
67. Hayashi J, Buckley JJ, Czogala E (1992) Direct fuzzification of neural network and fuzzified delta rule. In: Proc. Second Inter. Conf. Fuzzy Logic Neural Networks (IIZUKA 92). Iizuka, Japan, pp 73-76
68. Hayashi J, Buckley JJ, Gzogala E (1993) Fuzzy neural network with fuzzy signals and weights. J Int. J. of Intelligent Systems 8: 527-537

69. Hayes-Roth B, Van Gent R (1997) Story Making with Improvisational Agents. In: First International Conference on Autonomous Agents. New York, N.Y.: ACM, pp 1-7
70. Hayes-Roth B (1985) A Blackboard Architecture for Control. J Artificial Intelligence 26: 251-321
71. Hayes-Roth B (1995) An Architecture for Adaptive Intelligent Systems. J Artificial Intelligence 72: 329-365
72. Haynes Th, Sen S, Arora N, Nadella R (1997) An Automated Meeting Scheduling System. In: First International Conference on Autonomous Agents. New York, N.Y.: ACM, pp 308-315
73. Hampel R, Wagenknecht M, Chaker N (eds) (2000) Fuzzy Control. Theory and Practice. Physica-Verlag. A Springer-Verlag Company
74. Herrera F, Lozano M, Verdegay JL (1995) The use of fuzzy connectives to design real coded genetic algorithms. J Mathware and Soft Computing 1(3): 239-251
75. Horikawa S, Furuhashi T, Uchikava Y. (1992) On fuzzy modeling using fuzzy neural networks with the back-propagation algorithm. J IEEE Trans. on Neural Networks 3(5): 801-806
76. Hornic K (1991) Approximation capabilities of multiplayer feedforward network. J Neural Networks 4: 251-257
77. Ishibuchi H, Tanaka H (1991) Regression analysis with internal model by neural net. In: Inter. Joint Conf. Neural Networks (IJCNN 91). Seattle, pp 1594-1599
78. Ishibuchi H, Fujioka R, Tanaka H (1992) An architecture of neural networks for input vectors of fuzzy numbers. In: Proc. IEEE Inter. Conf. Fuzzy syst.(FUZZ-IEEE'92). San Diego, pp 1293-1300
79. Ishibuchi H, Okada H, Tanaka H. (1993) Fuzzy neural networks with fuzzy weights and fuzzy biases. In: ICNN 93. San Francisco, pp 1650-1655
80. Ishubuchi H, Morioka K, Tanaka H (1994) A Fuzzy Neural Network with Trapezoid Fuzzy Weights. J IEEE : 228-233
81. Jamshidi M (1997) Large-Scale Systems: Modeling, Control, and Fuzzy Logic. Prentice Hall, NJ
82. Jennings NR (1992) On being Responsible. In:Demazeau Y,Werner E (eds.) Decentralized AI, 3. Amsterdam, Holland: Elsevier, pp 93-102
83. Jong JSR (1992) Self learning fuzzy controllers based on temporal back-propagation. J IEEE Trans. on Neural Networks 3(5): 714-723
84. Jong Y, Liang W, Reza L (1997) Multiple Fuzzy Systems for Function Approximation. In: NAFIPS 97. Syracuse, New-York, pp 154-159
85. Kamel M, Chenniwa H (1994) Coordination of distributed intelligent systems In Soft Computing: Fuzzy Logic, Neural Networks, and Disributed Artificial Intelligence. Prentice Hall, pp 261-297
86. Karr CL, Stanley DA (1991) Fuzzy logic and genetic algorithms in time varying control problems. In: NAFIPS-91. pp 285-290
87. Kaufmann A, Gupta M M (1985) Introduction to Fuzzy Arithmetic. New York: Van Nostrand
88. Khan NA, Jain R (1985) Uncertainty Management in a Distributed Knowledge Base system. In: International Joint Conference on Artificial Intelligence. pp 318-320
89. Kinny D, Georgeff M (1996) Modelling and Design of Multi-Agent Systems. In: Muller JP, Wooldridge MJ, Jennings NR (eds) Intelligent Agents III: Agent Theories, Architectures, and Languages. Berlin, Germany: Springer Verlag, pp.1-20

90. Kiszka JB, Kochanska ME, Sliwinska DS (1985) The influence of some fuzzy implication operators on the accuracy of a fuzzy model, Part I. J Fuzzy Sets and Systems, vol 15: 111-128
91. Kusiak A (ed) (1988) Artificial Intelligence: Implication for CIM. Springer-Verlag, Berlin-Heidelberg, New-York-London-Paris-Tokyo
92. Lawry J (2001) A methodology for computing with words. J International Journal of Approximate Reasoning, 28, pp 51-89
93. Lee M, Lee SY, Park SH (1992) Neuro-fuzzy identification model of nonlinear dynamic systems. In: Second Inter. Conf. on Fuzzy Logic and Neural Networks, vol 1, pp 485-488
94. Lee M, Lee SY, Park SH (1994) Neuro-Fuzzy Identifiers and Controllers. J Intelligent and Fuzzy Systems, vol 2(1): 1-14
95. Lee MA, Takagi H (1993) Dynamic Control of Genetic Algorithms using Fuzzy Logic Techniques. In: Fifth Inter. Conf. on Genetic Algorithms. Urbana-Champaign, pp 76-83
96. Lee SC, Lee ET (1974) Fuzzy neural networks. J Cybernet 4: 23-103
97. Lee SC, Lee ET (1975) Fuzzy neural networks. J Math. Biosci 23: 151-171
98. Li YF, Lan CC (1989) Development of fuzzy algorithm for servo systems. J IEEE Contr. Syst. Mag.: 65-72
99. Luck M, D'Inverno M (1995) Engagement and Cooperation in Motivated Agent Modeling. In DIstibuted Artificial Intelligence.Architecture and Modeling. First Australian Workshop on DAI, Canberra, ACT, Australia, 70-84
100. Luck M, D'lnverno M (1995) A Formal Framework for Agency and Autonomy. In:First International Conference on Multi-Agent Systems. San- Francisco, CA: AAAI Press/MIT Press, pp 254-260
101. Maes P (1994) Agents that Reduce Work and Information Overload. J Communications of the ACM 37(7): 31-40, 146
102. Maes P (1995) Artificial Life Meets Entertainment: Life like Autonomous Agents. J Communications of the ACM 38(11): 108-114
103. Maes P (1995) Modeling Adaptive Autonomous Agents. In: Langton CG (ed) Artificial Life: An Overview. Cambridge, MA: The MIT Press, pp 135-162
104. Malone TW, Fikes RE, Grant KR, Howard MT (1988) Enterprise: A Market-like Task Scheduler for Distributed Computing Environments. In: Huberman B (ed) The Ecology of Computation. Amsterdam, Holland: North-Holland
105. Meredith DL, Kumar KK, Karr CL (1991) The use of genetic algorithms in the design of fuzzy logic controllers. In: Proc. WNN-AIND 91. pp 695-702
106. Michalewicz Z (1996) Genetic Algorithms+ Data Structures = Evolution Programs. Berlin: Springer-Verlag
107. Miller GF, Todd PM, Hedg SV (1989) Designing neural Networks using genetic algorithms. In: Schaffer JD (ed) Proc. of the Third Inter. Conf. on Genetic Algorithms. San Mateo, Calif.: Morgan Kaufmann, pp 379-384
108. Miller P (1986) Expert Critiquing Systems: Practice-Based Medical Consultation By Computer. Berlin, Germany: Springer-Verlag
109. Mohamad H Hassoun (1995) Fundamentals of Artificial Neural Networks. Cambridge, Massachusetts: MIT Press
110. Montana DJ, Davis L (1989) Training feedforward networks using genetic algorithms. In: Sridhara NS(ed) Eleventh Inter. Joint Conf. on Artifical Intelligence. San Mateo, Calif.: Morgan Kaufmann, pp 762-767

111. Moukas A, Giorgos Z (1997) Evolving a Multi-Agent Information Filtering Solution in Amalthaea. In: First International Conference on Autonomous Agents. New York, N.Y.: ACM, pp 394-403
112. Moulin B, Cloutier L (1994) Collaborative work based on multiagent architectures: a methodological perspective In: Aminzadeh F, Jamshidi M (eds) Soft Computing: Fuzzy Logic, Neural Networks, and Disributed Artificial Intelligence. Prentice Hall, pp 261-297
113. Mukaidono M, Yamaoka M (1992) A learning method of fuzzy inference rules with neural networks and its application. In: Proc. of the Inter. Conf. on Fuzzy Logic and Neural Networks, pp 185-187
114. Munday C, Dangedej J, Cross T, Lukose D (1995) Motivation and Perception Mechanisms in Mobile Agents for Electronic Commerce. In: Distributed Artificial Intelligence. Architecture and Modeling. First Australian Workshop on DAI, Canberra, ACT, Australia, pp 144-158
115. Nakamati G, Freitas R, Predo J, Gomide F (1994): "Fuzzy Distributed Artificial Intelligent Systems", IEEE, pp. 462-467
116. Neves MC, Oliveira E (1997) A Control Architecture for an Autonomous Mobile Robot. In: First International Conference on Autonomous Agents: New York, N.Y.: ACM, pp 193-200
117. Nwana HS, Ndumu DT (1997) An Introduction to Agent Technology. In:. Nwana HS, Azarmi N (eds) Software Agents and Soft Computing. Berlin, Germany: Springer-Verlag, pp 3-26
118. O'Brien PD, Wiegand ME (1997) Agents of Change in Business Process Management. In: Nwana HS, Azarmi N (eds) Software Agents and Soft Computing). Berlin, Germany: Springer- Verlag, pp132-145
119. Park D, Kandel A, Langholz G (1997) Genetic-based new fuzzy reasoning models with application to fuzzy control. J IEEE Transactions on Systems, Man, and Cybernetics - Part B: Cybernetics 24(1): 39-47
120. Park D, Kandel A, Laugholz G (1994) Genetic-based new fuzzy reasoning models with application to Fuzzy Control. J IEEE Trans. Systems, Man, and Cybernetics: 39-46
121. Pratihar DK, Deb K, Grosh A (1999) A genetic-fuzzy approach mobile robot navigation among moving obstacles. J International Journal of Approximate Reasoning 20: 145-172
122. Procyk TJ, Mamdani EH (1979) A linguistic self-organizing process controller. J Automatica, vol 15: 15-30
123. Rao AS, R Georgeff MP (1995) BDI Agents: From Theory to Practice. In: First International Conference on Multi-Agent Systems. San-Francisco, CA: AAAI Press/MIT Press, pp 312-319
124. Reeves CR (1993) Using genetic algorithms with small populations. In: Forrest S (ed) Proc. of the Fifth Inter. Conf. on Genetic Algorithms. San Mateo, Calif.: Morgan Kaufmann, pp 92-99
125. Requena I, Delgalo U (1992) A model of fuzzy neuron. In: Second Inter. Conf. Fuzzy Logic Neural Networks (IIZUKA 92). Iizuka, Japan, pp 793-796
126. Rosenschein JS, Genesereth MR (1985) Deals Among Rational Agents. In: International Joint Conference on Artificial Intelligence. pp91-99
127. Russel SJ, Norvig P (1995) Artificial Intelligence: A modern Approach. Upper Saddle River, N.J.: Prentice Hall

128. Sanchez E (1992) Genetic algorithms, neural networks and Fuzzy Logic systems. In: Second Inf. Conf. on Fuzzy Logic and neural networks (IIZUKA'92), vol 1. Japan, pp 17-19
129. Sanchez E, Shibata T, Zadeh L(eds) (1997) Genetic Algoritms and Fuzzy Logic Systems, Soft Computing Perspectives. Singapore: World Sceientific
130. Schaffer JD, Caruana RA, Eshelman LJ, Das R (1989) A study of control parameters affecting online performance of genetic algorithms for function optimization. In: Schaffer JD(ed) Third Inter. Conf. on Genetic Algorithms and their Application. San Mateo, Calif.: Morgan Kaufmann, pp 51-60
131. Scharf EM, Mandic NJ (1989) The application of a fuzzy controller to the control of a multi-degree-freedom robot arm. J Industry Application of Fuzzy Control: 41-62
132. Selker T (1994) A Teaching Agent that Learns. J Communications of the ACM 37(7): 92-99
133. Setnes M, Cross V (1997) Compatibility-based ranking of fuzzy numbers. In: NAFIPS 97. Syracuse, New-York
134. Shao S (1988) Fuzzy self-organizing controller and its application for dynamic processes. J Fuzzy Sets and Systems 26: 151-164
135. Shoham Y (1993) Agent Oriented Programming. J Artificial Intelligence 60: 51-92
136. Sikora R, Shaw MJ (1998) A Multi-Agent Framework for the Coordination and Integration of Information Systems. J Management Science 44(11): S65-S78
137. Singh MP (1994) Multi-Agent Systems: A Theoretical Framework for Intentions. J Know How and Communication. Berlin, Germany: Springer Verlag
138. Smith DC, Cypher A, Spohrer J (1994) Programming Agents without a Programming Language. J Communications of the ACM 37(7): 55-67
139. Stephen T, Wong E (1989) COSMO: A communication scheme for cooperative knowledge based systems. J IEEE Transaction on System, Man and Cybernetics 23(3): 809-822
140. Sugeno M, Kang YT (1988) Structure identification of fuzzy model. J Fuzzy Sets and Systems 28: 15-33
141. Takagi T, Sugeno M (1985) Fuzzy identification of systems and its applications to modeling and control. J IEEE Trans. Systems, Man, and Cybernetics 15(1): 116-132
142. Tanscheit R, Scharf EM (1988) Experiments with the use of a rule-based self-organizing controller for robotics applications. J Fuzzy Sets and Systems 26: 195-214
143. Tokagi H (1990) Fusion technology of fuzzy theory and neural networks. In: Inter. Conf. on Fuzzy Logic and Neural Networks, pp 13-26
144. Tokoro M (1996) Agents: Towards a Society in which Humans and Computers Cohabitate. Distributed Software Agents and Applications. In: 6th European Workshop on Modeling Autonomous Agents in Multi-Agent World, pp1-10. Berlin, Germany: Springer-Verlag
145. Tsuji T, Jazidie A, Kaneko M (1997) Distributed trajectory generation for cooperative multi-arm robots via virtual force interactions. J IEEE Transactions on Systems, Man, and Cybernetics - Part B: Cybernetics 27(5): 862-867
146. Tsuji T, Nakayama S, Ito K (1997) Parallel and distributed trajectory generation of redundant manipulators through cooperation and competition among subsystems. J IEEE Transactions on Systems, Man, and Cybernetics - Part B: Cybernetics 27(3): 498-509.
147. Vojdani N (1997) A Fuzzy Contract Net Approach to Manufacturing Control. Application of Fuzzy Logic. In: Jamshidi M, Titli A, Zadeh L, Bevrie S (eds) Towards

High Machine Intelligence Quotient Systems. Prentice Hall PTR, Upper Sadle River, New Jersey, pp. 289-299
148. Wang P (ed.) (2001) Computing with Words. John Wiley & Sons, Inc.
149. Werner E (1989) Cooperative agents: A unified theory of communication and social structure. In: Gasser L, Huhns MN (eds) Distributed Artificial Intelligence. Pitman morgan Kaufman, vol 2, pp 3-36
150. Whinston A (1997) Intelligent Agents as a Basis for Decision Support Systems. J Decision Support Systems 20(1): 1
151. Whitley D, Hanson T (1989) Optimizing Neural Networks using faster, more accurate genetic search. In: Schaffer JD (ed) Third Inter. Conf. on Genetic Algorithms. San Mateo, Calif.: Morgan Kaufmann, pp 391-396
152. Wooldridge M, Jennings N (1995) Agent Theories, Architectures, and Languages: a C 6l Survey. In: Wooldridge M, Jennings NR (eds) Intelligent Agents. Berlin, J Germany: Springer-Verlag, pp 1-22
153. Wooldridge M, Jennings N (1995) Intelligent Agents: Theory and Practice. J Knowledge Engineering Review 10 (2): 115-152
154. Yabuchi Y, Watada J, Tatsumi K (1994) Fuzzy regression analysis of data with error. Japanese J of Fuzzy Theory and Systems 6(6): 673-685
155. Yager R (1997) Intelligent Agents for World Wide Web Advertising Decisions. J International Journal of Intelligent Systems 12: 379-390
156. Yamakawa T, Fukurawa U (1992) A design algorithm of membership functions for a fuzzy neuron using example-based learnng. In: IEEE Inter. Conf. Fuzzy Syst. San Diego, pp 943-948
157. Yamakawa T, Tomoda S (1989) A fuzzy neuron and its application to pattern recognition. In: Proc. Third IFSA Conf. Seattle, pp 30-38
158. Yamakawa T (1990) Pattern recognition hardware system employing a fuzzy neuron. In: Inter. Conf. Fuzzy Logic. Iizuka, Japan, pp 943-948
159. Yamakawa T, Uchino E, Miki T, Kusanagi H (1992) A fuzzy neuron and its application to system identification and prediction of the system behavior. In: Second Inter. Conf. Fuzzy Logic and Neural Networks (IIZUKA 92). Iizuka, Japan, pp 477-483
160. Zadeh LA (1973) Outline of a new approach to the analysis of complex systems and decision processes. J IEEE Transactions on Systems, Man, and Cybernetics SMC3: 28-44.
161. Zadeh L A (1988) Fuzy Logic. J IEEE Computer April 1988: 83-93
162. Zadeh LA (1965) Fuzzy Sets. J Information and Control 8: 338-353
163. Zadeh LA (1975) Calculus of fuzzy restrictions. In Fuzzy Sets and Their Applications to Cognitive and Decision Processes. Zadeh LA, Fu KS, Shimura M (eds), New York: Academic, pp 1-39
164. Zadeh LA (1979) Theory of approximate reasoning. J Machine Intelligence Hayes J, Michie D, Mukilich LI (eds). New York: Halstead, pp 149-194
165. Zadeh LA (1986) Outline of a computational approach to meaning and knowledge representation based on a consept of a generalized assignment statement. In: Int. Seminar Artificial Intell.Man-Mach. Syst., Thoma M, Wyner A (eds). Heidelberg: Springer-Verlag, pp 198-211
166. Zadeh LA (1978) PRUF – A meaning representation language for natural languages. J Int. J. Man – Mach. Studies, vol:10, pp 395-460

167. Zadeh LA (1981) Test-score semantics for natural languages and meaning representation via PRUF, Empirical Semantics. Rieger B (eds). Germany: Brockmyer, pp 198-211
168. Zadeh L A (2001) From computing with numbers to computing with words – from manipulation of measurements to manipulation of perception. Wang P (eds) Computing with Word. John Wiley & Sons, Inc. pp 35-67
169. Zadeh LA (1997) Toward a theory of fuzzy information granulation and its centrality in human reasoning and fuzzy logic. J Fuzzy Sets and Systems 90, pp 111-127
170. Zadeh LA (1998) From computing with numbers to computing with words – from manipulation of measurements to manipulation of perception. In: Third International Conference on Application of Fuzzy Systems and Soft Computing. Aliev R, Bonfig KW, Aliew F (eds). Weisbaden, Germany October 5-7, pp 35-67
171. Zadeh LA (1986) Outline of a computational approach to meaning and knowledge representation based on the concept of a generalized assignment statement. In: Int. Seminar on Artificial Intelligence and Man-Machine Systems. Thoma M, Wyner A (eds). Springer – Verlag, Heidelberg, pp 198-211
172. Zadeh LA (2000) From Computing with Numbers of Computing with Words-From Manipulation of Measurements to Manipulation of Perceptions. In: Hampel R, Wagenknecht M, Chaker N (eds) Fuzzy Control. Theory and Practice. Physica-Verlag. A Springer-Verlag Company, pp 3-38
173. Zhang Ch (1992) Cooperation Under Uncertainty in Distributed Expert Systems. J Artificial Intelligence 56: 21-69
174. Zhang M, Zhang Ch (1995) Neural Network Strategies for Solving Synthesis Problems in Non-Conflict Cases in Distributed Expert Systems. In: Distributed Artificial Intelligence. Architecture and Modeling. First Australian Workshop on DAI, Canberra, ACT, Australia, pp 174-188

4 Soft Computing Technologies in Business and Economic Forecasting

4.1 Neuro-Computing and Forecasting

One of the first explicit uses of neural networks for time series analysis was in 1987 when Lapedes and Farber demonstrated that feed-forward neural networks could be used for modeling deterministic chaos [15].

It is known that one of the most spread forecasting methods is time-series analysis [26]. Time series analysis is based on the idea that data relating to past demand can be used to predict future demand. Past data may include several components, such as trend, seasonal, or cyclical influences.

Formally, the universe time series problem (for the case of simplicity we neglect the effect of noise) can be represented as

$$x_t = g(x_{t-1}, x_{t-2}, \dots, x_{t-n}) \tag{4.1}$$

where x_t is the current response (dependent variable), $x_{t-1}, x_{t-2}, \dots, x_{t-n}$ are past values of the series and g is the relationship (target function) between current and past data to be estimated.

In most real-word problems, few a priori assumptions can be made about the functional form of g. Since a parametric class of functions is not usually known, one resorts to non-parametric regression approaches, such as MLP, where an estimate $\hat{g} = f$ for g is constructed from a large class of functions F.

As it was mentioned in section 2.2 of this book, neural networks are universal approximators [21] and hence can be used to construct $\hat{g} = f \in F$. In a simple case we need a neural network with one hidden layer, n input and one output nodes, which can express the relationship:

$$\tilde{x}_t = \phi\left[\theta_0 + \sum_{i=1}^{n} w_i' \phi\left\{\sum_{i=1}^{n} w_{ij} x_{i-j} + \theta_i\right\}\right], \tag{4.2}$$

where w_{ij} is the weight, between unit j and hidden node i. θ_i is the bias for i-th node, w_i' is the weight connecting the hidden node i to the output node, ϕ is the activation function.

The problem now is in adjusting the weight vectors to minimize a cost function $E(\hat{x}_t, x_t)$, defined on the basis of (4.1) and (4.2) and representing the distance measure between the neural network output and the desired output pattern. In the general case, the parameterization of equation (4.2) is affected by some form of gradient descent over weight landscape in order to minimize the cost function for the complete training set. For time series analysis this has the effect of windowing through the data, creating input/output instances according to the order of the model given in equation (4.1).

In order to generate a successful neural network model for approximation of (4.1) it is required to choose the set of relevant parameters. Let's consider these parameters.

Data selection and processing

Analyzing and manipulating the data in preparation for neural network modeling is very important. For a given target process many possible causal effects may be relevant in the construction of a model.

Data preprocessing stage is to find out which parameters, such as scaling, smoothing, etc., are required to apply to the data. At this stage it is more important to know what is the appropriate forecast horizon as well as the selected and treated data.

Architecture of the neural network

It is necessary to determine the number of lags and correspondingly the number of input nodes of neural network. Also it is necessary to determine the number of hidden nodes and type of activation function providing the required forecasting accuracy.

Learning rule and style

Depending on the type and architecture of used neural network, we need to determine what rule should be used to control the weight update during training?

We also need to determine what style of neural network training algorithm should be used: batch versus online. The convergence feature of the developed neural network is to be paid much attention.

Neural network validation

The trained network should achieve good generalization, and that validation results are to accurately reflect the performance of the trained network.

4.2 Fuzzy Time Series Based Forecasting

In 1993 for solving forecasting problems, Song and Chissom offered a model of fuzzy time series [27,28,29]. In 1996 S.M.Chen et al [6], [11] offered new methods related to application of fuzzy time series, which help improve the quality of forecasting under conditions of uncertainty.

Fuzzy time series and forecasting

Let's assume, that U is the universal set and is given as follows[30]:

$$U = \{u_1, u_2, ..., u_n\}$$

The fuzzy set A of universal set U is defined as follows:

$$A = \propto_A (u_1)/u_1 + \propto_A (u_2)/u_2 + ... + \propto_A (u_n)/u_n$$

Let's assume that $Y(t)$ $(t = 0,1,2,3,...)$ is a subset of set R^1. In set $Y(t)$ the fuzzy set $\propto_i (t)$ $(t = 1,2,...)$ are defined and $Y(t)$ is universal set for these fuzzy sets. Let's define set $A(t)$, which consists of $\{\propto_i (t)$ $(t = 1,2,...)\}$, i.e. $A(t)$ is set of fuzzy sets. Then $A(t)$ is a fuzzy time series in universal set $Y(t)$ $(t = 0,1,2, ...)$. If we accept $A(t)$ as linguistic variable, then $\propto_i (t)$ will be accepted as possible linguistic values of $A(t)$. It can be seen that the function $A(t)$ changes with time, and this means that, for different times t, function $A(t)$ gets different values.

Let's assume that, $A(t)$ is a fuzzy time series. If for any t it holds $A(t) = A(t-1)$, and the fuzzy time series has final number of elements, then $A(t)$ is called an invariant fuzzy time series, otherwise it is called a time variant fuzzy time series.

Song and Chissom [27,28] applied fuzzy time series models for solving the problem of forecasting enrollments for university of Alabama. Their method can be briefly described as follows:

Step 1. Define universal set U;

Step 2. Divide universal set U into some intervals of equal length;

Step 3. Define linguistic values;

Step 4. Fuzzify the historical enrollment data;

Step 5. Choose appropriate parameter w>1 to calculate the matrix of fuzzy relations $R^w(t, t-1)$ and to forecast the enrollment for the next year;

Step 6. Defuzzify the obtained result.

Below we will consider the forecasting algorithm in more detail. For definition of universal set U, the variation of historical enrollments at university of Alabama for each consecutive year was used. The same approach was applied in [11]. The considered algorithm consists of the following steps:

Step 1. Using the historical enrollments data presented in Table 4.1, it is necessary to calculate a variation of enrollments between all consecutive years. The variation of historical enrollment data for the current year is calculated as a difference of the enrollments of the current and past year. For example, for 1972 the variation is calculated as:

"Variation for 1972" = 13563-13055=508. In Table 4.1 the variation for all years is given. In order to round up the obtained data using $D_1 = 45$ and $D_2 = 109$ it is possible to define the universal set U : $U = [D_{min} - D_1, D_{max} + D_2]$. Here D_1 and D_2 are the appropriate positive numbers, $D_{min} = -955$, $D_{max} = 1291$, $D_1 =$ = $= 45$, $D_2 = 109$. Hence the universal set U will be: $U = [-1000, 1400]$.

Step 2. To divide universal set U into some intervals of equal length. Here, the universal set U is divided into 6 intervals of equal length: $u_1 = [-1000, -600]$, $u_2 = [-600, -200]$, $u_3 = [-200, 200]$, $u_4 = [200, 600]$, $u_5 = [600, 1000]$, and $u_6 = [1000, 1400]$. The average values of these intervals are:

$$u_{av}^1 = -800, u_{av}^2 = -400, u_{av}^3 = 0, u_{av}^4 = 400, u_{av}^5 = 800, u_{av}^6 = 1200 \tag{4.3}$$

Later on, these numbers will be used to fuzzify the variations.

Step 3. To define fuzzy sets in universal set U . First of all for performance of variations it is necessary to define some linguistic values. Further, these linguistic values will be described with fuzzy sets. Here 6 fuzzy sets were considered: A_1=(big decrease), A_2=(decrease), A_3=(no change), A_4=(increase), A_5=(big increase), A_6=(too big increase). Fuzzy sets $A_1, A_2, ..., A_6$ are defined as follows:

$$\propto(u_i) = \frac{1}{1 + [C * (u_i - u_{av}^i)]^2} \tag{4.4}$$

Here u_{av}^i are average points of the appropriate intervals; C is a constant number. If in the formula (4.4) as the value of variable u_i we take the average values of the appropriate intervals, then the fuzzy sets $A_i\ (i = \overline{1,6})$ will be as follows:

$$\begin{aligned}
A_1 &= 1/u_1 + 0{,}86/u_2 + 0{,}61/u_3 + 0{,}41/u_4 + 0{,}28/u_5 + 0{,}20/u_6 \\
A_2 &= 0{,}86/u_1 + 1/u_2 + 0{,}86/u_3 + 0{,}61/u_4 + 0{,}41/u_5 + 0{,}28/u_6 \\
A_3 &= 0{,}61/u_1 + 0{,}86/u_2 + 1/u_3 + 0{,}86/u_4 + 0{,}61/u_5 + 0{,}41/u_6 \\
A_4 &= 0{,}41/u_1 + 0{,}61/u_2 + 0{,}86/u_3 + 1/u_4 + 0{,}86/u_5 + 0{,}61/u_6 \\
A_5 &= 0{,}28/u_1 + 0{,}41/u_2 + 0{,}61/u_3 + 0{,}86/u_4 + 1/u_5 + 0{,}86/u_6 \\
A_6 &= 0{,}20/u_1 + 0{,}28/u_2 + 0{,}41/u_3 + 0{,}61/u_4 + 0{,}86/u_5 + 1/u_6
\end{aligned} \tag{4.5}$$

The graphic representation of these fuzzy sets is given in Fig.4.1.

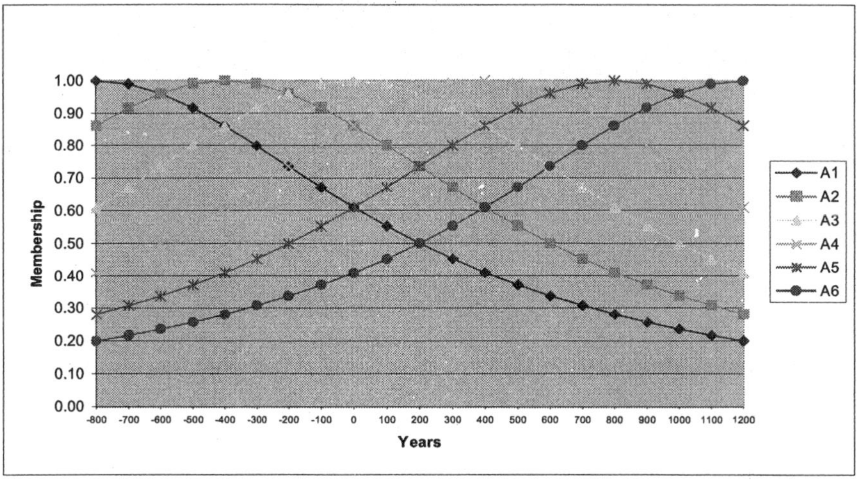

Fig.4.1. Memberships functions

Step 4. To fuzzify the historical enrollments data. Here the membership functions are calculated in accordance with formula (4.4). The results are presented in Table 4.1.

Table 4.1. Variation of actual enrollments

Years	Fact	Variation	Fuzzified variation
1971	13055		
1972	13563	+508	$A^2_4 = 0,37/u_1 + 0,55/u_2 + 0,79/u_3 + 0,99/u_4 + 0,92/u_5 + 0,68/u_6$
1973	13867	+304	$A^1_4 = 0,45/u_1 + 0,67/u_2 + 0,92/u_3 + 0,99/u_4 + 0,80/u_5 + 0,55/u_6$
1974	14696	+829	$A^1_5 = 0,27/u_1 + 0,40/u_2 + 0,59/u_3 + 0,84/u_4 + 1/u_5 + 0,88/u_6$
1975	15460	+764	$A^2_5 = 0,29/u_1 + 0,42/u_2 + 0,63/u_3 + 0,88/u_4 + 1/u_5 + 0,84/u_6$
1976	15311	-149	$A^1_3 = 0,70/u_1 + 0,94/u_2 + 0,98/u_3 + 0,77/u_4 + 0,53/u_5 + 0,35/u_6$
1977	15603	+292	$A^3_4 = 0,46/u_1 + 0,68/u_2 + 0,92/u_3 + 0,99/u_4 + 0,79/u_5 + 0,55/u_6$
1978	15861	+258	$A^4_4 = 0,47/u_1 + 0,70/u_2 + 0,94/u_3 + 0,98/u_4 + 0,77/u_5 + 0,53/u_6$
1979	16807	+946	$A^3_5 = 0,25/u_1 + 0,36/u_2 + 0,53/u_3 + 0,77/u_4 + 0,98/u_5 + 0,94/u_6$
1980	16919	+112	$A^2_3 = 0,55/u_1 + 0,79/u_2 + 0,99/u_3 + 0,92/u_4 + 0,68/u_5 + 0,46/u_6$
1981	16388	-531	$A^1_2 = 0,93/u_1 + 0,98/u_2 + 0,78/u_3 + 0,54/u_4 + 0,36/u_5 + 0,25/u_6$
1982	15433	-955	$A^1_1 = 0,98/u_1 + 0,76/u_2 + 0,52/u_3 + 0,35/u_4 + 0,25/u_5 + 0,18/u_6$
1983	15497	+64	$A^3_3 = 0,57/u_1 + 0,82/u_2 + 1/u_3 + 0,90/u_4 + 0,65/u_5 + 0,44/u_6$
1984	15145	-352	$A^2_2 = 0,83/u_1 + 1/u_2 + 0,89/u_3 + 0,64/u_4 + 0,43/u_5 + 0,29/u_6$
1985	15163	+18	$A^4_3 = 0,60/u_1 + 0,85/u_2 + 1/u_3 + 0,87/u_4 + 0,62/u_5 + 0,42/u_6$
1986	15984	+821	$A^4_5 = 0,28/u_1 + 0,40/u_2 + 0,60/u_3 + 0,85/u_4 + 1/u_5 + 0,87/u_6$
1987	16859	+875	$A^5_5 = 0,26/u_1 + 0,38/u_2 + 0,57/u_3 + 0,82/u_4 + 0,99/u_5 + 0,90/u_6$
1988	18150	+1291	$A^1_6 = 0,19/u_1 + 0,26/u_2 + 0,37/u_3 + 0,56/u_4 + 0,81/u_5 + 0,99/u_6$
1989	18970	+820	$A^6_5 = 0,28/u_1 + 0,40/u_2 + 0,60/u_3 + 0,85/u_4 + 1/u_5 + 0,87/u_6$
1990	19328	+358	$A^5_4 = 0,43/u_1 + 0,64/u_2 + 0,89/u_3 + 1/u_4 + 0,84/u_5 + 0,59/u_6$
1991	19337	+9	$A^5_3 = 0,60/u_1 + 0,86/u_2 + 1/u_3 + 0,87/u_4 + 0,62/u_5 + 0,41/u_6$
1992	18876	- 461	$A^3_2 = 0,90/u_1 + 1/u_2 + 0,82/u_3 + 0,57/u_4 + 0,39/u_5 + 0,27/u_6$

Step 5. To calculate the fuzzy operation matrix $O^w(t)$ for $w > 1$ and the criterion matrix $C(t)$. Accepting $w = 5$, actually we use the data for 6 previous years.

For example, we begin to predict enrollment for 1977. In this case, the operation matrix $O^5(t)$ and criterion matrix $C(t)$ are determined as follows:

$$O^5(1977) = \begin{bmatrix} \text{Fuzzy variation enrollments for 1975} \\ \text{Fuzzy variation enrollments for 1974} \\ \text{Fuzzy variation enrollments for 1973} \\ \text{Fuzzy variation enrollments for 1972} \end{bmatrix} \begin{matrix} A_5^2 \\ A_5^1 \\ A_4^1 \\ A_4^2 \end{matrix}$$

(big decrease)	(decrease)	(no change)	(increase)	(big increase)	(too big increase)
0,29	0,42	0,63	0,88	1	0,84
0,27	0,40	0,59	0,84	1	0,88
0,45	0,67	0,92	0,99	0,80	0,55
0,37	0,55	0,79	0,99	0,92	0,68

$C(1977)$ = Fuzzy variation enrollments for 1976 = $[A_3^1]$

(big decrease)	(decrease)	(no change)	(increase)	(big increase)	(too big increase)
0,70	0,94	0,98	0,77	0,53	0,35

The relation matrix $R(t)$ is calculated: $R(t)[i, j] = O^w(t)[i, j] \cap C(t)[i, j]$, here $1 \le i \le 4$ and $1 \le j \le 6$.

(big decrease) (decrease) (no change) (increase) (big increase) (too big increase)

$$R(1977) = \begin{Vmatrix} 0,29 & 0,42 & 0,63 & 0,77 & 0,53 & 0,35 \\ 0,27 & 0,40 & 0,59 & 0,77 & 0,53 & 0,35 \\ 0,45 & 0,67 & 0,92 & 0,77 & 0,53 & 0,35 \\ 0,37 & 0,55 & 0,79 & 0,77 & 0,53 & 0,35 \end{Vmatrix}$$

The final result for the forecast enrollment for 1977 will be shown as follows:

	(big decrease)	(decrease)	(no change)	(increase)	(big increase)	(too big increase)
F(1977)=	0,45	0,67	0,92	0,77	0,53	0,35

For other years forecasting results will be calculated similarly. For clearness once again it is necessary to note that for calculation of fuzzy relation matrixes R(1977) [11,27] the operation ($\cap$) Min is used, and for calculation of F(1977) the operation ($\cup$) Max is used. The results of calculations are summarized in Table 4.2.

Table 4.2. Forecasting of a variation for $w = 5$

Year	Value of membership functions of forecasted variations					
	u_1	u_2	u_3	u_4	u_5	u_6
1977	0,45	0,67	0,92	0,77	0,53	0,35
1978	0,46	0,68	0,92	0,99	0,79	0,55
1979	0,47	0,70	0,94	0,98	0,77	0,53

1980	0,25	0,36	0,53	0,77	0,98	0,84
1981	0,55	0,79	0,98	0,92	0,68	0,46
1982	0,55	0,79	0,78	0,54	0,36	0,25
1983	0,93	0,76	0,52	0,35	0,25	0,18
1984	0,57	0,82	0,99	0,90	0,65	0,44
1985	0,83	0,98	0,89	0,64	0,43	0,29
1986	0,60	0,85	1,00	0,87	0,62	0,42
1987	0,28	0,40	0,60	0,85	0,65	0,44
1988	0,26	0,38	0,57	0,82	0,99	0,87
1989	0,19	0,26	0,37	0,56	0,81	0,90
1990	0,28	0,40	0,60	0,85	1,00	0,87
1991	0,28	0,40	0,60	0,85	0,84	0,59
1992	0,43	0,64	0,89	0,87	0,62	0,41

Step 6. To defuzzify results obtained at Step 5. Defuzzification is done as follows:

$$V(t) = \frac{\sum_{i=1}^{6} \propto(u_i) \cdot u_i}{\sum_{i=1}^{6} \propto(u_i)}$$

Here $\propto(u_i)$ are calculated values of membership functions; u_i -are average values of intervals. For example, as it is seen in Table 4.3 for $F(1977)$ after calculations the below-mentioned value is found:

$$V(1977) =$$
$$= \frac{(-800)\cdot 0,45 + (-400)\cdot 0,67 + 0\cdot 0,92 + 400\cdot 0,77 + 800\cdot 0,53 + 1200\cdot 0,35}{0,45 + 0,67 + 0,92 + 0,7 + 0,53 + 0,35} =$$
$$= 143$$

Then to find the forecasted value for 1977 it is possible to perform the following calculation: $F(1977) = 15311 + 143 = 15454$. All forecasted values for the basis of $w = 5$ are presented in Table 4.3.

The graphic chart presenting the actual and forecasted values for each year for $w = 5$ is illustrated in Fig. 4.2.

The considered fuzzy forecasting method briefly can be described as follows:

$$R(t) = O^w(t) \otimes C(t) = \begin{bmatrix} R_{11} & R_{12} & \dots & R_{1m} \\ R_{21} & R_{22} & \dots & R_{2m} \\ \dots & \dots & \dots & \dots \\ R_{w1} & R_{w2} & \dots & R_{wm} \end{bmatrix}$$

$$A(t) = [Max(R_{11}, R_{21}, ..., R_{w1}), Max(R_{12}, R_{22}, ..., R_{w2}), ..., Max(R_{1m}, R_{2m}, ..., R_{wm})] =$$
$$= [Q_1, Q_2, ..., Q_m],$$

Here $A(t)$ is forecasting value for year t, represented by fuzzy set; $O^w(t)$ is operation matrix; $R(t)$ is fuzzy relation matrix.

Table 4.3. Results of the forecasting for w = 5

Years	Forecasted values	The obtained results
		Errors
1977	15454	0,95%
1978	15843	0,11%
1979	16086	4,29%
1980	17280	2,13%
1981	17081	4,23%
1982	16403	6,29%
1983	15265	1,50%
1984	15638	3,26%
1985	15118	0,30%
1986	15284	4,38%
1987	16296	3,34%
1988	17323	4,56%
1989	18701	1,42%
1990	19422	0,49%
1991	19703	1,89%
1992	19529	3,46%

In all above calculations w =5 is used. For other w the calculations are performed in similar way. The results of calculations for different values of w (from 2 up to 9) are presented in Table 4.4

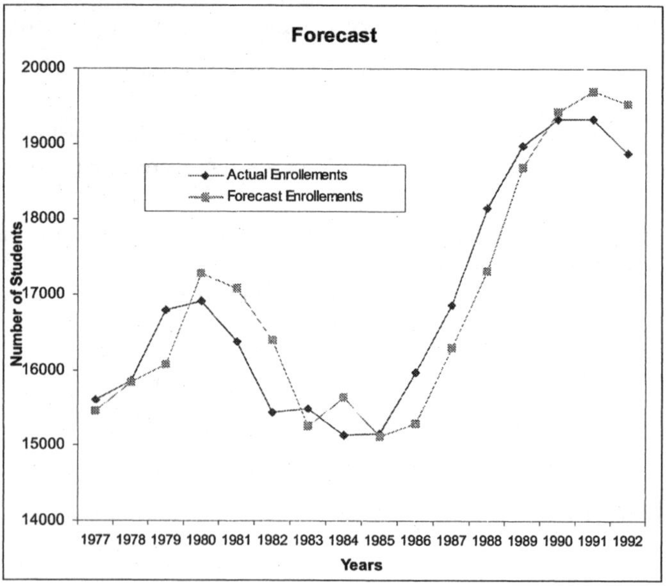

Fig. 4.2. Enrollments forecasting

Average forecasting error for these values of w is given in Table 4.5. The considered forecasting method was first suggested in [27,28,29], and then improved in [6] and [11].

Table 4.4. Forecasting enrollments

Forecasting enrolments for University of Alabama

Years	Actual	Variation	Forecast w =2	Forecast w =3	Forecast w =4	Forecast w =5	Forecast w =6	Forecast w =7	Forecast w =8	Forecast w =9
1971	13055									
1972	13563	508								
1973	13867	304								
1974	14696	829	14160							
1975	15460	764	15062	15103						
1976	15311	-149	15907	15892	15892					
1977	15603	292	15562	15562	15454	15454				
1978	15861	258	15743	15839	15839	15843	15843			
1979	16807	946	16094	16086	16086	16086	16086	16086		
1980	16919	112	17189	17196	17196	17280	17288	17288	17288	
1981	16388	-531	17268	17112	17112	17081	17081	17081	17081	17081
1982	15433	-955	16403	16403	16403	16403	16350	16350	16350	16350
1983	15497	64	15265	15265	15265	15265	15265	15265	15265	15265
1984	15145	-352	15416	15502	15638	15638	15638	15638	15638	15638
1985	15163	18	15190	15134	15118	15118	15118	15118	15118	15118
1986	15984	821	15199	15284	15284	15284	15284	15284	15284	15284
1987	16859	875	16286	16286	16296	16296	16296	16306	16434	16434
1988	18150	1291	17323	17323	17323	17323	17323	17323	17323	17329
1989	18970	820	18701	18701	18701	18701	18701	18701	18701	18701
1990	19328	358	19514	19434	19422	19422	19422	19422	19422	19422
1991	19337	9	19703	19703	19703	19703	19593	19593	19594	19594
1992	18876	-461	19529	19529	19529	19529	19529	19456	19454	19454

Table 4.5. Average forecasting error

	Basis of model							
	w =2	w =3	w =4	w =5	w =6	w =7	w =8	w =9
Error	2,77%	2,60%	2,71%	2,66%	2,72%	2,87%	2,70%	2,74%

The results of the comparative analysis evidently show that the results obtained using considered here improved method are better than those using other methods (see Table 4.6).

Table 4.6. Results of different forecasting methods

Years	Actual en-rollments	Method of Song-Chissoma (w =4)	Method of S.M. Chena (w =4)	Improved method (w =4)
1972	13563			
1973	13867			
1974	14696			
1975	15460	14700		
1976	15311	14800	16260	15892
1977	15603	15400	15511	15454
1978	15861	15500	16003	15839
1979	16807	15500	16261	16086
1980	16919	16800	17407	17196
1981	16388	16200	17119	17112
1982	15433	16400	16188	16403
1983	15497	16800	14833	15265
1984	15145	16400	15497	15638
1985	15163	15500	14745	15118
1986	15984	15500	15163	15284
1987	16859	15500	16384	16296
1988	18150	16800	17659	17323
1989	18970	19300	19150	18701
1990	19328	17800	19770	19422
1991	19337	19300	19928	19703
1992	18876	19600	19537	19529

4.3 Fuzzy Delphi Method

When a relevant data base is not available, as in the case of a new product being considered for market introduction, the conventional statistics forecasting methods and models cannot be used. The forecaster must then rely upon other evidence, usually in the form of subjective or judgmental information, as the basis for a market forecast. Forecasting models that do not rely on a historical data base are called qualitative forecasting models.

An important representative of qualitative forecasting models is Delphi Method. The essence of the Delphi Method is the following. A panel of experts are requested to give their opinion about a certain event. The responses of the experts to the first questionnaire are used to generate a second questionnaire. The same experts or new experts may be used to respond to the second questionnaire. Their responses are then used to generate questions for a third questionnaire, and so forth. This process could be repeated until the outcome converges to a reasonable solution. Delphi Method has the following characteristics [25]: anonymity of experts; iterative surveys; a structured feedback in a statistically analyzed format for the panel of experts for the next estimation.

The processing of judgment information by statistical tools does not allow the problems of ambiguity and uncertainty to be taken into account. Therefore a generalization of the classical Delphi method to consider these problems by using fuzzy sets theory is a challenge for researchers. Fuzzy Delphi methods (FDM) have been investigated by different researchers [4,13,22]. In [22] traditional Delphi procedure is combined with a "group average" membership functions. In [13] the max-min fuzzy Delphi Method and based on fuzzy integration a Delphi procedure are considered. Both methods rely on a single iteration of survey and utilize special kind of questions for each survey item.

Reference [4] considers the application of fuzzy Delphi Method for time estimation for technical realization of an innovative product. The work [4] also considers the use of fuzzy Delphi Method for fuzzy PERT for project management.

In [17] another more complete procedure of FDM is proposed. In this procedure, the fuzzy set theory is used by asking the participants to give a three-point estimate (i.e., the pessimistic, moderate, and optimistic values). On that basis, triangular fuzzy numbers (TFNs) (a_i, b_i, c_i), i=1,.,n, are formed and the mean, (a^m, b^m, c^m), is computed. In the next round, the feedback information, ($a^m - a_i$, $b^m - b_i$, $c^m - c_i$), is provided to individual experts. At the end of each round, if the distances between all individual experts' TFNs and the mean TFNs for all survey items satisfy a given criterion (value), the process is completed. The mean TFNs are the final forecasts.

A new fuzzy Delphi method is proposed in [25]. The method employs the fuzzy statistics and technique of the conjugate gradient search to fit membership functions. Membership functions besides triangles may be derived for the fuzzy forecasts.

In [19] the modified fuzzy Delphi Method is suggested. Below we present main steps of fuzzy Delphi Method adopted from [19], which will be widely used in next chapters of the book.

Step 1. Determine the qualitative data, say events to formulate the first questionnaire;

Step 2. Determine the triangular-shaped membership functions for each event. Here the parameters of each event is determined by three values: $\min(\min_j)$, $\max(\max_j)$, and $\text{mean} = \sqrt[n]{\prod \text{mean}_j}$, where j is sample number; n is the number of experts;

Step 3. Add the three event parameter values found in Step 2 into the next questionnaire and give to experts again;

Step 4. Do statistical tests to see whether the membership functions have converged or not;

Step 5. If membership functions have converged, then find the fuzzy number to describe each linguistic term.

For example, if qualitative model include k factors (political, economical, etc.), then a corresponding triangular membership function should be determined for every factor.

4.4 Soft Computing Based Forecasting Complex Time Series

Forecasting activities on the basis of prediction result of time series play an important role in area of economic and business planning, stock trading, inventory, demand and production control, and in many other fields. Many signals from financial, economic, technical and physical systems are linguistic values or words with fuzzy meaning. This type of time series significantly differs from traditional time series and the methods of the latter are not applicable in this case. So we deal with a new class of time series, a fuzzy time series, whose values are linguistic values. There are several approaches to modeling fuzzy time series. In this report we will use models based on fuzzy relational equations and fuzzy recurrent neural networks. We will analyze and compare several methods of times series analysis among which there exist our two methods: first—based on fuzzy relational equations with more appropriate reasoning mechanism; second—method based on recurrent neural networks. Using benchmark on forecasting of the enrollments of the University of Alabama, it will be shown that the forecasting error of the latter two methods is significantly smaller than that of existing fuzzy time series approaches. Also in this report it will be discussed the application of fuzzy times series to forecasting of power system load. Also, it will be discussed fuzzy-chaotic approach for time series prediction.

Introduction in [27,28,29,30], fuzzy time series models are proposed and their applications to forecasting problems are considered. They have proposed several definitions and theorems related to some properties of fuzzy time series. Some extension of fuzzy relational model based time series is analyzed in [7]. Applications of fuzzy time series are considered in [11,20,31]. Modification of fuzzy time series models for forecasting of university enrollments is discussed in [16]. It should be noted that almost in all these works the task used as the benchmark is forecasting enrollments of the University of Alabama.

In [23] a fuzzy fractal method for forecasting financial and economic time series is described. All the works on fuzzy time series mentioned above are based on fuzzy relational equations [32]. In this section we consider improved fuzzy time series model with appropriate fuzzy reasoning, neuro-fuzzy-genetic based time series and fuzzy-chaotic analysis of time series in fuzzy environment. The results are compared with those of existing methods [7,16,27,28,29].

Fuzzy Relational Equation Based Time Series

Definition. Let $U(t),(t=0,1,2,...)$ be the universe of discourse on which fuzzy sets $A_i(t),(i=1,2,...)$ defined, then the totality of $A_i(t)$ regarded as linguistic variable $A(t)$ is called a fuzzy time series on $U(t)$.

Definition. Let $A(t)$ is affected by $A(t-i), i=1,2,...,k$, where $k>0$, then fuzzy relational equation

$$A(t) = (A(t-1) \cdot A(t-2) \cdot \ldots \cdot A(t-k)) \circ R(t, t-k) \tag{4.6}$$

is called the k-th order model of $A(t)$.

Solution of the equation (4.6) significantly depends on fuzzy relation $R(t, t-k)$ and optimal value of k, where k denotes the number of time steps before t.

Here we consider forecasting enrollments of University of Alabama with the use of model (4.6). The historical data are adopted from [16]. We have considered different fuzzy reasoning methods and different numbers of inputs for calculating the value of the time series $A(t)$.

The experimental analysis has shown that a more optimal fuzzy reasoning mechanism is based on the logic suggested in [1]. $k=3$, that means forecasting using 3 years before is more effective. In the Table 4.7 we compare the actual forecasting errors of existing and offered methods.

Table 4.7. A comparison of average forecasting errors

	Time step							
	k=2	k=3	k=4	k=5	k=6	k=7	k=8	k=9
Average forecasting error of method of Song-Chissoma	3.15%	3.89%	4.87%	4.41%	4.49%	4.35%	4.45%	4.23%
Average forecasting error of method of Chen	2.99%	2.94%	3.12%	2.92%	3.01%	3.08%	2.89%	2.79%
Average forecasting error of the offered method	2.77%	2.60%	2.71%	2.66%	2.72%	2.87%	2.70%	2.74%

Neuro-fuzzy-genetic approach for forecasting time series

In this section of the book we use a fuzzy neural network learning by genetic algorithms for the implementation of fuzzy time series. It is known that recurrent neural networks are essentially more powerful than feed forward neural networks [1]. Therefore we use a recurrent fuzzy neural network to get better accuracy and effectiveness. The problem here is lack of learning methods for recurrent fuzzy neural networks. As shown earlier considered time series is described as

$$A(t) = f[A(t-1), A(t-2), \ldots, A(t-k)] \tag{4.7}$$

The structure of the recurrent fuzzy neural network for the implementation of this fuzzy time series is shown in Fig. 4.3. The input neurons are fed by data from previous time steps, the output neurons calculate the value of fuzzy time series $A(t)$. This technology has been used for forecasting the enrollment of University of Alabama. The fuzzy neural network consists of 3 input neurons (i.e. $k=3$ for time series), 10 hidden neurons and 1 output neuron. The comparison of the forecasted enrollments is shown in Fig. 4.4.

The results of the Computer Simulation have shown that the root mean square error on using this approach is significantly reduced. For example, the root mean square error when forecasting enrollments of University of Alabama for the Chen's method is 624.23, for the Song and Chusson method is 421.32, and for the Tsai and Wu method (by using high-order fuzzy time series) is 199.32, the offered method's error is 116.21, considerably smaller than the above mentioned methods. Hence, the feasibility of fuzzy-neuro-genetic approach to fuzzy time forecasting is evident. This results show improvement in forecasting under fuzzy data significantly.

Fuzzy-chaotic analysts of fuzzy time series

Many financial, economic, and technical time series are complex dynamic systems and usually has chaotic behavior. So we need to determine if the considered time series is predictable, i.e. to search the prediction strategies: is it of short-term or long-term character or is it unpredictable. It should be noted that the aim of chaotic analysis is not to make predictions, but to give information about the behavior of prediction models. In particular, the correlation dimension of the considered time series gives neural predictor the Information on how many past data from the time series should be processed, i.e. determines the number of input neurons of neural networks in the basis of the neural predictor.

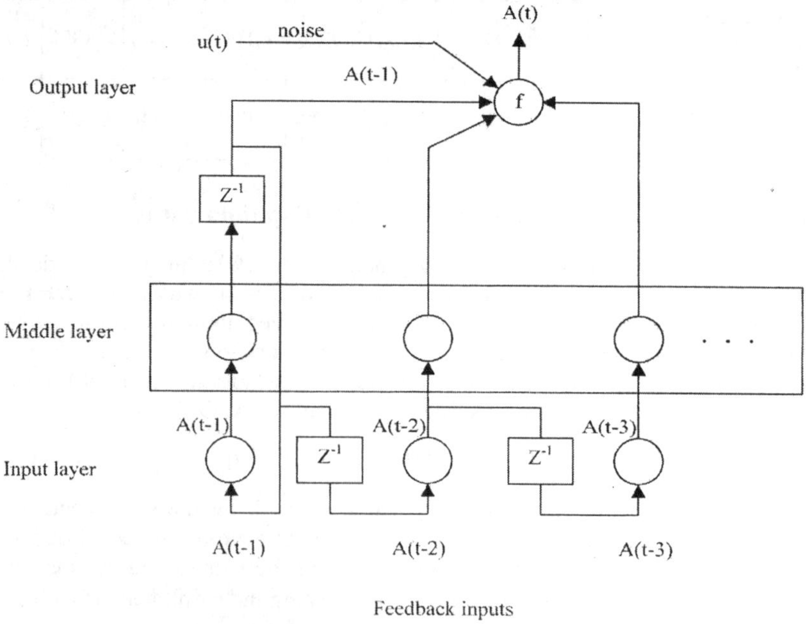

Fig. 4.3. Structure of the Recurrent Fuzzy Neural Network based Predictor

We describe in this section the application of a fuzzy-chaotic approach for time series prediction. This approach combines fuzzy logic with fractal theory to achieve the goal of forecasting complex dynamical systems [23].

Here we will consider only the calculation of fuzzy fractal dimension of fuzzy time series.

Definition. Let $A(t), t = 0, 1, 2, \ldots$ be an arbitrary fuzzy time series which is clustered into m objects $Y_1, Y_2, \ldots, Y_m$. Geometrical complexity of the objects Y_1, $Y_2, \ldots, Y_m$ measured by fuzzy fractal dimension is defined as

$$(d_1, \alpha_1),\ (d_2, \alpha_2), \ldots, (d_m, \alpha_m), \tag{4.8}$$

where $d_i, i = \overline{1, m}$, is a numerical value of the fractal dimension calculated as shown in [1], $\alpha_i, i = \overline{1, m}$ is the membership function for the object $Y_i, i = \overline{1, m}$.

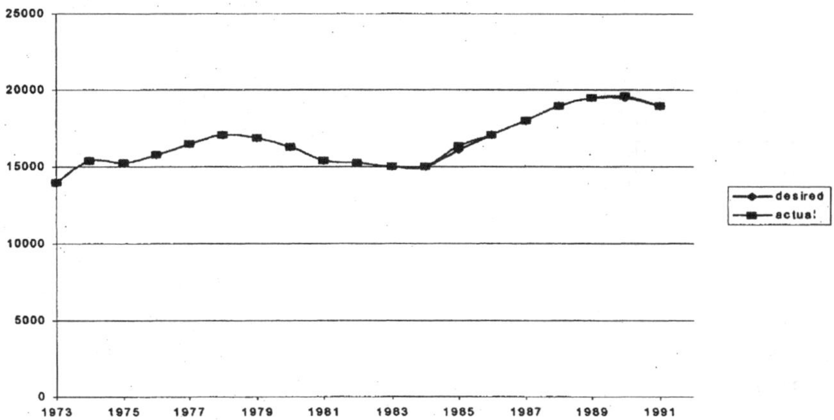

Fig. 4.4. Actual enrollments and enrollments forecasted by Fuzzy-Genetic Neural Networks

Considering the aforesaid the fuzzy time series forecasting model can be described as follows

$$\begin{array}{l} \text{IF } d_1 \text{ AND } A^1(t\text{-}l) \text{ AND } A^1(t\text{-}2) \text{ AND} \ldots \text{ AND } A^1(t\text{-}k) \text{ THEN } Y_1, \\ \text{IF } d_2 \text{ AND } A^2(t\text{-}l) \text{ AND } A^2(t\text{-}2) \text{ AND} \ldots \text{ AND } A^2(t\text{-}k) \text{ THEN } Y_2 \\ \ldots\ldots \\ \text{IF } d_m \text{ AND } A^m(t\text{-}l) \text{ AND } A^m(t\text{-}2) \text{ AND} \ldots \text{ AND } A^m(t\text{-}k) \text{ THEN } Y_m \end{array} \tag{4.9}$$

This method has been considered for forecasting products of an oil refinery plant and shown effective results. It should be noted that the experimental analysis shows that the application of fractal dimension in combination with other data in the fuzzy production model gives better effectiveness demonstrated by a substantial decrease in the forecasting error [23].

Fuzzy Time Series Based Forecasting of Electricity

Actual data of electricity production is shown in Table 4.8.

Table 4.8. Actual historical data on electricity production

Years	Actual
1982	1764
1983	3651
1984	4619
1985	6689
1986	7174
1987	8172
1988	9154
1989	9562
1990	10613
1991	11832
1992	9679
1993	8650
1994	8650
1995	8384
1996	9076
1997	8110
1998	10090
1999	10157
2000	10863

The calculations were performed on the basis of fuzzy time series using fuzzy relational model with fuzzy reasoning given in [1]. Actual and forecasting values of the electricity production are shown in Fig. 4.5.

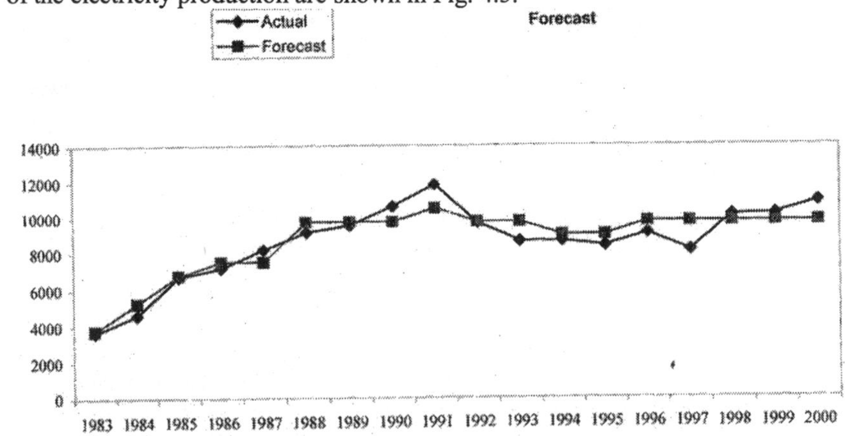

Fig. 4.5. Actual and forecasting values of the electricity production

The suggested methods based on both relational equations and neuro-fuzzy-genetic approach, have shown better results of fuzzy time series forecasting as compared with existing approaches. The use of fractal dimension along with other inputs in fuzzy production model substantially increase the accuracy of forecasting. The results of application of fuzzy time series forecasting method for forecasting of electricity production have confirmed the effectiveness of the suggested methods.

4.5 Soft Computing Based Prediction Ensemble for Forecasting in Chaotic Time Series

Forecasting activities on the basis of prediction of time series play an important role in area of economic and business planning, stock trading, inventory, demand and production control, and in many other fields [8,9,12,18].

Various predictors of time series have been introduced for accurate forecasting [8,14,24]. Among them are: stochastic, neural network based, fuzzy, hybrid etc. It is a very important task to decide which of these predictors to use for a specific application area to get the maximum of performance. This problem is usually solved through thoroughly examining different kinds of predictors upon a set of criteria (e.g. prediction error), selected in advance. Usually it is made by an ad-hoc procedure. Furthermore, in real-world problems, the character of data entering the system may vary. This may cause the situation when a predictor initially seemed to be optimal, in new changed conditions becomes less effective than other one. In this case we come across a problem of frequent selection of effective type of predictors by a human expert's subjective decision and behavioral knowledge. It is impossible to solve such a problem without an intelligent and automatic selection of suitable type of predictors in on-line mode. Thus the problem arises of creating an intelligent decision maker to perform the function of determining an optimal (near-optimal) predictor with the highest predictive accuracy at each step of forecasting. In the present section such an intelligent decision maker is created as a fuzzy module which performs "competition" among an ensemble of predictors, calculates membership functions recursively according to the predictive accuracy of each predictor and selects "winner" prediction module. The structure of Multi-Module Forecasting Time Series System is shown in Fig. 4.6 [10].

Many signals from financial, economic, technical and physical systems are chaotic, so the proposed system includes analyzer that determines if the considered time series is predictable. Then it determines the prediction strategies [8].

The chaotic analyzer identifies the character of prediction on the basis of the considered time series: whether it is of short-term or long-term character, or whether it is unpredictable or not, i.e. time series are random. It should be noted that the aim of chaotic analysis is not to make predictions, but to give information about the behavior of prediction models. In particular, the correlation dimension of the considered time series gives neural predictor the information on how many past data from the time series should be processed.

The outputs of the chaotic analyzer include, in particular, fractal dimensions (FR) which allow the determination of the character of prediction and Lyapunov exponents (LE) which allows the determination of predictability of the considered time series.

The predictors $P_i, i=\overline{1,n}$ are the main elements of the system. Different types of predictors can be used for forecasting. Among them are statistical forecasting time series models, neural models, fuzzy forecasting models, hybrid models, logistic-form models and others (Fig. 4.6).

$$Y_t = F_P(Y_{t-1}, Y_{t-2}, ..., Y_{t-q}), \tag{4.10}$$

where F_P is some non-linear mapping, q is the order of the predictor (number of past time series values used).

For the instant $t-1$ the observation Y_{t-1} becomes available and using the value $Y_{t-1}^{P_i}$, predicted for time $t-1$, the prediction module generates prediction error (for simplicity one-step prediction error is considered)

$$e_{t-1}^{i} = Y_{t-1} - Y_{t-1}^{P_i}, \; i=\overline{1,n} \tag{4.11}$$

Prediction error of each predictor is received by the fuzzy Decision Maker and on the basis of these errors the competitive computation of $\propto_{t-1}^{P_i}$ is performed. Thereby the “winner” predictor is determined (which takes the maximum value of $\propto_{t-1}^{P_i}$) which is used to predict X_t^P for time instant t. It should be noted, however, that all other predictors calculate $Y_t^{P_i}$ to use these values to determine $e_t^{P_i}$ at time t.

As it was shown, above the main element of the proposed multi-module forecasting time-series system is fuzzy decision maker.

The operation of the considered system implies two-stage design method. The first stage is off-line: on the basis of collected data each predictor is trained. At this time all data are divided into two parts: 2/3 of them are used to construct prediction modules; and 1/3 – for testing of created predictors.

The second stage is on-line. Here the fuzzy decision maker determines which type of predictors can predict Y_t on the basis of $Y_1, Y_2, ..., Y_{t-1}$ with better accuracy. We represent “the type of predictors with the highest predictive accuracy” as fuzzy set [24]

$$A\ (P_i, \propto_A (P_i, t) | P_i \in P), \tag{4.12}$$

where P_i is the i-th predictor, $\propto_A (P_i, t)$ is membership degree of P_i to A, P is the universe of predictors. $\propto_A (P_i, t)$ may change with time depending on new observations. The problem consists in calculation of membership grade of each predictor to fuzzy set A when every new data is available and in finding the predictor with maximum grade through performing competition among the predictors. The latter will be the “winner” for forecasting. Its forecasting will be considered to be near-optimal for next stage. To determine $\propto_A(P_i, t)$ at time instant t the following formula, performing fuzzy inference, is used [24]:

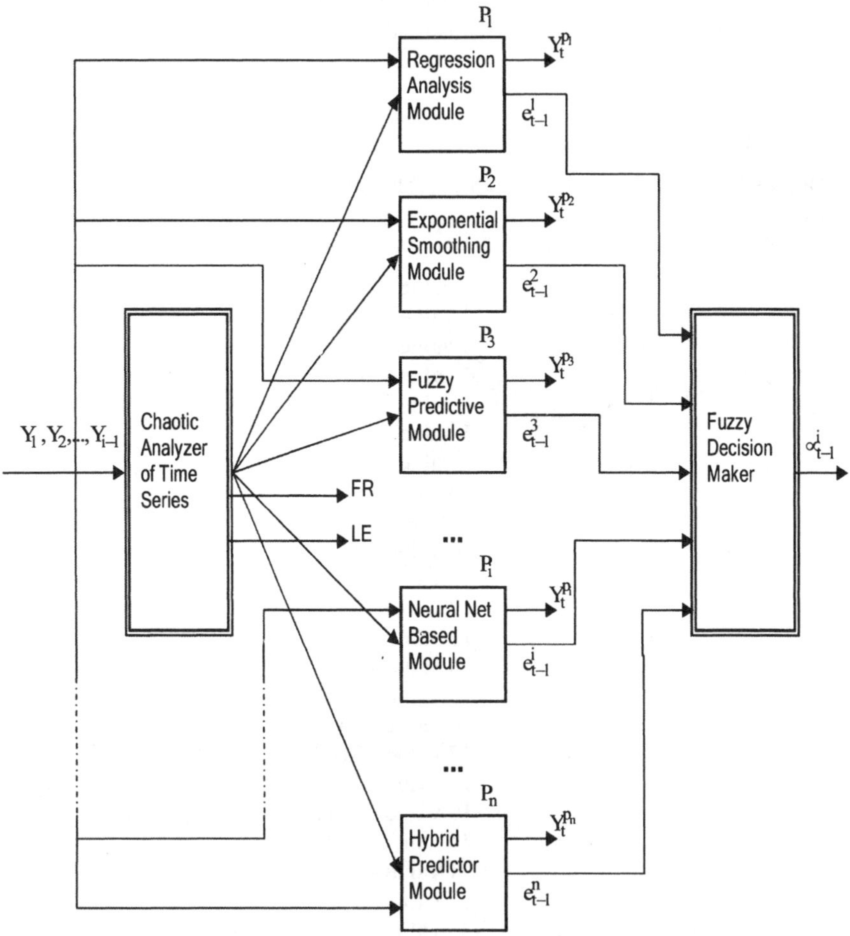

Fig.4.6. Architecture of the multi-module forecasting time series system

$$\propto_t^i = \frac{\propto_{t-1}^i e^{-(e_t^i/\sigma)^2}}{\sum_{i=1}^{n} \propto_{t-1}^i e^{-(e_t^i/\sigma)^2}} \tag{4.13}$$

where $\propto_t^i$ the membership grade of the i -th predictor to fuzzy set A at time t, σ is standard deviation of prediction in training stage.

After the determination of $\propto_t^i$ by using the defuzzification procedure, the "winner" predictor is found.

Experimental investigations

"Demand" time series is shown in the Table 4.9.

Table 4.9. "Demand" time series

1 value - 2217
2 value - 1748
3 value - 2010
4 value - 2417
5 value - 2383
6 value - 1890
7 value - 2401
8 value - 3132
9 value - 2875
10 value - 2375
11 value - 2583
12 value - 2730
13 value - 2883
14 value - 2577
15 value - 2677
16 value - 3035
17 value - 3148
18 value - 2459
19 value - 2841
20 value - 3501
21 value - 3534
22 value - 2798
23 value - 3109
24 value - 3621
....
....
....

Chaotic analysis of the considered time series shows that the given time-series is predictable. Four types of predictors are chosen: linear regression module, nonlinear regression module; exponential smoothing module and feed-forward neural-network module. The training and testing of all modules have been performed. Assume that at time instant $t=1$ the demand is $y(1)=2720$. For this time the following predictions have been generated: $y^{P_1}(1)=3475$ (linear regression); $y^{P_2}(1)=2715$ (nonlinear regression); $y^{P_3}(1)=3309$ (exponential smoothing); $y^{P_4}(1)=3047$ (neural networks module). Prediction error for each predictor is calculated:

$$e^{P_1}(1)=755; \quad e^{P_2}(1)=5; \quad e^{P_3}(1)=589; \quad e^{P_4}(1)=327.$$

Standard deviation obtained is $\sigma=385$. The initial value of $\propto_{t=0}$ was taken to be 1/n=1/4=0.25.

Using (4.13) we calculate $\propto_1^{P_i}$:

$$\propto_1^{P_1}=0.0134; \propto_1^{P_2}=0.6235; \propto_1^{P_3}=0.0600; \propto_1^{P_4}=0.3031$$

Because max ($\propto_1^{P_i}$) =0.6235, the 2-nd predictor is "winner" and is selected for the forecasting for the instant $t=2$.

Let's assume that the system receives data: $y(2)=2680$, $y(3)=2630$, $y(4)=2570$, $y(5)=2515$. Repeating the above procedure we calculate $\propto_2^{P_i}, i=\overline{1,4}$ (Fig. 4.7) and always $\propto^{P_2}$ becomes maximum. Then the system receives the data: $y(6)=3150$, $y(7)=3174$, $y(8)=2972$, $y(9)=3318$. As can be seen in Fig. 4.7 starting from $t=6$ $\propto_1^{P_4}$ is increasing constantly. Because $\propto_1^{P_4}$ is maximum, the fourth (neural network) predictor is selected for the forecasting. It should be noted that the selection of necessary predictor and switch from one predictor to another is performed in an on-line automatic mode without any involvement of a human expert.

The proposed system can produce minimal prediction errors on the basis of the membership function of the entire time-series and fuzzy inference technique. Main features of this system are: 1) chaotic analysis of the considered time series which determines predictability and prediction strategies. This analysis can also be used for predictors` module construction; 2) all predictors operate independently and in parallel and send the predictions to the Fuzzy decision maker. It is more effective to implement the proposed system on a parallel machine; 3) selection of the best predictor is performed by competition among predictors for a share of the membership degree according to the predictive accuracy of each module. The simulation results show the effectiveness of the proposed Soft Computing based multi-module forecasting system.

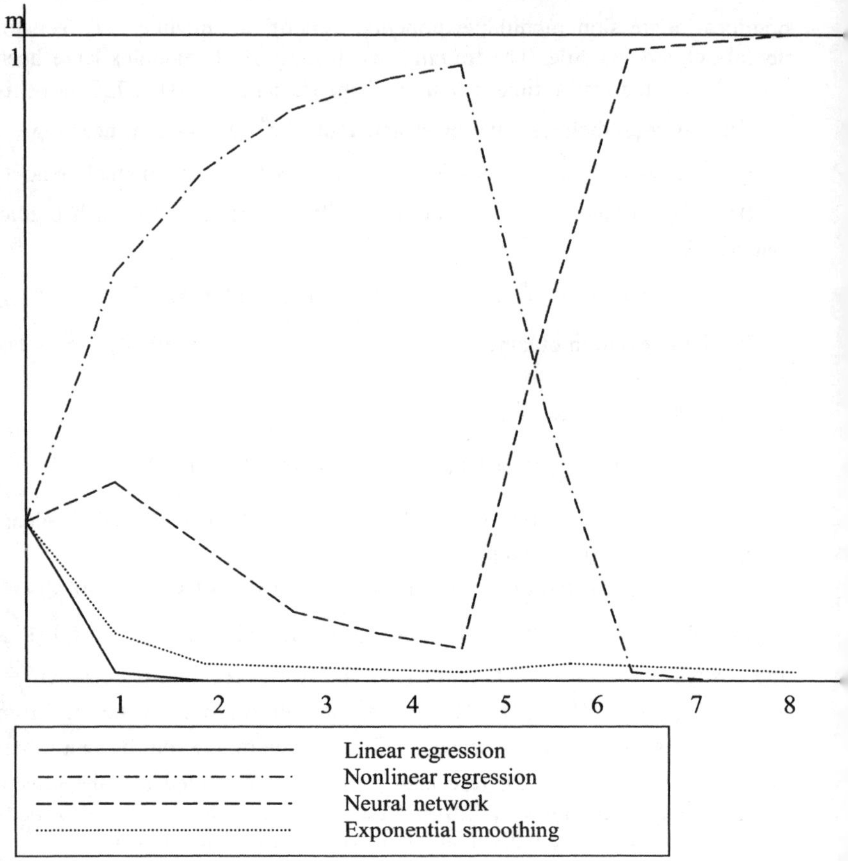

Fig. 4.7. Graphic representation of the forecasting procedure

References

1. Aliev RA, Aliev RR (2001) Soft Computing and its Application. World Scientific, New Jersey, London, Singapure, Hong Kong
2. Aliev RA, Shachnazarov MM, Gulko DE (1988) Scheduling expert systems (in English and Russian). J News of Academy of Sciences of USSR, Tech. Cybernetics 5: 118-128
3. Aliev RA, Shakhnazarov MM, Abdikeev NM (1990) Intelligent control systems. Moscow: Radio I Svyaz (in Russian)
4. Bojadziev G, Bojadziev M (1997) Fuzzy logic for business, finance, and management. World Scientific
5. Buckley JJ, Hayashi Y (1998) Application of fuzzy chaos to fuzzy simulation. J Fuzzy Sets and Systems '99: 151-157

6. Chen SM (1996) Forecasting enrollments based on fuzzy time series. J Fuzzy Sets and Systems 81: 311-319
7. Chao-Chih Tsai, Shun-Jyh Wu (2000) Forecasting enrollments with high-order fuzzy time series. In: IEEE 19n International Conference of the North American Fuzzy Information Processing Society , pp 196-200
8. Deboeck G (ed) (1994) Trading on the Edge. NY: John Wiley and Sons
9. Fazlollahi B, Aliev R (1998) Multi-agent systems in finance. In: Third Inter. Conf. on Application of Fuzzy Systems and Soft Computing. Wiesbaden, Germany, pp 65-62
10. Fazlollahi B, Aliev RR (2000) Forecasting time series system with fuzzy decision maker. In: World Conference on Intelligent Systems for Industrial Automation WCIS'2000. Tashkent, Uzbekistan, pp 111-114
11. Hwang JR, Chen S.M, Lee CH (1996) A new method for handling forecasting problems based on fuzzy time series. In: 7tln Internat. Conf. On Information Management. Chungli, Taoyuan, Taiwan, ROC, pp 312-321
12. Hwang JR, Chen ShM, Lee ChH (1998) Handling forecasting problems using fuzzy time series. J Fuzzy Sets and Systems 100: 217-228
13. Ishikawa A, Amagasa M, Shiga T, Tomizawa G, Tatsuta R, Mieno H (1993) The max-min Delphi method and fuzzy Delphi method via fuzzy integration. J Fuzzy Sets and Systems 55: 241-253
14. Jang JR, Sun C (1993) Predicting chaotic time series with fuzzy IF-THEN rules. In: Second IEEE Inter. Cons. Fuzzy systems. San Francisco, CA, Mar., vol 2, pp 1079-1084
15. Jason Kingdon (1998) Intelligent Systems and Financial Forecasting. Springer
16. Jeng-Ren Hwang, Shyi-Ming Chen, Chia-Hoang Lee (1998) Handling forecasting problems using fuzzy time series. J Fuzzy Sets and Systems 100: 217-228
17. Kaufmann A, Gupta (1988) MM Fuzzy Mathematical Models in Engineering and Management Science. North-Holland, Amsterdam
18. Kim D, Kim Ch (1997) Forecasting Time Series with Genetic Fuzzy Predictor Ensemble. J IEEE Transaction on Fuzzy Systems vol 5(4): 523-535
19. Kuo RJ (1998) A decision support system for the stock market through integration of fuzzy neural networks and fuzzy Delphi. J Applied Artificial Intelligence 12: 501-520
20. Li Zuoyoung, Chen Zhenpei, Li Jitao (1988) A model of weather forecast by fuzzy grade statistics. J Fuzzy Sets and Systems 26: 275-281
21. Mohamad H Hassoun (1995) Fundamentals of Artificial Neural Networks. A Bradford Book
22. Murray TJ, Pipino LL, JP van Gigch, (1985) A pilot study of fuzzy set modification of Delphi. J Human Systems Mgmt.5: 76-80
23. Oscar Castillo, Patricia Melin (2001) A New-Fractal Approach for Forecasting Financial and Economic Time Series. J IEEE: 929-934
24. Petridis V, Kehagias A (1997) Predictive modular fuzzy systems for time-series classification. J IEEE Transactions on Fuzzy Systems vol 5(3): 381-397
25. Ping-Teng Chang, Liang-Chih Huang, Horng-Jiun Lin (2000) The fuzzy Delphi method via fuzzy statistics and membership function fitting and an application to the human resources. J Fuzzy Sets and Systems 112: 511-520
26. Richard B, Chace Nicholas, Aquuilano J, F Robert Jacobs (1998) Production and Operations Management. Manufacturing and Services. Irwin/McGrow-Hill
27. Song Q, Chissom BS (1993) Forecasting enrollments with fuzzy time series — part I. J Fuzzy Sets and Systems 54: 1-9

28. Song Q, Chissom BS (1994) Forecasting enrollments with fuzzy time series - part II. J Fuzzy Sets and Systems 62: 1-8
29. Song Q, Chissom BS (1993) Fuzzy time series and its models. J Fuzzy Sets and Systems 54: 269-277
30. Sullivan J, Woodall WH (1994) A comparison of fuzzy forecasting and Markov modeling. J Fuzzy Sets and Systems 64: 279-293
31. Tsai CC, Wu SJ (1999) A Study for Second-order Modeling of Fuzzy Time Series. In: IEEE International Fuzzy Systems Conference. Seoul, Korea
32. Zadeh L (1975) The concept of a linguistic variable and its application to approximate reasoning. J Information Sciences 8: 43-80

5 Soft Computing Based Decision Making and DSS

5.1 Fuzzy Linear Programming

Statement of fuzzy linear programming

A classical (crisp) linear programming problem can be written as

$$\begin{aligned} &\text{maximize} \quad Z = cx \\ &\text{subject to} \ \ Ax \leq b, \quad x \geq 0, \end{aligned} \tag{5.1}$$

where Z denotes the objective, $Ax \leq b$ represents the set of (rigid) constraints and x are the structural variables.

In order to understand where and how fuzziness is applied in the linear programming problem, let us consider a simple example.

Manufacturing company must determine the mix of its commercial products B and C to be produced next year. The company produces two product lines, the B and the C. The average profit is \$400 for each B and \$800 for each C.

Fabrication and assembly are limited resources. There is a maximum of 5,000 hours of fabrication capacity available per month (each B requires 3 hours and each C requires 5 hours). There is a maximum of 3,000 hours of assembly capacity available per month (each B requires 1 hour and each C requires 4 hours).

How many of each products should be produced each month in order to maximize profit?

The objective function is

maximize $Z = 400x_1 + 800x_2$

where $Z =$ the monthly profit from B and C; $x_1 =$ the number of B produced each month; $x_2 =$ the number of C produced each month. We have following constraints

$B(x_1)$	$C(x_2)$		
Required Time/Unit	Required Time/Unit	Available Time/Month	
3	5	5,000	Fab
1	4	3,000	Assy

The company manager formulates the production scheduling problem as follows:

$$3x_1 + 5x_2 \leq 5{,}000 \qquad \text{Fab} \tag{5.2}$$

$$x_1 + 4x_2 \leq 3{,}000 \qquad \text{Assy} \tag{5.3}$$

$$x_1, x_2 \geq 0 \qquad \text{Nonnegativity}$$

However, the total available labor hours may not be that precise, because the company manager can ask workers to work overtime. Therefore, some tolerance would be put on the constraints. So, the of fuzziness appears in the specification of resources.

Since the market changes constantly, it may not be certain that the per unit profit of B and C are 400 and 800, respectively. The numbers could only be viewed as the most possible values. Therefore, it is reasonable to represent the objective coefficients as fuzzy numbers.

Because of the inconsistence of human workers, value of labor hours in making B and C could only be viewed as around 3 and 5 of materials. Therefore, it is reasonable to represent these coefficients by fuzzy numbers.

If we assume that the Linear Programming (LP) decision has to be made in fuzzy environment, then there exist several modification of (5.1). Now different models have been suggested which allow fuzziness in the objective and/or constraints in traditional LP. Some of those are considered below.

1. LP problem with fuzzy goal:

$$\text{maximize } Z = \tilde{c}x$$

$$\text{s.t.} \quad Ax \leq b,$$

$$x \geq 0.$$

2. LP problem with fuzzy resources:
 This version of the LP problem is

$$\text{maximize } Z = cx$$

$$\text{s.t.} \quad Ax \underset{\sim}{\leq} b,$$

$$x \geq 0.$$

3. LP problem with fuzzy constraint coefficients:

$$\text{maximize } Z = cx$$

$$\text{s.t.} \quad \tilde{A}x \leq b,$$

$$x \geq 0.$$

where $\tilde{A}$ is an n x m fuzzy matrix.

4. LP problem with all fuzzy coefficients:

$$\text{maximize } \tilde{Z} = \tilde{c}x$$

$$\text{s.t.} \quad \tilde{A}x \le \tilde{b},$$

$$x \ge 0.$$

5. LP problem with fuzzy variables:

$$\text{maximize } \tilde{Z} = c\tilde{x}$$

$$\text{s.t.} \quad A\tilde{x} \le b,$$

$$\tilde{x} \ge 0.$$

6. LP problem with fuzzy goal, fuzzy constraints and fuzzy variables:

$$\text{maximize } \tilde{Z} = \tilde{c}\,\tilde{x}$$

$$\text{s.t.} \quad \tilde{A}\,\tilde{x} \le \tilde{b},$$

$$\tilde{x} \ge 0.$$

Solution methods of fuzzy linear programming problem

The first approach to solving the Fuzzy Linear Programming (FLP) problem was introduced by Bellman and Zadeh [10].

Let X be a set of possible alternatives. Then fuzzy objective and fuzzy constraints are fuzzy sets $\propto_G$, $\propto_C \in F(X)$ respectively, where $F(X)$ is a set of fuzzy sets on X.

Thus a fuzzy decision is the fuzzy set $\propto_D(x) \in F(X)$ defined by

$$\propto_D(x) = \propto_G(x) \wedge \propto_C(x), \quad x \in X.$$

$$\propto_D(x) \in F(X)$$

If we want to optimize the solution, then we should to find the best $x \in X$ which simultaneously satisfies both objective and constraint (or constraints).

Let it be noted that in this basic framework of a FLP problem, there are no additional assumptions about the nature of the objectives and constraints.

Let G_i, $i = \overline{1,n}$ be fuzzy goals and $C_j, j = \overline{1,m}$ fuzzy constraints.

When goals and constraints have the same importance then a fuzzy decision D on X is defined as

$$D = (\underset{i}{\cap} G_i) \cap (\underset{j}{\cap} C_j)$$

where $x \in X$, $\propto_D(x) = \min\left\{\min_i \propto_{G_i}(x), \quad \min_j \propto_{C_j}(x)\right\}$

The final decision x_f can be chosen from the set

$$M_f = \{ x_f, \propto_D(x_f) \geq \propto_D(x) , \qquad \forall x \in X\} ,$$

where M_f is called the maximal decision set.

When criteria and constraints have inequal importance, membership functions can be weighted by x-dependent coefficients α_i and β_j such that

$$\forall x \in X, \sum_{i=1}^{m} \alpha_i(x) + \sum_{j=1}^{n} \beta_j(x) = 1$$

and

$$\propto_D(x) = \sum_{i=1}^{n} \alpha_i(x) \quad \propto_{G_i}(x) + \sum_{j=1}^{m} \beta_j(x) \quad \propto_{C_j}(x) .$$

Linear programming problem with fuzzy goals

In this section one considers a LP problem with a fuzzy objective that is

$$\begin{aligned} &\max \tilde{c}x \\ &\text{s.t. } Ax \leq b, \\ &x \geq 0, \end{aligned} \qquad (5.4)$$

where A is an $n \cdot m$ matrix of real numbers, $b \in R^m$ and $\tilde{c} = \{\tilde{c}_1, \tilde{c}_2, ..., \tilde{c}_n\}$ is an n-dimensional vector of fuzzy numbers, characterized by membership function

$$\propto_{\tilde{c}_j} = \begin{cases} 0 & \text{if} \quad c_j^U \leq x \text{ or } x \leq c_j^L, \\ h_j(x) & \text{if} \quad c_j^L \leq x \leq c_j , \\ g_j(x) & \text{if} \quad c_j \leq x \leq c_j^U, \end{cases} \qquad (5.5)$$

where c_j^u and c_j^L are upper and lower bound of $\tilde{c}_j$ and $h_j(x)$ and $g_j(x)$ are assumed to be strictly increasing and decreasing continuous functions, respectively.

To solve this problem three different approaches are described [23].

a) Multiobjective approach

Membership function for the goals in (5.4) will be supposed to be like (5.5).

By considering the $(1-\alpha)$ cut of every $\tilde{c}_j$ $\alpha \in [0,1]$

$$\forall x \in R, \qquad \propto_{\tilde{c}_j}(x) \geq 1 - \alpha$$

a fuzzy solution of (5.4) may be found from the parametric solution of the multiobjective parametric LP problem:

$$\max[c^1x, c^2x, ..., c^{2^n}x]$$
$$\text{s.t. } Ax \le b,\ x \ge 0$$
$$c^k \in E(1-\alpha),\quad \alpha \in [0,1],\quad k = 1,....,2^n,$$

where $E(1-\alpha)$ for each $\alpha \in [0,1]$, is the set of vectors in R^n such that each of its components is either in the upper bound, or in the lower bound of the respective $(1-\alpha)$-cut.

b) Possibilistic approach

Here it is assumed that the objective value

$$Z = \tilde{c}_1 x_1 + \tilde{c}_2 x_2 + ... + \tilde{c}_n x_n \tag{5.6}$$

when $\tilde{c}_j$ are fuzzy numbers with triangular membership functions $\Delta(c_j, c_j^L, c_j^U)$ $(j \in N = \{1,2,...,n\})$, is also a fuzzy number with

$$\alpha(y) = \begin{cases} 1-[|2y-(c^L+c^U)|x]/(c^U-c^L)x & \text{if } x>0 \\ 1 & \text{if } x=0,\ y=0 \\ 0 & \text{if } x=0,\ y\neq 0 \end{cases}$$

where $c^U = \left(c_1^U, c_2^U, ..., c_n^U\right)$, $c^L = \left(c_1^L, c_2^L, ..., c_n^L\right)$ and c_j^U and c_j^L similar to the ones in (5.5).

Maximization for a fuzzy set like (5.6) is done by means of

$$w_1 y^U x + w_2 y^L x \to \max,$$

where $w_1 + w_2 = 1$, $w_1, w_2 \in [0,1]$, $y^U = (c_1^U, c_2^U, ..., c_n^U)$, $y^L = (c_1^L, c_2^L, ..., c_n^L)$.

Then to solve (5.4) the auxiliary conventional LP problem is used

$$w_1 c^U x + w_2 c^L x \to \max$$

$$\text{s.t. } Ax \le b,$$

$$x \ge 0,$$

where (w_1, w_2) is a pair of values that the decision maker must choose according to his wishes.

c) Stratified piecewise reduction approach [23].

In this approach the vagueness of the goals is modeled by means of nested intervals, each of which has a membership or possibilistic degree $\alpha_i \in [0,1]$, $i \in M = \{1,2,...,m\}$ that is each $\tilde{c}_j$ in the objective function is defined by means of the set

$$\tilde{c}_j = \{[c_j^L, c_j^U]^i / a_i\ ;\ i \in M\},\quad \forall j \in N \tag{5.7}$$

verifying

$$\forall \alpha_1,\ \alpha_2 \in [0,1], \quad \alpha_1 \geq \alpha_2 \Rightarrow [c_j^L, c_j^U]^1 \subseteq [c_j^L, c_j^U]^2 .$$

Every fuzzy cost given by (5.5) can be formulated in the form (5.7) by using its α-cuts.

For each α_i, $i \in M$, the following auxiliary problem is used to solve (5.4):

$$\begin{aligned} &\max \lambda, \\ &\text{s.t. } f_1[c^{L\alpha} x] \geq \lambda, \ f_2[c^{U\alpha} x] \geq \lambda, \\ &Ax \leq b, \ x \geq 0, \lambda \geq 0, \end{aligned} \tag{5.8}$$

where

$$f_1[c^{L\alpha} x] = \frac{c^{L\alpha} x - z^{U\alpha}_{min}}{z^{*\alpha}_{min} - z^{U\alpha}_{min}} \quad \text{if} \quad z^{U\alpha}_{min} \leq c^{L\alpha} x \leq z^{*\alpha}_{min},$$

$$f_2[c^{U_\alpha} x] = \frac{c^{U\alpha} x - z^{U\alpha}_{max}}{z^{*\alpha}_{max} - z^{U\alpha}_{min}} \quad \text{if} \quad z^{U\alpha}_{max} \leq c^{U\alpha} x \leq z^{*\alpha}_{max}.$$

On the other hand, by definition [21]

$$z^{*\alpha}_{min} = c^{L\alpha}(x^*_{min}) = \max\{c^{L\alpha} x | x \in X\} \ ; \quad z^{U\alpha}_{min} = c^{L\alpha} x^*_{max},$$

$$z^{*\alpha}_{max} = c^{U\alpha}(x^*_{max}) = \max\{c^{U\alpha} x | x \in X\} \ ; \quad z^{U\alpha}_{max} = c^{U\alpha} x^*_{min}$$

with $X = \{x \in R^n | Ax \leq b, \ x \geq 0\}$.

Now, the solution of (5.4) may be found as the intersection of the solutions obtained by solving (5.8) for each $\alpha_i, i \in M$.

Linear programming problem with fuzzy resources

This problem is formulated as follows:

$$\begin{aligned} &cx \rightarrow \max, \\ &Ax \lesssim b, \\ &x \geq 0, \end{aligned} \tag{5.9}$$

where $c = (c_1, ..., c_n)$, $A = (a_1, ..., a_m)$, $a_i = (a_{i1}, ..., a_{in})$.

Here is considered the case in which a decision maker assumes he or she can tolerate violations in the accomplishment the constraints, i.e. he permits of the constraints to be satisfied "as well as possible". Each fuzzy constraint $(Ax)_i \lesssim b_i$, in the constraint set can be defined as a fuzzy set i by means of a membership function:

$$\alpha_i(x) = \begin{cases} 1 & \text{if } (Ax)_i \le b_i, \\ 1-[(Ax)_i - b_i]/\bar{b}_i & \text{if } b_i \le (Ax)_i \le b_i + \bar{b}_i, \\ 0 & \text{if } (Ax)_i \ge b_i + \bar{b}_i. \end{cases}$$

This membership function expresses that the decision maker tolerates violations in each constraint up the value $b_i + \bar{b}_i, i = 1, 2, ..., m$

Let it be noted that the functions α_i are defined for every $x \in X$ and give the accomplishment degree of the i-th constraint for $x \in R^n$. Therefore, the problem becomes to find x such that cx and $\alpha_i(x)$ for $i = 1,2,...,m$ are maximized. This is a multiple objective optimization problem.

We will use the following method to solve this problem [54,93]. First, the following two standard LP problems are solved:

$$\begin{aligned} &\text{maximize } cx \\ &\text{s.t. } (Ax)_i \le b_i, \ i = \overline{1,m}, \\ &x \ge 0. \end{aligned} \tag{5.10}$$

$$\begin{aligned} &\text{maximize } cx \\ &\text{s.t. } (Ax)_i \le b_i + \bar{b}_i, \ i = \overline{1,m}, \\ &x \ge 0. \end{aligned} \tag{5.11}$$

If x^0 and x^1 are solutions of (5.10) and (5.11), respectively, and $z^0 = cx^0$ and $z^1 = cx^1$. The degree of optimality can be characterized by:

$$\alpha_0(Ax) = \begin{cases} 1 & \text{if } cx > z^1, \\ 1 - \dfrac{z^1 - cx}{z^1 - z^0} & \text{if } z^0 \le cx \le z^1, \\ 0 & \text{if } cx < z^0. \end{cases}$$

The optimization problem becomes:

$$\max_{x \ge 0} \min[\alpha_0(x), \alpha_1(x), ..., \alpha_m(x)]$$

or equivalently

$$\begin{aligned} &\text{maximize } \alpha \\ &cx \ge z^1 - (1-\alpha)(z^1 - z^0) \\ &(Ax)_i \le b_i + (1-\alpha)\bar{b}_i, \ i = \overline{1,m} \\ &\alpha \in [0,1], \ x \ge 0. \end{aligned}$$

It is necessary to comment fuzzy multi-objective linear programming (FMOLP) in this context at least because of a very often use in business decision making. FMOLP can be presented as the following model [6,85]:

$$\begin{aligned} & c_i x \underset{\sim}{\geq} g_i \ , \quad i = \overline{1, \ q} \\ & c_i x \underset{\sim}{=} h_i \ , \quad i = \overline{q+1, \ k} \\ & a_i x \underset{\sim}{\geq} b_i \ , \quad i = \overline{k+1, \ m} \\ & x \geq 0 \end{aligned} \tag{5.12}$$

The solution to (1) is an x that maximizes a global measure of fulfillment of goals and constraints

$$F = \left(\propto_1 (x), \ \propto_2 (x), \ \ldots, \ \propto_m (x) \right). \tag{5.13}$$

For a simpler case (2) can be rewritten as:

$$F = \sum_{i=1}^{m} w_i \propto_i (x) \ , \quad w_i \geq 0 \ , \qquad \sum_{i=1}^{m} w_i = 1 \tag{5.14}$$

where w_i is the degree of importance of i-th constraint in (5.12).

Having considered the abovementioned the model can be transformed to the following form:

$$\begin{aligned} & \sum w_i \gamma_i \to \max \\ & \text{s.t} \quad \propto_i (x) = \gamma_i \quad , \quad i = \overline{1, m} \\ & \gamma_i \in [0 \ , \ 1] \ , \ \sum_{i=1}^{m} w_i = 1 \\ & x \geq 0 \quad , \quad w_i \geq 0 \end{aligned} \tag{5.15}$$

The membership functions $\propto_i$ for goals and constraints will be determined on the basis of information about the tolerance margin with respect to the target and right-hand side values.

For the case when $\propto_i$ is linear, we can consider the following [5,6].

For inequalities of type $c_i x \underset{\sim}{\geq} g_i$

$$\propto_i (x) = \begin{cases} 1 \Leftrightarrow c_i x - p_i = g_i, & \text{where } \ p_i \geq 0 \\ \gamma_i \in [0,1] \Leftrightarrow c_i x + n_i = g_i, & \text{where } \ n_i = (1 - \gamma_i) d_i \\ 0 \Leftrightarrow n_i \geq d_i \Leftrightarrow c_i x + d_i + r_i = g_i, & \text{where } \ r_i \geq 0 \end{cases} \tag{5.16}$$

Here d is the left tolerance margin, n_i and p_i are the target negative and positive deviations, respectively, r_i is the amount by which the constraint is strongly unfulfilled.

For $c_i x \underset{\sim}{=} h_i$, considering both tolerance margins d_i and t_i

$$\propto_i(x) = \begin{cases} c_i x + n_i = h_i, \text{ where } n_i = (1-\gamma_i)d_i \\ \text{or} \\ c_i x - p_i = h_i, \text{ where } p_i = (1-\gamma_i)t_i \\ \\ c_i x + d_i + r_i = h_i, \text{ where } r_i \geq 0 \\ \text{or} \\ c_i x - t_i - v_i = h_i, \text{ where } v_i \geq 0 \end{cases} \tag{5.17}$$

Here v_i is the amount by which the i-th constraint is strongly violated.

For inequalities $a_i x \underset{\sim}{\leq} b_i$

$$\propto_i(x) = \begin{cases} 1 \Leftrightarrow a_i x + n_i = b_i \;, \text{ where } n_i \geq 0 \\ \gamma_{ii} \in [0 \;, \; 1] \Leftrightarrow a_i x - p_i = b_i \;, \text{ where } p_i = (1-\gamma_i)t_i \\ 0 \Leftrightarrow a_i x - t_i - v_i = b_i \;, \text{ where } v_i \geq 0 \end{cases} \tag{5.18}$$

Taking into account the above considerations, the model (5.15) can be transformed [6] to the following one:

$$\left[\sum_{i=1}^{q} w_i \frac{n_i}{d_i} + \sum_{i=q+1}^{k} w_i \left(\frac{n_i}{d_i} + \frac{p_i}{t_i}\right) + \sum_{i=k+1}^{m} w_i \frac{p_i}{t_i}\right] \to \min \sqrt{b^2 - 4ac} \tag{5.19}$$

s.t.

$$c_i x + n_i - p_i = g_i, i = \overline{1,q}$$
$$c_i x + n_i - p_i = h_i, i = \overline{q+1,k}$$
$$a_i x + n_i - p_i = b_i, i = \overline{k+1,m}$$
$$n_i \leq d_i \;, \; i = \overline{1,k} \;, \; p_i \leq t_i \;, \; i = \overline{q+1,m} \;, \; x \geq 0 \;, \; p_i \geq 0 \;, \; n_i \geq 0$$

Linear programming problem with fuzzy constraints coefficients

If the decision maker does not know exactly the values of the coefficients involved in the problem (in our case a_{ij}'s or the b_j's because the costs c_j are supposed fixed throughout) he can model those inexact values by means of fuzzy numbers.

In this situation we are considering a FLP problem which may be represented as

$$\begin{aligned} &\max\ Z = cx \\ &\text{s.t. } \tilde{A}x \le b \\ &x \ge 0 \end{aligned} \tag{5.20}$$

The main feature of (5.20) is that there are no fuzzy constraints like (5.17). Here the fuzziness is in the coefficients but the decision maker does not permit any violation in the accomplishment of the constraints.

To solve (5.20) we assume that the fuzzy coefficients of matrix $\tilde{A}$ are L-R fuzzy numbers.

For simplicity and without loss of much generality, we assume that $\tilde{A} = [\tilde{a}_{ij}]$ consists of triangular fuzzy numbers. Then the problem becomes

$$\begin{aligned} &\max\ cx \\ &\text{s.t. } ((A\text{-}\underline{A})x, Ax, (A + \overline{A})x) \le b \\ &x \ge 0 \end{aligned}$$

Using the most-likely criterion [47,54] we can convert this problem into the following standard LP problem:

$$\begin{aligned} &\max\ cx \\ &\text{s.t. } \frac{6A\text{-}\underline{A} + \overline{A}}{6} x \le b \\ &x \ge 0 \end{aligned}$$

4th type of fuzzy linear programming in accordance with above given classification is LP problem where all the coefficients are fuzzy numbers (see page 245)

$$\begin{aligned} &\max\ \tilde{Z} = \tilde{c}x \\ &\text{s.t. } \tilde{A}x \le b \\ &x \ge 0. \end{aligned}$$

Assume that $\tilde{c}, \tilde{A}$ and $\tilde{b}$ are triangular fuzzy numbers. Then considered fuzzy LP problem can be converted into the following multiple objective LP problem:

$$\begin{aligned} &\min\ z_1 = \underline{c}\,x \\ &\max\ z_2 = c\,x \\ &\max\ z_3 = \overline{c}\,x \\ &\text{s.t. } (A - \underline{A})\,x \le b - \underline{b},\ A\,x \le b,\ (A + \overline{A})\,x \le b + \overline{b} \\ &x \ge 0 \end{aligned}$$

The method to solve this problem is known.

Linear programming problem with fuzzy variables

Let us consider the following fuzzy linear programming problem:

$$\begin{aligned} &\max\ Z = c\tilde{x} \\ &A\tilde{x} \le \tilde{b}, \\ &\tilde{x} \ge 0, \end{aligned} \tag{5.21}$$

where the variable $\tilde{x}$ is fuzzy.

For solving this problem the matrix A and vector b are divided into submatrices A^1, A^2, A^3 and subvectors b^1, b^2, b^3 in order to extract the groups of constraints of types"$\le$ ", "$\ge$" and "=".

The analog of the system (5.21) will have the following form:

$$\max\ Z = \begin{cases} cx & \text{when } Ax \le b, \\ c(x-\underline{x}) & \text{when } A(x-\underline{x}) \le b-\underline{b}, \\ c(x+\overline{x}) & \text{when } A(x+\overline{x}) \le b+\overline{b}, \end{cases}$$

subject to

$$\begin{aligned} &A^1x + A^1\overline{x} \le b^1 + \overline{b}^1 \\ &A^1x - A^1\underline{x} \le b^1 - \underline{b}^1 \\ &A^2x + A^2\overline{x} \ge b^2 + \overline{b}^2 \\ &A^2x - A^2\underline{x} \ge b^2 - \underline{b}^2 \\ &A^3x + A^3\overline{x} = b^3 + \overline{b}^3 \\ &A^3x - A^3\underline{x} = b^3 - \underline{b}^3 \\ &A^1x \le \overline{b}^1, A^2x \ge \underline{b}^2, A^3x = b^3, \\ &\overline{x} \ge 0, \underline{x} \ge 0, x \ge 0. \end{aligned} \tag{5.22}$$

Denote $x' = x - \underline{x}$, $x'' = x + \overline{x}$ and decompose the system (5.22) into three systems, which include the constraints on average and border values

$$\begin{aligned} &\max F = cx, \\ &A^1x \le b^1, \\ &A^2x \ge b^2, \\ &A^3x = b^3, \\ &x \ge 0. \end{aligned} \tag{5.23}$$

$$\max F' = cx', \tag{5.24}$$

$$A^1x' \le b^1 - \underline{b}^1,$$
$$A^2x' \le b^2 - \underline{b}^2,$$
$$A^3x' = b^3 - \underline{b}^3,$$
$$x \ge 0.$$

$$\max F'' = cx''$$
$$A^1x'' \le b^1 + b^{-1},$$
$$A^2x'' \ge b^2 + b^{-2}, \quad (5.25)$$
$$A^3x'' = b^3 + b^{-3},$$
$$x'' \ge 0.$$

By solving the system (5.23), we obtain the value x^*. Then, in order to satisfy L-R representation principle, it is necessary to add the following constraints to the systems (5.24) and (5.25):

$$x' = x - \underline{x} \le x^*,$$
$$x'' = x + \overline{x} \ge x^*.$$

Assume, that the systems (5.24) and (5.25) have been solved and the corresponding solutions are x'^* and x''^*. Then the solution of the system (5.20) will be the fuzzy vector of L-R type (Fig.5.1):

$$\tilde{x}^*_{LR} = (x^*, x^* - x'^*, x''^* - x^*)$$

i.e.

$$\propto_{\tilde{x}^*}(x) = \begin{cases} 0 & \text{if} \quad x \le x' \vee x \ge x''^* \\ \dfrac{x - x'^*}{x^* - x'^*} & \text{if} \quad x'^* < x < x^*, \\ -\dfrac{x - x''^*}{x''^* - x^*} & \text{if} \quad x^* < x < x''^*, \\ 1 & \text{if} \quad x = x^* \end{cases}$$

If the deviations $\underline{b}$ and $\overline{b}$ are proportional to the mean value the algorithm is simplified.

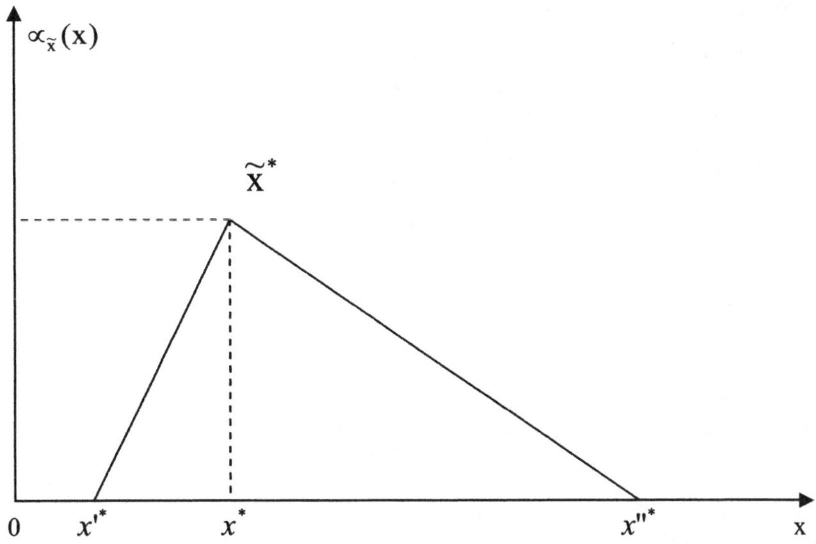

Fig. 5.1. L-R type fuzzy solution

Theorem: If the following conditions are satisfied:

$$\underline{b} = k_1 b \quad \text{if } k_1 \leq 1, \tag{5.26}$$

$$\overline{b} = k_2 b \quad \text{if } k_2 \leq 1, \tag{5.27}$$

and the system (5.23) has the optimal solution, then the systems (5.24) and (5.25) also have the optimal solutions x'^* and x''^*, moreover:

$$x'^* = (1 - k_1)x^*, \tag{5.28}$$

$$x''^* = (1 + k_2)x^*, \tag{5.29}$$

Proof. (See [2]).

Due to derived decomposition we can avoid the growth of dimension of precise analog of the system (5.29), that is the relevant advantage of the suggested approach.

The resulting solution of fuzzy L-R type vector

$$\tilde{x}_{LR} = (x^*, \; x^* - x'^*, \; x''^* - x^*)$$

is presented to the person, who makes decision.

Such representation of the solution allows to estimate the variation of the output $x \in [x'^*, x''^*]$, given the variation of input $\tilde{b} \in [b - \underline{b}, \; b + \overline{b}]$.

5.2 Evolutionary Algorithm Based Fuzzy Linear Programming

Let's consider the following fully fuzzified linear programming problem where all the parameters and variables are fuzzy (6th LP problem on classification given in section 5.1.):

$$\max \tilde{Z} = \tilde{c}\tilde{x} \tag{5.30}$$

$$\tilde{A}\tilde{x} \leq \tilde{b} \tag{5.31}$$

$$\tilde{x} \geq 0 \tag{5.32}$$

where $\tilde{c} = (\tilde{c}_1, \tilde{c}_2, ..., \tilde{c}_n)$, $\tilde{x} = (\tilde{x}_1, \tilde{x}_2, ..., \tilde{x}_n)$, $b = (\tilde{b}_1, \tilde{b}_2, ...\tilde{b}_n)$, , are vectors of fuzzy numbers (for example, triangle, trapezoidal) and $\tilde{A} = (\tilde{a}_{ij})$ is an $m \cdot n$ matrix of fuzzy numbers. The determination of an optimal solution $\tilde{x}_{opt.}$ to (5.30)-(5.32) is a very difficult problem and the use of the above described methods for searching an optimal solution is accompanied with heavy computational needs. Our experience in solving the problem of linear and non-linear fuzzy regression problems (section 7.4) as well as the evolving in the literature evolutionary algorithm based approaches to fuzzy problems [14] show that it is purposeful to use the directed search methods to find approximate solutions to fuzzy linear programming. For this reason in this section we consider the use of a directed search algorithm such as an evolutionary algorithm where the direction is given by the fitness function to solve to mentioned problem.

Before we describe the algorithm to solve the problem (5.30)-(5.32) it is necessary to note the following. First we need to find out which of the methods is more purposeful to use for computing $\max \tilde{Z}$ in formula (5.30). Secondly we must specify how to evaluate the inequality (5.31). As shown in [14] 23 different ranking methods are presented in [17]. None of these methods guarantees that $\tilde{Z}$ is bounded. In reality, for all real world problems $\tilde{Z}$ is bounded almost always.

Below we describe an evolutionary algorithm based procedure to solve the fuzzy linear programming problem.

1. Constructing fitness function F using fuzzy objective function $\tilde{Z}$.
2. Initializing the population (set of individuals) randomly according to the constraints (5.30) and (5.32). Each individual of the population is described by a set of fuzzy numbers (triangular, trapezoidal, Gaussian, etc.) $\tilde{x}_i, i = \overline{1,n}$, which is a feasible solution to the (5.30)-(5.32) linear programming problem. It is necessary to note that an individual is taken from the initial population only if it satisfies the constraints (5.31) and (5.32).
3. Evaluating each individual (chromosome) in the initial population in terms of fitness function F, if the termination conditions are met, then stop the evaluation.

4. Creating a new generation using the selection operator in accordance with the fitness values of the chromosomes. Note the importance to choose k fittest chromosomes from the initial population with l individuals to build an offspring of the next generation, where $k \leq l$.
5. Creating new individuals by applying a crossover operator to the k fittest elements of the previous generation according to a user specified probability of crossover P_i. The resulting offspring replaces the original chromosomes in the previous generation. It is necessary to note that the new individuals are taken into the new generation only if they satisfy the constraints (5.31) and (5.32).
6. Applying the mutation operator with some user-specified probability of mutation P_m to elements of the temporary generation. If the new element (chromosome) satisfies the constraints (5.31) and (5.32), it is taken into the new generation. The evaluation including these three processes continues until the fitness of individual is greater than the positive value or the maximum number of generations is reached.
7. Stop, return the best individual and translate it into a fuzzy feasible solution to the problem (5.30)-(5.32).

The development of an effective evolutionary algorithm based solution to the product mix problem is considered in detail in [14]. Let's briefly consider this example.

A company produces three products A, B, and C each of which must be processed through three departments. The approximate times the products are in departments and selling price for each product can vary (for example due to small discounts). So the following fuzzy linear programming problem is formulated [14]:

$$\max \tilde{Z}=(5.8,6,6.2)\,\tilde{x}_1+(7.5,8,8.5)\,\tilde{x}_2+(5.6,6,6.4)\,\tilde{x}_3 \tag{5.33}$$

$$\text{s.t. } (5.6,6,6.4)\,\tilde{x}_1+(7.5,8,8.5)\,\tilde{x}_2+(2.8,3,3.2)\,\tilde{x}_3 \leq (283,288,293), \tag{5.34}$$

$$(11.4,12,12.6)\,\tilde{x}_1+(7.6,8,8.4)\,\tilde{x}_2+(5.7,6,6.3)\,\tilde{x}_3 \leq (306,312,318), \tag{5.35}$$

$$(1.8,2,2.2)\,\tilde{x}_1+(3.8,4,4.2)\,\tilde{x}_2+(0.9,1,1.1)\,\tilde{x}_3 \leq (121,124,127), \tag{5.36}$$

$$\tilde{x}_1,\tilde{x}_2,\tilde{x}_3 \geq 0, \tag{5.37}$$

where $\tilde{x}_1$, $\tilde{x}_2$, and $\tilde{x}_3$ are triangular fuzzy numbers for the amount to produce each product per week.

In [14] the determination of max $\tilde{Z}$ boils down to a multi-objective optimization problem. The solution of the problem depends also on the kind of inequality relation $\leq$ used. When using Chen's inequality [14], an "optimal" solution of (5.33)-(5.37) is determined as $\tilde{x}_1 = \{63, 63, 100\}$, $\tilde{x}_2 = \{80, 80, 120\}$ $\tilde{x}_3$ $\{65,65,202\}$. The corresponding value of $\tilde{Z}=\{1325,1410,2940\}$. In reality $\tilde{Z}$ is a triangle shaped fuzzy number very close to being a triangle.

5.3 Fuzzy Chaos Approach to Fuzzy Linear Programming Problem

For generality again we will consider a fuzzy linear programming problem with fuzzy goal, fuzzy constraints, and fuzzy variables

$$\max \tilde{Z} = \tilde{c}\tilde{x} \tag{5.38}$$

$$A\tilde{x} \leq \tilde{b} \tag{5.39}$$

$$\tilde{x} \geq 0 \tag{5.40}$$

There is no straight way for solving the problem (5.38)-(5.40) as opposed for the case of crisp linear programming problem. Even if there exists an "optimal" value of $\tilde{x}_i, i=\overline{1,n}$, there is no algorithm available to calculate an exact value of the optimal $\tilde{x}_i, i=\overline{1,n},$. In this section we consider an approximate solution to (5.38)-(5.40) problem using chaotic generation of fuzzy numbers $\tilde{x}_i, i=\overline{1,n},$.

Preserving generality, but for simplicity, we represent $\tilde{x}_i, i=\overline{1,n}$, as fuzzy triangular numbers. The algorithm for solving the problem (5.38)-(5.40) on the base of chaos approach is described below:

1. Generate chaotically an initial set of fuzzy numbers $\tilde{x}_i, i=\overline{1,n},$. Test $\tilde{x}_i, i=\overline{1,n}$, to see if they are triangular fuzzy numbers (or trapezoidal, if we describe $\tilde{x}_i, i=\overline{1,n}$, by trapezoidal fuzzy numbers).
2. Check if the test set of generated fuzzy numbers satisfies the constraints (5.39) and (5.40). If yes, then go to step 3, otherwise chaotically generate a new set of fuzzy numbers $\tilde{x}_i, i=\overline{1,n},$.
3. Compute $\tilde{Z}$ and save this current value of objective function.
4. Generate chaotically a new set of fuzzy numbers $\tilde{x}_i, i=\overline{1,n}$, test this set of fuzzy numbers in accordance with step 2.
5. Compute the current value of $\tilde{Z}^c$ and compare to the previous value of $\tilde{Z}$. If $\tilde{Z}^c$ is better than previous $\tilde{Z}$, then replace the previous "best" value of $\tilde{Z}$ by $\tilde{Z}^c$.
6. Continue the process until the objective function does not further improve.

The application of fuzzy chaos to the optimization, fuzzy programming, fuzzy maximum flow is considered in [13], which also discusses on the ways of generating chaotic fuzzy sets. In the present work we use the following procedure to generate series of fuzzy numbers. The following chaotic function is used to generate chaotic numbers:

$$f(x) = \lambda x(M - x)/M, \tag{5.41}$$

where f maps [0,M] into [0,M].

Assume we want to generate a chaotic sequence of triangular fuzzy number set $\tilde{x}_n = (a_n, b_n, c_n)$ in [0,M]. If $\tilde{x}_{n+1} = (a_{n+1}, b_{n+1}, c_{n+1})$, then choosing $\lambda_i, i = \overline{1,3}$ such that f is chaotic and considering that $f_i(x) = \lambda_i x(M - x)/M, i = \overline{1,3}$, the values of $a_{n+1}, b_{n+1}, c_{n+1}$ can be determined from the following relationships:

$$a_{n+1} = \min\{f_1(a_n), f_2(b_n), f_3(c_n)\}$$

$$c_{n+1} = \max\{f_1(a_n), f_2(b_n), f_3(c_n)\}$$

$b_{n=1}$ is the middle value of $\{f_1(a_n), f_2(b_n), f_3(c_n)\}$.

5.4 Fuzzy–Probabilistic Scheduling for Oil Refinery

In this section nature of the various existence units in the process of the oil Refinery plant and also the nature of the desired plant is investigated, which in result two different types, probability and fuzzy, uncertainties and its sources are recognized.

Considering these uncertainties, principles of probability and fuzzy sets theories are applied in order to formulate various appropriate mixed mathematical models for each unit. For more illustration, the mathematical model with mixed: fuzzy, stochastic and deterministic constraints on the basis of observation datas for Vacuum unit of the oil plant are introduced.

The subject of decision making is included of choosing an alternative from among a number of available alternatives which best provides the organization's goals.

Decision may be taken on the basis of intuitive approaches or on the basis of quantitative procedures.

In complex situations which are involved a considerable number of variable and also, possibility of providing the following factors: availability of relevant data, determination objectives in quantitative terms, and availability of applicable models, the experienced and successful managers use quantitative approaches. Scheduling modeling is a suitable quantitative mathematical tool for dealing decision making problems. Scheduling process is involved of selection an alternative plan from among several available alternative plans and allocation limited resources.

On large-scale oil refinery plant designing an appropriate and reasonable scheduling model for the different reasons such as: there are a number of various units included in the plant process, existence a number of different feasible schedules, the relevant process involved complex interrelationships among actiyities, due to multi-stage character of scheduling, existence different kinds economic and technical constraints, coordination current decisions with an integral one, due to partially structured both from formulating and from solution techniques points of view, is quite complicated and difficult tasks .for planners and schedulers. Today, nearly almost all organizations make high efficient use of scheduling techniques in various fields for dealing and solving their decision making problems.

We shall first discuss the sources that cause different types of uncertainty. To understand the need for dealing with uncertainty one must identify the sources of uncertainty and how the uncertainty accumulated within System, one way of classifying and summarizing the sources of uncertainty maybe expressed as follows.

Information deficiency

Uncertainty involved in problem-solving situations is a result of imperfect and incomplete information. Usually, collection of total information about the situation of the problem seems impossible, so, lack of information induces uncertainty, also, information may be fragmentary, and not fully reliable.

Natural language

Using natural language for purpose of describing the problem and transforming the information causes other kind of uncertainty, namely, vagueness.

Natural language contains, to a considerable amount the characteristic of vagueness. Vague information can be considered as opposed of error-free, certain, and crisp data for the case under analysis.

Measurement

The result of measurements may contain errors, measurement error produce uncertain information in practice, especially in the petroleum industries and manufacturing processes. Our measurements due to applied methods or measuring instruments are uncertain. Conflicting, erroneous, inconsistency, and inaccuracy are other sources of uncertainty, in real context the conflicting information is quite common, the data used to derive a decision tree and rules may contain inconsistencies. Contradictory observations induces error data or lack of consistent rules in real problems, input data which is provided to a system for solving specific problem may not be accurate, so, inaccurate data is another source of uncertainty, indeed, uncertainty is impossibility to predict occurrences of events, in general case, various information deficiencies may result in different types of uncertainty, similar to the sources of uncertainty, the types of uncertainty is a wide range concept, therefore, to give a comprehensive illustration don't seem applicable in this study, basically, various kinds of uncertainty can be recognized in practice which may be classified as follows.

Fuzziness (or vagueness)

Fuzziness results from imprecise boundaries of fuzzy sets, fuzziness is different from ambiguity, it results from the lack of sharpness of relevant distinctions. Fuzziness can be found in various areas such as in engineering and manufacturing, there is considerable difference between fuzziness and imprecision, in that the latter can be described by crisp tolerance interval which is resulted from lack of information about a value of a specific parameter, where as fuzziness exists in the case when the relevant interval contains no sharpness boundaries.

Finally, the terms such as: unclearness, cloudiness, vagueness can be considered as the relevant concepts of fuzziness. Ambiguity may be assumed as an other type of uncertainty, ambiguity results from the lack of certain distinction characterizing an object, ambiguity are involved two higher uncertainty types, namely, non specificity (or imprecision) and strife (or discord) impossibility of exactness observation induce imprecision and it is relevant to the sizes of alternative sets, strife (or discord) arises in the situations which there is conflicting among various sets of alternatives, these strife occurs in the conditions which it remains unclear which of the different alternatives must be selected as the genuine one.

Investigations and estimation the values of the effective parameters on the basis of collected data leads to recognizing three different types of decision making environments in the activities conditions of the desired refinery plant as follows:
a) decision making under conditions of certainty
b) decision making under situations of risk (or probability uncertainty)
c) decision making under conditions of uncertainty (or fuzziness).

In case (a), availability of complete, perfect, and crisp information leads to formulating appropriate deterministic model, the decision making problem becomes maximizing the utility function.

According to case (b), there is possibility of allocation probability values to each states of nature, therefore, the problem can be considered as maximizing the expected utility function. Applied measurement methods and due to used measuring instruments are caused this type of uncertainty in the nature of the plant, considering the case (c), the assignment of probabilities on the states of nature due to lack of sufficient information in not possible, so, this type of uncertainty is the prime domain for fuzzy decision making, arising fuzzy uncertainty in the nature of the desired plant is a result of imperfect and incomplete information, decision making under uncertainty is perhaps the most important class of decision making problem, formulating a model for incorporating different kinds of uncertainty is not an easy task for analysts. We need a quantitative method for taking into account different types of uncertainty, namely, probabilistic and fuzziness, which is occurred in the plant process, therefore, we focus on the mathematical modeling aspects of these various kinds of uncertainties. Basically, type of information dictates the suitable and appropriate mathematical tools, while probabilistic situations have been modeled by probability theories, fuzzy sets theories attempt to deal with the fuzziness inherent in formulating of constraints and objectives. Considering above discussion we have proposed and designed hybrid scheduling model with mixed, deterministic, stochastic, and fuzzy constraints for the plant. The above mentioned peculiarities are formulated in the following scheduling model with three different types, deterministic, stochastic, and fuzzy constraints.

$$\text{Maximize} \quad \sum_{t=1}^{N}\sum_{j=1}^{n_t} c_j^t x_j^t \tag{5.42}$$

$$\text{Subject to: } \sum_{j=1}^{n_t} a_{ij}^t x_j^t \le b_i^t; \quad t=\overline{1,N}, \quad i=\overline{1,m_t}; \tag{5.43}$$

$$P\left[\sum_{t=1}^{N}\sum_{j=1}^{n_t} a_{ij}^t x_j^t \le b_i^t(\omega)\right] \ge \beta_i, \qquad \beta_i \in [0,1]; \tag{5.44}$$

$$\sum_{j=1}^{n_t} \tilde{b}_{ij}^t x_j^t \le \tilde{\beta}_i^t; \quad i=\overline{1,m_t}; \quad t=\overline{1,N}; \tag{5.45}$$

$$x_j^t \ge 0; \; j=\overline{1,n_t}; \; t=\overline{1,N} \tag{5.46}$$

Here (5.42) is the goal function, (5.43) are deterministic constraints on t-th sub-interval, (5.44) are stochastic constraints on the whole horizon, P[] denotes probability mesaure, β_i are positive constants, (5.45) are fuzzy constraints on t-th sub-interval, x_j^t are decision variables, β_i are reliability level of i-th constraint.

Vacuum tower unit mathematical model

For illustration purpose, we choose vacuum tower unit and express its mathematical model which is contained mixed stochastic, fuzzy, and deterministic constraints. The state of the considered unit in the plant process including inputs and outputs data is shown in Fig 5.2.

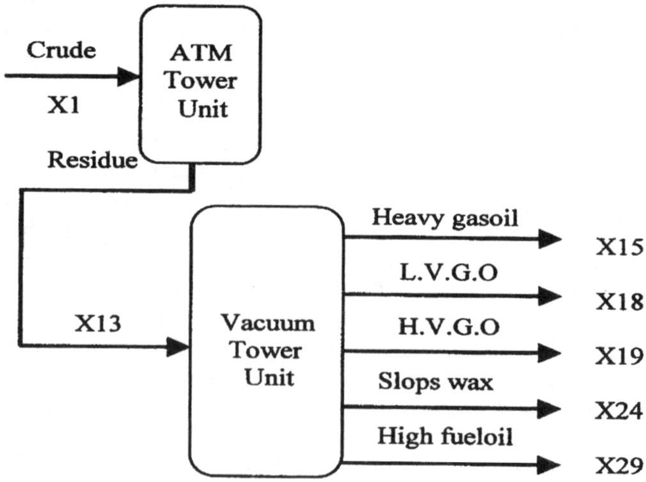

Fig. 5.2. The states of the vacuum tower unit in the plant process .

Fuzzy constraints

The constraint associated with the operating capacity considering in Fig 5.2, can be written as:

$$x_{13} - \tilde{B}_{ij} x_{14} = 0 \tag{5.47}$$

In Eq. (5.47), x_{13} denotes the residue which is assumed as output from ATM tower unit and as a input to the vacuum unit and x_{14} denotes the operating days in month, the employed coefficient $(\tilde{B}_{ij})$ represents the operating capacity coefficient. The estimation is resulted that the coefficient $(\tilde{B}_{ij})$ is characterized as a fuzzy number with the following relevant values: lower bound=4500, mean value=7769, upper bound=10936.

Fuzziness involved in the coefficient $(\tilde{B}_{ij})$ is a result of imperfect and incomplete information. For simplicity, it is assumed that the coefficient $(\tilde{B}_{ij})$ is a triangular fuzzy number with membership function which is shown in Fig 5.3

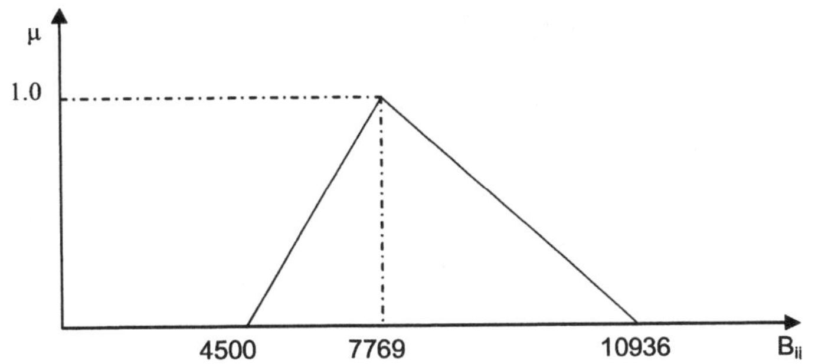

Fig. 5.3. Triangular fuzzy number employed in Eq. (5.47)

Stochastic Constraints

According to the Fig 5.2, consider the output variable x_{15} (heavy gas oil), the observation values in tons on 100 days is expressed in Table 5.1.

Table 5.1. The observation values of heavy gasoil on 100 days

890	549	950	547	810	539	845	529	514	665
840	514	830	521	850	544	880	527	800	635
710	502	777	513	860	524	895	525	782	612
671	485	526	512	732	523	721	577	765	528
630	690	499	487	720	495	667	575	659	492

610	670	602	470	668	493	666	590	657	475
851	649	560	661	601	501	619	920	680	492
522	545	580	660	595	515	615	785	700	589
530	648	535	791	532	520	645	735	712	570
542	647	541	795	540	820	538	510	715	514

Estimation shows that the variable x_{15} (heavy gasoil) is a random variable with mean value =631.5 and standard deviation = 121.78.

Due to applied measurements methods and also used measuring instruments have been caused this probability uncertainty in the nature of the mentioned unit, the percentages of heavy gasoil (x_{15}) on the basis of operating capacity of the vacuum unit is calculated and expressed in Table 5.2.

So the relevant stochastic constraint can be written as:

$$P\left[a_{ij}x_{13} - x_{15} = 0\right] \geq B_i \tag{5.48}$$

In (5.48), the coefficient a_{ij}, is a random variable, by calculating the mean and Standard deviation of the random variable a_{ij}, We obtain $\bar{a}_{ij} = 9.36$, $S.d\,a_{ij} = 2.45$

Table 5.2. The percentages of heavy gasoil on 100 days

14.86	10.83	12.59	9.45	13.7	6.73	13.86	6.95	6.96	9.58
13.92	6.94	16.31	8.99	13.98	9.49	14.64	7.	13.38	10.26
9.88	6.85	10.25	7.	14.26	7.07	15.14	6.97	10.37	9.61
9.68	6.39	7.	7.	9.68	7.02	9.57	8.86	10.08	6.99
9.98	9.34	6.76	6.54	9.66	6.65	8.82	8.8	9.43	6.57
9.52	9.75	9.35	6.26	8.86	6.73	9.72	9.12	9.21	6.31
8.67	7.79	8.51	8.9	9.32	6.79	9.75	15.92	9.21	6.58
6.97	9.49	8.94	8.98	9.21	7.14	9.7	10.45	9.87	9.09
6.86	12.73	6.85	12.99	6.89	7.11	10.45	9.71	9.81	8.71
9.38	10.55	9.18	13.17	9.29	14.11	6.74	6.61	9.73	6.94

Now, Consider another output variable x_{18} (L.V.G.O), the percentage of L.V.G.O (x_{18}) during 100 days on the basis of operating capacity of the vacuum unit is calculated and shown in Table 5.3.

Table 5.3. The percentages of L.V.G.O on 100 days

6.63	5.72	7.76	4.19	8.14	6.04	6.38	5.92	5.62	4.51
7.92	7.26	5.78	4.31	7.99	6.03	6.32	6.24	5.98	4.32
5.99	4.41	5.5	4.5	6.19	9.23	5.84	7.87	5.64	5.95
7.61	6.11	5.29	4.4	4.43	6.35	7.63	5.8	4.49	4.12

5.9	4.59	5.77	6.14	4.57	7.41	7.62	4.56	4.53	4.57
6.18	4.38	8.06	6.11	5.62	5.82	5.99	4.55	5.94	4.53
5.89	4.18	8.46	6.12	5.86	5.77	5.72	4.58	4.29	4.53
6.07	5.73	6.1	7.73	4.21	8.15	5.45	4.12	5.73	4.47
5.42	4.48	5.91	5.77	6.01	4.45	5.57	4.17	4.06	5.79
4.76	6.34	3.99	6.41	4.35	4.47	6.07	7.76	5.77	5.47

Therefore, the relevant probabilistic constraint can be written as:

$$P\left[l_{ij}x_{13} - x_{18} = 0\right] \geq B_i \tag{5.49}$$

where, the coefficient (l_{ij}) is a random variable with mean value of $\bar{l}_{ij} = 5.73$ and s.d $l_{ij} = 1.2$ the third output variable from the vacuum unit is determined by x_{19} (H.V.G.O), the percentages of H. V.G.O (x_{19}) during 100 days is calculated on the basis of collected data and according to the operating capacity of the desired unit which is expressed in Table 5.4.

The associated stochastic constraint can be written as:

$$P\left[d_{ij}x_{13} - x_{19} = 0\right] \geq B_i \tag{5.50}$$

In Eq. (5.50), the employed coefficient $\left(d_{ij}\right)$, is a random variable with mean value of $\overline{d_{ij}}$ =22.83 and standard deviation of, s.d $d_{ij} = 4.43$

Table 5.4. The percentages of H.V.G.O on 100 days

29.12	19.21	25.56	22.31	29.09	34.17	22.51	23.33	23.43	34.67
22.54	23.21	25.48	22.46	18.19	25.36	18.75	31.52	29.49	17.67
25.36	30.59	18.57	22.23	25.26	24.57	23.1	22.86	21.88	17.52
16.64	23.57	17.25	17.02	18.73	18.64	21.83	23.77	22.87	16.73
18.64	23.78	18.05	18.58	22.37	18.95	29.26	22.68	18.87	17.21
22.4	24.39	18.22	17.94	22.09	22.42	17.26	18.46	22.66	22.05
22.64	18.86	22.6	22.98	17.57	29.69	17.75	25.08	18.63	16.73
21.57	27.7	24.08	24.12	22.72	30.35	24.04	18.83	23.32	22.58
17.51	22.56	28.66	23.93	24.22	18.92	23.56	18.49	23.92	17.65

Now, consider x_{24} (slops wax), as another output variable from vacuum unit, the relevant probabilistic constraint can be expressed as:

$$P\left[e_{ij}x_{13} - x_{24} = 0\right] \geq B_i \tag{5.51}$$

Here, the estimation on the basis of observation data shows that the coefficient e_{ij} is a random variable with mean value of $\bar{e}_{ij} = 8.65$ and standard deviation of s.d $e_{ij} = 2.02$. Finally, another output variable from desired unit is denoted by x_{29} (high fueloil), the percentages of high fueloil during 100 days is calculated on the basis of operating capacity of the vacuum unit which the results are expressed in Table 5.5

Table 5.5. The percentages of high fuel oil (x_{29}) on 100 days

29.55	29.99	35.14	18.13	22.4	22.43	22.56	18.96	25.3	18.51
20.95	22.67	22.8	22.85	36.21	22.75	30.22	17.92	18.46	23.41
20.7	18.52	24.8	21.44	28.51	18.4	22.61	21.51	30.29	34.
17.84	17.97	24.7	22.52	18.45	22.48	22.67	23.05	17.72	18.59
22.34	18.07	18.5	24.38	17.82	22.81	18.76	23.23	18.15	33.06
23.03	18.67	23.84	23.78	22.59	35.36	22.79	29.61	18.74	18.62
30.1	22.95	18.43	17.86	21.61	24.17	21.22	36.6	24.88	31.5
18.57	28.81	37.11	18.43	23.66	24.4	18.76	22.99	18.67	23.47
18.27	22.02	20.87	17.7	26.15	22.63	18.77	25.07	18.37	18.62
23.32	18.59	22.47	24.21	25.07	22.47	18.74	28.9	23.36	25.21

The associated stochastic constraint can be written as:

$$P\left[r_{ij}x_{13} - x_{29} = 0\right] \geq B_i \tag{5.52}$$

The estimation is proved that, in Eq. (5.52), the employed coefficient (r_{ij}), is a random variable with mean value of $\bar{r}_{ij} = 23.07$ and standard deviation of $\text{s.d}\, r_{ij} = 4.82$.

Summarizing the previous discussion, Eqs. (5.47) to (5.52) and in addition to other deterministic constraints, represent the mixed mathematical model of the vacuum unit as follows:

$$x_{13} - \tilde{Z}_{ij}x_{14} = 0 \tag{5.53}$$

$$x_{14} \leq 30 \tag{5.54}$$

$$P\left[a_{ij}x_{13} - x_{15} = 0\right] \geq B_i \tag{5.55}$$

$$x_{15} - x_{16} - x_{17} = 0 \tag{5.56}$$

$$P\left[l_{ij}x_{13} - x_{18} = 0\right] \geq B_i \tag{5.57}$$

$$P\left[d_{ij}x_{13} - x_{19} = 0\right] \geq B_i \tag{5.58}$$

$$P\left[e_{ij}x_{13} - x_{24} = 0\right] \geq B_i \tag{5.59}$$

$$P\left[r_{ij}x_{13} - x_{29} = 0\right] \geq B_i \tag{5.60}$$

Investigations and also estimation the values of parameters which affect the situation of existence units on the basis of observations is proved that similar to the vacuum unit, two different types of uncertainty, namely, probabilistic and fuzziness, are involved in the nature of the units. Considering these probability and fuzzy uncertainty, mathematical models with mixed constraints are designed and formulated for each unit. Finally by combining these various mathematical models, the hybrid scheduling model with mixed constraints is created for the plant.

Architecture of the system

The architecture of the suggested hybrid system shown below Fig 5.4 includes fuzzy expert system, database, model base, neural network shell, interpreter and interfaces. Briefly the system functions in the following way. Information about current states of plant and its environment enters the database of the system. The database consists of information concerning the main technological parameters, the quality of raw materials, and semi-finished products. The database also includes subjective information reflecting human observation and monitoring of the plant and its units. The database keeps the main part of the information that is necessary for managerial analysis and running Expert System (ES). We differentiate market based data, plant based data, and technological unit based data. Knowledge base includes plant level rules and technological unit level rules. Scheduling model base allows utilizing depending on the state of environment fuzzy model, stochastic model, and model with mixed (fuzzy, stochastic and crisp) constraints.

Problem solution

We suggest a two-step process of solving such problems under uncertain environment.

Step one: Preliminary solution of the problem by means of fuzzy ES.

The aim of this step is to determine a satisfactory rough solution that is based on experts' knowledge. At this step the decision maker has a possibility to get information from different external and internal sources. If necessary, the neural network technique for short-term forecasting of delivery can be used for estimations of supplying resources. It might be necessary to get values of attributes of fuzzy objects or to evaluate fuzzy relations between different objects for updating and extending the knowledge base. Such information can be obtained via processing the fuzzy database.

The use of fuzzy database and fuzzy expert system enables a user to get feasible and more reliable preliminary solution. This solution may be used for obtaining more precise final solutions by means of the various Operations Research (OR) models in step two. Here depending on the states of the plant and its environment one can modify the parameters, constraints and structure of the scheduling model and priorities. For example, in accordance to rules 5, 6 and 7

RULE 5
IF *UNIT=U15*
AND *U15_ST=NORMAL*

THEN *U15_CAP=NORMAL*

RULE 6
IF *UNIT=U16*
AND *UN16_ST=BELOW NORMAL*
THEN *U16_CAP=BELOW_NORMAL*

RULE 7
IF *PLANT_CAP=INSUFFICIENT*
AND *U15_CAP=NORMAL*
AND *U16_CAP=BELOW_NORMAL*
THEN *INCREASE U15_CAP SLIGHTLY*

we have to replace the constraint
x_{15}<18,000
by the constraint x_{15}<l 8,500

RULE 8
IF *UNIT=U17*
AND *U17_ ST=NORMAL*
THEN *U17_CAP=NORMAL*
IF *PLANNED U17 CAP=NORMAL*
AND *AVAILABLE BENZINE LESS THAN REQUIRED*
THEN *INCREASE OUTPUT OF BENZINE FROM U15 AND U!6*

RULE 9
IF *UNIT=MJ16*
AND *U16_ST= NORMAL*
THEN *U16_CAP= NORMAL*

RULE 10
IF *UNIT=U16*
AND *U16_ST= NORMAL*
AND *U15_ ST - SHUT_ DOWN*
AND *DAILY PLANNED CRUDE CONSUMPTION CONSIDERABLE GREAT THAN AVAILABLE PLANT CAPACITY*
THEN *TEMPRORARY INCREASE U16_ CAP CONSIDERABLE*

In this case as a result we have to replace constraint
$x_{16} < 18,00$
by the constraint
$x_{16} < 20,00$

Initial value of the output coefficients for the UNIT15 (U15) are equal $a_{11} = 0.08$; $a_{12} = 0.14$; $a_{13} = 0.22$; $a_{21} = 0.10$; $a_{22} = 0.22$; $a_{23} = 0.08$. The output coefficients of the light products should be subject to the constraints

$$0.39 < a_{11} + a_{12} + a_{13} < 0.44$$
$$0.38 < a_{21} + a_{22} + a_{23} < 0.43$$

The outputs of the U15 can be described by equations

$$x_2 - 0.\tilde{0}8x_1 = 0$$
$$x_3 - 0.\tilde{1}4x_1 = 0$$
$$x_4 - 0.\tilde{2}2x_1 = 0$$

These equations implicitly take into account a condition that total content of light products in crude is approximately equal 44% and content of the benzene is equal 14%. If one wants to change the output of benzene he has to change the outputs of the fraction 85°C and kerosene simultaneously taking into consideration total output of the light products. Also there is an opportunity of changing of total output of light product within striated range. All above mentioned knowledge generate a number of additional constraints.

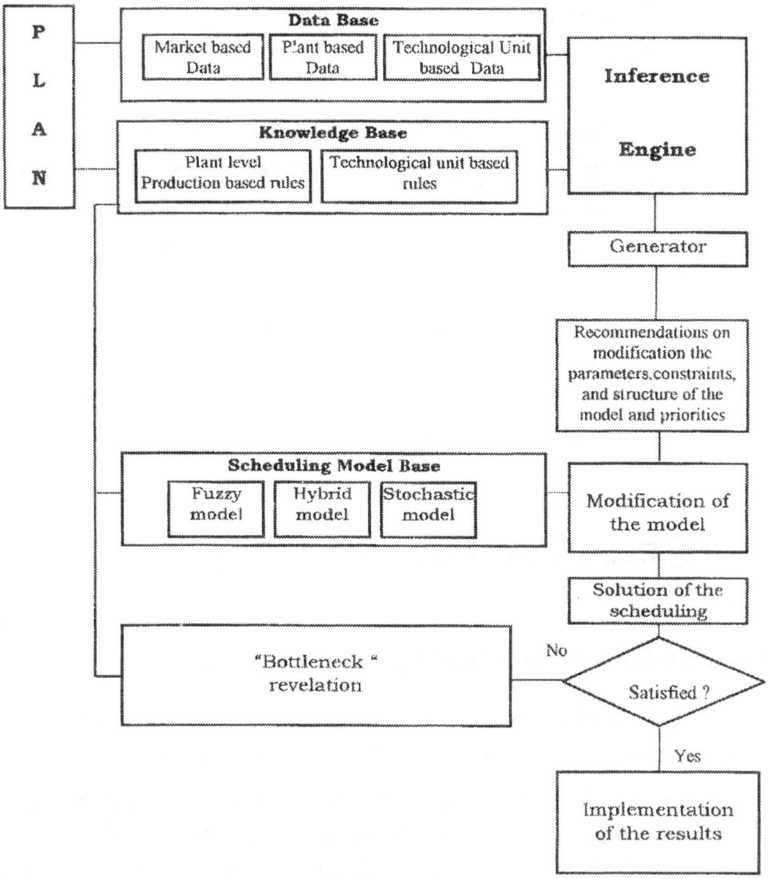

Fig. 5.4. Hybrid Scheduling System

Step two: Solution of the problem via OR model.

At the second step the decision maker chooses a suitable model from the model base and obtains more adequate solution taking into consideration objectives, uncertainty of the environment, and states of the plant. The solution of the problem found in the first step by means of ES is used for identifying the ways for elimination of deviations.

The above approach has been used for solving the oil-refinery scheduling problem. The results of the solution of the scheduling problem are presented in the Table 5.6.

The hybrid system has high level of flexibility and allows to make various decisions depending on the state of environment and goals.

Table 5.6. Results of the solution

	α =1.0		α = 0.5		α =0.8
Crude	427016.7	510698.5	944982.5	493633.6	945408.
Gasoline	25000.	25000.	38707.6	25000.	25000.
Kerosene	64448.2	101316.7	211708.4	83030.2	200078.
Gasoil	49500.	63934.6	172500.	49920.9	172500.
Sulfur	931.3	1063.7	3000.	970.5	3000.
Treated Gases	4159.2	6525.4	13408.3	5056.5	12556.8
Fuel oil	123861.	156670.6	158248.	157119.4	158248.
Asphalt	16074.	26438.4	140000.	19966.2	135596.
A.W.402	2530.	2530.	3180.	2530.	3180.
L.S.R.G.	9804.2	14371.3	23250.	15667.8	23250.
H.S.R.G.	7100.	7100.	7100.	7100.	7100.

5.5 Fuzzy Decision Making

Alternatives selection under condition of uncertainty

There is a set of alternatives $A = \{a_1, a_2, ..., a_m\}$, which is characterized by n criteria C_1, C_2, C_n. Among alternatives A the best alternative is such one, that answers satisfies to all criteria C_i, $i = \overline{1,n}$ simultaneously. It is known that all rules for best alternative selection may be written as [11, 96]:

$$D = C_1 \cap C_2 \cap ... \cap C_n \tag{5.61}$$

If we define the estimation of alternative a_j on criterion C_i through $\propto_{C_i}(a_j) \in [0,1]$ then one may write

$$C_i = \{\propto_{C_i}(a_j)/a_j\}, i = \overline{1,n}, j = \overline{1,m} \tag{5.62}$$

The expression (5.61) with taking (5.62) into account may be represented as

$$\propto_D(a^*) = \max_j \min_i \propto_{C_i}(a_j), \tag{5.63}$$

where a^* is the best alternative.

Selection of manager for factory consider the example [11], when the director of an enterprise union selects the manager for subordinate factory from five candidates to post:

a_1 - chief engineer of aforesaid factory,
a_2 - director of smaller subordinate factory,
a_3 - research institute collaborator,
a_4 - third deputy director,
a_5 - young talented engineer.

The candidates are estimated in accordance with the following criteria:

C_1 - professional skills,
C_2 - organization capabilities,
C_3 - experience in such work,
C_4 - authority,
C_5 - ability in human relations,
C_6 - age.

After determining the conformity degree of each candidate to criteria C_i we derive the following sets:

$C_1 = \{0.9/a_1,\ 0.9/a_2,\ 0.6/a_3,\ 0.8/a_4,\ 0.5/a_5\}$,
$C_2 = \{0.8/a_1,\ 0.9/a_2,\ 0.5/a_3,\ 0.7/a_4,\ 0.5/a_5\}$,
$C_3 = \{0.7/a_1,\ 0.9/a_2,\ 0.8/a_3,\ 0.5/a_4,\ 0.3/a_5\}$,
$C_4 = \{0.9/a_1,\ 0.8/a_2,\ 0.5/a_3,\ 0.6/a_4,\ 0.5/a_5\}$,
$C_5 = \{0.9/a_1,\ 0.9/a_2,\ 0.4/a_3,\ 0.7/a_4,\ 0.6/a_5\}$,
$C_6 = \{0.9/a_1,\ 0.4/a_2,\ 0.8/a_3,\ 0.7/a_4,\ 0.5/a_5\}$.

Using (5.63) we derive

$$\propto_D(a^*) = \{\max[\min(0.9,0.8,0.7,0.9,0.9,0.9)/a_1,$$
$$\min(0.9,0.9,0.9,0.8,0.9,0.4)/a_2, \min(0.6,0.5,0.8,0.5,0.4,0.8)/a_3,$$
$$\min(0.8,0.7,0.5,0.6,0.7,0.7)/a_4, \min(0.5,0.6,0.3,0.5,0.6,0.5)/a_5]\} =$$
$$= \{\max[0.7/a_1, 0.4/a_2, 0.4/a_3, 0.5/a_4, 0.3/a_5\} = \{0.7/a_1\}$$

The best alternative is a_1, i.e. chief engineer of aforesaid factory.

We have considered the task of alternative selection, when they have equal importance.

If criteria C_i have different importance $\alpha_i \geq 0$, then expression (5.61) is transformed to the form

$$D = C_1^{\alpha_1} \cap C_2^{\alpha_2} \cap ... \cap C_n^{\alpha_n} \quad (5.64)$$

$$\frac{1}{n}\sum_{i=1}^{n}\alpha_i = 1. \quad (5.65)$$

Locating chemical plant using fuzzy logic. Let's consider the following problem [11]. It is necessary to choose the best site among alternatives a_1, a_2, a_3 and a_4 for building of chemical plant based on the following criteria:

C_1 - proximity to consumers,

C_2 - proximity to raw material sources,

C_3 - the availability of labor force in the city.

Every criterion has the coefficient of the relative importance, determined on the base of procedure of criteria comparison. In this case the array B is formed, in which the elements b_{ij} are determined on the scale of importance evaluation:

equal importance $b_{ij} = 1$,

slightly more important $b_{ij} = 3$,

more important $b_{ij} = 5$, and so on,

and satisfied the following conditions

$$b_{ii} = 1,\ b_{ji} = \frac{1}{b_{ij}}.$$

After that the eigenvector W of array B corresponding to maximal eigenvalue λ_{max}, is found:

$$BW = \lambda_{max} W.$$

Value α_i is derived from

$$\alpha_i = nw_i,$$

where w_i are elements of eigenvector W.

Array B for aforesaid problem has the form

$$B = \begin{array}{c} \\ C_1 \\ C_2 \\ C_3 \end{array} \begin{array}{c} \begin{array}{ccc} C_1 & C_2 & C_3 \end{array} \\ \left\| \begin{array}{ccc} 1 & 5 & 1/3 \\ 1/5 & 1 & 1/9 \\ 3 & 9 & 1 \end{array} \right\| \end{array}.$$

The elements of eigenvector W are derived as $w_1 = 0.06$; $w_2 = 0.27$; $w_3 = 0.67$, and coefficients of relative importance of criteria α_i are defined as $\alpha_1 = 3 \cdot 0.06 = 0.18$, $\alpha_2 = 3 \cdot 0.27 = 0.81$, $\alpha_3 = 3 \cdot 0.67 = 2.01$

Fuzzy sets characterizing alternative variants on criteria C_i without taking into consideration the coefficients of relative importance have the following form:

$$C_1 = \{0.5/a_1, 0.7/a_2, 0.3/a_3, 0.6/a_4\},$$
$$C_2 = \{0.5/a_1, 0.4/a_2, 0.8/a_3, 0.4/a_4\},$$
$$C_3 = \{0.2/a_1, 0.1/a_2, 0.6/a_3, 0.9/a_4\}$$

Considering α_i they are transformed to the form:

$$C_1^{0.18} = \{0.5^{0.18}/a_1, 0.7^{0.18}/a_2, 0.3^{0.18}/a_3, 0.6^{0.18}/a_4\} =$$
$$= \{0.88/a_1, 0.94/a_2, 0.81/a_3, 0.91/a_4\}$$
$$C_2^{0.81} = \{0.57/a_1, 0.48/a_2, 0.83/a_3, 0.48/a_4\}$$
$$C_3^{2.01} = \{0.04/a_1, 0.01/a_2, 0.36/a_3, 0.81/a_4\}$$

Therefore according to (5.64)

$$\propto_D(a^*) = \{\max(0.04/a_1, 0.01/a_2, 0.36/a_3, 0.48/a_4)\} = \{0.48/a_4\}$$

Hence it follows that a_4 is the site, where one need build the chemical plant.

Multicriteria decision making on the basis of the fuzzy inference rule

The decision set is characterized by linguistic criteria C_i, $i = \overline{1,n}$ on the basis sets $U_1, U_2, ..., U_n$ accordingly. Totality of criteria determines the acceptable decision, characterized by linguistic variable "satisfaction measure" denoted by S. In the general case the proposition's are represented as

$$d_j : \text{if } C_1 = C_{j1}^* \text{and } C_2 = C_{j2}^* \text{ and...} C_n = C_{jn}^* \text{then } S = S_j^*, j = \overline{1,q} \tag{5.66}$$

Expression

$$\propto_{C_j^*}(\upsilon) = \min_{\upsilon \in V} (\propto_{C_{j1}^*}(u_1), \propto_{C_{j2}^*}(u_2), ..., \propto_{C_{jn}^*}(u_n))$$

corresponds to intersection of sets $A_j = \bigcap_{i=1}^{n} C_{ji}$.

Here $V = U_1 \cdot U_2 \cdot ... \cdot U_n$, $\upsilon = (u_1, u_2, ..., u_n)$ is the membership value of elements u_j to fuzzy set C^*_{jl}.

Then (5.66) one may write in the form

$$d_j : \text{If } C = C^*_j \quad \text{then} \quad S = S^*_j \tag{5.67}$$

If we define the basis set U or V through W and consequently adopt that S^*_j is fuzzy subset of W, and S is fuzzy subset of interval I, then implication of fuzzy sets (5.67) may be written in the form

$$\propto_H(w, i) = \min_{w \in W}(1, (1 - \propto_{C^*}(w) + \propto_S(i))$$

Here H is fuzzy subset on $W \cdot I$, $w \in W$, $i \in I$.

For the set $D = \bigcap_{j=1}^{q} H_j$ and for every $(w, i) \in W \cdot I$ we can write

$$\propto_D(w, i) = \min_{w \in W}(\propto_{H_j}(w, i)), \quad j = \overline{1, q}.$$

Satisfaction measure of alternatives which are described by fuzzy subset A from W are determined on the base of compositional rule of inference:

$$G = A \circ D$$

or

$$\propto_G(i) = \max_{w \in W}(\min(\propto_A(w), \propto_D(w, i))$$

Alternatives matching is performed with utilization of pointing estimations, values of which for the set A is determined as

$$F(A) = \frac{1}{\alpha_{max}} \int_0^{\alpha_{max}} M(A_\alpha) d\alpha, \tag{5.68}$$

where $M(A_\alpha)$ is the middle number of elements of α-level set A_α, and α_{max} is value under which the A has maximal value.

Alternatives selection under decision making is executed by means of determining the satisfaction measure and the corresponding pointing-estimation for every ones. The decision is the alternative with maximal value of pointing estimation.

Example [11]. Choosing best candidate for vacant post. It is necessary to choose the best candidate from five candidates $A = \{a_1, a_2, a_3, a_4, a_5\}$ for a vacant faculty post.

Five criteria is used for decision making:

C_1 - capabilities for investigations,

C_2 - industrial service record,

C_3 - experience in teaching technical subjects,

C_4 - experience in teaching the fuzzy sets theory,

C_5 - ability to work with customers from firms.

We will measure these variables on the basis set of alternatives.

There is the following knowledge which describe cause-effect relationships between criteria values and satisfaction measure of decision:

d_1 :If C_1 = educated and C_2 = some experience and C_3 = good then S = satisfactory ,

d_2 :If C_1 = educated and C_2 = some experience and C_3 = good and C_4 = able then S = more than satisfactory ,

d_3 :If C_1 = educated and C_2 = some experience and C_3 = good and C_4 = able and C_5 = capable then S = irreproachable ,

d_4 :If C_1 = educated and C_2 = some experience and C_3 = good and C_5 = capable, then S = very satisfactory ,

d_5 :If C_1 = very educated and C_2 = has not experience and C_3 = good and C_5 = capable, then S = satisfactory ,

d_6 :If C_1 = not educated and C_3 = not able to teach then S = non satisfactory .

The variable S is given on the set $J = \{0, 0.1, 0.2, ..., 1\}$ and has the following linguistic terms:

S_1^* - satisfactory,

S_2^* - more than satisfactory,

S_3^* - irreproachable,

S_4^* - very satisfactory,

S_5^* - non satisfactory.

Here "satisfactory" is defined as $\propto_S(c) = c$, $c \in J$, "very satisfactory" - as $\propto_S(c) = c^2$, "non satisfactory" - as $\propto_S(c) = 1 - c$, "more than satisfactory" - as $\propto(c) = \sqrt{c^3}$, "irreproachable" – as

$$\propto_S(c) = \begin{cases} 1 & \text{if } c = 1, \\ 0 & \text{if } c \neq 1. \end{cases}$$

There are the following estimations of each candidate:

C_1^* = educated investigator = $\{0.8/a_1, 0.6/a_2, 0.5/a_3, 0.1/a_4, 0.3/a_5\}$,

C_2^* = industrial experience = $\{0.5/a_1, 1.0/a_2, 0.0/a_3, 0.5/a_4, 1.0/a_5\}$,

C_3^* = good ability for teaching = $\{0.6/a_1, 0.9/a_2, 1.0/a_3, 0.7/a_4, 1.0/a_5\}$

C_4^* = ability for teaching the fuzzy sets theory =
$=\{1.0/a_1, 0.3/a_2, 1.0/a_3, 0.0/a_4, 0.0/a_5\}$

C_5^* = ability for work with customers =
$=\{0.0/a_1, 0.5/a_2, 1.0/a_3, 0.8/a_4, 0.1/a_5\}$

Using these estimations the above-mentioned knowledge may be represented in the form:

d_1 : If $C = C_1^*$ and C_2^* and C_3^* then $S = S_1^*$;

d_2 : If $C = C_1^*$ and C_2^* and C_3^* and C_4^* then $S = S_2^*$;

d_3 : If $C = C_1^*$ and C_2^* and C_3^* and C_4^* and C_5^* then $S = S_3^*$;

d_4 : If $C = C_1^*$ and C_2^* and C_3^* and C_5^* then $S = S_4^*$;

d_5 : If $C =$ very C_1^* and not C_2^* and C_3^* and C_5^* then $S = S_1^*$;

d_6 : If $C =$ not C_1^* or not C_3^* then $S = S_5^*$.

We will measure these variables on the basis set of alternatives.

Applying rule (5.66) we derive

for $d_1 : \propto_{M_1}(a) = \min(\propto_{C_1^*}(a), \propto_{C_2^*}(a), \propto_{C_3^*}(a))$,

$M_1 = \{0.5/a_1, 0.6/a_2, 0/a_3, 0.1/a_4, 0.3/a_5\}$,

for $d_2 : \propto_{M_2}(a) = \min(\propto_{C_1^*}(a), \propto_{C_2^*}(a), \propto_{C_3^*}(a), \propto_{C_4^*}(a))$,

$M_2 = \{0.5/a_1, 0.3/a_2, 0/a_3, 0/a_4, 0/a_5\}$,

for $d_3 : \propto_{M_3}(a) = \min(\propto_{C_1^*}(a), \propto_{C_2^*}(a), \propto_{C_3^*}(a), \propto_{C_4^*}(a), \propto_{C_5^*}(a))$,

$M_3 = \{0/a_1, 0.3/a_2, 0/a_3, 0/a_4, 0/a_5\}$,

for $d_4 : \propto_{M_4}(a) = \min(\propto_{C_1^*}(a), \propto_{C_2^*}(a), \propto_{C_3^*}(a), \propto_{C_5^*}(a))$,

$M_4 = \{0/a_1, 0.5/a_2, 0/a_3, 0.1/a_4, 0.1/a_5\}$,

for $d_5 : \propto_{M_5}(a) = \min(\propto_{C_1^{*2}}(a), 1-\propto_{C_2^*}(a), \propto_{C_3^*}(a), \propto_{C_5^*}(a))$,

$M_5 = \{0/a_1, 0/a_2, 0/a_3, 0.01/a_4, 0/a_5\}$,

for $d_6 : \propto_{M_6}(a) = \min(1-\propto_{C_1^*}(a), 1-\propto_{C_3^*}(a))$,

$M_6 = \{0.4/a_1, 0.4/a_2, 0.5/a_3, 0.9/a_4, 0.7/a_5\}$.

With taking M_i $(i=\overline{1,6})$ into account we can write:

d_1 : If $C = M_1$ then $S = S_1^*$,

d_2 : If $C = M_2$ then $S = S_2^*$,

d_3 : If $C = M_3$ then $S = S_3^*$,

d_4 : If $C = M_4$ then $S = S_4^*$,

d_5 : If $C = M_5$ then $S = S_1^*$,

d_6 : If $C = M_6$ then $S = S_5^*$.

Using rule (5.66) of implication transformation " If $C = M$ then $S = S^*$ " in expression $\propto_D(a, i) = \min(1, 1 - \propto_M(a) + \propto_{S^*}(a))$, for every pair $(a, i) \in A \cdot I$ we derive fuzzy subset D_i $(i = \overline{1, 6})$ from $U \cdot I$. For example,

$D_1 =$	0	0.1	0.2	0.3	0.4	0.5	0.6	0.7	0.8	0.9	1
a_1	0.5	0.6	0.7	0.8	0.9	1	1	1	1	1	1
a_2	0.4	0.5	0.6	0.7	0.8	0.9	1	1	1	1	1
a_3	1	1	1	1	1	1	1	1	1	1	1
a_4	0.9	1	1	1	1	1	1	1	1	1	1
a_5	0.7	0.8	0.9	1	1	1	1	1	1	1	1

The general solution will be

$$D = D_1 \cap D_2 \cap D_3 \cap D_4 \cap D_5 \cap D_6$$

or

$$\propto_D(a, i) = \min_j \{\propto_{D_j}(a, i)\}$$

i.e.

$D =$	0	0.1	0.2	0.3	0.4	0.5	0.6	0.7	0.8	0.9	1.0
a_1	0.5	0.53	0.59	0.66	0.75	0.85	0.96	0.9	0.8	0.7	0.6
a_2	0.4	0.5	0.54	0.59	0.66	0.7	0.7	0.7	0.7	0.7	0.6
a_3	1	1	1	1	1	1	0.9	0.8	0.7	0.6	0.5
a_4	0.9	0.91	0.9	0.8	0.7	0.6	0.5	0.4	0.3	0.2	0.1
a_5	0.7	0.8	0.9	0.99	0.9	0.8	0.7	0.6	0.5	0.4	0.3

For determining the satisfaction measure for each alternative the compositional rule of inference (5.68) is used, otherwise

$$\propto_{G_k}(i) = \max(\min(\propto_{A_k}(a), \propto_{D_k}(a,i)))..$$

Here $\propto_{G_k}(i) = \propto_D(a_k, i)$, i.e. G_k is k-th line in matrix D.

For the first alternative

$$G_1 = \{0.5/0,\ 0.53/0.1,\ 0.59/0.2,\ 0.66/0.3,\ 0.75/0.4,\ 0.85/0.5,\ 0.96/0.6,\ 0.9/0.7,\ 0.8/0.8,\ 0.7/0.9,\ 0.6/1\}.$$

Compute α-level sets G_j^α and their power $M(G_j^\alpha)$

$$M(G_j^\alpha) = \sum_{i=1}^{n} \frac{a_i}{n}.$$

When $0 \le \alpha \le 0.5$ and $d\alpha = 0.5$ then

$G_1^\alpha = \{0, 0.1, 0.2, 0.3, 0.4, 0.5, 0,6, 0.7, 0.8, 0.9, 1\}$ and $M(G_1^\alpha) = 0.5$.

When $0.9 < \alpha \leq 0.96$ and $d\alpha = 0.6$ then $G_1^\alpha = \{0.6\}$ and $M(G_1^\alpha) = 0.6$.

Pointing estimation G_1 is determined according to formula (5.68):

$$F(G_1) = \frac{1}{0.96} \int_0^{0.96} M(G_1^\alpha) d\alpha = \frac{1}{0.96}(0.5 \cdot 0.5 + 0.55 \cdot 0.03 + 0.6 \cdot 0.06 +$$

$$+ 0.65 \cdot 0.04 + 0.6 \cdot 0.05 + 0.6 \cdot 0.05 + 0.65 \cdot 0.05 + 0.65 \cdot 0.05 + 0.6 \cdot 0.06) = 0.554$$

Analogously we derive

$$F(G_2) = 0.555, \; F(G_3) = 0.425, \; F(G_4) = 0.298, \; F(G_5) = 0.391.$$

Pointing estimation of satisfaction measure for alternative a_i are

$$a_1 = 0.554, \; a_2 = 0.555, \; a_3 = 0.425, \; a_4 = 0.298, \; a_5 = 0.391.$$

The best alternative is a_2.

Multicriteria method of decision making on the basis of fuzzy preference relation

The entity method may be defined as building the set of non-dominated alternatives on the base of fuzzy preference relation [11]. There are set of alternatives A and the selection is done according to criteria $C_j, \quad i = \overline{1, m}$. Every criterion generates a preference relation R_j on alternatives set. One need choose the alternative from set $\{A, R_1, R_2, \ldots, R_m\}$.

Let relations R_j are described by given functions of utility f_j which characterize the numerical evaluation of alternative A on j-th criterion. The decision making problem is to select the alternative that has maximal estimations on all criteria. Such alternative is efficient one. The problem's solution is a set of all efficient alternatives. It is known that for determination of this set in place of the relations set $R_j, \quad j = \overline{1, m}$ may be utilized their intersection

$$Q_1 = \bigcap_{j=1}^{m} R_j.$$

Hence we can find the set of non-dominated alternatives in the set (A, Q_1). Q_1 may be determined according to the following formula:

$$\propto_{Q_1}(a_i, a_j) = \min(\propto_1(a_i, a_j), \ldots, \propto_m(a_i, a_j)). \quad (5.69)$$

A fuzzy subset of non-dominated alternatives in the set $(a, \propto_{Q_1})$ may be determined as

$$\propto_{Q_1}^{ND}(a_i) = 1 - \sup_{a_j \in A} (\propto_{Q_1}(a_j, a_i) - \propto_{Q_1}(a_i, a_j)) \qquad (5.70)$$

If given relations are different by importance, then (5.69) may be rewritten as

$$\propto_{Q_1}(a_i, a_j) = \min(w_1 \propto_1 (a_i, a_j), ..., w_m \propto_m (a_i, a_j)).$$

Consider another form of relations R_j convolution

$$\propto_{Q_2}(a_i, a_j) = \sum_{k=1}^{m} w_k \propto_k (a_i, a_j).$$

Let's build the fuzzy subset of non-dominated alternatives of the set $(A, \propto_{Q_2})$:

$$\propto_{Q_2}^{ND}(a_i) = 1 - \frac{1}{m} \sup_{a_j \in A_k} \sum_{k=1}^{m} \propto_k (a_j, a_i) - \propto_k (a_i, a_j)), \quad a_i \in A. \qquad (5.71)$$

$\propto_{Q_2}^{ND}$ regulates alternatives in accordance with degree of their ability to be non-dominated. As a result of intersection sets $\propto_{Q_1}^{ND}$ and $\propto_{Q_2}^{ND}$ we derive

$$\propto^{ND}(a) = \min(\propto_{Q_1}^{ND}(a_i), \propto_{Q_2}^{ND}(a_i))$$

or

$$A^{ND} = \left\{ a_i \mid a_i \in A, \propto^{ND}(a_i) = \sup_{a_i' \in A} (\propto_A^{ND}(a_i') \right\}$$

from which one needs to choose a necessary alternative. An alternative, that has maximal degree of ability to be non-dominated, will be the rational solution.

Example [11]. Selection of appropriate way for project development in a institute during development of robot a problem, not related to the main work, appeared. The deadline for project development requires quick execution at this stage.

The director has three possibilities:

a_1 - train his collaborator,

a_2 - hire a collaborator for this work, who is able to perform required functions,

a_3 - make contract with another firm for the execution of this work.

The director decides in accordance with the following criteria:

C_1 - speed of work execution,

C_2 - expenditure for execution,

C_3 - quality of work execution.

As pointed out above each criterion generates the preference relation $R_j, j = \overline{1,3}$:

R_1 - a_1 is equal to a_2 by preference,
a_3 is more preferable than a_2 by criterion C_1;

R_2 - a_1 is more preferable than a_2 and a_3,
a_2 is more preferable than a_3 by criterion C_2;

R_3 - a_1 is equal to a_2 by preference,

a_3 is more preferable than a_1 by criterion C_3.

On this basis we form relation matrices R_j, for that we will utilize

$$r_{ij}^k = \begin{cases} 1 & \text{if } i-\text{th alternative is preference than } j-\text{th one by criterion } k, \\ 0 & \text{if alternatives are equal by preference or } i-\text{th alternative is} \\ & \text{worse than } j-\text{th one by criterion } k. \end{cases}$$

$$\propto_1 = \begin{matrix} & a_1 & a_2 & a_3 \\ a_1 & 1 & 1 & 0 \\ a_2 & 1 & 1 & 0 \\ a_3 & 0 & 1 & 1 \end{matrix}, \quad \propto_2 = \begin{matrix} & a_1 & a_2 & a_3 \\ a_1 & 1 & 1 & 1 \\ a_2 & 0 & 1 & 1 \\ a_3 & 0 & 0 & 1 \end{matrix}, \quad \propto_3 = \begin{matrix} & a_1 & a_2 & a_3 \\ a_1 & 1 & 1 & 0 \\ a_2 & 1 & 1 & 0 \\ a_3 & 1 & 0 & 1 \end{matrix}$$

Determine the intersection of relations R_j, i.e. $Q = R_1 \cap R_2 \cap R_3$:

$$\propto_{Q_1}(a_i, a_j) = \min(\propto_1(a_i, a_j), \propto_2(a_i, a_j), \propto_3(a_i, a_j)) = \begin{matrix} & a_1 & a_2 & a_3 \\ a_1 & 1 & 1 & 0 \\ a_2 & 0 & 1 & 0 \\ a_3 & 0 & 0 & 1 \end{matrix}.$$

According to formula (5.70) we determine the subset of non-dominated alternatives:

$$\propto_{Q_1}^{ND}(a_1) = 1 - \sup(\propto_{Q_1}(a_2,a_1) - \propto_{Q_1}(a_1,a_2),$$
$$\propto_{Q_1}(a_3,a_1) - \propto_{Q_1}(a_1,a_3)) = 1 - \sup(0-1, 0-0) = 1;$$
$$\propto_{Q_1}^{ND}(a_2) = 1 - \sup(\propto_{Q_1}(a_1,a_2) - \propto_{Q_1}(a_2,a_1),$$
$$\propto_{Q_1}(a_3,a_2) - \propto_{Q_1}(a_2,a_3)) = 1 - \sup(1-0, 0-0) = 0;$$
$$\propto_{Q_1}^{ND}(a_3) = 1 - \sup(\propto_{Q_1}(a_1,a_3) - \propto_{Q_1}(a_3,a_1),$$
$$\propto_{Q_1}(a_2,a_3) - \propto_{Q_1}(a_3,a_2)) = 1 - \sup(0-0, 0-0) = 1.$$

Thus, $\propto_{Q_1}^{ND} = \begin{matrix} a_1 & a_2 & a_3 \\ 1 & 0 & 1 \end{matrix}$.

The relation Q_2 is defined as

$$\propto_{Q_2}(a_i,a_j) = \sum_{j=1}^{m} w_j m_j(a_i,a_j) = \frac{1}{3}(\propto_1(a_i,a_j) + \propto_2(a_i,a_j) + \propto_3(a_i,a_j)),$$

$$\propto_{Q_2}(a_1,a_1) = \frac{1}{3}(1+1+1) = 1,$$

$$\propto_{Q_2}(a_1,a_2) = \frac{1}{3}(1+1+1) = 1,$$

$$\propto_{Q_2}(a_1,a_3) = \frac{1}{3}(0+1+0) = \frac{1}{3},$$

$$\propto_{Q_2}(a_2,a_1)=\frac{1}{3}(1+0+1)=\frac{2}{3},$$

$$\propto_{Q_2}(a_2,a_2)=\frac{1}{3}(1+1+1)=1,$$

$$\propto_{Q_2}(a_2,a_3)=\frac{1}{3}(0+1+0)=\frac{1}{3},$$

$$\propto_{Q_2}(a_3,a_1)=\frac{1}{3}(0+0+1)=\frac{1}{3},$$

$$\propto_{Q_2}(a_3,a_2)=\frac{1}{3}(1+0+0)=\frac{1}{3},$$

$$\propto_{Q_2}(a_3,a_3)=\frac{1}{3}(1+1+1)=1,$$

$$\propto_{Q_2}(a_i,a_j)=\begin{array}{c} \\ a_1 \\ a_2 \\ a_3 \end{array}\begin{array}{ccc} a_1 & a_2 & a_3 \\ \left\|\begin{array}{ccc} 1 & 1 & 1/3 \\ 2/3 & 1 & 1/3 \\ 1/3 & 1/3 & 1 \end{array}\right\| \end{array}$$

With the help of formula (5.71) we find the subset of non-dominated alternatives of the set $(A, \propto_{Q_2})$:

$$\propto_{Q_2}^{ND}(a_1)=1-\sup(\propto_{Q_2}(a_2,a_1)-\propto_{q_2}(a_1,a_2)),$$

$$\propto_{Q_2}(a_3,a_1)-\propto_{Q_2}(a_1,a_3)=1-\sup(\frac{2}{3}-1,\frac{1}{3},-\frac{1}{3})=1;$$

$$\propto_{Q_2}^{ND}(a_2)=1-\sup(\propto_{Q_2}(a_1,a_2)-\propto_{Q_2}(a_2,a_1)),$$

$$\propto_{Q_2}(a_3,a_2)-\propto_{Q_2}(a_2,a_3)=1-\sup(1-\frac{2}{3},\frac{1}{3}-\frac{1}{3})=\frac{2}{3};$$

$$\propto_{Q_2}^{ND}(a_3)=1-\sup(\propto_{Q_2}(a_1,a_3)-\propto_{Q_2}(a_3,a_1)),$$

$$\propto_{Q_2}(a_2,a_3)-\propto_{Q_2}(a_3,a_2)=1-\sup(\frac{1}{3}-\frac{1}{3},\frac{1}{3}-\frac{1}{3})=1;$$

$$\propto_{Q_2}^{ND}(a_i)=\begin{array}{c} a_1 \quad a_2 \quad a_3 \\ \|1 \quad 0 \quad 1\| \end{array}.$$

The solution choice is done from the set of non-dominated alternatives and is derived as a result of the sets $\propto_{Q_1}^{ND}$ and $\propto_{Q_2}^{ND}$ intersection:

$$\propto_{Q_1}^{ND}\cap\propto_{Q_2}^{ND}=\{(101)\}\cap\{(101)\}=\{(101)\}$$

The solution is either alternative a_1 or a_3 which have maximal degree of ability non-domination.

Decision making under conditions of fuzzy states and outcomes

Let there is a set of alternatives $A = \{a_1, a_2,..., a_m\}$ and the choice of one from this set depends on environment state $X = \{x_1, x_2,..., x_m\}$.

For example, we make decision for choosing the clothes, that depends on weather state (sunny, nasty, rainy etc). There are alternatives: put clothes on for cold weather, take umbrella, etc.

The selection of some alternative from the set under different states of weather is characterized by corresponding use. The rational decision is made on the basis of this use.

If the environment state is $x_j \in X$ and the alternative a_i is selected, then utility of this alternative will be u_{ij}. In general form we derive an utility matrix U:

$$U = \begin{Vmatrix} u_{11}, & ..., & u_{1n} \\ . & ..., & . \\ u_{m1}, & ..., & u_{mn} \end{Vmatrix}.$$

Of course, under known state x_j it is necessary to choose an alternative proceeding from the following conditions

$$u_0 = \max_{i=1,m} u_{ij}$$

It should be noted that often we can't determine the system (environment) state and utility of alternative exactly. Therefore we will consider the method of decision making under such situations, when the system state is known inexactly and use of alternatives under different system's states is determined inexactly too.

Fuzzy state of a system

The system state is defined by fuzzy set

$$\tilde{X} = \bigcup_k \propto_{\tilde{X}}(x_k)/x_k,\ x_k \in X.$$

The utility of alternatives can be determined as

$$\tilde{U}_i^* = \bigcup_k \propto_{\tilde{U}_i}(u_k)/u_k,$$

where $u_k = u_{ik}$, $\propto_{\tilde{U}_i}(u_k) = \propto_{\tilde{X}}(x_k)$.

In order to choose an optimal alternative we will use the concept of maximizing set. The maximizing set $M(Y)$ of the set Y is fuzzy subset. The membership degree to this subset $y \in Y$ expresses the proximity of y to $\sup Y$.

Determine the set Y

$$Y = \bigcup_{i=1}^{m} S(U_i) \qquad (5.72)$$

The maximizing set for alternative $a_i \in A$ has the form:

$$\tilde{U}_{im} = \propto_{\tilde{U}_{im}}(u_k)/(u_k), \qquad (5.73)$$

where $\propto_{\tilde{U}_{im}}(u_k) = (u_k : u_{max})^n$; $u_{max} = \sup Y$.

After determining the maximizing set it is necessary to find an optimizing set on the base of intersection of fuzzy sets $\tilde{U}_{im}$ and $\tilde{U}_i$

$$\propto_{\tilde{U}_{i0}}(u_k) = \min\ (\propto_{\tilde{U}_{im}}(u_k), \propto_{\tilde{U}_i}(u_k)).$$

The set, that determines the optimal alternative, is found as

$$\propto_{\tilde{A}_0}(a_i) = \max_k \propto_{\tilde{U}_{i0}}(u_k).$$

One need select the the necessary alternative proceeding from the following conditions:

$$\propto_{\tilde{A}_0}(a_0) = \max_i \propto_{\tilde{A}_0}(a_i).$$

Fuzzy utility and state

The fuzzy utility of alternative a_i is given in the form

$$\tilde{U}_i^* = \bigcup_k \propto_{\tilde{U}_i^*}(\tilde{U}_k)/\tilde{U}_k \qquad (5.74)$$

where $\tilde{U}_k = \tilde{U}_{ik}$, $\propto_{\tilde{U}_i^*}(\tilde{U}_k) = \propto_{\tilde{X}}(x_k)$.

If

$$\propto_{\tilde{U}_i^*}(\tilde{U}_k)/\tilde{U}_k = \propto_{\tilde{U}_i^*}(\propto_{\tilde{U}_{ik}}(u_1)/u_1)/(\propto_{\tilde{U}_{ik}}(u_1))/u_1$$

then the membership degree of crisp value u_1 to fuzzy set $\tilde{U}_i^*$ is determined from membership of u_1 to $\tilde{U}_{ik}$ and from membership of $\tilde{U}_{ik}$ to $\tilde{U}_i^*$:

$$\propto_{\tilde{U}_i^*}(u_1) = \min(\propto_{\tilde{X}}(x_k), \propto_{\tilde{U}_{ik}}(u_1)). \qquad (5.75)$$

After repeating this procedure for all k the set $\tilde{U}_i^*$ has the form:

$$\tilde{U}_i^* = \propto_{\tilde{U}_i^*}(u_1)/u_1.$$

Example. There is a set of alternatives $A = \{a_1, a_2, a_3\}$ and system can be in one of states $X = \{x_1, x_2, \ldots, x_{10}\}$. An utility matrix is given as

$$U = \begin{Vmatrix} & x_1 & x_2 & x_3 & x_4 & x_5 & x_6 & x_7 & x_8 & x_9 & x_{10} \\ a_1 & VH & H & L & L & M & VL & H & VH & VH & M \\ a_2 & L & VL & H & VH & VH & H & H & M & M & VH \\ a_3 & H & M & L & M & M & H & VH & M & L & L \end{Vmatrix}$$

Values of the utility are represented by linguistic fuzzy sets:

Low : $L = \{0.4/1,\ 1.0/2,\ 0.5/3\}$,
Very Low : $VL = \{1.0/1,\ 0.4/2\}$,
Medium : $M = \{0.4/3,\ 0.7/4,\ 1.0/5,\ 0.7/6, 0.4/7\}$,
High : $H = \{0.5/7,\ 1.0/8,\ 0.5/9\}$,
Very High : $VH = \{0.5/9,\ 1.0/10\}$.

The system state is given as

$$\tilde{X} = \{0.4/x_3, 0.8/x_4, 1.0/x_5, 0.7/x_6, 0.3/x_7\}.$$

According to formula (5.76) we determine the utility of alternatives with taking into account the given system's state:

$$\tilde{U}_1^* = \{0.4/L,\ \ 0.8/L,\ \ 1.0/M,\ \ 0.7/VL,\ \ 0.3/H\} =$$
$$= \{0.4 + 0.8 - 0.4 \cdot 0.8)/L, 1.0/M, 0.7/VL, 0.3/H\} =$$
$$= \{0.88/L, 1.0/M, 0.7/VL, 0.3/H\},$$

$$\tilde{U}_2^* = \{0.864/H,\ 0.8/VH,\ 1.0/VL\},$$

$$\tilde{U}_3^* = \{0.4/L,\ 1.0/M,\ 0.7/H,\ 0.3/VH\},$$

The fuzzy set $\tilde{U}_1^*$ (considering the linguistic values of the use) is transformed as

$$\tilde{U}_1^* = \{0.88/\{0.4/1,\ 1.0/2,\ 0.5/3\}; 1.0/\{0.4/3,\ 0.7/4,\ 1.0/5,\ 0.7/6,\ 0.4/7\};$$
$$0.7/\{1.0/1,\ 0.4/2\};\ 0.3/\{0.5/7,\ 1.0/8,\ 0.5/9\}\}.$$

Taking into account

$$\propto_{U_i^*}(u_1) = \min(\propto_{\tilde{X}}(x_k), \propto_{\tilde{U}_{ik}}(u_1))$$

we can write

$$\tilde{U}_1^* = \{\min\ (0.88,\ 0.4)/1,\ \min\ (0.88,\ 1.0)/2,\ \min\ (0.88,\ 0.5)/3,$$
$$\min(1.0,\ 0.4)/3, \min(1.0, 0.7)/4, \min(1.0, 1.0)/5,$$
$$\min(1.0,\ 0.7)/6, \min(1.0, 0.4)/7, \min(0.7, 1.0)/1,$$
$$\min\ (0.7,\ 0.4)/2,\ \min\ (0.3,\ 0.5)/7,\ \min\ (0.3,\ 1.0)/8,$$
$$\min(0.3,\ 0.5)/9\} = \{0.4/1, 0.88/2, 0.5/3, 0.4/3, 0.7/4,$$
$$1.0/5, 0.7/6, 0.4/7, 0.7/1, 0.4/2, 0.3/7, 0.3/8, 0.3/9\} =$$
$$= \{0.82/1, 0.928/2, 0.7/3, 0.7/4, 1.0/5, 0.7/6, 0.58/7,\ 0.3/8, 0.3/9\},$$

Similarly

$$\tilde{U}_2^* = \{1.0/1,\ 0.4/2,\ 0.5/7,\ 0.864/8,\ 0.75/9,\ 0.8/10\},$$

$$\tilde{U}_3^* = \{0.4/1, 0.4/2, 0.64/3, 0.7/4, 1.0/5, 0.7/6, 0.7/7, 0.7/8, 0.65/9, 0.3/10\}.$$

Let's compute the set Y. Using (5.72) we derive

$$Y = \{1,2,3,4,5,6,7,8,9\} \cup \{1,2,7,8,9,10\} \cup \{1,2,3,4,5,6,7,8,9,10\} =$$
$$= \{1,2,3,4,5,6,7,8,9,10\}.$$

One can see that $u_{max} = 10$. With utilization of (5.71) and (5.72) we determine the maximizing sets:

$$\tilde{U}_{1m} = \{0.1/1, 0.2/2, 0.3/3, 0.4/4, 0.5/5, 0.6/6, 0.7/7, 0.8/8, 0.9/9\},$$
$$\tilde{U}_{2m} = \{0.1/1, 0.2/2, 0.7/7, 0.8/8, 0.9/9, 1.0/10\},$$
$$\tilde{U}_{3m} = \{0.1/1, 0.2/2, 0.3/3, 0.4/4, 0.5/5, 0.6/6, 0.7/7, 0.8/8, 0.9/9, 1.0/10\}.$$

Optimizing sets can be written as

$$\tilde{U}_{10} = \{(\min(0.1, 0.82)/1, \min(0.2, 0.928)/2, \min(0.3, 0.7)/3, \min(0.4, 0.7)/7, \min(0.8, 0.3)/8, \min(0.9, 0.3)/9\} = \{0.1/1, 0.2/2, 0.3/3, 0.4/4, 0.5/5, 0.6/6, 0.58/7, 0.3/8, 0.3/9\}$$
$$\tilde{U}_{20} = \{0.1/1, 0.2/2, 0.5/7, 0.8/8, 0.75/9, 0.8/10\},$$
$$\tilde{U}_{30} = \{0.1/1, 0.2/2, 0.3/3, 0.4/4, 0.5/5, 0.6/6, 0.7/7, 0.7/8, 0.65/9, 0.3/10\},$$

Determine the set $\tilde{A}_0$, in which one need search the optimal alternative:

$$\propto_{\tilde{A}_0}(a_1) = \max(0.1, 0.2, 0.3, 0.4, 0.5, 0.6, 0.58, 0.3, 0.3) = 0.6,$$
$$\propto_{\tilde{A}_0}(a_2) = 0.8, \quad \propto_{\tilde{A}_0}(a_3) = 0.7, \text{ i.e.}$$
$$\tilde{A}_0 = \{0.6/a_1, 0.8/a_2, 0.7/a_3\}.$$

It is obvious that we need to choose the alternative a_2.

Fuzzy multistage decision-making

In multistage decision making processes where the system is developed with regard to time as well as to space the pointed- out one-stage decision-making becomes useless. In such cases decision is made in finite steps in advance and the optimization problem becomes multistage (multi-step or dynamical). The main mathematical tool, which is used for solving multistage optimization problem, is dynamic programming method and the discrete principle of maximum. Nevertheless the further research shows that in real-world dynamic systems the state, solution, transition equations, mappings, control and also terminating time may be fuzzy. In other words, the system may completely or partially operate in fuzzy environment. In such cases existing crisp methods for optimization such systems are inapplicable. In such case multistage decision making problems become fuzzy and for its application the method described below may be applied.

In multistage decision-making problem in each stage there exist fuzzy goals and constraints. In other words there are n goals $G^1, \ldots, G^n$ and m constraints

$U_1,\dots,U_m$. Analogously to fuzzy mathematical programming problem fuzzy solution is defined by aggregation of all goals and constraints:

$$D=G^1 *\dots*G^n *U_1 *\dots*U_m \tag{5.76}$$

or

$$\propto_D(x)=\min\{\propto_{G^1}(x),\dots,\propto_{G^n}(x),\propto_{U_1}(x),\dots,\propto_{U_m}(x)\} \tag{5.77}$$

From the expression (5.71) it is obvious that in fuzzy decision-making process the goal and constraints are same-identical role.

Optimal solution is usual non-fuzzy subset $(u_1^*,\dots,u_m^*)$, which maximizes $\propto_D(x)$

$$\propto_D(x^*)=\max_x \propto_D(x)$$

Up to now we considered only case when the goal and constraints are defined on the same spaces. Now we consider more general case when the goal and constraints are fuzzy sets defined on different spaces.

Let the constraints defined as a fuzzy set $U_i=\{\propto_{U_i}(u_i)/u_i)\}$ defined on V and fuzzy goal $G^i=\{\propto_{G^i}(x_i)/x_i)\}$ defined on W.

Let f be mapping the space V to W: $f:X\to Y, y=f(x)$ .Then for given fuzzy sets $\underset{\sim}{G}^i\in Y$ the fuzzy set $\underset{\sim}{\overline{G}}^i\in X$ is given by the formula:

$$\propto_{\overline{G}^i}(x)=\propto_{G^i}(f(x))$$

When the goal and constraint are defined in the different spaces the fuzzy solution D is defined as follows:

$$D=\overline{G}^1 *\square *\overline{G}^n *U_1 *\square *U_m$$

or

$$\propto_D(x)=\min\{\propto_{\overline{G}^1}(f(x)),\dots,\propto_{\overline{G}^n}(f(x)),\propto_{U_1}(x),\dots,\propto_{U_m}(x)\} \tag{5.78}$$

The last solution often is called f-solution. This formula basically is used in multistage decision-making processes.

The maximizing solution will be

$$\propto_D(x^*)=\max_x \min\{\propto_{G^1}(x),\dots,\propto_{G^n}(x),\propto_{U_1}(x),\dots,\propto_{U_m}(x)\} \tag{5.79}$$

$$\propto_D(x^*)=\max_x \min\{\propto_{\overline{G}^1}(f(x)),\dots,\propto_{\overline{G}^n}(f(x)),\propto_{U_1}(x),\dots,\propto_{U_m}(x)\} \tag{5.80}$$

The most widely used method for solving the problem (5.79) or (5.80) is the fuzzy analogue of the dynamic programming and branch and bound methods [2].

5.6 Multi-Agent Distributed Intelligent System Based on Fuzzy Decision Making

Generally decision making for solving ill-structured problems in DSS takes place in uncertain situations. The main drawbacks of existing traditional DSS are inefficiencies associated with dealing with complex models and large databases. Usually a fuzzy DSS has many input variables and, hence, its knowledge base, containing the totality of fuzzy rules, is very large. Large rule base leads to disadvantages in speed, reliability, and complexity of DSS.

This section introduces an alternative concept for designing fuzzy DSS based on multi-agent distributed artificial intelligent technology and fuzzy decision making.

One of the main characteristics of DSS is the ability to represent and manipulate imprecise knowledge [81, 86], which is usually associated with the uncertainty intrinsic to ill-structured problems [27].

Fuzzy logic is used to handle imprecise knowledge. There is a number of examples of fuzzy DSS in the literature. In [72] an implementation of DSS is presented to solve multiple-attribute problems. The main features of this fuzzy DSS are: it accepts and manipulates goals of fuzzy multiple-attribute problems; its structure is flexible; it learns; it provides different types of inference. The suggested fuzzy DSS is based on the FRIL language. A fuzzy logic based DSS for medical diagnosis is presented in [30]. The representation of uncertain and imprecise knowledge is implemented using fuzzy sets and relations. The new generation of DSS, active DSS, with more advanced features than those of traditional DSS is suggested in [15]. This type of DSS uses hyper-knowledge and fuzzy logic and has significantly improved performance characteristics for problem solving and decision support.

The main drawbacks of the DSS mentioned above are inefficiency associated with dealing with complex mathematical models and large databases. Complex decision situations have large number of input variables and require a large knowledge base of fuzzy rules. A single agent DSS carrying the totality of fuzzy rules on computers with monolith software tools has speed, reliability, and complexity disadvantages. Furthermore, large rule bases lack high consistency and validity and are difficult to maintain and update.

The main idea of the proposed DSS is based on granulation of the overall system into cooperative autonomous intelligent agents. These agents compete and cooperate with each other in order to propose total solution to the problem and organize (combine) individual total solutions into the final solution. Determination of a total solution of DSS is based on multi-criteria fuzzy decision making with unequal objectives as proposed by Yager [96]. Each distributed intelligent agent is implemented as Fuzzy Knowledge Based system with modest number of rules.

Below we will consider fuzzy DSS aimed at solving multiple-attribute problems. Consider a set of decision alternatives $U = \{u_1, u_2, ..., u_N\}$. Each alternative is characterized by M criteria (attributes) $C = \{c_1, c_2, ..., c_M\}$ according to which the desirability of the solution is determined.

Let $\propto_{c_i}(u_j) \in [0,1]$ be rating estimation value (membership value) of alternative u_j on criteria c_i. The objective is to determine the optimal alternative u_j which is better than the others in terms of criteria c_i.

Architecture of Multi-Agent Distributed Intelligent DSS.

The architecture and acting principles of the proposed fuzzy DSS are based on multi-agent distributed artificial intelligence technology and a fuzzy multi-criteria decision –making procedure.

The structure of the considered Multi-Agent Distributed Intelligent DSS in Fig. 5.5 is similar to shown in the system is composed of N parallel agents. All agents receive the same fuzzy (or partly non-fuzzy) input information $x_1, x_1, ..., x_m$. Each agent, which is a knowledge-based system, performs inference and produces its own solution of the full problem (not partial problem as in classical distributed systems)

$$u_j = [u_{j1}, u_{j2,} ..., u_{jl}]^T, j = \overline{1,N},$$

Each agent Ag_j, $j = \overline{1,N}$ is an autonomous intelligent agent, capable of performing inference to produce solution. The agents are knowledge-based systems with modest number of fuzzy rules related through "ALSO":

$$\begin{aligned} &R^k: \text{IF } x_1 \text{ is } A_{k1} \text{ AND } x_2 \text{ is } A_{k2} \text{ AND ... AND } x_m \text{ is } A_{km} \text{ THEN} \\ &u_{k1} \text{ is } B_{k1} \text{ AND } u_{k2} \text{ is } B_{k2} \text{ AND ... AND } u_{kl} \text{ is } B_{kl},\ k = \overline{1,K} \end{aligned} \quad (5.81)$$

The architecture of agent is given in Fig. 5.6. Note that granulation fine, i.e., number of agents in DSS is determined by the system designer.

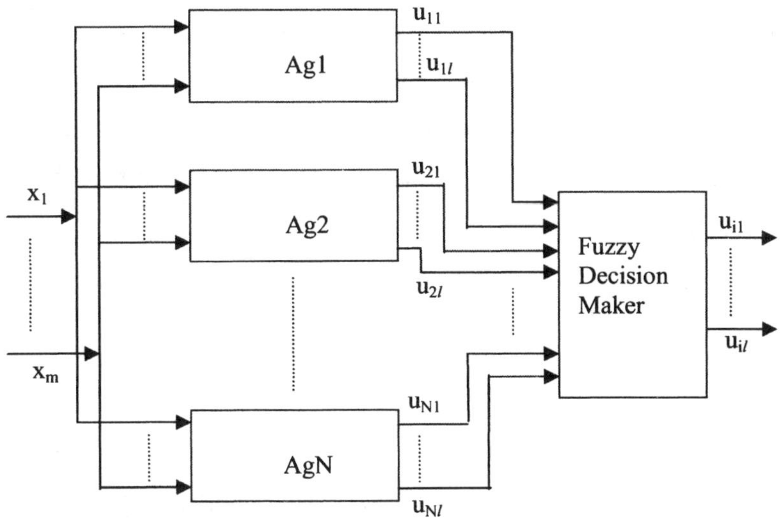

Fig. 5.5. Architecture of multi-agent DSS.

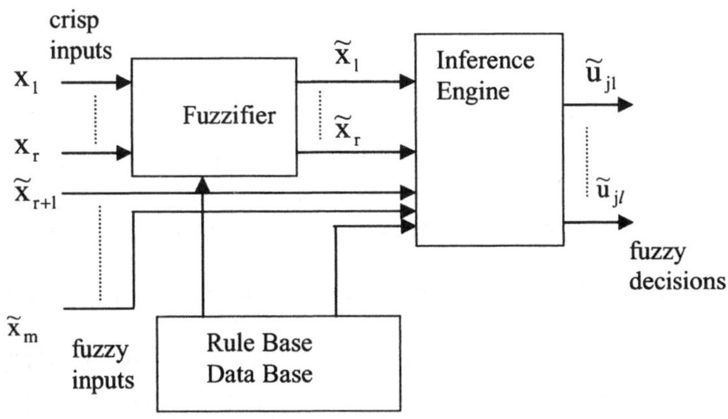

Fig. 5.6. Structure of a competitive agent.

where x_i, $i=\overline{1,m}$ and u_j, $j=\overline{1,l}$ are common input and individual agent output variables, A_{kj}, B_{kj} are fuzzy sets, and K is the number of rules. Note, that inputs $x_1, x_2, \ldots, x_m$ may be crisp or fuzzy variables. If input data are crisp, then fuzzifier (Fig. 5.6) will map these data into fuzzy sets. Decision of each agent is made by the composition rule, which is the basis of inference mechanism, as follows:

$$\widetilde{U}_j = R_j \circ \widetilde{X}, \qquad j = \overline{1, N}, \tag{5.82}$$

where $\widetilde{U}_j$ is fuzzy value of decision of j-th agent, R_j is the fuzzy relation corresponding to the fuzzy model (5.83), and $\widetilde{X}$ is total input information after fuzzification.

Every agent's solution is evaluated by the fuzzy decision maker on the basis of their achievements on M criteria C_i, $i = \overline{1, M}$. Among N agents the best agent ("winner" agent) is the one that satisfies criteria C_i, $i = \overline{1, M}$ best. The solution of this agent will be taken as total solution of the full system.

Decision Generation in Multi-Agent Distributed Intelligent DSS

A set of alternative solutions of DSS $\{u_{Ag\,1}, u_{Ag2}, \ldots, u_{AgN}\}$ is formed through fuzzy reasoning by parallel working agents. Each decision alternative is characterized by M criteria, $\{C_1, C_2, \ldots, C_M\}$, according to which the desirability of a solution is determined. The problem here is to determine optimal alternative (winner), which is better than other alternatives in terms of criteria C_i. This problem is alternative selection under condition of uncertainty, which was considered in [2,11, 45,95,96,98].

We will use the method suggested by Yager [96] for selection of the winner alternative. The proposals of agents are evaluated in accordance with the criteria C_i, $i = \overline{1, M}$. In real decision situation criteria C_i have different importance $\alpha_i, i = \overline{1, M}$. The relative importance of each criterion is determined using the criteria comparison procedure. For example, as it is shown in [96], equal importance $b_{ij} = 1$, weak importance $b_{ij} = 3$, strong importance $b_{ij} = 5$, ($b_{ij} = \frac{1}{b_{ji}}$) etc. are determined on the scale of importance evaluation. $\alpha_i, i = \overline{1, M}$ are obtained by using a matrix composed from b_{ij}. We define the estimation of alternative u_{Ag_i} on criteria C_i through $\propto_{C_i}(u_{Ag_j}) \in [0, 1]$ (for simplicity reasons we will denote $\propto_{C_i}(u_{Ag_j})$ as $\propto_{ij}$). After determining the conformity degree of each alternative $u_{Ag_j}, j = \overline{1, N}$, to criteria C_i ($\propto_{C_i}(u_{Ag_j})$) and $\alpha_i, i = \overline{1,M}$ we derive the following fuzzy sets:

$$C_1^{\alpha_1} = \{(\propto_{11})^{\alpha_1} / u_{Ag1}, (\propto_{12})^{\alpha_1} / u_{Ag2}, \ldots, (\propto_{1N})^{\alpha_1} / u_{AgN}\}$$
$$C_1^{\alpha_2} = \{(\propto_{11})^{\alpha_2} / u_{Ag1}, (\propto_{12})^{\alpha_2} / u_{Ag2}, \ldots, (\propto_{1N})^{\alpha_2} / u_{AgN}\} \quad (5.83)$$
$$\vdots$$
$$C_1^{\alpha_M} = \{(\propto_{M1})^{\alpha_M} / u_{Ag1}, (\propto_{M2})^{\alpha_M} / u_{Ag2}, \ldots, (\propto_{MN})^{\alpha_M} / u_{AgN}\}$$

It is known that optimal alternative selection may be performed by [96]:

$$\propto_D(u^*_{Ag}) = \max_j \min_i \propto_{C_i}^{\alpha_i}(u_{Ag_j}), \quad (5.84)$$

where u^*_{Ag} is the optimal alternative.

Application. Here we consider a marketing system in oligopolistic industry. In such industry the number of firms is limited to a few, where the actions of one firm have an impact on the industry demand. A number of firms compete for three products (*x*, *y*, and *z*). Each firm is managed by a team, which pursues profitability and market share goals. The teams make marketing, production, and financial decisions each quarter using DSS incorporating data and models. Marketing decisions include the decisions on mix and amounts of goods to produce, pricing, advertising expenses, and other variables. The firm's price and advertising strategies, as well as prices and advertising expenditures of the competitors are the major factors influencing profits, market share, and other marketing variables [74].

The proposed multi-agent distributed intelligent marketing DSS consists of 5 knowledge based agents, each of which has 9 fuzzy rules. Input information are fuzzy variables of average price $\tilde{x}_1$ (AvgPrice) and average advertising $\tilde{x}_2$ (Avg-Adv) of competitors and are the same for all 5 agents. Using fuzzy inference rule each agent produces its output solutions: firms own price ($\tilde{u}_{i1}$), $i = \overline{1,5}$, and firm's own advertising $\tilde{u}_{i2}$, $i = \overline{1,5}$. Fuzzy rules in knowledge base of each agent are of the following type (for example, for the first agent):

IF Avg Price is HIGH and AvgAdv is LOW, THEN
Price is HIGH and Advertising is MEDIUM
IF Avg Price is LOW and AvgAdv is HIGH , THEN
Price is MEDIUM and Advertising is HIGH
IF AvgPrice is MEDIUM and Avg Adv is MEDIUM, THEN
Price is HIGH and Advertising is MEDIUM

⋮

For the second agent:
IF AvgPrice is HIGH and Avg Adv is LOW, THEN
Price is HIGH and advertising is LOW
IF Avg Price is LOW and AvgAdv is HIGH, THEN
Price is MEDIUM and Advertising is HIGH
IF AvgPrice is MEDIUM and Avg Adv is MEDIUM, THEN

Price is HIGH and Advertising is LOW

⋮

For the third agent:
IF Avg Price is HIGH and AvgAdv is LOW, THEN
Price is LOW and Advertising is LOW
IF Avg Price is LOW and AvgAdv is HIGH, THEN
Price is LOW and Advertising is HIGH
IF AvgPrice is MEDIUM and Avg Adv is MEDIUM, THEN
Price is LOW and Advertising is MEDIUM

⋮

For the fourth agent:
IF Avg Price is HIGH and AvgAdv is LOW, THEN
Price is HIGH and Advertising is LOW
IF Avg Price is LOW and AvgAdv is HIGH, THEN
Price is LOW and Advertising is HIGH
IF AvgPrice is MEDIUM and Avg Adv is MEDIUM, THEN
Price is MEDIUM and Advertising is MEDIUM

⋮

Finally, for the fifth agent:
IF Avg Price is HIGH and AvgAdv is LOW, THEN
Price is HIGH and Advertising is MEDIUM
IF Avg Price is LOW and AvgAdv is HIGH, THEN
Price is LOW and Advertising is HIGH
IF AvgPrice is MEDIUM and Avg Adv is MEDIUM, THEN
Price is MEDIUM and Advertising is HIGH

⋮

The agents in the considered multi-agent distributed intelligent marketing DSS are characterized by the following criteria:

C_1 - conformity to situations;

C_2 - confidence factor (CF);

C_3 - track record.

Conformity to situations specifies the degree to which an agent's expertise is adequate for the situation at hand. For example, certain agent's expertise may be most adequate to the situations when there is excess inventories and low demand in the industry. Confidence factor measures an agent's confidence in the submitted proposals. Proposals with high degree of confidence tend to be superior to proposals with low degree of confidence. Track record indicates an agent's past performance. Agents whose proposals led to profitable decisions in the past acquire strong track record.

Fuzzy sets characterizing alternatives on criteria C_i have the following form:

$$C_1 = \{0.7/u_{Ag1}, 0.8/u_{Ag2}, 0.6/u_{Ag3}, 0.7/u_{Ag4}, 0.4/u_{Ag5}\}$$
$$C_2 = \{0.8/u_{Ag1}, 0.7/u_{Ag2}, 0.5/u_{Ag3}, 0.6/u_{Ag4}, 0.3/u_{Ag5}\}$$
$$C_3 = \{0.8/u_{Ag1}, 0.6/u_{Ag2}, 0.4/u_{Ag3}, 0.4/u_{Ag4}, 0.25/u_{Ag5}\}$$

The pair wise comparisons of the above criteria resulted in the following:

C_1 and C_2; C_1 is more important than C_2 with the relative importance 7;

C_1 and C_3; C_1 is more important than C_3 with the relative importance 5;

C_3 and C_2; C_3 is more important than C_2 with the relative importance 3;

This results in the following matrix:

$$\begin{bmatrix} 1 & 7 & 5 \\ 1/7 & 1 & 1/3 \\ 1/5 & 3 & 1 \end{bmatrix}$$

The corresponding eigenvector V with $\lambda = 3.065$ is:

$$\begin{bmatrix} 2.192 \\ 0.243 \\ 0.565 \end{bmatrix}$$

Therefore, we derive:

$$\begin{aligned} C_1^{2.192} &= \{0.458/u_{Ag1}, 0.613/u_{Ag2}, 0.326/u_{Ag3}, 0.458/u_{Ag4}, 0.134/u_{Ag5}\} \\ C_2^{0.243} &= \{0.947/u_{Ag1}, 0.917/u_{Ag2}, 0.845/u_{Ag3}, 0.883/u_{Ag4}, 0.746/u_{Ag5}\} \\ C_3^{0.565} &= \{0.882/u_{Ag1}, 0.749/u_{Ag2}, 0.596/u_{Ag3}, 0.596/u_{Ag4}, 0.457/u_{Ag5}\} \end{aligned} \tag{5.85}$$

On the basis of the formula (5.84), and using (5.82) and (5.85) we derive:

$$\begin{aligned} \propto_D(u^*_{Ag}) = \{\max[&\min(0.458, 0.947, 0.882)/u_{Ag_1}, \min(0.613, 0.917, 0.749)/u_{Ag_2}, \\ &\min(0.326, 0.845, 0.596)/u_{Ag_3}, \min(0.458, 0.883, 0.596)/u_{Ag_4}, \\ &\min(0.134, 0.746, 0.457)/u_{Ag_5}]\} \end{aligned}$$

$$\begin{aligned} \propto_D(u^*_{Ag}) = \max\ &[0.458/u_{Ag_1}, 0.613/u_{Ag_2}, 0.326/u_{Ag_3}, \\ &0.458/u_{Ag_4}, 0.134/u_{Ag_5}] = \{0.613/u_{Ag_2}\} \end{aligned}$$

The optimal alternative is u_{Ag_2}, So Ag_2 is the "winner" agent with the proposal as the solution of the system.

5.7 Soft Computing and Data Mining

According to most well-known definitions, data-mining is a process of extraction of implicit, previously unknown and potentially useful information (including rules, patterns, constraints, relationships, and regularities) from various structured and unstructured data sources (databases, data warehouses etc.). The problem is best stated by J.Naisbitt in his bestseller Megatrends: "We are drowning in information but starved for knowledge and intelligence". It should be noted that data mining is a much more difficult process than a simple machine learning process. The complication arises mainly due to the basic problem related to data mining: it has to deal with massive datasets with high dimensionality and missing and dynamic data. In contrast to simple machine learning process, the datasets for data mining is never complete, fully true, or sufficient. That means that it is necessary to modify learning algorithms for data mining to deal with large sets of noisy data.

To cope with this situation the Soft Computing technology can be suggested which combines the ability of Fuzzy Logic to represent and manage imprecise data, the ability of Neural Network to learn and find regularities and generalizations in noisy data, the ability of Genetic Algorithms to learn in case of high dimensionality of data and non-linearity in input/output relations.

Neural Networks are being used by financial industry to reveal patterns in vast pools of data that may help firms predict corporate bond ratings and bankruptcies. Many firms are presently using neural networks to predict security ratings, timing stock buying and selling, and other business related areas. Like human decision makers, neural networks can learn from examples. Figure 5.7. illustrates a simple example of a neural network that uses rules it learned from patterns in data to construct a hidden (black-box) logic model to identify good and bad credit risk [49].

Solutions to certain type of problems in optimization and related areas are appropriate for application of Genetic Algorithms. Many business problems require optimization: minimization of costs and maximization of profit, efficient scheduling and use of resources, etc. In a dynamic and complex environment with hundreds variables and formulas, Genetic Algorithms can be applied successfully, because they attack the problem from many directions at once ranking the strings in the population according to their level of desirability as possible solutions.

Fuzzy logic is found very useful for decision making and organizational control, because using it we can create rules that use approximate or subjective values and incomplete or ambiguous data and can represent and infer knowledge that is imprecise, uncertain, or unreliable.

Mining dependencies and association rules. Often in databases presence of a value of some attribute A implies the presence of another value of some other attribute B. This is called a functional dependency between the attributes: $A \Rightarrow B$. This kind of dependencies can be described by the use of fuzzy rules with confidence degrees denoting the strength of implication. Algorithms to find such rules can be found in [16].

Generalization of data. Databases often contain data detailed at primitive level. Generalization is used to provide compact descriptions for parts of data to represent them at a higher conceptual level. An approach to do this is called the "data cube" approach [16]. Aggregation functions are used for some dependent attribute according to all of the grouping possibilities of its determinant attributes. A problem when using this approach is that the size of the cube sometimes appears to be very large. For example, if we consider weight and height as grouping variables and set weight domain as integers from 0 to 300 and height domain as integers from 50 to 250, we will have 300· 200=60,000 entries in the table corresponding to (*weight, height*) dimension. The cube size will significantly decrease if we describe weight and height domains as linguistic label sets. For example for the weight domain we can use {very light, lightish, light-weight, heavy, very heavy, overweight} and for the height domain: {very small, small, medium height, tall, very tall}. These fuzzy sets used in the aggregation function computation make the generalization more effective.

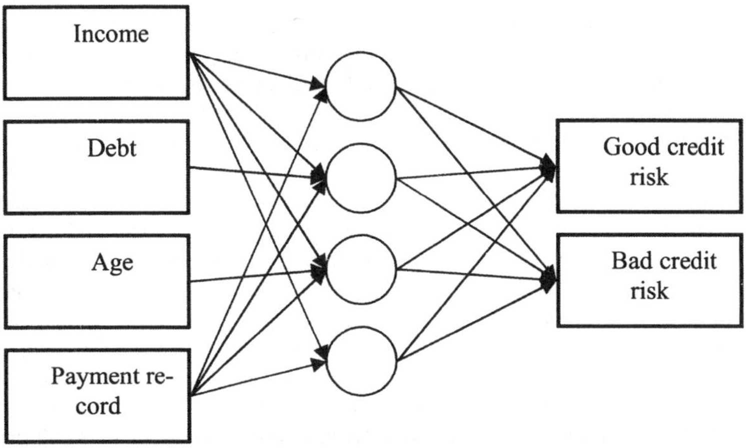

Fig. 5.7. A neural Network to identify credit risk

Moreover, fuzzy logic can be used for performing queries in databases including linguistic quantifiers and with smoothed versions of aggregation functions:

List all of young MBA students who have passed "Accounting" and "Economics" course with high marks.

Reference [89] describe a fuzzy logic based mechanism to compute the accomplishment degrees of sentences inside linguistic queries and implement corresponding inference.

In modeling fuzzy dependencies the equality (=) used in classical functional dependencies is replaced by fuzzy resemblance (similarity) relations. Using fuzzy logic allow to design a complete and effective method for the dependency extraction [22].

Data mining from large databases often involves pattern identification or clustering. Every clustering process is imprecise in nature. That is why the current studies in fuzzy clustering are focused mainly on the use of Fuzzy Logic, Neural Networks and Genetic Algorithms to develop different procedures to obtain fuzzy clusters. These approaches are especially suitable for clustering because they consider the imprecision in the fuzzy cluster concept presuming that the processed data are incomplete and uncertain, and also because of the computational power of Neural Networks and Genetic Algorithms to deal in the environment of large databases where conventional methods are unable to find an appropriate decision. Fuzzy rules can be used to describe regularities and patterns in databases and also to describe complex situations with fuzzy inputs and outputs. A simple example of such a rule for classification is [89]:

IF attitude is "positive" AND avg(marks) is "not low", THEN the student belongs to the class "good students".

As we mentioned above the source of data for discovery and information analysis is different kinds of large and unstructured databases and warehouses, containing credit card transactions, demographic data, purchase data from checkout counter scanners at supermarkets and retail stores, and also databases on the Internet. Many companies design their web pages so that they are able to extract user specific information when people access and interact with World Wide Web sites. The data of interest include specific customer data and preferences (behavior, tastes, interests etc.). The data are mined for information to create user profiles of web site visitors, which are then used to target individuals and common interest groups: the pages dynamically change the content, advertising, and presentation thereby making an appropriate reaction and maximizing the effect and profit. The patterns and hypotheses found from facts and business data can be used to predict future behavior, evaluate proposed business strategies, which in turn would improve competitiveness, improve profits, and transform business processes. Data mining from the Internet data bases and web pages is widely used in marketing to improve customer retention, cross-selling opportunities, market and pricing analysis, customer evaluation and segmentation analysis.

For example, mining from supermarket purchases databases might reveal the following fuzzy regularity:

WHEN potato chips are purchased, THEN soda is also purchased with the confidence degree 0.65.
WHEN potato chips are purchased AND there is a promotion THEN soda is purchased with the confidence degree 0.85.

This information may help firms make better promotion of their offered products.

Presently there exist a very large number of data mining applications. Such great companies as Oracle, Informix Software, Sybase and others are incorporating data mining tools into their products. Some of applications are: market segmentation (characteristics of customer who buy the same products), customer churn (predict customers which are likely to switch to competitors), fraud detection (identify fraudulent transactions), direct marketing (design mailing lists to increase response rate), market basket analysis (identify products commonly purchased together), trend analysis (difference of data on this month versus the previous) [80].

We can conclude that data mining as an information analysis tool is more discovery driven than data driven decision support systems, it uses statistical analysis tools as well as neural networks, fuzzy logic, genetic algorithms, or rule-based and other intelligent techniques. The types of information that can be revealed from data mining include associations, sequences, classifications, clusters, and forecasts.

5.8 Soft Computing Based Multi-Agent Marketing DSS

Here we consider a simplified marketing system in an oligopolistic industry. We suppose that the firm's price and advertising strategies dominate the marketing strategy [74]. The existing DSS are oriented to econometric models of the industrial history. In particular, as it is shown in [74] existing econometric models deal inadequately with the information about competitors future behavior assuming competitor actions are known. [74] presents procedure to model competitors behavior uncertainties as fuzzy information. Here multi-agent DSS is created using multi-agent distributed intelligent system methodology suggested in section 3.8 The proposed multi-agent distributed intelligent marketing DSS consist of 5 agents (GA included 5 agents from author pre-specified 7). Input information are fuzzy variables of average price $\tilde{x}_1$ (AvgPrice) and average advertising $\tilde{x}_2$ (AvgAdv) of competitors, and are the same for all 5 agents. Using fuzzy inference rule each agent produces its output solutions: firm's own price ($\tilde{u}_{i1}$), $i=\overline{1,5}$, and firm's own advertising $\tilde{u}_{i2}$, $i=\overline{1,5}$. Fuzzy rules in knowledge base of each agent is of the following type (for example for the first agent):

IF Avg Price is HIGH and AvgAdv is MEDIUM, THEN
Price is HIGH and Advertising is MEDIUM
⋮

For the second agent:
IF AvgPrice is HIGH and Avg Adv is MEDIUM, THEN
Price is HIGH and advertising is LOW
⋮

Using switching procedure of 6 types of fuzzy inference rule, each agent produces its output solution: firm's own price, and firm's own advertising.

Solutions of each agent for situation where AvgPrice is about three hundred dollars $325 and AvgAdv is about sixty thousand $60,000 are shown in Table 5.7.

Table 5.7. Solutions proposed by 5 agents

Agent #	Price	Advertising
Agent 1	$331.59	$80,000
Agent 2	$331.67	$53,000
Agent 3	$313.61	$60,000
Agent 4	$331.59	$53,000
Agent 5	$331.59	$70,000

Coordinating agent receives values of solutions (price and advertising) of each agent and total input fuzzy information on AvgPrice and AvgAdv and calculates fuzzy profit. The output of the estimator is 5 fuzzy profits associated with the solutions of each agent. The fuzzy profits for the above situation are shown in Table 5.8.

Table 5.8. Fuzzy expected profits

Agent #	Fuzzy profit (in $)
Agent 1	(505,858; 84,195; 52,467)
Agent 2	(401,376; 56,963; 98,182)
Agent 3	(434,827; 29,321; 76,352)
Agent 4	(410,508; 56,823; 98, 130)
Agent 5	(501,342; 84,115; 50,786)

The evaluator component of coordinating agent performs fuzzy ranking of fuzzy profits and provides total ordering of 5 fuzzy profits as follows:

$$\tilde{P}_{r_1} > \tilde{P}_{r_5} > \tilde{P}_{r_3} > \tilde{P}_{r_4} > \tilde{P}_{r_2}$$

First agents is the "winner" agent and its solution is accepted as solution for the given situation:

Price =$331.59
Advertising =$80,000

After optimization of relational matrixes, triangle membership functions, scaling factors and finding appropriate fuzzy inference mechanism the global solution of DSS (Price and Advertising for quarters 1, 2, 3, and 4) consists of the sequence of solutions of agent 1 for quarter 1, agent 5 for quarter 2, agent 5 for quarter 3, and agent 3 for quarter 4. All five agents competed for the total solution of the problem. In each quarter the most suitable agent for the situation was the "winner", while the other four agents were losers. For example, for situation in quarter 2 agent 5 was more suitable than the rest of the agents. Weights in objective function (3.42) are:

$$\text{Quarter1: } \alpha_1 = 1, \quad \alpha_2 = 0, \ \alpha_3 = 0, \ \alpha_4 = 0, \ \alpha_5 = 0;$$

$$\text{Quarter 2: } \alpha_1 = 0, \quad \alpha_2 = 0, \ \alpha_3 = 0, \ \alpha_4 = 0, \ \alpha_5 = 1;$$

$$\text{Quarter 3: } \alpha_1 = 0, \quad \alpha_2 = 0, \ \alpha_3 = 0, \ \alpha_4 = 0, \ \alpha_5 = 1;$$

$$\text{Quarter 4: } \alpha_1 = 0, \quad \alpha_2 = 0, \quad \alpha_3 = 1, \quad \alpha_4 = 0, \quad \alpha_5 = 0$$

respectively.

5.9 Hybrid DSS Based on Simulation and Genetic Algorithms

Many real-life ill-structured problems involve high uncertainty and complexity preventing application of analytical optimization techniques in building effective Decision Support Systems (DSS). These systems may employ simulation method and search for a "good" solution through "what-if" analysis. However, this method is very time consuming and often overlooks the consideration of many promising alternative solutions. A Genetic Algorithm (GA) automates the search for "good" solutions by finding near-optimal solutions and increases effectiveness of DSS. This session introduces a hybrid method based on the combination of Monte-Carlo simulation and genetic algorithms.

Introduction and Related work

Many real-life problems involve complexity, uncertainty, hard and soft constraints, as well as multiple and often conflicting objectives. Simulation models (e.g., Monte-Carlo simulation) may be the only legitimate choice for a decision maker to approach these problems [89]. In simulation models, the decision maker decides which promising alternatives to evaluate and performs "what-if" analysis. Decision support systems (DSS) often provide only the means for mechanical calculation for evaluating alternative solutions for these problems in terms of feasibility achievement of objectives. Little or no support is when to stop the search, or how to ensure the feasibility of trial solutions. Generally, as the problem complexities and uncertainties increase it becomes even more difficult for the decision maker to generate feasible promising alternatives that allow for uncertainties and do not violate the constraints. Consequently, the decision makers is likely to "satisfies" and thus avoid an exhaustive, time consuming, and occasionally futile search to discover and evaluate the optimal choice.

An effective DSS with increased participation in the decision process is needed to avoid the above shortcomings. The objective of such DSS is to help search for feasible "good" solutions, stop the search when benefits becomes negligible, and reduce the effect of user cognitive biases.

One way to improve simulation-based DSS is through the use of genetic algorithms. Monte-Carlo simulations allow modeling of complex uncertain response functions with little need for unrealistic simplifying assumptions. GA, on the other hand, are capable of finding near-optimal solutions of such functions. Combination of GA and simulations allows us to search for acceptable solutions without unnecessary simplifications required by traditional optimization models.

DSS has been one of the most popular research topics in Information Systems Research [20, 51]. Presently, there are no strong theory-based methods for relating DSS features and their effectiveness, and the issues of DSS design have not been adequately addressed [76]. Nevertheless, there are several research results that may help in pointing the way on how to increase the effectiveness of simulation model-based DSS. More effective DSS are implied through features that enable increased DSS participation in the decision process, active participation [1,57,62,69], and balanced restrictiveness.

Alter [3] developed a taxonomy that distinguishes between the DSS on the basis of level of participation in the decision, i.e., the degree to which the system determines the decision. Alter's taxonomy, although over 25 years old, is still the only taxonomy that systematically distinguishes between DSS. In his classification representational models, optimization models, and suggestion systems are in the order from a low to high degree of system participation in the decision. Moving from the representational model type of DSS to an optimization type DSS is equivalent to climbing Alter's taxonomy ladder towards higher DSS participation in determining the decision.

Optimization models establish enough structure to use models as analysis tools. Here the model is in control of the search process and it determines which feasible candidate decisions to evaluate on the way to obtaining optimal solutions (assuming it utilizes an iterative optimization procedure). Elevation from the representational models to the optimization models may result from discovering additional structure in the decision task and assigning more activities to DSS. All DSS require some human judgment in performing the decision task [44]. Increasing task structure implies less need for human judgment. Here the increased task structure is achieved through coupling of Monte-Carlo simulations with GA.

DSS effectiveness may be increased through active participation of DSS in the decision-making process. Traditional DSS offer a weak form of support, where the DSS user is assumed to have full knowledge of a system's capabilities and take the initiative in human-computer interaction. In this setup, the potential power of human-computer collaboration is significantly underutilized. Using GA, the DSS may participate more actively in the decision-making process by taking the initiative in search for promising solutions. Mirchandani and Pakath discuss a GA-driven adaptive DSS as an example of making decision support more active.

The concept of "restrictiveness" may also guide how to increase the effectiveness of simulation model-based DSS through improved DSS features. *System restrictiveness* is the degree to which, and manner in which, a DSS limits its users' decision-making processes to a subset of all possible processes. It is a quality but it is not measurable. A DSS with just the right amount of restrictiveness may lead to better decisions. A DSS that limits the available processes to potentially superior prescriptive and normative techniques is more restrictive and may provide more effective support. However, a DSS which reduces the need for simplifying heuristics and the effects of their "systematic cognitive biases", expands a decision maker's information processing capabilities and therefore is less restrictive and may also provide more effective support.

The conceptual model of restrictiveness does not specify the methodology for increasing restrictiveness. In particular, how would one go about manipulating restrictiveness in simulation-based DSS? One practical way to achieve this is to build effective simulation-based DSS through increasing restrictive-ness via genetic algorithms that can perform the search for promising alternatives, thus eliminating the necessity for manual search and restricting the user to the analysis rather than the search. The GA also reduces the cognitive bias for "satisfying" and the strategy of "anchor and adjust" which tend to settle for local optima. Also, hard and soft constraints may be incorporated in the GA component to further restrict the system and ensure the generation of feasible solutions.

There have been several attempts to incorporate GA into DSS. Wang and Leu [91,92] used GA for adaptive selection of rules for the DSS for trading at Taiwan stock market. They compared the performance of this adaptive design to that of the stationary rules to show that the former is better. Balakrishnan and Jacob [7] incorporate different techniques, including complete enumeration, heuristic dynamic programming and GA in their DSS for product design. The approach increases the confidence in the quality of decisions thus promising more effective decisions. These applications are encouraging contributions to the field of DSS. However, further investigation into the use of GA in DSS for problems with uncertainty and various types of constraints was needed.

Several research results report the application of the hybrid method to searching for near-optimal solutions. An example of finance application of Monte-Carlo in combination with GA includes finding optimal hedging strategies for trading options, where simulations were used to model the uncertain market factors [87]. In this application, the GA - generated hedging strategies were better than those generated by the traditional delta rule. However, this application did not incorporate constraints. Examples of applications of Monte-Carlo in combination with GA for engineering problems include optimal scheduling for flexible manufacturing systems [67], optimal scheduling of hot parts operating [73], and optimal tolerance allotment [50]. Some of these problems incorporated hard or *crisp* constraints; none considered the soft type of constraints.

Statement of the problem

Gatignon [37] defines the general model for marketing mix as follows:

$$y = f(X, Z, \beta, \gamma, \varepsilon) \tag{5.86}$$

where y is a measure of market response (i.e., sales, j market share, profit), X a set of marketing variables, Z a set of environmental variables, β the response parameters of marketing variables; γ the response parameters of environmental variables, and ε the disturbance term. In general, the functional form of the model is non-linear, and linearizing the model would make it inappropriate in most marketing contexts because of simplifying assumptions [36].

Different techniques for making marketing mix decisions that will optimize the response function have been proposed. However, the inherent simplifying assumptions of these techniques limit their usefulness in real situations. The Dorfman-Steiner theorem specifies the conditions for the optimality of marketing mix, which includes values of elasticities [25]. Similar conditions are specified by Lambin [48], The use of these results, however, is hampered by the difficulty in estimating the exact values of elasticities. Moreover, the techniques require specifications of quantities of products sold and these cannot be accurately estimated.

A number of marketing and joint marketing-production decision-making models emerged from the OR/MS area. In particular, the mathematical programming models were reviewed by Rangaswamy [70], and Eliashberg and Steinberg [28]. These models represent marketing problems in terms of objective function and a number of constraints. However, the use of mathematical programming tools becomes more difficult as the complexity of the objective function increases (e.g., it becomes non-linear, non-differentiable, and uncertain). Moreover, these tools are inappropriate for objective functions that have local minima.

The marketing mix problem belongs to a class of problems that, due to inherent complexity and uncertainty, cannot be adequately solved by traditional optimization techniques. In highly competitive and uncertain dynamic environments, external factors, such as the state of the economy, consumer preferences, as well as strategies and tactics of competing firms determine the dynamics of the market. In making marketing mix decisions on price, advertising expenses, quantity and other variables for multiple products, the firm must consider various constraints and uncertain environmental and competitive factors. In general, the objective function may be complex, non-differentiable, and uncertain. The uncertainty may arise from ignorance about competitor actions and other environmental factors. Constraints may include both crisp and "soft", or "fuzzy" ones. These characteristics of the problem prohibit the use of the traditional calculus-based and mathematical programming techniques.

Both Gatignon and Rangaswamy stress the use of simulation models as more appropriate for complex market response, than are traditional optimization tools. For complex market response patterns, the simulation models seem to be the best tools to analyze different alternatives. These models allow the decision maker to do "what-if" analysis and see how the key criteria change in response to changes in the input variables.

Simulation method overcomes the problems associated with most optimization methods: unrealistic assumptions, failure to deal with local optima, uncertainties, and non-differentiability of the objective function. The drawback of the simulation-based approach is that the user may spend significant time and effort in manually optimizing the solution and fail to consider many promising alternatives. This disadvantage grows exponentially as the number of decision variables increase, since each new variable leads to a considerably larger space of possible combinations.

Genetic algorithms (GA) may be used to identify a set of "good" alternatives for the marketing mix problem. They are new optimization tools that search for solutions based on the principles of evolution [34]. The technique is applicable to a wide class of problems, and GA are capable of dealing with more complex response functions than traditional optimization approaches. The potential applications of GA to marketing optimization problems are consumer behavior; segmentation, targeting and positioning, managing the marketing mix; and strategic marketing. Hurley et al. [41] discussed applications to site location analysis and segmentation (product-market structure) in detail. Terano and Ishino [83,84] used GA for extracting marketing decision rules from questionnaire.

A number of applications of GA to economic problems are reported in the literature including those reported by Holland and Miller [40] J. Arifovic [4] reported application of GA to setting the quantities of a product in an environment where multiple firms competed in a market for a single product. The firms were price-takers, and the production decisions were made before observing the market price. Also, Midgley et al. [60] reported an application of GA to the pricing problem. They used GA to determine optimal pricing for three major coffee brands in a regional US coffee market. The GA considered only four levels for price of coffee.

Monte-Carlo plus GA-based hybrid DSS

The overall architecture of the marketing DSS based on Monte-Carlo simulations plus GA is shown in Fig. 5.8. The GA module overcomes the problem of manual search for promising alternatives. It generates sets of candidate solutions, which are entered into the simulation module for evaluation. The simulation module then produces the outputs based on the known facts, estimates, and candidate solutions. The output of the model is finally used to evaluate the submitted candidate solutions.

A number of Monte-Carlo simulations are performed at each evolution step for each chromosome in the population to evaluate their fitness. The simulation module uses statistical models to calculate the outputs and returns average results of simulations. The database module holds the historical facts, estimates, and parameters of the models, as well as the parameters of GA. The database also keeps track of the evolution history, last updated genetic pool, the values of the best chromosome, and other useful information.

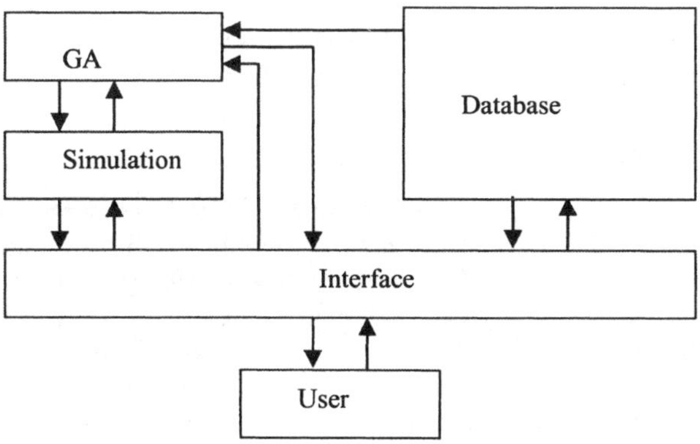

Fig. 5.8. Structure of the GA-based marketing DSS

The algorithm proceeds as follows:

1. The system updates the database with the relevant information (e.g., past facts, existing constraints, etc.) in interaction with the user.

2. The user specifies his/her estimates concerning uncertain factors (e.g., future actions of competitors).
3. The GA starts its work in interaction with the Monte-Carlo simulation module as follows:
   3.1. An initial population of candidate solutions is randomly generated.
   3.2. The fitness value is calculated for each chromosome (potential solution) in the population:

- Each is entered as a candidate solution to the simulation module.
- The module performs a number of Monte-Carlo simulations and returns the results to the GA module.
- The GA module computes . the fitness value for a chromosome based on the output of the simulations.

   3.3. The selection operator generates a new population from the old one based on the fitness values of its chromosomes.
   3.4. A crossover operator is applied to the population.
   3.5. A mutation operator is applied to the population.
   3.6. The fitness value is calculated for each chromosome in the population (as in step (3.2)).
   3.7. If the stopping criteria are met, then the process ends; otherwise it returns to step (3.3).
4. The results of GA, including the last generated pool of candidate solutions, the history of evolution of solutions, and other information is written to the database. The best generated solution is presented to the user.

The user can request other solutions from the database. He or she can then modify the proposed solution and directly interact with the Monte-Carlo simulation module to perform refining "what-if' analysis. The user can also change his/her estimates and re-run the GA to do sensitivity analysis and see how sensitive the solution is to the estimates of uncertain factors.

The experiments

For our experiments we used the SIMQ game, which is a system that simulates a hypothetical business environment; in this, a number of firms compete for three products (x, y, and z) in an olygopolistic market [19,74]. A team of students who generally decide to pursue profitability and market share goals, manages each firm. The teams make marketing, production, and financial decisions each quarter using DSS incorporating accounting and representational models. The decisions include the mix and amounts of goods to produce, pricing, advertising expenses, and production capacity. The DSS for marketing decisions incorporates a Monte-Carlo simulation model.

We formulate the general marketing mix problem as follows:

Optimize:

$$y = f(X, Z, \beta, \gamma, \varepsilon) \tag{5.87}$$

Subject to constraints:

$$g_1(X, Z) \in \Omega \tag{5.88}$$

where the parameters and variables in (5.87) and (5.88) were described previously, g_i and Q_i describe the ith constraint on the values of marketing variables in general form.

In the context of SIMQ game, the objective is to maximize the overall marketing value, which is aggregation of the individual marketing values for the three products:

$$MV = \sum_{i \in (x,y,z)} MV_i \tag{5.89}$$

Subject to:

$$\sum_{i \in \{x,y,z\}} Q_i \leq \text{Capacity}, \tag{5.90}$$

$$\text{Supply}_i \cong \text{Demand}_i, \quad i \in \{x, y, z\} \tag{5.91}$$

The objective function (5.89) is the total marketing value of the mix. There are crisp and soft constraints in the problem. Crisp constraint (5.90) limits the total production volume for three products by the capacity of firm's facilities. Soft constraints (5.91) require the probabilities of stocking out for the three products to be away from 0 and 1 to avoid large inventory buildups as well as shortages in delivering the products. In other words, soft constraints require that the supply should meet the demand to some extent.

Individual marketing values are affected by the decisions of the firm as well as external (possibly uncertain) factors:

$$MV_i = MV_i(P_i, A_i, P_i^{\,c}, A_i^{\,c}, C_i, Q_i, V_i, T_i), \quad i \in \{x, y, z\},$$

where P is the price of a product, A the advertising expenses, P^c the average price charged by competitors, A^c the average competitor advertising, C the cost of a product, Q the quantity produced, V the future value (estimated value of the product in the time frame beyond the current planning horizon) of a product, and T reflects the temporal pattern of dynamics of the demand for the product.

The expected marketing value (MV) represents the effectiveness of marketing decisions. It is based on the estimate of profit and future value of inventory for a given set of potential decisions.

The expected profit and inventory level depends on the estimate of sales, which, in turn depends on the estimate of demand. If demand is more than the quantity produced, then all the produced goods will be sold, otherwise, the level of demand will determine the sales volume [56]:

$$\text{IF } D_i \geq Q_i, \text{THEN } S_i = Q_i, I_i = 0,$$
$$\text{ELSE } S_i = D_i, I_i = Q_i - D_i$$

Here D is demand, 5 the sales, I the inventory level, and Q the quantity produced. The demand is influenced both by the firm's decisions and the decisions of competing firms. Uncertainty arises from the ignorance of competitor decisions on price and advertising, as well as the uncertainty inherent in demand forecasting.

Summarizing the above, the problem is defined as: find such a combination of decision variables (P, A, Q) that improves the total MV (5.89), subject to constraints (5.90) and (5.91). Making effective decisions is difficult: one has to consider all of the three products jointly in order to manage them properly. Moreover, the situation is complicated by uncertainties involved in competitors' decisions on price and advertising.

Solving the marketing mix problem: integration of Monte-Carlo simulation with GA and fuzzy set concepts

Monte-Carlo simulations are used to deal with the uncertainly in the described problem. The objective is to predict the level of the MV for a given set of decisions and environmental variables. Different sets of variables are involved in the simulations. The decision variables include: the quantities of products to produce, prices, and advertising expenses. The decision maker has control over the decision variables. The known environmental variables include historical industry and firm's prices, advertising expenses, R&D expenditures, and market share data. The uncertain variables include average industry price, advertising expenses, and R&D expenditures for the current quarter.

Basically, the uncertainty stems from lack of knowledge about competitors' future actions. Although the exact competitors' behavior can hardly be determined, judgment, as well as business intelligence information may be used [31, 32] to estimate the approximate intentions of competitors. Here, uncertainties are represented as triangular distributions with user-defined ranges and peak points for average industry price, advertising expenses, and R&D expenditures. During Monte-Carlo simulations the values for average industry price, advertising expenses, and R&D expenditures are generated from user-specified triangular distributions. The results from a number of runs, with different values for uncertain variables, are averaged. Also, the probability of stocking out is estimated as the ratio of cases when the demand exceeded the supply to the total number of cases.

In evaluating different decisions, the user avoids both the cases when the probability of stocking out is close to zero (zero inventory), or one (excessive inventory). In other words, the desired situation is where demand is met to some extent by supply. The good solutions would also generate higher levels of MV without violating the capacity constraint; i.e., the MV is the objective function. The general strategy for the user is to determine the quantities for the three products so that the probabilities of stock out are far from zero and one, and then perform local search for better prices and advertising expenditures until no further improvement is made. The user can then try other set of quantities.

Solving the marketing mix problem: integration of Monte-Carlo simulation with GA and fuzzy set concepts

Balakrishnan and Jacob point out that the ability to generate a set of alternative solutions from which a decision maker can make a choice is a strong advantage of GA. Here we use both Monte-Carlo and GA.

Solution representation

Each chromosome represents the set of decision variables. We included into consideration nine variables: quantities to produce, prices, and advertising expenses for three hypothetical products X, Y, and Z. We employed binary representation for chromosome encoding of our decision variables, 10 binary digits per variable, with the total length of the chromosomes being 90.

The structure of a chromosome is:

$$(P_x^b, P_y^b, P_z^b, A_x^b, A_y^b, A_z^b, Q_{x/y}^b, Q_{x/z}^b, Q_{yz}^b).$$

Here, P_x^b, P_y^b, P_z^b are binary string representations of prices for products x, y, and z; A_x^b, A_y^b, A_z^b are those for advertising expenditures; and $Q_{x/y}^b, Q_{x/z}^b, Q_{yz}^b$ are those encoding the mix of the products and trie total quantity of them. The conversion from the binary representation is made using the upper and lower limits for the variables shown in Table 5.9 . For example, the following partial chromosome

(0000111110 0100000110 0100111010

0101100101 0000001000 1100000001 ...)

Table 5.9. Ranges for variables

	Product		
	X	Y	Z
Min price ($)	310	410	250
Max price ($)	340	450	275
Min advertising ($)	70000	40000	20000
Max advertising ($)	120000	100000	80000

translates into the real-valued vector:

(324.55; 425.09;259.04; 102,551; 43,754; 50, 205;...)

representing the set of decisions that implies the price for product x be \$324.55; advertising expenditures of \$102,551, etc. In other words, the floating point chromosome explicitly represents the set of possible marketing decisions. The total quantity of products is restricted to 13,000.

Objective function

MV (5.84) is the objective function and the basis for constructing the fitness function for GA. A set of potential decisions is input to the Monte-Carlo simulation model along with the values of environmental variables. The Monte-Carlo simulation model calculates the average marketing value of the given set of decisions. In solving the marketing mix problem we are not concerned with mathematically exact location of the optimum, rather, we are trying to locate close-to-optimal solutions. GA can operate in noisy environments, therefore, use of stochastic models, such as Monte-Carlo simulation in combination with GA is appropriate.

Handling constraints

The capacity of firm's facilities limits the total quantity of products to be produced. The capacity requirement dictates the necessity of developing the way to handle constraints in GA. The crisp constraint is handled by careful implementation of the genetic operators. Instead of explicitly generating the quantities for the three products, we encode the *product mix* by taking one product (in our case the product *x*) as the base product and encoding the quantities of others through the ratio of the quantity of the base product to them. The numbers $Q^b_{x/y}$ and $Q^b_{x/z}$ are actually the angles representing the mix of products. The tangent of these angles gives the actual ratio of product x to the products y and z. Use of angles instead of actual ratios is preferred, since we can control the ranges for these angles easier than those for the ratios. The last substring Q^b represents the total actual quantity of the three products and is limited by the capacity limit of 13,000. Symbolically,

$$\frac{Q_x}{Q_y} = \tan Q_{x/y}, \quad \frac{Q_x}{Q_y} = \tan Q_{x/z},$$

$$Q_X + Q_Y + Q_Z = Q,$$

where $Q_{X/Y}$, $Q_{X/Z}$, Q are the real-number translations of the binary $Q^b_{X/Y}, Q^b_{X/Z}, Q^b$ and Q_X, Q_Y, Q_Z are the quantities of products *x*, y, and z. From the above formulas we can derive the quantities of the products as follows:

$$Q_X = \frac{Q}{1 + (1/\tan Q_{X/Y}) + (1/\tan Q_{X/Z})},$$

$$Q_Y = \frac{Q_X}{\tan Q_{X/Y}}, \quad Q_Z = \frac{Q_X}{\tan Q_{X/Z}}$$

This method guarantees that no invalid chromosomes (infeasible solutions) will be generated as a result of genetic operations.

The requirement that the supply to be close to the demand (5.92), translates into the "soft" or "fuzzy" constraint of probabilities of stocking out being away from zero and one. Here the term "fuzzy" implies that there is no strict constraint on the range of the probability of stocking out. Rather, it is transformed into a mathematical form from a linguistic definition, such as "the supply should be close to the estimated demand". Trying to make supply exactly meet the demand does not make much sense, because the demand cannot be predicted with accuracy.

Fuzzy Sets Theory, introduced by Zadeh [96] helps to deal with situations similar to this. We used the membership function for the term "the supply is not close to demand", defined as follows:

$$\propto_i = \begin{cases} \tan h(r_i \cdot (|D_i - Q_i| - \vartheta_i)) & \text{if } |Q_i - Q_i| \geq \vartheta_i, \\ 0, & \text{otherwise} \end{cases} \tag{5.92}$$

Here, $\propto_i$ is the membership function for ith product, D_i the demand for ith product, Q_i the quantity produced, ϑ_i the allowance for a gap between the demand and supply, i.e., considered to be insignificant, and r_i the coefficient that defines the steepness of the membership function beyond the range defined by the allowance. The shape of the membership function is represented graphically in Fig. 5.9.

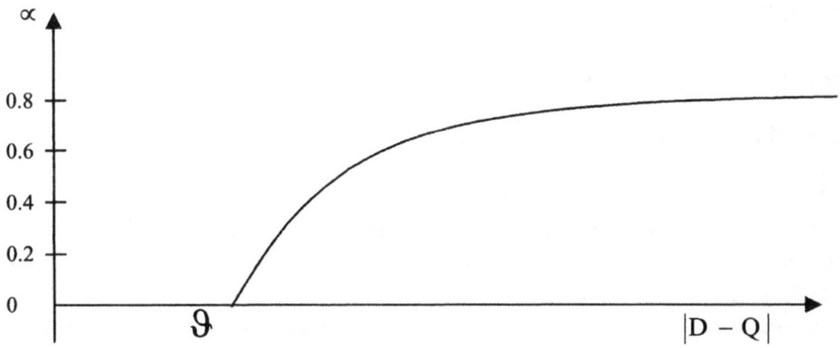

Fig. 5.9. The shape of membership function.

One way of dealing with constraints in GA employs penalty functions. If a candidate solution violates a constraint, then "penalty" is applied to the fitness function by decreasing its value for the corresponding chromosome. We incorporated the soft constraint based on the membership function (5.92) with the adjusted steepness coefficient r_i^j as a penalty in the following way:

$$\text{fitness} = \sum_{i \in \{x,y,z\}} MV_i - \sum_{i \in \{x,y,z\}} l_i \cdot \propto_i' \cdot \tag{5.93}$$

Here, MV_i is the marketing value of the ith product, $\propto_i'$ is derived from $\propto_i$, by using the adjusted coefficient of steepness r_i' instead of r_i and l_i is the maximum penalty that can be subtracted from the solution's marketing value for violating the constraint on supply being close to demand. The reason for using the adjusted shape of the original membership function is to be able to control the convergence of the GA. If the gap between demand and supply is within a pre-set interval, no penalty is imposed. As the gap increases, the penalty starts rising according to the tangent curve. There is an upper limit for a penalty since the membership function cannot be higher than one. Setting the parameter r_i' at low level makes the curve close to linear. It is impoitant to make the curve smooth enough, so, that the algorithm "feels" the direction in which the penalty is decreased. The values of the parameters l_i, r_i, ϑ_i were set to 20,000; 0.00001; and 20, respectively, for all three products. These values for parameters were found empirically.

Although, in many cases the penalty function approach does not produce good results, since it allows the presence of many invalid chromosomes in the population, in our case, because of the nature of soft constraint (5.91), such chromosomes are not invalid but are undesirable.

It may appear that the objective/fitness function is linear, However, the formulas only show the aggregation of different marketing values. The calculation of individual marketing values involves combined use of complex non-linear discontinuous functions (since there are decision rules involved to determine the sales level) and Monte-Carlo simulations. Hence, the relationship between the decision variables and the value of fitness function is complex and uncertain.

Results of empirical test of the hybrid DSS

We used a traditional DSS based on a Monte-Carlo simulation model and a hybrid DSS based on GA plus Monte-Carlo for our experiments.

We conducted experiments to test the performance of a Monte-Carlo + GA-based system as compared to that of a Monte-Carlo + human "what-if". The subjects included 25 upper division business students taking a course in generalized modeling techniques. We feel that these subjects are appropriate for the experiment, because they were knowledgeable of the business environment and modeling techniques, and, in particular, they were exposed to the use of Monte-Carlo simulation models in the course they were taking. All the subjects used the Monte-Carlo-based simulation DSS. The subjects had been told how to perform simulations. Their goal was to maximize the total MV of their decisions, while observing the crisp and soft constraints.

We ran GA 25 times, for 100 and 1000 generations. In each run GA produced slightly different results, because it uses stochastic component in the search process. The parameters are summarized in Table 5.10. We experimented with different rates for genetic j operators and finally chose those ones. Fig. 5.10. shows a typical example of the dynamics of the fitness function for 1000 generations of GA.

Table 5.10. Parameters of the GA

Representation	Binary
Chromosome length	90
Sampling	Elitist
Population size	50
Crossover rate	0.7
Mutation rate	0.007
Number of generations	100; 1000

The performance of the human "what-if" was compared with the performance of the GA-based system. We measured the performance of both approaches in terms of achieved MV, estimated probabilities of stocking out, and time elapsed. Table 5.11 compares the performance of the two approaches in terms of MV achieved.

The results indicate that Monte-Carlo 4- GA outperformed the Monte-Carlo + "what-if" approach after 100 and 1000 generations. The one-tail hetero-skedastic (since the variances of MV are obviously unequal for the two methods) t-tests for the difference of the means between the Monte-Carlo + "what-if" and both 100 and 1000 generations GA combined with Monte-Carlo was significant (see Table 5.12).

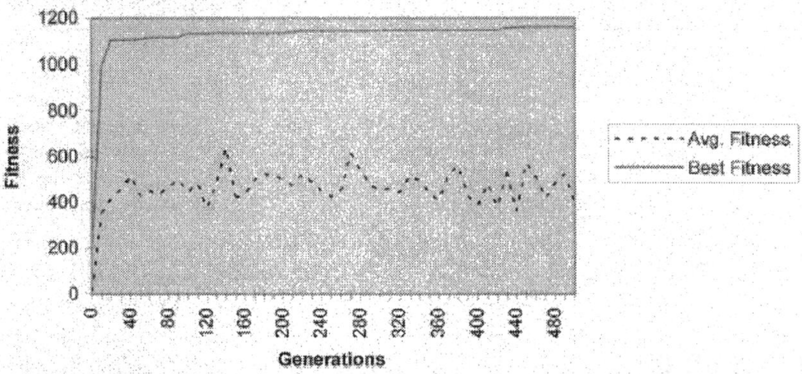

Fig. 5.10. A typical example of dynamics of evolution of population of marketing mix decisions

Table 5.12 reports the probabilities of stocking out for the three products. The standard deviation of these probabilities is an important measure of the performance, because even if the averages of the reported probabilities are neither close to zero, nor to one, their spread across the experiments may be wide. Therefore, low standard deviation would be the desired characteristic of these probabilities. In general, the standard deviations of the probabilities of stocking out for each of the three products was significantly lower for the decisions produced by GA, as confirmed by the F-test (see the p-values in Table 5.13). An exception is the probability of stock out of x for 100-iterations GA, where the p-value for the difference in standard deviations between the GA and human "what-if" is 0.053, and is, hence not significant at 0.05 level, but significant at 0.1 level.

Table 5.11. Comparison of marketing values produced by MC + GA vs. MC + human "what-if" (The figures in the table are based on simulations and hence do not reflect the accuracy of real world measures)

Method	Average	S.D.	Median	p-Value
MC+ "what-if"	1117173	41650	1133910	N/A
MC+GA(100 iterations)	1156956	11675	1158122	0.000042
MC+GA(1000 iterations)	1164106	6614	1164944	0.000004

The average time spent by humans was 75 min. It took approximately 15 s for GA to evolve 100 generations and 2.5 min to run 1000 generations.

The most common comments reported by the subjects included the difficulty of making joint decisions on multiple products while watching not to violate the crisp and soft constraints, as well as little guidance concerning the direction of the search. The results suggest that the method that combines genetic algorithms with simulation models produces better decisions in terms of marketing value and the probabilities of stocking out for the products than human "what-if". Furthermore, the time spent on search for "good" decisions is significantly reduced.

Table 5.12. Comparison of probabilities of stocking out produced by MC + GA vs. MC + human "what-if"

Method	Probability of stockout for x			Probability of stockout for y			Probability of stockout for z		
	Averag	S.D.	p-Value	Averag	S.D.	p-Value	Averag	S.D.	p-Value
MC+"what-if"	0.361	0.304	N/A	0.300	0.247	N/A	0.386	0.264	N/A
MC + GA(100)	0.378	0.104	0.000	0.272	0.177	0.053	0.389	0.055	0.000
MC + GA(1000)	0.403	0.009	0.000	0.269	0.032	0.000	0.384	0.029	0.000

5.10 Soft Computing Based Alternatives Generations by Decision Support Systems

An essential feature of active Decision Support Systems (DSS) is the ability to take the initiative in performing decision-related tasks. One possibility for providing active high cognitive level decision support is through facilitating alternative generation in DSS. The method considered in this section enables the generation of several diverse alternatives in a single run. The method relies on the principles of effective problem-solving/decision-making and facilitates divergent processes, the separation of alternative generation from evaluation, as well as the diminishing of human cognitive biases. A hybrid DSS based on Genetic Algorithms (GA) and fuzzy sets is used to operationalize the approach. The section outlines the design requirements for alternative generation in DSS and discusses the inadequacies of the "what-if" simulation and traditional optimization methods in light of these requirements. The section further elaborates on the appropriateness of GA as a tool for alternative generation in DSS for solving complex ill-structured problems.

Conventional DSS offer a weak form of support, where the potential power of human-computer collaboration is significantly underutilized. The process is user-directed, where the user must take the initiative to perform all necessary operations having full knowledge on where and how to use DSS tools. Researchers have recommended increasing effective utilization of DSS through a more active participation in decision-making [1, 18, 57, 69, 71], active support of high level cognitive processes [49], and stressing divergent processes [69].

One way to address the above concerns is to facilitate alternative generation in a DSS. In a conventional model-based DSS, the human performs alternative generation and the machine evaluates each alternative with simple calculations of the relevant criteria. This ignores the human yearning for more alternatives from which to choose. Thus, a conventional DSS supports primarily the choice phase of Simon's decision-making model [77]. It does not provide adequate support for the design phase, particularly for alternative generation. The drawbacks of a human-only alternative generation include:

- A search for promising alternatives may be very time-consuming, especially if the number of decision variables is large,
- Human biases, such as "anchor and adjust" [53], may lead to accepting inferior solutions,
- Because of the above drawbacks and the nature of "what-if" analysis, the generated alternatives may lack true diversity, and
- Since in traditional DSS alternative generation is usually immediately followed by evaluation, the well-known problem-solving principles of deferred judgment and divergence-convergence [64, 66] are inevitably violated.

In less-structured problems, the conditions indicating the existence of problems are not defined. There is no best methodology to approach solving these problems, and the criteria for choosing the best decisions are not clear [44], A problem is unstructured if any of the three stages of decision-making (intelligence, design, choice) is unstructured. According to another definition, if either the initial or desired state, or the set of transformations to cover the gap between the two, is not precisely known, the problem is less structured [55],

The area of problem-solving is very closely related to decision-making. In fact, some authors make no distinction between problem-solving and decision-making [55]. Researchers have made many prescriptions for improving problem-solving and decision-making effectiveness [88] including divergent/convergent thinking, deferred judgment, and control of human biases. Alex Osborn, the father of brainstorming, suggested that problem-solving consists of three stages: fact-finding, idea-finding, and solution-finding [64]. Osborn's model was further expanded by Fames, who suggested two additional stages and stressed the importance of divergent and convergent activities [65]. Divergent thinking is a creative input to problem-solving, while the convergent thinking deals with choice [32]. The divergence-convergence principle implies separation of evaluation and judgment from idea generation. This principle referred to as deferred judgment [64, 65], or ideation-evaluation [9] lies in the basis of creative problem-solving. It is also necessary to generate alternatives that are truly diverse, and not "clones" of one another. Otherwise this would lead to "myopic" alternative generation [12]. Generation of diverse promising alternatives may be hindered by human biases, such as "anchoring and adjusting" [38]. Therefore, one has to reduce the impact of such biases in a problem-solving process.

In a less structured problem there are normally several conflicting objectives that need to be achieved [44, 79]. One aspect of the lack of structure in such problems is the uncertainty about the desired states. Therefore, a desired characteristic of a DSS is support for the generation of diverse alternatives to achieve objectives.

Past research has focused on the evaluation of alternatives in terms of multiple objectives. [38] distinguishes compensatory and non-compensatory models as strategies for choice. Compensatory models include linear, additive difference, and ideal point models, while non-compensatory models include conjunctive, disjunctive, lexicographic, and elimination-by-aspects models. One widely used Operations Research approach to multi-objective problems is goal programming [42], which is a generalization of linear programming to the case of multiple objectives. Traditional use of optimization methods assumes that these techniques will generate the "best" solution. These methods are not well suited for ill-structured problems, where the desired objectives are not known in advance. They do not produce multiple alternatives in a single run. They do not consider the divergence - convergence principles for effective problem-solving discussed previously. A possible use of optimization methods in DSS is to incorporate them as part of the human-directed tool.

Ironically, traditional model-based DSS are not characterized with the above principles of problem-solving/decision-making. For example, "what-if' DSS ignores the guideline for the separation of alternative generation from evaluation as well as the divergence-convergence principle. However, research findings have recommended the inclusion of these principles in guidelines for building DSS. [29] argues that a computer algorithm for supporting problem-solving should generate "spread" [94] of alternatives preferably in a single run.

Recently, researchers have advocated a more proactive role for DSS in the man-machine collaboration. This may enable the DSS to support the user in abiding by the principles of problem-solving/decision-making. [57] introduced a notion of active DSS and proposed an architecture for constructing such DSS, where certain tasks can be performed by the system in an autonomous fashion [18] argued in favor of active support and proposed a framework for adaptive DSS. In [24] it is proposed an active DSS for econometric analysis. In [69] it is stressed the necessity to build more active DSS that would take the initiative in performing certain tasks, stress divergent processes, and engage in insightful conversations. As well as an architecture for active DSS and a prototype called JANUS, Rao proposed an architecture for active DSS based on intelligent agents [71].

Developments in active DSS may contribute positively to the problem of alternative generation. [75] and [81] argued that little attention had been paid in the past to the model analysis stage in model-based DSS. They proposed a system called INSIGHT for supporting model analysis based on multiple model instantiations. Through such analysis the decision-maker can get a better understanding into the situation by exploring the impact of different model parameter values on the solutions. In effect, this analysis facilitates alternative generation, since different parameters lead to different solutions. The work on the support for model building facilitates an easy construction of models, as well as an incorporation of the changes in those models to test different assumptions and generate alternatives [63]. There is also literature on interactive methods for alternative generation in decision support to elicit user preferences [43, 82]. The above methods, however, are sequential in manner. They do not aim at generating truly diverse alternatives with respect to the achievement of different objectives, and they are prone to human biases (e.g., anchoring and adjusting).

Advances in man-machine systems, where more activities can be delegated to machines necessitate development of advanced features/functions. [68] pointed out that decision support has been mainly restricted to the low cognitive level support. Typical functionality of a DSS included such tasks as data storage, retrieval, manipulation, consistency checking, and small calculation. In an attempt to support higher level tasks, the researchers advocated the use of artificial intelligence techniques [68], as well as hybrid methods [52, 53, 82], that combine various techniques (e.g., fuzzy logic, neural nets, etc.) in one DSS. The major rationale for employing the latter includes combining the strengths of the individual techniques for more effective decision support. One such relatively novel technique that can be used in DSS [7] is GA [21]. GA can deal with problems that incorporate nonlinearity, discontinuity, uncertainty, complexity, and other demanding features. These features make the application of traditional search and optimization methods inappropriate.

The developments in active and adaptive DSS could lead to building a DSS, which follows the principles of decision-making and problem-solving. For example, an active "what-if" DSS would suggest promising and diverse alternatives giving the user more options from which to choose. However, currently there is no conceptual basis for building such systems. There is a pressing need to develop the design methods for active DSS that would support high level cognitive functions, such as alternative generation to solve-less structured problems.

The alternative generation problem considered here suggests diverse alternatives, which promise different potential desired states for the DSS user. The major difference from the traditional multi-objective optimization methods is that in our approach, DSS also support the decision-maker's value judgment [38]. For example, in the case of marketing mix decisions, one alternative would lead to higher profits, while the other would promise higher market share. The user can then exercise judgment in making the final choice.

The alternative generation capability would benefit DSS in the following ways:

- ☐ It would facilitate divergent processes by offering a spread of alternatives,
- ☐ It would facilitate the separation of alternative evaluation from alternative generation,
- ☐ It would make DSS more active by enabling the exercise of initiative in searching for promising solutions and thus alleviate the user from a time-consuming manual search,
- ☐ It would reduce natural human biases, such as "anchor and adjust," for the above reasons,
- ☐ It would provide cognitively "higher" support. It goes beyond simple calculations and data retrieval/manipulation capabilities to create a higher degree of man-machine collaboration.

Problem of Alternative Generation

As we have stressed, traditional optimization methods of Operations Research (OR), including multi-criteria optimization methods (e.g., goal programming), strive for finding a single "best" solution to a problem at hand. These techniques require specification of the objective function and the set of constraints. Then, the optimization procedure identifies the solution that best fits the given specifications. However, for less structured problems with the presence of complexities and uncertainties, the notion of optimality may be fuzzy at best. For these problems, there is often some degree of uncertainty about the desired state. Hence, the objectives are not clear. The OR methods are inappropriate for making decisions in such situations. Human judgment is necessary for making a choice.

A more appropriate strategy for solving less structured problems is to use the methods that help generate promising candidate alternatives. These "good" candidates can then be passed to the decision-maker for selection, or, if necessary, modification. This relieves the decision-maker from the time-consuming task of locating promising candidates. Here, the computer-based procedure attempts to solve the alternative generation part of the original problem. Therefore, the procedure does not produce the solution, but a set of solutions. Each of these should be a "good" or viable alternative. The general formulation of the problem is given below.

Let $X = (x_1, x_2, ..., x_n)$ be a vector of decision variables and $C = (c_1, c_2, ..., c_m)$ be a vector of relevant criteria. Let $F=(f_1(\ ), f_2(\ ), \ldots, f_m(\ ))$ be a vector of functions mapping the decision variables into the set of criteria $F: X \rightarrow C$. Let $C_j^* = (c^*_{j1}, c^*_{j2}, ..., c^*_{jm})$ be a vector containing desirable values for criteria c_i, $i = \overline{1, m}$. We will call this vector the alternative goal (not to be confused with the term "alternative," which here means a potential solution). In traditional multi-objective optimization problems, one would attempt to find such values of X that would achieve C_j^* best. In general, however, there are a number of different combinations of desired values for C_j^* $j=\overline{1,r}$, that is, r alternative goals. Since all values for criteria should be reached in order to achieve a given alternative goal, we can write:

$$G_j = [f_1(x_1,x_2,...,x_n) \to c_{j1}^*] \wedge [f_2(x_1,x_2,..,x_n) \to c_{j2}^*] \wedge$$
$$\wedge ... \wedge [f_m(x_1,x_2,...,x_n) \to c_{jm}^*]$$

or

$$G_j = \bigwedge_{i=1}^{m} \left[f_i(x_1,x_2,...,x_n) \to c_{ji}^*\right] \tag{5.94}$$

As mentioned previously, the automated procedure for alternative generation should find a set of viable "solutions" where these solutions lead to multiple alternative goals. Therefore, we apply "OR" (disjunction) operation to the alternative goals in Equation (5.86) to obtain the global goal:

$$G = \bigvee_{j=1}^{r} G_j = \bigvee_{j=1}^{r} \bigwedge_{i=1}^{m} \left[f_i(x_1,x_2,...,x_n) \to c_{ji}^*\right] \tag{5.95}$$

Adding constraints in a very general form:

$$\Phi(X) \subset \Omega, \tag{5.96}$$

where Φ restricts the values of criteria associated with solution X to a feasible region Ω, we complete the formulation of the problem. The solutions to the problem (Equations 5.95 and 5.96) can be represented as a matrix r x n with r rows being alternative solutions, and n columns being values of different decision variables:

$$X^* = \begin{bmatrix} x_{11}^* & x_{12}^* & \cdots & x_{1n}^* \\ x_{21}^* & x_{22}^* & \cdots & x_{2n}^* \\ \vdots & \vdots & & \vdots \\ x_{r1}^* & x_{r2}^* & \cdots & x_{rn}^* \end{bmatrix} \tag{5.97}$$

such that

$$X_k^* \neq X_l^*, \quad k,l=\overline{1,r}; \quad k \neq l \tag{5.98}$$

Here, x_{ij}^* is ith alternative value of jth decision variable, and X_k^* is kth row of matrix (Equation 5.97), that is, kth alternative. The requirement Equation (5.98) specifies that no two alternatives should be the same.

Note two important aspects of the above formulation. First, as we already mentioned, the global solution is not a vector, but a matrix. This feature distinguishes the current approach from traditional approaches, where solutions are typically vectors. Second, the disjunction connective in the global goal will most likely introduce multiple optima. Although traditionally multiple optima are considered detrimental, in this problem formulation we *intentionally* introduce nondominated multiple optima and seek multiple candidate solutions. We try to avoid dominated local optima that may be present due to the complexities of the functions included in F.

We further advocate use of fuzzy sets in constructing the goal Equation (5.95). There are several reasons for doing this:

- Human decision-makers prefer linguistic, or "soft" information [42], and fuzzy-set theory allows representing such information.
- Preferences are formulated in fuzzy terms in DSS, and the use of fuzzy preferences makes DSS more robust [35].
- Since the membership function of fuzzy sets has a range [0,1], it provides a standard scale for measuring the degree of achievement of different objectives.
- It makes it easy to incorporate possible "soft" constraints in the problem (e.g., "avoid large inventory build-ups") into the overall objective.

With a little modification, we can rewrite Equation (5.87) as

$$\tilde{G} = \bigwedge_{j=1}^{r} \bigwedge_{i=1}^{m} [f_i(x_1, x_2, ..., x_n) \rightarrow \tilde{c}^*_{ji}] \tag{5.99}$$

where $\tilde{c}^*_{ji}$ is jth fuzzy desired value for the ith criterion. One way of aggregating the components of the above goal is through use of min and max operations for $\wedge$ and v, respectively [26]:

$$\propto_{\tilde{G}}(X) = \max_{j} \min_{i} \{\propto_{\tilde{c}^*_{ji}}[f_i(x_1, x_2, ..., x_n)]\} \rightarrow \max, \quad i = \overline{1,m}, \quad j = \overline{1,r} \tag{5.100}$$

where $\propto_{\tilde{G}}(X)$ membership function indicating the degree to which the fuzzy goal $\tilde{G}$ is achieved, and $\propto_{\tilde{c}^*_{ji}}()$ is that indicating the degree to which the fuzzy desired value $\tilde{c}^*_{ji}$ is achieved.

Equation (5.94) is of a general form. The alternative goals, including the fuzzy sets for the desired values, should be specified in accordance with the problem at hand. The most important guideline in specifying the alternative objectives is to try to set the diverse alternative goals. For example, in a strategic planning case the diverse alternatives could be "low cost leader", "differentiated approach," and "best value." Each of these translates into a different set of objectives.

Consider, for example, a simplified case of a marketing mix problem where two alternatives are described as achieving (5.96) high profit and moderate sales, or (5.89) moderate profit and high sales. Assume a candidate solution S achieves the objective "high profit" to degree 80 percent, the objective "moderate sales" to 70 percent, the objective "moderate profit" to 100 percent, and the objective "high sales" to degree 90 percent. Then, the alternative goal (5.96) would be achieved to 70 percent (min(80%, 70%)), and alternative goal (5.97) would be achieved to 90 percent (min(100%, 90%)). The total goal would be then achieved to 90 percent (max(70%, 90%)).

In [32] authors conclude that GA are applicable for the problem of alternative generation. In particular, the hybrid approach with the combination of GA as an alternative generation procedure and fuzzy sets for expressing diverse preferences for evaluating the alternatives, promises to make DSS a more active and advanced collaborator in a decision-making process.

Now we consider generating Alternatives for Marketing Mix Problem Using Genetic Algorithms. The marketing mix problem involves making decisions on price, promotion, and other important variables for various products. Traditionally, this problem was considered an optimization problem [37], and there have been attempts to optimize some specified objective function [25, 28, 48, 70]. However, the inherent simplifying assumptions of these approaches limit their usefulness [36, 70]. In highly competitive and uncertain dynamic environments, external factors, such as the state of the economy, consumer preferences, as well as strategies and tactics of competing firms determine the dynamics of the market. [36] and [70] stress the use of simulation models, rather than traditional optimization as an appropriate decision-support tool for these problems. These models allow the decision-maker to evaluate alternatives and see how the key criteria change in response to changes in the input variables. The simulation method overcomes the problems associated with most optimization methods: unrealistic assumptions, failure to deal with local optima, uncertainties, non-differentiability of the objective function, and the necessity to aggregate multiple criteria into a single objective function. Simulation may serve as a basis for "what-if" type of DSS. The drawbacks of the simulation-based approach include:

- the user may waste significant time and effort examining inferior (in a Pareto sense) solutions,
- the user is prone to different biases (e.g., "anchor and adjust"), generates "clone"
  alternatives (because of "anchoring," or settling on one solution), and fails to consider many promising solutions,
- the generated alternatives are immediately evaluated, which may lead to premature termination of search.

Genetic algorithms identify a set of "good" alternatives for the marketing mix problem better than humans do. GA-directed search is automated and is not prone to human biases. Furthermore, the evaluation by the decision-maker is done after the set of alternatives has been generated.

The potential applications of GA to marketing optimization problems are: consumer behavior; segmentation, targeting and positioning, managing the marketing mix; and strategic marketing [41]. [41] discuss applications to site location analysis, and segmentation (product-market structure) in detail. [83, 84] used GA for extracting marketing decision rules from a questionnaire. [7, 8] used GA to support the product design process. The above approaches treat marketing problems in a traditional optimization sense, assuming that there is a single "best" solution. Our approach employs GA as an alternative generation tool (rather than conventional optimization tool) for generating promising alternatives for ill-structured problems.

Let us reproduce Equation (5.95):

$$G = \bigvee_{j=1}^{r} \bigwedge_{i=1}^{m} [f_i(x_1, x_2, ..., x_n) \rightarrow c]$$

With respect to the marketing mix problem, vector C consists of important marketing criteria, including profit, market share, sales volume, marketing expenditures, and other important indicators. As we stressed earlier, we will use fuzzy criteria in our formulation. This makes sense, since specifying precise target values for sales, market share, and other variables would not be adequate in complex and imprecise environments. Vector X consists of decision variables, such as price, advertising expenditures, quantity, and other controllable variables, for different products. Note, that some decision variables can also be part of criteria, for example, advertising expenditures. Function vector F maps the values of decision variables into the values of criteria. If there are intrinsic uncertainties in the problem formulation, the simulation model can be used to determine the expected values of criteria. Constraints (5.90) may contain both "hard" (crisp) and "soft" (fuzzy) constraints. We advocate the use of special methods to ensure that no invalid (in a sense of hard constraints) chromosomes are produced and the inclusion of soft constraints into the fitness function as a fuzzy objective. In the next section we will discuss handling constraints in more detail using an example.

In general, the values of decision variables will need to be translated into the chromosome representation. Genetic algorithm then proceeds as follows:

1. Initial population of candidate solutions is randomly generated.
2. Fitness value is calculated for each chromosome (potential solution) in the population (this may require running simulations and passing the results to the GA).
3. Selection operator generates new population from the old population based on fitness values of the chromosomes.
4. Crossover operator is applied to the population.
5. Mutation operator is applied to the population.
6. Fitness value is calculated for each chromosome in the population (similar to step 2).
7. If stopping criteria are met, then the process ends, otherwise it goes back to step 3. The stopping criteria may include the maximum number of generations, or some goal attainment based measure.
8. The set of the best generated alternative solutions for the alternative goals and the key criteria are presented to the user.

Example Application

For our experiments we used SIMQ Business simulation software, where a number of firms compete for three products (x', y', and z') in an oligopolistic market of a hypothetical business environment [19, 72]. The teams make marketing, production, and financial decisions each quarter using DSS incorporating accounting and representational models. The decisions include a mix and quantity of goods to produce, pricing information, advertising expenses, as well as production capacity. The DSS for marketing decisions incorporates a Monte-Carlo simulation model. This model allows estimating the distributions of important output variables (e.g., sales, market share, profits, inventory levels) given the trial decisions and other input information. The decision-makers use these criteria as a guide to analyze different alternatives.

In searching for promising decisions, the decision-makers have to watch for hard and soft constraints in the problem. The hard constraint includes the requirement that the total production quantity of the three products should be limited by the capacity of the production facilities. The soft constraint is to avoid large inventory buildups, and still have some level of inventory. We call it "soft," since there are no "crisp" boundaries for inventory levels. Even if such boundaries were to be specified, slight violation of them would not lead to infeasible solutions.

In our prototype, we tried to alleviate the user's search process by delegating GA the task of generating a set of promising alternatives. We used the variables Volume Sales, Profit, and Advertising Expenditures as criteria to evaluate the alternatives. We generated the following set of situation descriptors to specify "good," or "promising" alternatives.

1. High Volume Sales And Moderate Profit And Somewhat High Advertising (to capture market share and yet make substantial profit);
2. High Profit And Somewhat High Volume Sales And Moderate Advertising (to make high profit and have some inventory to avoid stock outs);
3. Moderate Volume Sales And Somewhat High Profit And Somewhat Low Advertising Expenditures;
4. Moderate Volume Sales And Some Profit And Low Advertising.

The above items form the basis for alternative goals. In reality, one could use multiple experts in a brainstorming session to specify a set of diverse specifications for desired goals. Note that the target values of the criteria are expressed in fuzzy linguistic terms, rather than precise values, which is more appropriate for human experts and decision-makers.

We used binary-coded chromosomes for encoding the prices, advertising expenditures, and quantities for the three products. Here GA interacts with Monte-Carlo simulation model to obtain the expected values for the above criteria, and then uses membership functions to obtain the value of the fitness function. To calculate the fitness of a trial solution (chromosome), the conditions specified above are aggregated through the "OR" operation to form a single clause in accordance with

Equation (5.89). Since we are using fuzzy terms, the actual fitness function is similar to Equation (5.94).

We handled the capacity constraint in the following way. Instead of explicitly generating the quantities for the three products, we encode the *product mix* by taking one product (in our case, the product *x*, which is one of the product quantity decision variables) as the base product and encoding the quantities of others through the ratio of the quantity of the base product to them. We encode these using $Q_{x'/y'}$ and $Q_{x'/z'}$, which are actually the angles representing the mix of products. The tangent of these angles gives the actual ratio of product x' to the products y' and z'. Use of angles instead of actual ratios is preferred since we can control the ranges for these angles easier than those for the ratios. The variable *Q* represents the total actual quantity of the three products and is limited by the capacity limit of 15,000. Symbolically,

$$\frac{Q_{x'}}{Q_{y'}} = \tan Q_{x'/y'}, \quad \frac{Q_{x'}}{Q_{z'}} = \tan Q_{x'/z'}, \quad Q_{x'} + Q_{y'} + Q_{z'} = Q,$$

where $Q_{x'}$, $Q_{y'}$, $Q_{z'}$ are the quantities of products x', y and z'. From the previous formulas we can derive the quantities of the products as follows:

$$Q_{x'} = \frac{Q}{1 + \dfrac{1}{\tan Q_{x'/y'}} + \dfrac{1}{\tan Q_{x'/z'}}}, \quad Q_{y'} = \frac{Q_{x'}}{\tan Q_{x'/y'}}, \quad Q_{z'} = \frac{Q_{x'}}{\tan Q_{x'/z'}}$$

This method guarantees that no invalid chromosomes (infeasible solutions) will be generated as a result of genetic operations. This method seems ad hoc in our case. It can potentially be generalized to include more complex constraints. However, our interest here is in generating diverse alternatives and not in developing general constraint handling procedures.

The soft constraint concerning inventory buildups is directly incorporated into the multi-objective function as a fuzzy term. To this end, we added the term "Some Inventory" to our alternative goals specified earlier.

To complete the specification of the fitness function we specified the membership functions used for fuzzy terms. Membership functions reflect the degree to which an item can be described by a given fuzzy term. Fuzzy set theory provides the experts with the flexibility in specifying fuzzy terms they use in their respective fields [46]. Experts also use their judgment in specifying the parameters of the membership functions. For example, if market share is described by a fuzzy term "medium" and a triangular function is chosen for representation, the experts will have to specify what is the smallest and highest values for market share to be considered medium at all (membership 0), and what is the most representative value that describes "medium" market share (membership 1). With these parameters specified, the membership of any crisp value for market share to the fuzzy set "medium" can be calculated.

The chromosomes encoding candidate solutions are represented in GA as follows:

$$\left(P^b_{x'}, P^b_{y'}, P^b_{z'}, A^b_{x'}, A^b_{y'}, A^b_{z'}, Q^b_{x'/y'}, Q^b_{x'/z'}, Q^b\right).$$

Here, $P^b_{x'}$, $P^b_{y'}$, $P^b_{z'}$ are binary string representations of prices for products x', y' and z'; $A^b_{x'}, A^b_{y'}, A^b_{z'}$ are those for advertising expenditures; and $Q^b_{x'/y'}, Q^b_{x'/z'}, Q^b$ are those encoding the mix of the products and the total quantity of the products.

In our experiments GA used elitist selection in which the chromosomes best satisfying the described conditions are always preserved. The chromosome length was 90, the population size was set to 80, number of generations in a single run was 300, the probability of mutation was 0.008, and the probability of crossover was 0.7. The results of 50 runs are presented in Table 5.13.

The marketing mix problem requires making decisions on price, promotion, and other variables in order to achieve "best" outcome in terms of key marketing criteria (e.g., profit and market share). However, because of the ill-structured nature of the problem, determination of what makes the outcome "best" is largely judgmental. The considered method makes a thorough search of the solution space and presents several promising alternative solutions to the decision-maker. The method avoids the drawbacks of "what if"-type simulation that often leads to waste of time and effort in examining inferior solutions (in Pareto sense), fails to consider many promising alternatives, and may terminate the search process prematurely.

Table 5.13. Results of the Simulations

	Sales Volume, units		Profit, $		Advertising,		Inventory, units	
Condition #	Mean	St.D.	Mean	St.D.	Mean	St.D.	Mean	St.D.
1	13,720	313	844,127	23,691	215,235	24,240	80	49
2	12,562	474	1,097,802	32,927	149,525	33,310	27.6	25.7
3	11,897	699	942,212	21,377	108,121	6,039	82	40.4
4	8,867	321	762,757	14,320	38,575	3,730	109	116

The tabulated results (Table 5.13) show that diverse alternative goals lead to diverse alternatives in terms of the key criteria. The first goal emphasizes high volume sales, whereas the second one stresses high profit and some inventory. The third and the fourth goals lead to the somewhat low to low advertising expenditures with moderate to high profit and some inventory.

Fig.5.11 plots the alternatives 1 through 4 along the Profit and Sales Volume dimensions. As shown in the figure, alternative 1 attempts to maximize sales volume, while alternative 2 strives to maximize profit. The alternatives 3 and 4 lead to decreased Sales Volume and Profit, but also imply lower Advertising' Expenditures as one can see from Fig. 5.12 and Fig. 5.13.

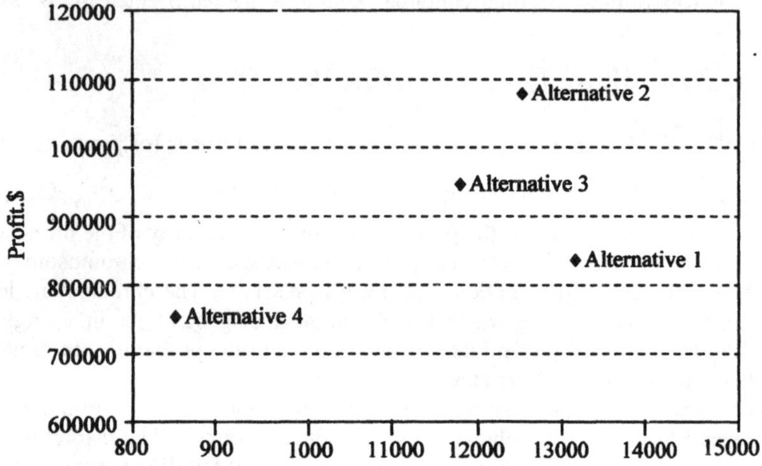

Fig. 5.11. Alternatives 1 Through 4 Mapped on Sales Volume and Profit Dimensions

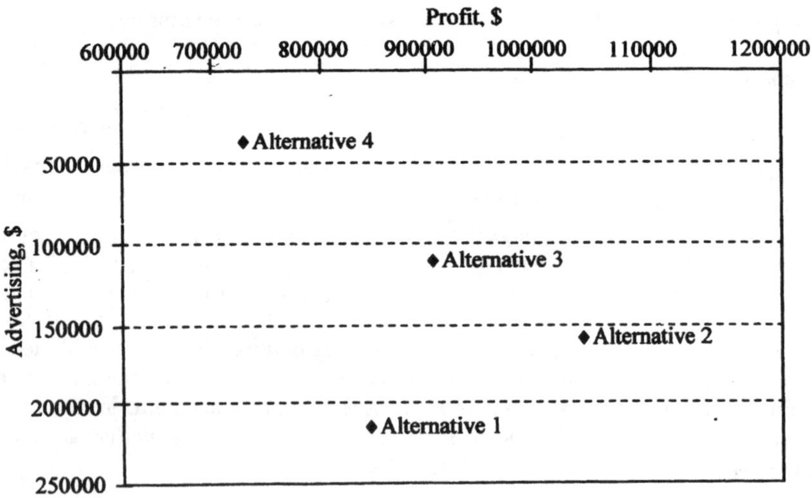

Fig. 5.12. Alternatives 1 Through 4 Mapped on Sales Volume and Advertising Dimensions

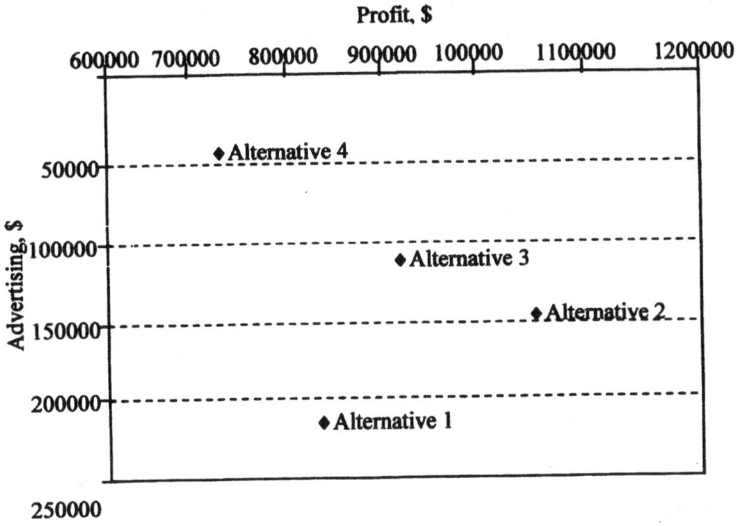

Fig. 5.13. Alternatives 1 Through 4 Mapped on Profit and Advertising Dimensions

The divergence of alternatives was assessed using MANOVA. In MANOVA, only Profit, Sales Volume, and Advertising Expenditures were included, since the Inventory Level is not a criterion, but is used to impose a soft constraint. We used MANOVA since we were interested in detecting significant differences among the groups of observations measured by multiple criteria. The analysis of the results of the procedure rejected the hypothesis that the generated alternatives are the same with the p-value of 0.000. This provides some evidence in favor of our expectation concerning the diversity of alternatives.

The results showed that:

- DSS can be built for a manageable number of promising diverse alternatives,
- The alternatives were diverse in terms of the achievement of objectives,
- Each alternative informs the decision-making process on making trade-offs between conflicting objectives,
- The alternative generation is completed before the evaluation of alternatives by the decision-maker.

In the context of the marketing mix problem, the method generates alternative decisions tied to the key marketing criteria (e.g., profits, sales, etc.)- This allows a decision-maker to examine diverse and promising scenarios in making marketing mix decisions. Furthermore, the decision-maker would also acquire a better insight into the problem as he analyzes the impact of different alternatives on the above criteria. Possible difficulties related to the choice of various parameters required by the method can be addressed by: setting those parameters automatically (e.g., rate of crossover and mutation), or eliciting them from the user (if these relate directly to the marketing problem domain, e.g., parameters for fuzzy values of profits, inventories, etc.) in a user-friendly fashion. In the latter case, the user may be asked to define his perception of fuzzy terms using numeric input or interactive graphics. The outlined approach is best suited for the less structured problems. Specifically, it is most appropriate for the situations where the desired objectives are not clear. In many such cases the alternative goals can only be rationally specified in vague terms, thus, leading to the use of fuzzy sets. The GA-based DSS can recommend different promising alternatives for further analysis by the decision-maker who can then analyze/modify the alternatives and assess their implications and consequences for the final choice. Use of GA is preferred, not only because they operate on a population of solutions and produce multiple alternatives in a single run, but also because they can cope with complex problems and are less prone to inferior local minima as compared to local search methods.

References

1. Ahgehrn AA (1991) Computers that criticize you: stimulus-based decision support systems. J Interfaces 23 (3): 3-16
2. Aliev RA, Mamedova GM, Aliev RR (1993) Fuzzy Sets and its Application. Tabriz University. Tabriz
3. Alter SA (1977) Taxonomy of decision support systems. Sloan Mgmt. Rev. 19 (1): 39-56
4. Arifovic J (1994) Genetic algorithm learning and the Cobweb model. J Economic Dynamics and Control 18 (1): 3-28
5. Arenas M, Bilbao A, Jimenez M, Rodrigue (2000) On considering flexible constaints of different importance in goal programming problems. The Fourth International Conference on Multiobgective Puming and Coal Programming: Theory & Applications. (Ustron, Poland.9-june 1,)
6. Arenas M,.Bilbao A, Uria R (2001) A Fuzzy Goal Programming Model for Evaluating a Hospital Service Performance. In: Fuzzy Sets in Management, Economics and Marketing. Zopounidis C, Pardalos PM, Baourakis G (eds), World Scientific
7. Balakrishnan PV, Jacob VS (1995) Triangulation in decision support systems: algorithms for product design. J Decision Support Systems 14 (4): 313-327
8. Balakrishnan PV, Jacob VS (1996) Generic algorithms for product design. J Management Science 42(8): 1105-1117
9. Basadur M (1994) Managing the creative process in organizations. In: Runco MA (ed) Problem Finding. Problem Solving and Creativity. Norwood. NJ: Ablex Publishing. pp 237-268
10. Bellman RE, Zadeh LA (1970) Decision making in a fuzzy environment. J Management Science 17 : B141-B164

11. Borisov AN, Krumberg OA, Fedorov IP (1984) Decision making on the basis of fuzzy models. Examples for utilization. Riga. Zinatne
12. Brightman HJ (1980) Problem Solving: A Logical and Creative Approach. Atlanta: Georgia State University. College of Business Administration. Business Publishing Division
13. Buckley J, Hayashi Y (1998) Applications of fuzzy chaos to fuzzy simulation. J Fuzzy Sets and Systems 99: 151-157
14. Buckley J, Thomas Feuring (2000) Evolutionary algorithm solution to fuzzy problems: Fuzzy linear programming. J Fuzzy Sets and Systems 109: 35-52
15. Carlsson C, Sebastian HJ (1995) Active DSS: Theory and Methodology for a New DSS Technology. In:Third European Congress on Intelligent Techniques and Soft Computing. Aachen. Germany. vol 2.pp 1202-1208
16. Chen M, Han J, Yu PS (1996) Data mining: An overview from a database perspective. J IEEE Trans. on Knowledge and Data Engineering 8: 866-883
17. Chen SJ, Hwahg CL (1992) Fuzzy multiple attribute decision making: Methods and application.Springer. New York
18. Chuang T, Yadav SB (1998) The development of an adaptive decision support system. J Decision Support Systems. 24(2): 73-87
19. Churchill G (1992) Applied Decision Sciences. Alphagraphics. Atlanta
20. Claver E, Gonzales R, Llopis J (2000) An analysis of research in information systems (1981-1997). J Information and Management 37: 181-195
21. Coello CAC (1999) A comprehensive survey of evolutionary-based multiobjective optimization techniques. J Knowledge & Information Systems 1(3): 269-308
22. Cubero JC, Medina JM , Pons O, Vila MA (1997) Extensions of a resemblance relation. J Fuzzy Sets and Systems 86: 197-212
23. Delgado M, Verdegay JL, and Vila MA (1990), Relating different approaches to solve linear programmimg problems with imprecise costs, J Fuzzy Sets and Systems. 37: 39-42
24. Dolk DR, Kridel DJ (1991) An active modeling system for econometric analysis. J Decision Support Systems 7(4): 315-328
25. Dorfman R, Steiner PO (1954) Optimal advertising and optimal quality. J American Economic Review 44: 826-836
26. Dubois D, Prade H (1988) Possibility Theory: An Approach to Computerized Processing of Uncertainty. New York: Plenum Press
27. Dutta A The Explicit Support of Human Reasoning in Decision Support System. J Advances in Computers: 1-45
28. Eliashberg J, Steinberg R(1993) Market production joint decision making. In: Eliashberg Lilien GL (eds) Handbooks in Operations Research and Management Science. Marketing. vol 5. Norch Hoiland. Amsterdam. pp 827-880
29. Evans JR (1990) Creative Thinking in the Decision and Management Sciences. Cincinnati. OH: South-Western Publishing Co
30. Fathi Torbaghan M, Meyer D (1994) DSMD - A Decision Support System for Medical Diagnosis. In: Second European Congress on Intelligent Techniques and Soft Computting. Aachen. Germany. vol 3. pp 1509-1515
31. Fazlollahi BB, Vahidov RM (1995) Applicability of FRIL to business intelligence. In: Proceedings of the Third European Congress on Intelligent Techniques and Soft Computing. Aachen. Gennany. pp 121-125

32. Fazlollahi BB, Vahidov RM (1995) Intelligent neural networks in business intelligence. In: Proceedings of the World Congress on Neural Networks. Washington. vol II. pp 721-724
33. Flood RL (1995) Solving Problem-Solving: A Potent Force for Effective Management. New York. John Wiley & Sons
34. Fogel LJ, Owens AJ, Walsh MJ (1966) Artificial Intelligence through Simulated Evolution. Wiley. New York
35. Garavelli AC, Gorgoglione M, Scozzi B (1999) Fuzzy logic to improve the robustness of decision support systems under uncertainty. J Computer & Industrial Engineering 37: 1-2. 477-480
36. Gatignon H, Hanssens DM (1987) Modeling marketing interactions with application to sales force effectiveness. J Marketing Research 24 (3): 247-257
37. Gatignon H (1993) Marketing mix models. J Eliashberg, GL Lilien (eds). Hand books in Operations Research and Management Science. Marketing. North-Holland. Amsterdam. vol 5: pp 697-731
38. Hogarth R (1987) Judgement and Choice. New York: John Wiley & Sons
39. Holland JH (1975) Adaptation in Natural and Artificial Systems. Ann Arbor: University of Michigan Press
40. Holland JH, Miller JH (1991) Artificial adaptive agents in economic theory. American Economic Review. In: Papers of Proceedings of the 103rd Annual Meeting of the American Economic Association. pp 365-370
41. Hurley S, Moulinho L, Stephens NM (1995) Solving marketing optimization problems using genetic algorithms. J European Journal of Marketing 29 (4): 39-56
42. Ignizio JP (1976) Goal Programming and Extensions. Lexington. MA: Lexington Books
43. Kalu TC (1999) An algorithm for systems welfare interactive goal programming modeling. J European Journal of Operational Research. 116(3): 508-529
44. Keen PGW, Scott Morton MS (1978) Decision Support Systems: An Organizational Perspective. Reading. MA: Addison-Wesley
45. Klir CJ, Yuan B (1995) Fuzzy Sets and Fuzzy Logic: Theory and Applications. Prentice Hall. PTR
46. Klir GJ, Clair US, Yoan B (1997) Fuzzy Set Theory. Upper Saddle River. NJ: Prentice Hall
47. Lai YJ, Hwang CL (1992) Fuzzy mathematical programming, Springer-Verlag,NY.
48. Lambin JJ (1970) Optimal allocation of competitive marketing efforts: an empirical study. J Journal of Business 43 (4): 468-484
49. Laudon KC, Laudon JP (2000) Management Information Systems. Organization and Technology in the Networked Enterprise. 6th edition. Prentice Hall
50. Lee J, Johnson JE (1993) Optimal tolerance allotment using a genetic algorithm and truncated Monte-Carlo simulation. J Computer Aided Design 25 (9): 601-611
51. Lee Z, Gosain S, Im I (1999) Topics of interest in IS: evolution of themes and differences between research and practice. J Information and Management 36: 223-246

52. Lenard MJ, Madey GR, Alam, P (1998) The design and validation of a hybrid information system for the auditor's going concern decision. J Management Information Systems. 14(4): 219-237
53. Li S (2000) The development of a hybrid intelligent system for developing marketing strategy. J Decision Support Systems. 27(4): 395-409
54. Li Xin Wang (1997) A course in fuzzy systems and control. Prentice Hall PTR.
55. MacCrimmon KR, Taylor RN (1976) Decision making and problem-solving. In: MD Dunnette (ed). Handbook of Individual and Organizational Psychology. Chicago: Rand- McNally. pp 1397-1453
56. Maddala CS, Nelson ED (1974) Maximum likelihood methods for models for markets in disequilibrium. J Economelrica 42: 1013-1030
57. Manheim M (1988) An architecture for active DSS. In: Proceedings of the 21st Hawaian International Conference on Systems Sciences. vol III. pp 356-365
58. McHaney R, Cronan TP (2000) Toward an empirical understanding of computer simulation implementation success. J Information and Managment 37: 135-151
59. Michalewicz Z (1992) Genetic Algorithms + Data Structures = Evolution Programs. Berlin: Springer-Verlag
60. Midgley DE, Marks RE, Cooper LG (1997) Breeding competitive strategies. J Management Science 43 (3): 257-275
61. Mintzberg H (1973) The Nature of Managerial Work. New York: Harper and Row
62. Mirchandani D, Pakath R (1999) Four models for a decision support system. J Information and Management 35 (1): 31-42
63. Murphy FH, Stohr EA, Ma P (1992) Composition rules for building linear programming models from component models. J Management Science. 38(1): 948-963
64. Osborn A (1963) Applied Imagination. New York: Charles Scribner's Sons
65. Parnes SJ (1967) Creative Behavior Guidebook. New York: Charles Scribner's Sons
66. Parnes SJ, Noller RB, Biondi AM (1977) Guide to Creative Action. New York: Charles Scribner's Sons
67. Rabelo L, Yih Y, Jones A, Tsai ShJ (1993) Intelligent scheduling for flexible manufacturing systems. In: Proceedings of the IEEE International Conference on Robotics and Automation 3. pp 810-815
68. Radermacher FJ (1994) Decision support systems: Scope and potential. J Decision Support Systems. 7 (4-5): 315-328
69. Raghavan SA (1991) JANUS: a paradigm for active decision support. J Decision Support Systems 7: 379-395.
70. Rangaswamy A (1993) Marketing decision models: From linear programs to knowledge-based systems. In: Eliashberg J, Lilien GL (eds). Handbooks in Operations Research and Management Science. Marketing. Amsterdam and New York: North-Holland vol 5. pp 733-771
71. Rao H, Raghav, Sridhar R, Narain S (1994) An active intelligent decision support system. J Decision Support Systems. 12(1): 79-91

72. Ribeiro RA, Baldwin JF (1994) A Fuzzy Decision Support System in FRIL. In: Second European Congress on Intelligent Techniques and Soft Computing. Aachen. Germany. vol 1. pp 404-409
73. Sakawa M, Utaka J, Inoiguchi M, Shiromaru I, Suginohara N, Inoue T (1993) Hot parts operating schedule of gas turbines by genetic algorithms. In: Proceedings of the International Joint Conference on Neural Networks and Fuzzy Satisfying Methods. vol I. pp 746-749
74. Schott B, Whalen T (1994) Fuzzy uncertainty in imperfect competition. J Information Sciences 76(3-4): 339-354
75. Sharda R, Steiger DM (1996) Inductive model analysis systems: Enhancing model analysis in decision support systems. J Information Systems Research. 7(3): 328-341
76. Silver M (1991) Systems that Support Decision Makers: Description and Analysis. Wiley. New York
77. Simon H (1960) The New Science of Management Decisions. New York: Harper
78. Sprague RH, Watson HJ (1986) Decision Support Systems: Putting Theory into Practice. Prentice Hall
79. Stabell C (1994) Towards a theory of decision support. In: Gray P(ed.). Decision Support and Executive Information Systems. Englewood Cliffs. NJ: Prentice Hall. pp 45-57
80. Stair R, Reynolds G (1999) Principles of Information Systems. 4th edition. Course Technology
81. Steiger DM (1998) Enhancing user understanding in a decision support system: A theoretical basis and framework. J Journal of Management Information Systems. 15(2): 199- 220
82. Sun M, Stam A, Steuer RE (2000) Interactive multiple objective programming using Tchebycheff programs and artificial neural networks. J Computers & Operations Research. 27(7-8): 601-620
83. Terano T, Ishino Y (1995) Marketing data analysis using inductive learning and genetic algorithms with interactive and automated phases. In:Proceedings of Fourth IEEE International Conference on Evolutional Computing. Piscataway. NJ: IEEE. pp 771-776
84. Terano T, Ishino Y (1996) Knowledge acquisition from questionnaire data using simulated breeding and inductive learning methods. J Expert Systems with Applications. 11(4): 507-518
85. Torra I, Reventos V (1998) On considering constraints of different importance in fuzzy constraints satisfaction problems. J International Journal of Uncertainty, Fuzziness and Knowledge-Based Systems. vol.6, №5: 489-491
86. Turban E (1988) Decision Support and Expert Systems. Macmillan Publishing Company
87. Vacca L (1997) Managing options risk with genetic algorithms. In: Proceedings of the IEEE/IAFE 1997 Computational Intelligence for Financial Engineering. New York. pp 29-32
88. Van Gundy AB (1987) Creative Problem-Solving: A Guide for Trainers and Management. New York: Quorum Books

89. Vila MA, Cubero JC, Medina JM, Pons O (1997) Soft Computing: a new perspective for some data mining problems. J Vistas in Astronomy. Elsevier Science Ltd vol 41(3): 379-386
90. Wagner HM (1975) Principles of Operations Research. Prentice-Hall. Englewood Cliffs. NJ
91. Wang JH, Leu JY (1996) Dynamic trading decision support system using rule selector based on genetic algorithms. In: Proceedings of the 1996 IEEE Signal Processing Society Workshop. pp 119-128
92. Wang JH, Leu JY (1997) Stock trading decision support system using rule selector based on sliding window. In: Proceedings of the 1997 IEEE International Conference on Systems, Man, and Cybernetics. vol I. pp 559-564
93. Werner B (1987) An interactive fuzzy programming system. J Fuzzy Sets and Systems. 23: 131-147
94. Woolsey G (1978) Two essays on model motivation: With this sign optimize & the shekels of silver solution. J Interfaces 9(1): 13-17
95. Yager RR (1977) Multiple Objective Decision Making Using Fuzzy Sets. J International Journal of Man-Machine Studies. 9 (4): 375-382
96. Yager RR (1978) Fuzzy Decision Making Including Unequal Objectives. J Fuzzy Sets & Systems. 1: 87-95
97. Zadeh L (1965) Fuzzy sets. J Information and Control 8: 338-353
98. Zimmermann HJ (1990) Fuzzy Sets Theory and its Application. Kluwer Academic Publishers

6 Soft Computing in Marketing

6.1 Marketing Analysis of a Customer's Purchasing Behavior

Marketers often are interested in the decision rules the customers use when choosing and making a purchase. On evaluating the products the customers rely on various attributes such as price, quality etc. In other words a customer's purchasing behavior frequently comes along with multiple-attribute decisions. On the other hand, the analysis of a customer's purchasing behavior enables the possibility for customers to find better ways to offer products and improve marketing mixes [17].

The value based on which the consumer shows his preferences of a product often contain fuzziness and imprecision. That is why section of the book will consider the problem of development of fuzzy multiple-model to analyze decision making in a consumer purchasing behavior study.

Such a model can be used in different markets to identify the preferences of customers, make effective advertisement emphasizing them and to produce product with demanded specifications. The guidelines can be prepared for marketers to make successful business in one particular market.

The considered model [17] is based on a combination of multi-attribute utility model [5] (MAUM) and fuzzy regression analysis, in particular, in [14]. Let's consider the statement of fuzzy multiple-attribute decision making problem when selecting products to buy. These products are always characterized by a multitude of attributes, e.g. conformance quality, durability, reliability, reparability, price etc [12].

Most products can be offered with varying features, characteristics that supplement the product's basic function. Being the first to introduce valued new features is one of the effective ways to compete. Most products are established at one of four performance levels: low, average, high, or superior. Performance quality refers to the level at which the product's primary characteristics operate. Buyers expect products to have a high conformance quality, which is the degree to which all the produced units are identical and meet the promised specifications. Durability, a measure of the product's expected operating life under natural or stressful conditions, is a valued attribute for certain products. Buyers normally will pay a premium for more reliable products. Reliability is a measure of the probability that a product will not malfunction or fail within specified time period. Buyers prefer products that are easy to repair. Reparability is a measure of the ease of fixing a product when it malfunctions or fails. Buyers are normally willing to pay a premium for products that are attractively styled [12].

Let's denote these attributes as x_i, $i=\overline{1,n}$ and assume that a product attributes x_i are usually based on subjective cognition and therefore represented as fuzzy numbers. The differentiation of products is determined by the view of products bought by customers. This resulting parameter designated here as *y* is determined by set of attributes x_i. y here is also a fuzzy number. In order to find how the attributes x_i influence on the value of y we use the following fuzzy linear regression model [14]:

$$y=a_i x_i,\ i=\overline{1,k} \tag{6.1}$$

where

$$a_i=(a_0,a_1,...,a_n),\ x_i=(1,x_{i1},...,x_{in}) \tag{6.2}$$

Here the fuzzy input data x_{ij}, the fuzzy output data y_i, and the fuzzy parameters a_j are often symbolically represented by

$$x_{ij}=(x_{ij},d_{ij})_L;\ y_i=(y_i,l_i)_L;\ a_j=(a_j,c_j)_L \tag{6.3}$$

The creation of the regression equation in [17] with the use of α -cut procedure is converted to a multi-objective programming problem. α $(0<\alpha\le 1)$ in [8] represents the degree of conformity between the fuzzy output y_i and the fuzzy linear regression model $a_i x_i$. As the objective function we use the degree of fuzziness of the fuzzy linear regression model.

Since the fuzziness of the model and the fuzzy threshold α conflict with each other, the usual optimal solution concept for scalar optimization problems cannot be applied to the formulated multi-objective programming problems. Alternatively, for the multi-objective programming problems, the Pareto optimal solution concept is introduced, where any improvement of one objective function can be achieved only at the expense of another [14].

Given an appropriate level for α, the decision maker can derive a satisfying solution from among the Pareto optimal solution set. Thus we can determine fuzzy weights of the attributes and how α influences the chosen attributes.

A very interesting example of application of fuzzy multi-attribute utility model is considered in [17] for automobile purchasing behavior. The authors focus on the five automobile brands: Nissan, Toyota, Mitsubishi, and Honda, which are all manufactured in Taiwan.

The car attributes used as purchase decision rule were chosen on the basis of a survey. 250 questionnaires (survey 1) were distributed among car buyers, salespersons, and various employees at car selling companies. 169 of them were used to identify the attributes. The survey 1 results (chosen attributes) were included in survey 2 questionnaires. Then 200 questionnaires of survey 2 were distributed to find the fuzzy attribute preference portion. Among them 126 were considered usable. Only attributes chosen by at least 65% of respondents were used in the fuzzy MAUM study (Table 6.1)

Table 6.1. Car Attributes

Customers with no car	Customers having a car
Safety structure	Safety structure
Price	Operation ability
Gas consumption	Maintenance fees
Operation ability	Price
Car style	Gas consumption

Fuzzy Linear Regression Model [14] was used with alpha value from 0.1 to 0.9. This allowed estimating the influence of alpha (estimated fuzzy weight) on the values of parameters. The results of sensitivity analysis made for the attributes (Table 6.1) show that fuzzy weights have overlapped (for example, when α =0.1, fuzzy weights for the attributes price and car style have overlapped). The overlapping areas represent the indifference of emphasis that customers put on the car style and car price. This means that these attributes have vagueness in customers' minds. The analysis has shown that first-time car buyers have different opinions about price. This conclusion is drawn because the fuzzy price spread value is 0.169 while the car style spread value is 0.05. That means they can accept wide price range while having strict cognition for the car style. The results of fuzzy sensitivity analysis show also that when α =0.3÷0.6 consumers have different viewpoints about the attributes of operation ability, gas consumption and car style.

The application of fuzzy multiple-attribute utility model can help marketers to make a market analysis for designing successful marketing strategy. Using this model, marketers can estimate fuzzy attributes and their importance to customers and consensus in market. They can emphasize important attributes in advertisements and use this information for designing new products.

6.2 Customer Credit Evaluation

Recently credit risk has been attracting a great attention. Ability to percept the counterparts' creditworthiness is important to avoid the risk of loss of valuable assets. Shareholders are not satisfied with rough forecasts of a company's expected returns – they are looking for effective risk-management technique for measuring credit risk.

Consider banks granting credit to customers. The information about customers can be gained through customer questionnaires, by examining the history of the bank's relation to the customer, from external databases. All these data reports are fed into a scoring model. The output is a numerical score predicting bank's risk as a borrower.

In conditions when the number of applicants tremendously increases, customer credit evaluation becomes impossible without advanced intelligent technique.

Methodologies to measure the credit are diverse. Among statistical tools the most widely applied are Linear Discriminant Analysis and Logistic Regression Model. Recently for credit scoring there was offered also an approach called Classification and Regression Tree [8, 16].

Although statistical tools are well-studied and straight-forward, sometimes they require data which are not always available; they are too computationally intensive and thereby slow. They do not give good results when finding analogies in huge pool of data containing incomplete or contradictory entries. Neural networks can be used instead to more effectively learn the relationships in these data.

Neural network based approach.

Data for learning the neural model are taken from real credit bank databases. It is important to include data on both positive and negative credit banking decisions. The input data can be of both continuous and categorical types. Usual input variables in a customer evaluation neural network system may be: credit history, credit owed, length of credit, level of delinquency on account, employment status, personal information, age, housing, job type, etc. All the available data are first filtered and scaled to a predefined data range suitable for a specific neural network type. Then the data are divided into two groups of approximately the same size: a group for training (developing the model) and a group for testing.

The system's output should detect the level of creditworthiness of the client whose data are applied to the inputs. The neural network can have two or one output. In case of one output, the value of this output shows the value of creditworthiness in some range (usually it is set to 0-100, where 100 displays the maximum possible creditability). During the testing phase, the threshold is adjusted. Commonly used rule is to consider all the clients who evaluate below 50% as non-creditworthy.

Artificial neural networks (ANNs) approach give better results in case of complex relationships among dependent and independent variables. ANNs show a very good performance when doing analysis for nonlinear problems. The predictability of ANNs appears to be superior over Logistic Regression Models.

However, the problem with ANNs credit scoring model is the lack of sufficient and accurate training information. Hence we may need to provide ANN with additional information to further improve the performance.

The experiments show that the results given by neural network can be significantly improved if we can combine them with some statistical method (e.g. the Classification and Regression Tree approach). In this case, the input data are applied first to a statistical model. The most appropriate known model for evaluating the creditability has been the Classification and Regression Tree model described in [6]. Then, the output from this model is fed to the neural network based system as one of its inputs. The integration neural network is illustrated in Fig.6.1.

In the presented system one of the outputs to the neural network is the predicted output provided by Classification and Regression Tree (CART) model. Reference [6] considers an example of applying a neural network based approach to measure customer's creditworthiness. Data for training the system were collected from a bank in Taiwan for 1995 to 2000. Data consists of information about 2250 credit applicants with the positive decision about granting the credit and 750 applicants with negative credit granting decision. The system used 9 input variables (2 continuous and 7 categorical variables) describing credit history, employment status, personal information, age, housing, and job type. The training data for estimating the parameters in the credit scoring model were divided into two groups: data for training (1980=1485+495) and data for testing (1020).

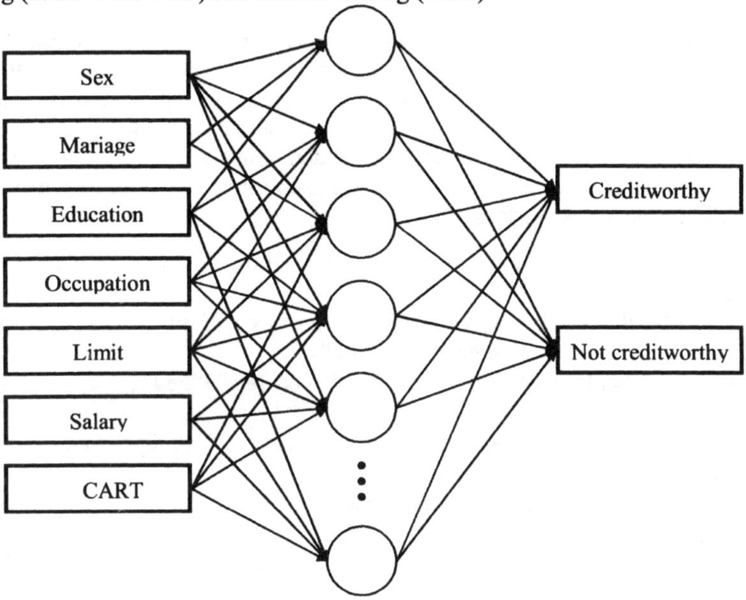

Fig. 6.1. Combined CART ANN system for credit scoring

The Classification and Regression Tree model was constructed and applied to the training and testing sets, which for the case of creditworthy customer gave 88.43% correct classification for training set and 88.22% for testing set, and for the case of non-creditworthy customer gave 58.27% and 56.27% correct classification for training and testing sets, respectively.

The comparison results of the pure CART model and combined neural model are presented in Table 6.2 [6].

Table 6.2. Performance comparison between pure CART and combined system for credit scoring

	Pure CART	CART ANN
Creditworthy customer	56.27%	72.24%
Non-creditworthy customer	88.22%	90.22%

Fuzzy approaches

The weakness of most datasets when using ANNs is that the database contains more "good" than "bad" creditors as the training data come from the bank and the banks normally reject the obviously bad customers. An ANN trained only on this database will not find obviously bad customers, because of no information available. In this case Neuro-Fuzzy approach is used which allows datasets to be added with appropriate rules [13].

It is important to note also that the data given by customer in questionnaires may contain wrong information. Also the business relationships between the bank and the customer may not be long enough to build a reliable database. Hence a more effective alternative to predict risk in a dynamic world and an imprecise environment which includes subjective considerations for credit scoring is Fuzzy query and ranking. For example, below we present the pieces of information and fuzzy linguistic terms used for credit cards:

- Amount owed on accounts = {"Too low", "Low", "Average", "High" "Too high", "Extremely high", "Not care"};
- Level of delinquency on account = {"Too low", "Low", "Average", "High" "Too high", "Extremely high", "Not care"};
- Number of bank revolving accounts = {"Too few", "Few", "Some", "Many", "Too many", Not care"};
- Number of bank or national revolving accounts = {"Too few", "Few", "Some", "Many", "Too many", Not care"};
- Availability of recent installment loan information = {"Lacking", "Not enough", "Enough", "Not care"};
- Number of accounts with balances = {"Too few", "Few", "Some", "Many", "Too many", Not care"};
- Number of consumer finance accounts = {"Too few", "Few", "Some", "Many", "Too many", Not care"}
  etc. [9].

The credit scores are evaluated as series of "Excellent", "Very good", "Not good", "Not bad", "Bad", and "Very bad". Then Fuzzy similarity and ranking is used to rank the new customer and define the associated credit score.

The reference [9] introduces fuzzy query and fuzzy aggregation for ranking and predicting the risk for credit scoring in an imprecise environment which includes subjective consideration.

6.3 Soft Computing Based Fraud Detection

Combating different kind of fraud, including credit card fraud, is one of the most important problems in service sector [7]. The volume of fraudulent actions in some businesses is so high that even modest improvement might lead to significant financial savings. However the task of detecting fraud is very difficult because of a number of factors:

- huge volume and rapid turnover of transactions that should be monitored continuously;
- the transactions can contain uncertain, imprecise, missing, contradictory data;
- new styles of fraud are continually invented and no specific set of patterns can be specified in advance.

For example, London stock exchange process 80000 bargains daily and each transaction consists of more than 100 data entries.

Technical difficulties in processing such transactions may be so high that sometimes overpass the saved money, making it technically unfeasible when applying traditional technique.

Fuzzy, neural and genetic algorithms based approaches in many cases permit to design and implement systems that make it possible to combat fraud effectively. The factors that make the application of these techniques particularly appropriate for detecting fraud include their tolerance for imprecision and flexibility and adaptability associated with the developed systems. Learning from past experience is extremely useful in changing business environment (and dynamic forms of fraud). The system can adapt and adjust its response to accord the changes. The system also can discover new patterns of fraud itself making its own associations between the events it experiences and creating new relationships.

The system is trained to recognize fraudulent activity by comparing legitimate transactions with known cases of fraud. The comparisons to existing models are made by using patterns based on the data entries such as date and time of transaction, the frequency of transaction, the volume of transaction, the type of transaction, etc. If the system considers a pattern of behavior potentially fraudulent, it generates on its output an alert signal. The system can also assign a score (ranging, say, 0 to 100) to every processed transaction, with the highest score marking the highest chance of fraud.

A rule based system to detect case of fraud can use, for example, the following rule:

IF Time> T_1 AND Time< T_2 THEN Score= S_0

IF Cost> C_0 THEN Score=Score+ S_1 ;

IF Type= T_0 OR Type= T_1 THEN Score=Score+ S_2 ;

IF Frequency> F_0 THEN Score=Score+ S_3

...

IF Score>50 THEN report the case of potential fraud

The system can use various parameters matching them to many fuzzy values, e.g. a rule can say:

IF Cost is "unusual" THEN ...

The training of the parameters (various S_i and the threshold 50 in our simple example) can be done through neural or evolutionary cycles performed on the historical data with indicated fraud cases.

Below we will shortly consider a Neuro-Fuzzy model that is capable of extracting the knowledge rules from pool of credit card based data and transaction records.

The fuzzy rules are presented in usual way:

$$\text{IF } x_1 = A_1^1 \text{ AND } x_2 = A_2^1 \text{ AND} \ldots \text{AND } x_n = A_n^1 \text{ THEN } y = B^1$$

$$\text{IF } x_1 = A_1^2 \text{ AND } x_2 = A_2^2 \text{ AND} \ldots \text{AND } x_n = A_n^2 \text{ THEN } y = B^2$$

...

$$\text{IF } x_1 = A_1^k \text{ AND } x_2 = A_2^k \text{ AND} \ldots \text{AND } x_n = A_n^k \text{ THEN } y = B^k$$

...

$$\text{IF } x_1 = A_1^m \text{ AND } x_2 = A_2^m \text{ AND} \ldots \text{AND } x_n = A_n^m \text{ THEN } y = B^m,$$

Where x_i are fuzzy input variables, y is a fuzzy output variable, A_i^k are input fuzzy constant values, B^k are output fuzzy constant values.

The values of A_i^k and B^k are described by bell-shaped membership functions (Fig.6.2):

$$\propto_{A_i^k} = e^{-\left(\frac{x_i - a_i^k}{\sigma_i^k}\right)^2}$$

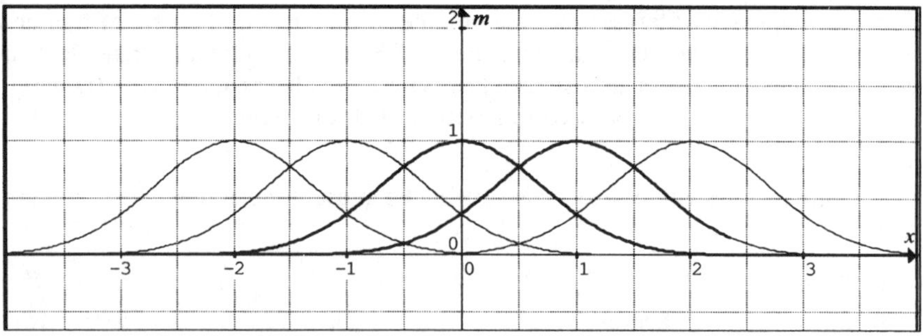

Fig. 6.2. Bell-shaped membership functions

The inputs are fed to a parallelized network structure, the nodes and connections of which are designed to perform the appropriate processing, inference, and defuzzification. Here for simplicity we consider a one output model. A detailed description of defuzzification and learning algorithm of an appropriate Neuro-Fuzzy network can be found in [19]. The authors of [15] conducted a lot of experiences with a version of such a network called a granular neural network implemented on a very powerful parallel computer.

The input data are divided into three groups: first group is for training, second group is for prediction, and third is for fraud detection. The network is trained using the training data. Then the second set of data is used for prediction, and later the third set of data is used for the fraud detection.

The credit card fraud detection procedure is implemented as follows:

1. Mine the principal data from the customer database;
2. Normalize data and prepare the groups for training, prediction, and detection;
3. Train on the first data set;
4. Predict on the second data set;
5. Detect fraud for the third data set.

An example of application of the granular neural network for credit card detection is illustrated in [15]. Different sets of transaction amounts for 50 days were used for training, prediction, and detection. Fuzzy training, prediction, and detection processes were done using a parallel fuzzy neural network implemented on a 24 processors based system. The transactions were considered as really fraudulent if they give the output between 0.5 and 1. In this experiment, 6 transactions were reported as suspicious to be fraudulent: the outputs were 0.060, 0.885, 0.635, 0.505, 0.085, and 0.009. For quality testing purpose, the results given by the system were compared to the actual status of the transactions. Three of these cases were actually fraudulent, which means a 50% detection rate.

6.4 Fuzzy Evaluation of Service Quality

In highly competitive service companies, advantages are mainly specified by service quality. In contrast to goods quality, which may be evaluated on the basis of measurable indicators, service quality can be evaluated only on the basis of perception information, because the perceived quality is the customer's objective judgment about an entity's superiority. Therefore, the evaluation fuzzy measure should be used, not a precision measure [2,8]. As shown in [2], service quality, as perceived by customers, can be measured by an evaluation analysis which results from a comparison between customers' expectations and experiences. In fact, service experiences are perceptions of reality, in which prior expectations are inherent. In this section we will discuss on how to use perception based information and fuzzy logic for evaluation of service quality.

The most widely used customer perceived service quality model is perhaps the Gap Analysis and SERVQUAL model [10,11]. Despite its validation in concept, there is no quantitative yardstick available.

In [2] the evaluation of customer-perceived service quality based on multiple service attributes is discussed. Because the best known formulation of attitude models is probably the Fishbein's multiple attribute model [3,4], the authors use this methodology to construct models for service quality evaluation problem. The comparative evaluation of service quality is formulated as a fuzzy multi-attribute decision making problem that requires cardinal ranking, in which customer-perceived service attribute weights and performance ratings are represented with fuzzy numbers.

The fuzzy model of measuring transportation service quality is discussed in [18]. Authors use fuzzy number to denote the consumer fuzzy rating of the perception performance and the expectation performance. They use cognitive threshold to develop the concept of measuring service quality, and develop the process of measuring service quality. Having [18] as a basis, let's suggest a methodology for evaluating the service quality.

For each attribute the perceived quality is determined as

$$Q = P - E, \tag{6.4}$$

where P and E are the fuzzy ratings representing by fuzzy numbers (P is consumer's mental evaluation of object (service) for each factor, E is desires of the customer.

To calculate service quality Q by formula (6.4) we need any fuzzy distance concept to measure the service quality (see section 2.1). In this case we use Hamming distance between fuzzy numbers P and $E: d_{ij}(P,E)$, where d_{ij} is the distance between P and E for i-th attribute and j-th customer. When P is very close to E, then it is said to be the best. The importance of each attribute is characterized by relative weight w_i. Then i-th attribute service quality rating will be determined as.

$$S_i = \frac{\sum_{j=1}^{N} d_{ij}(P,E)/w_i}{N} \tag{6.5}$$

Obviously, the greater the S_i, the better is service quality. Let's now describe the procedure to evaluate service quality attributes.

First, using the factor analysis we choose necessary q attributes that define the service quality. Then we calculate the fuzzy relative weights for every attribute

$$w_i = \frac{\sum_{j=1}^{N} w_{ij}}{N} \tag{6.6}$$

Then, using formula (6.5) we can calculate $S_i, i = \overline{1,q}$, i.e. the i-th attribute of the service quality.

After the calculation of $S_i, i = \overline{1,q}$, the manager of the company on the basis of the following rules can choose the service quality management strategy.

If P_i is greater than E_i and $S_i \leq \beta_i^u$, where β_i^u is the upper level of the user specified threshold, then the service quality rating for i -th attribute remains as P_i.

If P_i is less than E_i and $S_i < \beta_i^u$, then the i-th attribute should be improved. It is so because S_i is below the lower control limit.

Below we will consider an example from [18]. This example is related to measuring the air transportation service quality. The author interviews the passengers in the Kaohsiung domestic airport. The questionnaire return is 300, the effective questionnaire is 251, and the effective return rate is 83.67 %.

As a result of factor analysis of data, 11 attributes are chosen with the following defuzzified relative weights: flight safety, 0.118; manner and service attitude of flight attendants, 0.116, service attitude of ground employees, 0.115, on-time flight, 0.101, comfort of in-flight seat, 0.087, convenience choice of flight schedule, 0.085, simplicity of making reservation, 0.082, in-flight air circulation and air conditioning, 0.078, simplicity of purchasing the ticket, 0.075, in-flight cleanliness & layout, 0.073, convenience of entering & exiting the airplane, 0.070.

For each of considered 6 airlines A1-A6 the values of P, E, and S are calculated for every attribute, for example, for airline A1, flight safety has $P=6.38$, $E=7.91$, and $S=0.22$, while on-time flight has $P=6.81$, $E=7.97$, and $S=0.30$. And the simplicity of making reservations has $P=6.67$, $E=7.44$, and $S=0.52$.

The obtained results allow constructing improvement priority of service quality. On the basis of the fuzzy model of measuring of service quality, it has been found that for Airline A1: Flight safety=1, manner and service attitude of flight attendants=7, service attitude of ground employees=2, on-time flight=3, comfort of in- flight seat=4, convenience choice of flight schedule=4, simplicity of making reservation=11, in-flight air circulation and air conditioning=9, simplicity of purchasing the ticket=8, in-flight cleanliness & layout=10, convenience of entering & exiting the airplane=6. The latter allows taking a decision on how to mitigate the "bottleneck" in the service.

6.5 Application of Fuzzy Programming to Hospital's Service Performance Evaluating

Hospitals are highly complex institutions with uncertain operation environment seeking to provide high quality of care on the one hand and generate profits on the other. Frequently, goals and constraints in decision making for planning and control are conflicting. A Decision Maker (DM) deals with the problem of minimizing the costs, meeting community medical service needs, and maintaining job allocation simultaneously. Generally the DM is highly interested in achieving some degree to which the goals and constraints related to the above mentioned decision making situation are reached. These satisfaction degrees of objectives and constraints in the decision process are formulated by DM and usually are imprecise or uncertain. In this case the use of fuzzy approach to evaluating hospital service performance can look to be more suitable and adequate.

Below, using section 5.1 of the book, we will consider evaluating the performance of surgical service at the hospital. To be strict, we will consider evaluating a plan for one year ahead. The problem of finding a near-optimal plan will be considered as the problem of fuzzy multi-objective programming. To this end we will consequently consider formation of goals and constraints of the hospital's plan model [1].

Let's formulate staff and spatial requirements constraints. Vector x in model (5.12) include the following components (variables): x_1 – scheduled operation with admission; x_2 – emergency operation with no admission; x_3 – ambulatory scheduled operation; x_4 – ambulatory nonscheduled operation; x_5 – first appointments; x_6 – following appointments; x_7 – ambulatory emergency; x_8 – hospitalization. Let's assume that staff requirements characterize with the following values (for both physicians and nurses):

Constraints to time resources are given as: g^{Ph} – for physicians; g^{HN} – for hospitalization nurses; g^{CN} – for consulting nurses; g^{SN} – for surgical nurses.

Constraints for spatial requirements are defined by the values: beds g^{B}; consulting rooms g^{CR}; operating theatres g^{OT}.

Assume that the DM can use the available time for physicians and nurses with a tolerance threshold m% (for example 3% or 5%). With this account, the given constraints can be written as:

$$\sum_{i=1}^{8} c_i^{Ph} x_i \mathrel{\tilde{=}} \left(g^{Ph}, \frac{m}{100} g^{Ph}, \frac{m}{100} g^{Ph} \right) \text{ (for physicians)} \tag{6.7}$$

$$\sum_{i=1}^{4} c_i^{N} x_i \mathrel{\tilde{=}} \left(g^{SN}, \frac{m}{100} g^{SN}, \frac{m}{100} g^{SN} \right) \text{ (for surgical nurses)} \tag{6.8}$$

$$\sum_{i=1}^{7} c_i^{N} x_i \mathrel{\tilde{=}} \left(g^{CN}, \frac{m}{100} g^{CN}, \frac{m}{100} g^{CN} \right) \text{ (for consulting nurses)} \tag{6.9}$$

$$c_8^{N} x_8 \mathrel{\tilde{=}} \left(g^{HN}, \frac{m}{100} g^{HN}, \frac{m}{100} g^{HN} \right) \text{ (for hospitalization nurses)} \tag{6.10}$$

Assume that from the viewpoint of DM, the available spatial resources can be increased up to n%. With the account of the constraints for spatial resources we will have the form (in minutes):

$$\sum_{i=5}^{6} a_i x_i \mathrel{\tilde{\leq}} \left(b^{CR}, \infty, \frac{n}{100} b^{CR} \right) \tag{6.11}$$

$$\sum_{i=5}^{4} a_i x_i \lesssim \left(b^{OT}, \infty, \frac{n}{100} b^{OT} \right) \tag{6.12}$$

where b^{CR}, b^{OT} are spatial requirements on consulting rooms and operating theatres, respectively.

As was mentioned above, in hospital's plan, service costs must not exceed a predefined amount and they should be kept as minimum as possible. In order to consider the above condition, in the model for the optimization of plan one need hospital's cost evaluation to allocate it into each of variables $x_i, i = \overline{1,8}$. In particular, with this purpose, it is possible to use a cost assignment technique. Assume that as the result we have got the following values of costs corresponding to each $x_i, i = \overline{1,8}$:

$x_1 - a_9; x_2 - a_{10}; x_3 - a_{11}; x_4 - a_{12}; x_5 - a_{13}; x_6 - a_{14}; x_7 - a_{15}; x_8 - a_{16}$

Assuming that DM formulates the tolerance threshold for service costs as "service cost does not exceed the desired level by more than f%" we can represent the costs constraint as:

$$a_9 x_1 + a_{10} x_2 + a_{11} x_3 + a_{12} x_4 + a_{13} x_5 + a_{14} x_6 + a_{15} x_7 + a_{16} x_8 \lesssim \lesssim \left(b^{act}, \infty, \frac{f}{100} b^{act} \right) \tag{6.13}$$

The constraints related to waiting lists, which mainly determine the quality of service in hospitals can be rewritten as:

$$x_i \cong \left(x_i^d, x_i^d - x_i^{act}, \frac{S}{100} x_i^d \right) \quad i = \overline{1,8} \tag{6.14}$$

Here x_i^d, x_i^{act} are desirable and actual activity levels, s is the value of the tolerance threshold given by DM as "activity level can be exceeded by as much as s %".

It is known that the quality service in hospitals are measured in the same way as the activity indexes such as Occupation Index (OI), Turnover Index (TI), and average length of stay (ALS) [1]. These indexes are determined as: OI=(Patient-days/Annual average of beds utilized· 365)/100; TI=Number of admission/Number of beds; ALS=(Patient-days/Number of admission (scheduled on emergency). The information about current minimum and maximum values of these activity indexes is available. The constraint related to these indexes can be written as:

$$TI^{min} \le (x_1 + x_2)/v \le TI^{max} \tag{6.15}$$

With the account of the maximum and minimum degrees of satisfaction, the above can be rewritten as:

$$x_1 + x_2 \lesssim \left(vTI^{max}, (vTI^{max})^{dev}, 0 \right) \tag{6.16}$$

where $(vTI^{max})^{dev}$ is the deviation of value vTI^{max}. Now we can form the complete fuzzy model of hospital's service performance evaluation. To this end, we need one more piece of information – weighting up the different goals and constraints by DM. Assume that DM gives the following information. For the first group of constraints (human and spatial resources) – w_1. Members (goals) in this group have the following weights: physicians- w_{11}; surgical nurses- w_{12}; consulting nurses- w_{13}; hospitalization nurses- w_{14}; consulting room- w_{15}; operating room- w_{16}. For the second group, weight is w_2, for waiting list – is w_3, and for fourth group is w_4. On the basis of equation (5.19) and considering (6.7)-(6.16), the full model can be written as:

$$
\begin{aligned}
& w_1\left[w_{11}\left(\frac{n_1}{0.01mg^{Ph}}+\frac{p_1}{0.01mg^{Ph}}\right)+w_{12}\left(\frac{n_2}{0.01mg^{SN}}+\frac{p_2}{0.01mg^{SN}}\right)+\right.\\
& \left.+w_{13}\left(\frac{n_3}{0.01mg^{CN}}+\frac{p_3}{0.01mg^{CN}}\right)+w_{14}\left(\frac{n_4}{0.01mg^{HN}}+\frac{p_4}{0.01mg^{HN}}\right)\right]+\\
& +w_1\left(w_{15}\frac{p_5}{0.01nb^{CR}}+w_{16}\frac{p6}{0.01nb^{OT}}\right)+w_2\frac{p_7}{0.01fb^{act}}+\frac{w_3}{8}\left(\frac{n_8}{x_1^d-x_1^{act}}+\right.\\
& +\frac{p_8}{0.01sx_1^d}+\frac{n_9}{x_2^d-x_2^{act}}+\frac{p_9}{0.01sx_2^d}+\frac{n_{10}}{x_3^d-x_3^{act}}+\frac{p_{10}}{0.01sx_3^d}+\frac{n_{11}}{x_4^d-x_4^{act}}+\\
& +\frac{p_{11}}{0.01sx_4^d}+\frac{n_{12}}{x_5^d-x_5^{act}}+\frac{p_{12}}{0.01sx_5^d}+\frac{n_{13}}{x_6^d-x_6^{act}}+\frac{p_{13}}{0.01sx_6^d}+\frac{n_{14}}{x_7^d-x_7^{act}}+\\
& \left.+\frac{p_{14}}{0.01sx_7^d}+\frac{n_{15}}{x_8^d-x_8^{act}}+\frac{p_{15}}{0.01sx_8^d}\right)+w_4\frac{n_{16}}{(vTI^{max})^{dev}}\rightarrow \min
\end{aligned}
\tag{6.17}
$$

S.t.

$$\sum_{i=1}^{8} c_i^{Ph}x_i+n_1-p_1=g^{Ph};\sum_{i=1}^{4} c_i^{N}x_i+n_2-p_2=g^{SN};$$

$$\sum_{i=5}^{7} c_i^{N}x_i+n_3-p_3=g^{CN};c_8^{N}x_8+n_4-p_4=g^{IIN};$$

$$\sum_{i=5}^{6} a_i x_i+n_5-p_5=b^{CR};\sum_{i=1}^{4} a_i x_i+n_6-p_6=b^{OT};$$

$$a_9x_1 + a_{10}x_2 + a_{11}x_3 + a_{12}x_4 + a_{13}x_5 + a_{14}x_6 + a_{15}x_7 + a_{16}x_8 + n_7 - p_7 =$$
$$= b^{act};\ x_1 + n_8 - p_8 = x_1^d;\ x_2 + n_9 - p_9 = x_2^d;\ x_3 + n_{10} - p_{10} = x_3^d;$$
$$x_4 + n_{11} - p_{11} = x_4^d;\ x_5 + n_{12} - p_{12} = x_5^d;\ x_6 + n_{13} - p_{13} = x_6^d;$$
$$x_7 + n_{14} - p_{14} = x_7^d;\ x_8 + n_{15} - p_{15} = x_8^d;\ x_1 + x_2 + n_{16} - p_{16} = vTI^{max};$$
$$n_1 < 0.01mg^{Ph};\ p_1 < 0.01mg^{Ph}; n_2 < 0.01mg^{SN};\ p_2 < 0.01mg^{SN};$$
$$n_3 < 0.01g^{CN}; p_3 < 0.01g^{CN}; n_4 < 0.01g^{HN};\ p_4 < 0.01g^{HN}; n_5 < 0.01nb^{CR};$$
$$p_5 < 0.01nb^{CR};\ n_6 < 0.01nb^{OT};\ p_6 < 0.01nb^{OT};\ p_7 < 0.01fb^{act};$$
$$n_8 < {x_1}^d - {x_1}^{act}; p_8 < 0.01s{x_1}^d;\ n_9 < {x_2}^d - {x_2}^{act};\ p_9 < 0.01s{x_2}^d;$$
$$n_{10} < {x_3}^d - {x_3}^{act}; p_{10} < 0.01s{x_3}^d;\ n_{11} < {x_4}^d - {x_4}^{act}; p_{11} < 0.01s{x_4}^d;$$
$$n_{12} < {x_5}^d - {x_5}^{act}; p_{12} < 0.01s{x_5}^d;\ n_{13} < {x_6}^d - {x_6}^{act};\ p_{13} < 0.01s{x_6}^d;$$
$$n_{14} < {x_7}^d - {x_7}^{act}; p_{14} < 0.01s{x_7}^d; n_{15} < {x_8}^d - {x_8}^{act};\ p_{15} < 0.01s{x_8}^d;$$
$$n_{16} < (vTI^{max})^{dev}$$

With all variables non-negative.

Having solved the problem, DM on the basis of the obtained information can apply necessary changes to the organization of hospital to improve the operational results.

Example [1]. In [1] the authors have developed a method and concrete model for the COT (Traumatology and Orthopedic Surgery) service of a local state owned hospital. Numerical model for one-year plan was created which is described as:

$$0.35\left[0.20\left(\frac{n_1}{70000} + \frac{p_1}{70000}\right) + 0.15\left(\frac{n_2}{17940} + \frac{p_2}{17940}\right) + 0.10\left(\frac{n_3}{10920} + \right.\right.$$
$$\left.\left. + \frac{p_3}{10920}\right) + 0.10\left(\frac{n_4}{71760} + \frac{p_4}{71760}\right)\right] + 0.35\left(0.20\frac{p_5}{25260} + 0.25\frac{p_6}{6300}\right) +$$
$$+0.22\frac{p_7}{64571} + 0.18\frac{1}{8}\left(\frac{n_8}{90} + \frac{p_8}{26.4} + \frac{n_9}{294} + \frac{p_9}{7.2} + \frac{n_{10}}{103} + \frac{p_{10}}{6.5} + \frac{n_{11}}{12} + \frac{p_{11}}{0.4} + \right.$$
$$\left. + \frac{n_{12}}{200} + \frac{p_{12}}{4187} + \frac{n_{13}}{400} + \frac{p_{13}}{2663} + \frac{n_{14}}{200} + \frac{p_{14}}{6539} + \frac{n_{15}}{417} + \frac{p_{15}}{1042.5}\right) + 0.16\frac{n_{16}}{164}$$

S.t.

$$
\begin{aligned}
&200x_1+200x_2+90x_3+90x_4+20x_5+10x_6+32x_7+15x_8+n_1-p_1=\\
&=1400000;\\
&200x_1+200x_2+90x_3+90x_4+n_2-p_2=358800;\\
&6.66x_5++3.33x_6+n_3-p_3=218400;\ 70.28x_8+n_4-p_4=1435200;\\
&20x_5+10x_6+n_5-p_5=505200;\\
&100x_1+100x_2+45x_3+45x_4+n_6-p_6=126000;\\
&19437.57x_1+19437.57x_2+9718.78x_4+9719.47x_5+5830.5x_6+\\
&+11663.25x_7+38875.15x_8+n_7-p_7=1291423;\\
&x_1+n_8-p_8=1320;\ x_2+n_9-p_9=360;\ x_3+n_{10}-p_{10}=130;\\
&x_4+n_{11}-p_{11}=20;\ x_5+n_{12}-p_{12}=10000;\ x_6+n_{13}-p_{13}=20000;\\
&x_7+n_{14}-p_{14}=10000;\ x_8+n_{15}-p_{15}=20850;\\
&x_1+x_2+n_{16}-p_{16}=1465;\ n_1<70000;\ p_1<70000;\ n_2<17940;\\
&p_2<17940;\ n_3<10920;\ p_3<10920;\ n_4<71760;\ p_4<71760;\\
&p_5<25260; p_6<6300;\ p_7<64571;\ n_8<90;\ p_8<26.4; n_9<294;\\
&p_9<7.2; n_{10}<103;\ p_{10}<6.5; n_{11}<12;\ p_{11}<0.4; n_{12}<200;\ p_{12}<4187;\\
&n_{13}<400;\ p_{13}<2663;\ n_{14}<200;\ p_{14}<6539; n_{15}<417;\ p_{15}<1042.5;\\
&n_{16}<164
\end{aligned}
\tag{6.18}
$$

It is shown that this problem is unfeasible [1]. In order for this problem to be feasible, the DM in an interactive manner should apply some iteration by changing the tolerance margin. For example, using the model (6.18) DM should increase the left tolerance margin of consulting nurses and right tolerance margin of desirable activity level of x_6 and x_7. By some reformulation of the investigated model near optimal solution was obtained as:

Table 6.3. Near-optimal solution

	Variable	Desirable	Activity level	Our approach solution
Scheduled operation with admission	x_1	1320	1230	1235
Emergency operation with no admission	x_2	360	66	66
Ambulatory scheduled operation non-scheduled	x_3	130	27	27

Ambulatory operation non-scheduled	x_4	20	8	20
First Appointments	x_5	10000	14187	14187
Following Appointments	x_6	20000	22663	22663
Ambulatory Emergency	x_7	10000	16539	10000
Hospitalization	x_8	20850	20655	20850
	Target deviation	Tolerance Margin		
Physicians	p_1	70000		75550
	n_1	70000		0
Surgical nurses	n_2	17940		94370
Consulting nurses	n_3	10920		48446
Hospitalization nurses	p_4	71760		30138
Consulting rooms	n_5	25260		0
	p_5	25260		5170
Operating theatres	n_6	6300		0
	p_6	6300		6215
Cost	n_7	∞		68734
Turnover	n_{16}	164		164

One year hospital service plan described above is more balanced and resources are used more intensively.

References

1. Arenas M, Bilbao A, Uria R (2001) A Fuzzy Goal Programming Model for Evaluating a Hospital Service Performance. In: Fuzzy Sets in Management Economics and Marketing. Zopounidis C, Paradalos PM, Baourakis G (eds), World Scientific
2. Chung-Hsing Yeh A (2002) Preference-Based Approach to Fuzzy Multiattribute Evaluation of Service Quality. J IEEE: 7803-7280
3. Engel JF, Blackwell RD, Miniard PW (1995) Consumer Behavior (Eighth Edition). Forth Worth, TX. The Dryden Press
4. Fishbein M, Ajzen I (1975) Belief attitude. Intention and Behavior. An Introduction to Theory and Research. Reading. Mass: Addio-Wesley
5. Fishbein M (1963) An Investigation of the Relationships Between Beliefs About an Object and the Attitude Toward That Object. J Human Relations vol 16: 233-240
6. Kao LJ (2001) Mining the customer credit by using the neural network model with classification and regression tree approach. 0-7803-7078-3/01, J IEEE: 923-928
7. Kingdon J (1997) Intelligent systems for fraud detection. In: Sanchez E, Shibata T, Zadeh L (eds) Genetic Algorithms and Fuzzy Logic Systems. Soft Computing Perspectives. Advances in Fuzzy Systems -- Applications and Theory vol 7, pp 133-153

8. Maher J, Sen TK (1997) Predicting bond ratings using neural networks: a comparison with logic regression. J Intelligent Systems in Accounting, Finance and management vol 6: pp. 59-72
9. Nikravesh M (2001) Credit scoring for billions of financial decisions. 0-7803-7078-3. J IEEE:191-196
10. Parasuraman A, Zeithaml VA, Berry LL (1985) A Conceptual Model of Service Quality and its Implications for Future Research. J Journal of Marketing. vol. 49:4 pp 44-50
11. Parasuraman A, Zeithaml VA, Berry LL (1988) Servqual: A multiple-item scale for measuring consumer perception of service quality. J Journal of Retailing. vol.64:1:12-40
12. Philip Kotler (2000) Marketing Management. The Millennium Edition, Northwestern University
13. Rast M (1997) Application of Fuzzy Neural Networks on Financial Problems. In: Annual Meeting of the North American Fuzzy Information Processing Society NAFIPS'97, pp 347-349
14. Sakawa M (1992) Multiobjective Fuzzy Linear Regression Analysis for Fuzzy Input-output Data. J Fuzzy Sets and Systems vol.47: 173-181
15. Syeda M, Zhang Y-Q, Pan Y (2002) Parallel Granular Neural Network for Fast Credit Card Fraud Detection. 0-7803-8/02, J IEEE: 572-577
16. Thomas LC (2000) The surway of credit and behavioral scoring: forecasting financial risk of lending to customers. J International Journal of Forecasting vol 16: 149-172
17. Tsuen-Ho Hsu, Monle Lee (2000) Fuzz & Automobile Curtains Produce Clarity of View to the Adept Marketer. J CR vol 10(2): 87-96
18. Tsuen-Ho Hsu (2001) Applying Fuzzy Set Theory to Measuring Air Transportation Service Quality. In: IEEE: 0-7803-7078-3/01/(c)
19. Zhang Y-Q, Kandel A (1998) Compensatory genetic neural networks and their applications. J Series in Machine Perception Artificial Intelligence vol 30. Singapore: World Scientific

7 Soft Computing Applications in Operations Management

7.1 Application of Fuzzy Logic in Transportation Logistics

Problems in transportation logistics, such as routing, assignment, dispatching etc. are very complex combinatorial optimization problems and their solution is related with certain difficulties.

Logistics problems in practice are very difficult to solve because additional constraints should be considered. Many of the constraints, cost coefficients and relations are in reality not crisp but fuzzy [53].

Fuzzy technology has already been used widely in different areas of logistic primarily for evaluation of traffic flows, for data mining in traffic management, for data fusion in traffic supervision and others because it is more flexible and adaptive to quickly changing conditions.

Fuzzy Logic and Approximate Reasoning are successfully applied in traffic management: traffic supervision, traffic situation forecasting and evaluation, individual traffic behavior; selection of optimal routes and traffic control etc. [23,45,53].

Fuzzy sets are also used in fleet scheduling and dispatch. The central control of the vehicle fleet allows many haulers to economically cope with transport orders and to succeed in cost reductions [53].

The real time dispatcher poses additional problems which are solved by using fuzzy approaches.

In [53] approximate reasoning in container terminals is considered. Below we will discuss the application of Fuzzy Logic for selection of vehicles. The transport process is the complex technological object. Its operation happens in conditions of uncertainty and unmonitored traffic-climatic perturbations, the nature of which varies in time and in space. Absence of full prior information on behavior of the transport process and physical impossibility of direct measurement of all parameters, necessary for efficient control, suffices compounds of distinctive features.

Application of conventional mathematical methods for optimization of the transport processes is either inconvenient or almost impossible for some reasons. At the same time the simplification of model, by abstracting some parameters, makes it inadequate to real transport processes [9]. It is because the solution of some tasks of the transport process, the conventional deterministic and stochastic models quite often appear to be inadequate to the operating environment, leave out uncertainty of actual conditions and, generally, do not consider fuzzy nature of the state of units of the transport process.

For the rise of efficiency of the transportation process on motor transport, it is often necessary to solve the problem of choice of the rational vehicle. Problems bound with the decision making at the choice of rational dimensions of vehicle is now based on usage of normative values (average speeds, execution time of handling operations, etc.), which do not reflect actual exploitation conditions. Only on the basis of some estimation and matching of different alternate variants, with allowance for uncertainties, it is possible to solve the considered problems [1].

Let's have a set n of alternatives of vehicles $A=\{a_1, a_2, ..., a_n\}$ which can be utilized for delivery of given packages from the supplier to customers.

The choice of a vehicle is made on the basis of estimations on a number of criteria R_i, namely:

C_1- weight-lifting ability of vehicle;

C_2 - traveling speed of vehicle when delivery of package;

C_3 - time of loading of a package onto a vehicle;

C_4 - time of unloading of a package from a vehicle.

Then for each criterion the fuzzy set can be considered

$$C_i = \{\propto_{C_i}(a_1)/a_1, \propto_{C_i}(a_2)/a_2, ..., \propto_{C_i}(a_n)/a_n\}, \quad i=\overline{1,n}. \tag{7.1}$$

Where $\propto_{C_i}(a_j) \in [0.1]$ is the estimation of alternative a_j by criterion C_i, describing the degree of correspondence of alternative a_j to criterion C_i.

The rational vehicle A* is selected in accordance (7.2)

$$\propto_D(A^*) = \max_{j=1,n} \propto_D(A_i) \tag{7.2}$$

where $\propto_D(A_i)$ is defined by (5.53).

The relation which describes the satisfaction degree of criteria C_i to alternative a_i is given as

$$\begin{matrix} C_1 \\ C_2 \\ \cdots \\ C_m \end{matrix} \begin{bmatrix} c_{11} & c_{12} & \cdots & c_{1n} \\ c_{21} & c_{22} & \cdots & c_{2n} \\ \cdot & \cdot & \cdot & \cdot \\ c_{m1} & c_{m2} & \cdots & c_{mn} \end{bmatrix} \tag{7.3}$$

Set of alternatives A and set of criteria C characterize the situation of solutions. All elements of this matrix are real numbers of correspondence [0,1] and each coefficient c_{ij} expresses the degree with which criterion C_i satisfies the alternative a_j.

On the basis of the calculated data the fuzzy sets characterizing alternatives on criteria C_i are described as follows:

$$
\begin{aligned}
C_1 &= \{\ 0.3/A_1,\ 0.5/A_2,\ 0.75/A_3,\ 0.9/A_4\} \\
C_2 &= \{\ 0.8/A_1,\ 0.7/A_2,\ 0.65/A_3,\ 0.4/A_4\} \\
C_3 &= \{\ 0.6/A_1,\ 0.5/A_2,\ 0.25/A_3,\ 0.2/A_4\} \\
C_4 &= \{\ 0.7/A_1,\ 0.5/A_2,\ 0.40/A_3,\ 0.3/A_4\}
\end{aligned} \tag{7.4}
$$

On the basis of expert estimations the importance of criteria has been ordered

C_1 is more relevant C_2 with relative importance 4
C_2 is more relevant C_3 with relative importance 6
C_1 is more relevant C_4 with relative importance 5
C_3 is more relevant C_2 with relative importance 3
C_4 is more relevant C_2 with relative importance 5
C_4 is more relevant C_3 with relative importance 2

The matrix B is formed on the basis of relative importance criteria

$$
\begin{array}{cc}
 & \begin{array}{cccc} C_1 & C_2 & C_3 & C_4 \end{array} \\
B = & \begin{bmatrix} 1 & 4 & 6 & 5 \\ 1/4 & 1 & 1/3 & 1/5 \\ 1/6 & 3 & 1 & 1/2 \\ 1/5 & 5 & 2 & 1 \end{bmatrix} \begin{array}{c} C_1 \\ C_2 \\ C_3 \\ C_4 \end{array}
\end{array} \tag{7.5}
$$

where $b_{ii} = 1, b_{ji} = 1/b_{ij}$.

Eigenvalues of this matrix $\lambda_1, \lambda_2, \lambda_3, \lambda_4$ are determined with the use of characteristic equation. The equation has a solution

$\lambda_1 = 4{,}3896$; $\lambda_2 = -0{,}1728 + 1{,}2904i$; $\lambda_3 = -0{,}1728 - 1{,}2904i$; $\lambda_4 = 0{,}0439$.

Thus, $\lambda_{max} = 4.3896$.

The eigenvector appropriate to the greatest eigenvalue $\lambda_1 = 4{,}3896$ is vector $w = (w_1, w_2, w_3, w_4)$

$$w = [\,0.9234,\ 0.1052,\ 0.1856,\ 0.3191\,] \tag{7.6}$$

The required values of coefficient of relative importance are obtained from the condition

$$
\begin{gathered}
\alpha_i = n w_i\,, \\
\alpha_1 = 3.6936, \quad \alpha_2 = 0.4208, \quad \alpha_3 = 0.7424, \quad \alpha_4 = 1.2764\,.
\end{gathered} \tag{7.7}
$$

We modify (7.4)

$$
\begin{aligned}
C_1^{\alpha^1} &= \{\ 0.3^{\alpha_1}/A_1,\ 0.5^{\alpha_1}/A_2,\ 0.75^{\alpha_1}/A_3,\ 0.9^{\alpha_1}/A_4\} \\
C_2^{\alpha^2} &= \{\ 0.8^{\alpha_2}/A_1,\ 0.7^{\alpha_2}/A_2,\ 0.65^{\alpha_2}/A_3,\ 0.9^{\alpha_2}/A_4\} \\
C_3^{\alpha^3} &= \{\ 0.6^{\alpha_3}/A_1,\ 0.5^{\alpha_3}/A_2,\ 0.25^{\alpha_3}/A_3,\ 0.2^{\alpha_3}/A_4\} \\
C_4^{\alpha^4} &= \{\ 0.7^{\alpha_4}/A_1,\ 0.5^{\alpha_4}/A_2,\ 0.40^{\alpha_4}/A_3,\ 0.3^{\alpha_4}/A_4\}
\end{aligned} \tag{7.8}
$$

Having substituted values of coefficients of relative importance of criteria we obtain the following

$$
\begin{aligned}
C_1^{3.6936} &= \{0.3^{3.6936}/A_1, 0.5^{3.6936}/A_2, 0.75^{3.6936}/A_3, 0.9^{3.6936}/A_4\} \\
C_2^{0.4208} &= \{0.8^{0.4208}/A_1, 0.7^{0.4208}/A_2, 0.65^{0.4208}/A_3, 0.4^{0.4208}/A_4\} \\
C_3^{0.7424} &= \{0.6^{0.7424}/A_1, 0.5^{0.7424}/A_2, 0.25^{0.7424}/A_3, 0.2^{0.7424}/A_4\} \\
C_4^{1.2764} &= \{0.7^{1.2764}/A_1, 0.5^{1.2764}/A_2, 0.40^{1.2764}/A_3, 0.3^{1.2764}/A_4\}
\end{aligned}
\tag{7.9}
$$

or

$$
\begin{aligned}
C_1^{3.6936} &= \{\ 0.0117/A_1,\ 0.0773/A_2,\ 0.3456/A_3,\ 0.6776/A_4\} \\
C_2^{0.4208} &= \{\ 0.9104/A_1,\ 0.8606/A_2,\ 0.8342/A_3,\ 0.6801/A_4\} \\
C_3^{0.7424} &= \{\ 0.6844/A_1,\ 0.5977/A_2,\ 0.3573/A_3,\ 0.3028/A_4\} \\
C_4^{1.2764} &= \{\ 0.6343/A_1,\ 0.4128/A_2,\ 0.3105/A_3,\ 0.2151/A_4\}
\end{aligned}
\tag{7.10}
$$

Now we can choose a rational vehicle

$$
\begin{aligned}
\propto_D(A^*) = \{&\max[\min(0.0117;\ 0.9104;\ 0.6844;\ 0.6343)/A_1, \\
&\min(0.0773;\ 0.8606;\ 0.5977;\ 0.4128)/A_2, \\
&\min(0.3456;\ 0.8342;\ 0.3573;\ 0.3105)/A_3, \\
&\min(0.6776;\ 0.6801;\ 0.3028;\ 0.2151)/A_4]\}
\end{aligned}
\tag{7.11}
$$

As the rational vehicle alternative A^* is selected, having the greatest value of the membership function

$$
\begin{aligned}
\propto_D(A^*) &= \max[0.0117/A_1,\ 0.0773/A_2,\ 0.3105/A_3,\ 0.2151/A_4] \\
\propto_D(A^*) &= \{0.3105/A_3\}
\end{aligned}
\tag{7.12}
$$

Thus the best alternative is the alternative A_3.

7.2 Scheduling Fuzzy Expert Systems with Probabilistic Reasoning for Oil Refineries

Structure of the system

In this section some aspects of the shell of expert systems ESPLAN [2,3], are considered. The problems connected with fuzzy knowledge representation in knowledge base of this system and the inference organization will be discussed. The mathematical description of knowledge in the shell of expert systems ESPLAN is based on fuzzy interpretation of antecedents and consequents in production rules.

The common structure of developed expert systems is shown in Fig. 7.1.

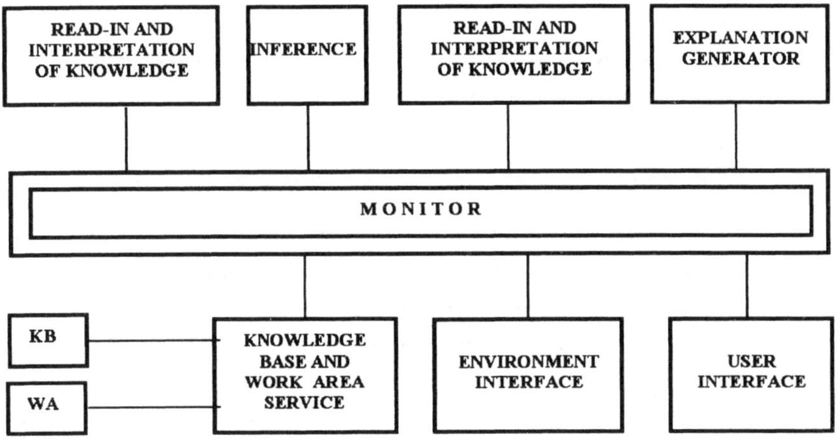

Fig. 7.1. Structure of the expert system

Knowledge representation

For the knowledge representation in the ESPLAN shell the antecedent of each rule contains a conjunction of binary logical connectives like (Fig.7.2): <name of object> $\left\{ \begin{matrix} = \\ \neq \end{matrix} \right\}$ <linguistic value> named elementary antecedent (for instance, "IF quantity of petroleum = great AND demanded output of residual oil $\neq$ little AND ...").

The consequent of the rule is a list of imperatives, among which may be some operator-functions (i.e. input and output of objects' values, operations with segments of a knowledge base, etc). Each rule may be complemented with a confidence degree $cf \in [0,100]$ and with the author's comments on the rule. For example, let us take a rule from the knowledge base (Fig.7.2).

IF the remainder of local petroleum = great
THEN the load of 17th unit = great
CONFIDENCE 90
BECAUSE "if there is great remainder, then the 17th unit may be completely loaded"

For appropriate interpretation of linguistic values in ESPLAN the fuzzy sets theory is used. Each linguistic value has a corresponding membership function which is built using parametric LR-representation (Fig.7.3).

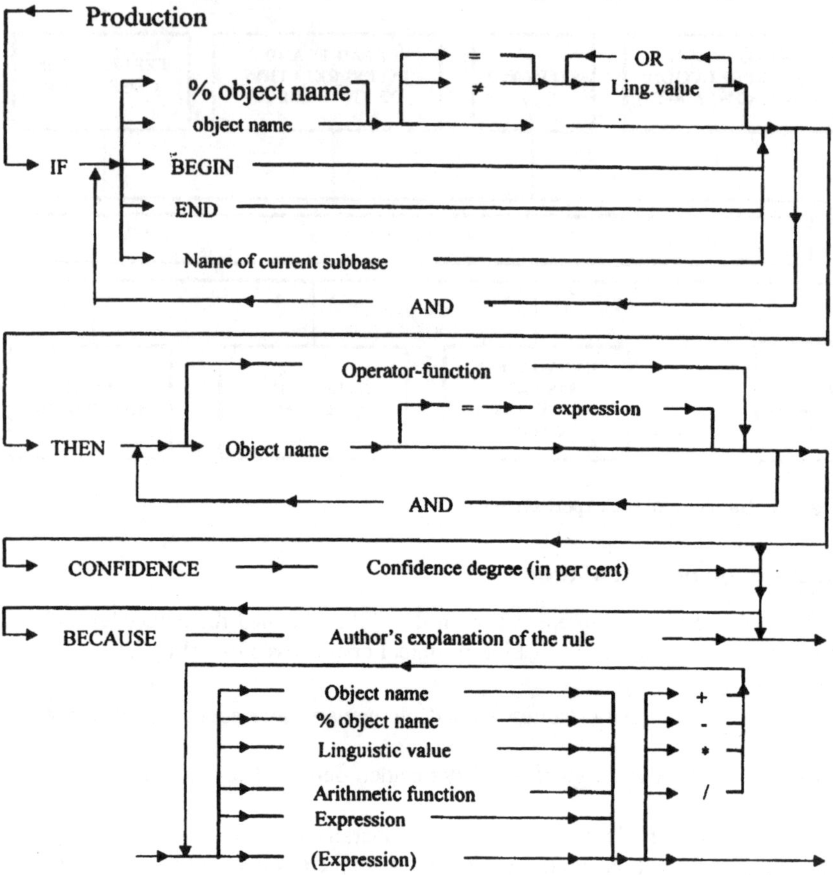

Fig. 7.2. Production rules

So, each membership function is defined by 4 parameters: α_l - left deviation, a_l – left peak, a_2 – right peak, α_r – right deviation, i.e. $\alpha_a(x) = (\alpha_1, a_1, a_2, \alpha_r)$.

So we have an analytic form of membership function [4]:

$$\alpha_a(x) = \begin{cases} 1 - \dfrac{a_1 - x}{\alpha_l}, & \text{if } a_1\text{-}\alpha_l \le x \le a_l \\ 1, & \text{if } a_l \le x \le a_2 \\ 1 - \dfrac{x - a_2}{\alpha_r}, & \text{if } a_2 \le x \le a_2 + \alpha_r \\ 0, & \text{in other cases} \end{cases}$$

The graphical representation of membership function is a trapezoid with the tops a_l and a_r and plumb deviations α_l and α_r, respectively. Such representation allows the definition of unimodal membership functions, but knowledge representation language of ESPLAN also allows definition of polimodal membership functions using construction "OR". For instance, "low OR near 50". The result of interpretation is a disjunction (max) of membership functions of composite elements.

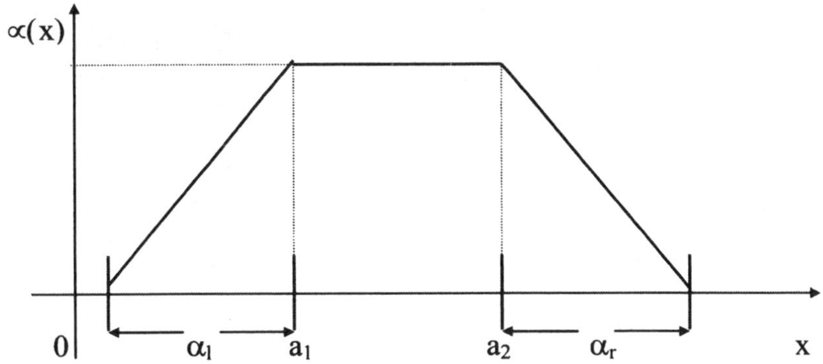

Fig. 7.3. Membership function

The subsystem of fuzzy arithmetic and linguistic values processing (see Fig.7.1) provides automatic interpretation of linguistic values like "high", "low", "OK", "near...", "from ... to ..." and so on; i.e. for each linguistic value this subsystem automatically computes α_l, a_1, a_2 and α_r using universes of corresponding variable. The user of the system may define new linguistic values, modify built-in ones and explicitly prescribe a membership function in any place where linguistic values are useful.

Inference

Efficiency of inference engine considerably depends on the knowledge base internal organization. That is why ESPLAN realizes paradigm of "network of production rules" similar to semantic network. Here the nodes are rules and vertexes are objects. Inference mechanism acts as follows. First, some objects take some values (initial data). Then, all production rules, containing each of these objects in antecedent, are chosen from the knowledge base. For these rules the truth degree is computed (in other words, the system estimates the truth degree of the fact that current values of objects correspond to values fixed in antecedents). If the truth degree exceeds some threshold then imperatives from consequent are executed. At that time the same objects as well as a new one take new values and the process continues till work area contains "active" objects ("active" object means untested one).

The assigned value of the object is also complemented by a number, named confidence degree, which is equal to the truth degree of the rule.

A truth degree of a rule's antecedent is calculated according to the following algorithm [3,5,6].

Let us consider an antecedent of a rule in the form:

$$\text{IF ... AND } w_i \begin{Bmatrix} = \\ \neq \end{Bmatrix} a_{ji} \text{ AND ... AND } w_k \begin{Bmatrix} = \\ \neq \end{Bmatrix} a_{jk} \text{ AND ...}$$

Confidence degree of the rule $cf \in]0,100]$.

Objects w_i, w_k etc. have current values of the form (v, cf) in the work area (here v is linguistic value with its membership function, $cf \in]0,100]$ is confidence degree of the value v). Truth value of k-th elementary antecedent is:

$r_k = \text{Poss}(v_k | a_{jk}) * cf_k$, if the sign is "="
and $r_k = (1 - \text{Poss}(v_k | a_k)) * cf_k$, if the sign is "≠"
$\text{Poss}(v|a) = \max_u \min(\alpha_v(u), \alpha_a(u)) \in [0,1]$.

For the LR-representation of membership function (see section 2.1):

$$\text{Poss}(v|a) = \begin{cases} 1 - \dfrac{v_l - a_2}{\gamma_l + \alpha_r}, & \text{if } 0 < v_l - a_2 < \gamma_l + \alpha_r \\ 1, & \text{if } \max(a_1, v_1) \le \min(a_2, v_2) \\ 1 - \dfrac{a_1 - v_2}{\alpha_l + \gamma_r}, & \text{if } 0 < a_1 - v_2 < \alpha_l + \gamma_r \\ 0, & \text{otherwise} \end{cases}$$

The truth degree of the rule:

$$R_j = (\min_k r_k) * \frac{cf_j}{100}$$

After the inference is over, the user may obtain for each object a list of its values with confidence degree which are accumulated in the work area. The desirable value of the object may be obtained using one of the developed algorithms:

$$w_i : (v_i^n, cf_i^n), \quad n = \overline{1,S},$$

S is total number of values.

Calculation of resulting value:

I. Last - v_i^s

II. The value with maximum confidence degree - $v_i^m / cf_i^m = \max_n cf_i^n$

III. The value $\underline{v}_i = \wedge_n (v_i^n * cf_i^n)$, or $\overline{v}_i = \vee_n (v_i^n * cf_i^n)$

IV. The average value $\overline{v}_i = \dfrac{\sum_n v_i^n * cf_i^n}{\sum_n cf_i^n}$

IF $x_1 = a_1^j$ AND $x_2 = a_2^j$ AND ... THEN $y_1 = b_1^j$ AND $y_2 = b_2^j$ AND ...

IF ... THEN Y_1 = AVRG (y_1) AND Y_2 = AVRG (y_2) AND ...

ESPLAN has a built-in function AVRG which calculates the average value. This function simplifies the organization of compositional inference with possibility measures. As a possibility measure here a confidence degree is used. So, the compositional relation is given as a set of production rules like:

$$\text{IF } x_1 = a_1^j \text{ AND } x_2 = a_2^j \text{ AND ... THEN } y_1 = b_1^j \text{ AND } y_2 = b_2^j \text{ AND ...},$$

where j is a number of a rule (similar to the row of the compositional relation matrix). After all these rules have been executed (with different truth degrees) the next rule (rules) ought to be executed:

IF ... THEN Y_1 = AVRG (y_1) AND Y_2 = AVRG (y_2) AND ...

The approach based on construction of compositional relations is often used for synthesis of control systems for technological processes. Particularly, it was used for the construction of a knowledge base for the control of initial oil-refinery unit.

The expert system for current scheduling of oil-refinery production

Here we consider some aspects of system's application for solving the tasks of scheduling oil-refinery production. The process was partitioned into 4 main subprocesses - initial oil-refinery, catalytic cracking, catalytic reforming and petrol compounding.

```
{Distribution of petroleum among units 15, 16, 17}
.............................................................
{Description of objects: shortname, fullname, minimal
value, maximal value, unit of measurement}
OB(S_100, "Plan on local petroleum", 10,30, "thous.tons");
OB(P_106, "Plan on output of residual oil",1,3,"thous.tons");
OB(R_106, "Real output of residual oil",1,3,"thous.tons");
OB(IN_42_106,"Input of residual oil at the unit 42",0.5,1.5, "thous.tons");
OB(E_17_106, "Yield efficiency for residual oil at the unit 17");
. . . . . . . . . . . . . . . . . . . . . .
IF . . .
THEN . . .
AND P3=(P_106+IN_42_106)/E_17_106
AND P31=TP_17-P3
BECAUSE "Computing necessary load of 17-th unit to provide the
plan with respect to the residual oil and the load unit 42";
IF P31=Less 0
THEN IN_17_100=TP_17 AND OUT_17_106=IN_17_100*E_17_106
AND R_106=OUT_17_106-IN_42_106
AND DISPLAY("Power of 17-th unit is insufficient for the load")
AND DISPLAY("of the unit 42 and demanded plan on residual oil")
BECAUSE "Necessary load of 17-th unit is greater than the top
```

capacity of the unit"

Fuzzy hypotheses generating and accounting systems

Using ESPLAN one may construct hypotheses generating and accounting systems. Such system contains the rules:

IF <condition$_j$ > THEN X=A_j CONFIDENCE CF_j

Here "X=A_j" is a hypothesis that the object X takes the value A_j.Using some preliminary information, this system generates elements X=(A_j,R_j), where R_j is a truth degree of j-th rule. In order to account the hypothesis (i.e. to estimate the truth degree that X takes the value A_0) the recurrent Bayes-Shortliffe formula (see Chapter 2), generalized for the case of fuzzy hypotheses, is used [40]:

$$P_0 = 0$$

$$P_j = P_{j-1} + cf_j^* Poss(A_0 / A) * (1 - \frac{P_{j-1}}{100}$$

This formula is realized as a built-in function BS :
IF END THEN $P = BS(X, A_0)$.

7.3 Detection and Withdrawal of Defect Parts in the Computer Aided Manufacturing of Evaporators

Here we consider the process of decision making by an industrial sorter-robot (SR) for on-line detection and rejection of defect parts, which is installed in the heat exchanger production shop. Practically, the decision making process is being executed by an on-board computer and two stages of this process should be remarked:

- the qualitative estimation of parts on the base of input information;
- the definition of the manipulator's turning angle using the results of previous stage.

The input information of the part is picked off in the form of the set $x = \{x_i\}$, $i = \overline{1,n}$. Here $x_i \in [x_{i\,min}, x_{i\,max}]$, $i = \overline{1,n}$ where $x_{i\,min}$ and $x_{i\,max}$ are minimal and maximal values of the i- th parameter. The parameters of the set { x_i }, $i = \overline{1,n}$ are considered to be fuzzy sets, forming the linguistic variables in the form of triplets:

$$\overset{\circ}{X}_i = \{< X_i^j, U_{X_i}, \widetilde{X}_i >\}, \quad X_i^j \in T_i^*(u), \quad i = \overline{1,n}, \quad j = \overline{1,10}$$

where $T_i^*(u)$ is the extended term-set of the linguistic variable "Parameter I"; $\widetilde{X}_i$ is a normal fuzzy set of the kind:

$$X_i = \int_{U_{X_i}} \propto_{X_i}(u)/u \qquad (7.13)$$

$U_{X_i} = \{0,1,2,...,10\}$ is the universe. The values of the linguistic variable "Parameter i" are given in the Table 7.1.

Table 7.1. The values of the linguistic variable "Parameter i"

The terms of variables		$u_j \in U_{Xi}$
"Parameter i"	"Quality 1"	$v_i \in V_y$
Negligible	very low	0
almost small	almost low	1
small	low	2
more than small	better than low	3
almost average	almost average	4
average	average	5
more than average	better than average	6
almost big	almost high	7
big	high	8
more than big	better than high	9
extreme	highest	10

As a mapping $q_i : \tilde{x}_i \to U_{X_i}$, $i = \overline{1,n}$ we use the expression (7.14)

$$U_j = \mathrm{Ent}\left\{ (\mathrm{Card} U_{X_i} - 1) * \left[\frac{x_i^c - x_{i\,min}}{x_{i\,max} - x_{i\,min}} \right]^{\alpha} \right\}, \quad j = \overline{1,10}; \quad i = \overline{1,n} \qquad (7.14)$$

Membership functions are defined on the base of

$$\propto(u_j) = 1 - \frac{1}{\mathrm{Card}\ U_{X_i} - 1} \left| u_j - \mathrm{Ent}\left\{ (\mathrm{Card}\ U_{X_i} - 1) * \left[\frac{x_i^c - x_{i\,min}}{x_{i\,max} - x_{i\,min}} \right]^{\alpha} \right\} \right| \qquad (7.15)$$

Calculated estimations of membership functions in singletons permit to apply the L-R representation of the membership functions and the equality,

$$\propto(u_j) = 1$$

takes the place in the point $u_j^* \in U$, which is defined using (7.14).

Assume the output parameter (the quality of the part) in the form of a linguistic variable $\overset{0}{Y}$:

$$\overset{0}{Y} = \{< Y_j, V_Y, \tilde{Y} >\}, \quad Y_j \in T^*(v)$$

where $T^{*}(v)$ is the extended term-set of the linguistic variable "Quality 1"; $\tilde{Y}$ is the normal fuzzy set of the kind:

$$\tilde{Y} = \int_{V_y} \propto_Y (v)/v \tag{7.16}$$

$V_Y = \{0,1,\ldots,10\}$ is the universe. The values of the linguistic variable "Quality 1 are given in the Table 7.2.

Table 7.2. The values of the linguistic variable "Quality 1"

The terms of variables		$w_i \in W_{Y_i}$
" Quality 2"	"Angle"	$w_{Z_j} \in W_Z$
hard defect	small	0
repeated processing is possible	average	1
ready for the next operation	big	2

For the definition of the output parameter the rule of fuzzy conditional inference (see Chapter 2) is used. For the correspondent membership functions the rule of fuzzy conditional inference is taken as (see Chapter 2) and is calculated using the conditional proposition:

P1="IF ($\tilde{X}_1$ is A_1) AND ($\tilde{X}_2$ is A_2) AND ...AND ($\tilde{X}_n$ is A_n) THEN $\tilde{Y}$ is B OR $\tilde{Y}$ is non B",

where $A_i \subseteq U_{X_i}$, $i = \overline{1,n}$, $B \subseteq V_Y$ are fuzzy sets, corresponding to the terms of linguistic variables (see Table 7.1). At the same time, the current output fuzzy set may be defined using the rule of compositional inference of Zadeh [51]:

$$\begin{aligned}\tilde{Y}^c &= R(A_2(y)) = R(A_1(x)) \circ R_1(A_1(x), A_2(y)) = \\ &= \left[\bigcap_1 \tilde{X}_1^c\right] \circ R_1(A_1(x), A_2(y))\end{aligned} \tag{7.17}$$

where $R(A_1(x))$ and $R(A_2(y))$ are unary relations, corresponding to the current values of input and output parameters, respectively.

The subnormal fuzzy set

$\tilde{X}^c = \bigcap_1 \tilde{X}_1^c$ from (7.15) ought to be normalized:

$$\tilde{X}_n^c = \mathrm{NORM}(\tilde{X}^c) \int_{U_{X_i}} [\frac{\wedge_i \propto_{X_i}(u)}{\sup_{U_{X_i}}\left(\wedge_i \propto_{X_i}(u)\right)}]/u \tag{7.18}$$

The input information about the turning angle of the manipulator is represented in form of the linguistic variable "Quality 2":

$$\overset{\circ}{Y}_1 = \{< Y_{1i}, W_{Y1}, \widetilde{Y}_1 >\} \quad , \qquad Y_{1i} \in T^*(w)$$

where $T^*(w)$ is the extended term-set of the linguistic variable "Quality 2"; $\widetilde{Y}_1$ is the normal fuzzy set:

$$\widetilde{Y}_1 = \int_{W_{Y_1}} \propto_{Y_1}(w)/w$$

$W_{Y1} = \{0,1,2\}$ is the universe. Estimations of membership functions in singletons $\propto_{Y1}(w_i)/w_i$ have the L-R form. Note that,

$$\propto_{Y1}(w_i) = 1 \quad \text{if} \quad w_i = w^*$$

The turning angle of the manipulator is represented by the linguistic variable "Angle": $\overset{\circ}{Z} = \{< Z_i, W_Z, \widetilde{Z} >\}$ and correspondent estimations are calculated as

$$\propto_Z(w_{Z_j}) = 1 - \frac{1}{\text{Card } w_Z - 1}\left| w_{Z_j} - \text{Ent}\left\{(\text{Card } w_Z - 1)\cdot\left[\frac{\psi_1 - 90}{180}\right]\right\}\right|, \; j = \overline{1,3} \tag{7.19}$$

In order to construct the binary relation $R_2(A_2(y_1), A_2(z))$ the following conditional proposition P2 is used:

P2="IF $\widetilde{Y}_1$ is B_1 THEN $\widetilde{Z}$ is C OR $\widetilde{Z}$ is non C"

Here, as a binary relation we use the rule of fuzzy conditional inference or, for the corresponding membership functions,

$$R_2(A_1(Y_1), A_2(z)) =$$

$$= \int_{W_{Y_1} * W_Z} (\propto(w_{Y_1}) \underset{s}{\rightarrow} \propto(w_Z)) \wedge ((1-\propto(w_{Y_1})) \underset{s}{\longrightarrow} (1-\propto(w_Z))/(w_{Y_1}, w_Z) = \tag{7.20}$$

$$= \begin{cases} 1, & \propto(w_{Y_1}) = \propto(w_Z) \\ 0, & \propto(w_{Y_1}) \neq \propto(w_Z) \end{cases}$$

The current value of the fuzzy set $\widetilde{Z}^c$ may be defined, as in previous case, using the Zadeh compositional rule of inference:

$$\widetilde{Z}^c = R(A_2(z)) = R(A_1(Y)) \circ R(A_1(Y_1), A_2(z)) = \widetilde{Y}_1^c \circ R(A_1(Y_1), A_2(z)) * \tag{7.21}$$

Further, we define $w_z^* \in W_Z$ and taking into account that card $W_Z = 3$, we obtain an expression for the value of the turning angle:

$$\psi_i = \frac{\pi}{2} \cdot w_Z^* + \frac{\pi}{2} \tag{7.22}$$

Thus, if, for example, w_Z^*=0, then $\psi_i = \pi/2$; and if w_Z^*=1, then $\psi_i = \pi$ and so on (note here that in practice the value of the turning angle vary in the range $\psi_i \in [\pi/2, 3\pi/2]$).

Example [2]

Let $x = \{x_1, x_2\}$, $x_1 \in [30, 80]$ $x_2 \in [100,185]$, Card U_{X_1} =Card V_Y =11, and Card W_{Y1} =Card W_Z =3. Assume also the proposition P1:

P1= "IF $\widetilde{X}_1$ is negligible AND $\widetilde{X}_2$ is small THEN $\widetilde{Y}$ is highest OR $\widetilde{Y}$ is very low".

Then, from the Table 7.2 we define u_1 =0, u_2 =2 and using (7.14)

$$x_1 = 30,\ x_2 = 117,\ \alpha = 1$$

On the base of (7.15) we obtain,

$\widetilde{X}_1 \cong$ negligible $\cong$ {1.0/0 + 0.9/1 + 0.8/2 + 0.7/3 + 0.6/4 + 0.5/5 + +0.4/6 + 0.3/7 + 0.2/8 + 0.1/9 + 0.1/10}

$\widetilde{X}_2 \cong$ small $\cong$ {0.8/0 + 0.9/1 + 1.0/2 + 0.9/3 + 0.8/4 + 0.7/5 + 0.6/6+ +0.5/7 + 0.4/8 + 0.3/9 + 0.2/10}

By analogy, we obtain $\widetilde{Y}$,

$\widetilde{Y}$ highest $\cong$ {0.1/0 + 0.1/1 + 0.2/2 + 0.3/3 + 0.4/4 + 0.5/5 +0.6/6 + + 0.7/7 + 0.8/8 + 0.9/9 + 1.0/10}

Further, using (7.15) and (7.16) the fuzzy relations matrix is constructed, i.e.,

$$R_1(A_1(x), A_2(y)) = \\ = \begin{vmatrix} 0.1 & 0.1 & 0.2 & 0.3 & 0.4 & 0.5 & 0.6 & 0.7 & 0.8 & 1.0 & 0.0 \\ 0.1 & 0.1 & 0.2 & 0.3 & 0.4 & 0.5 & 0.6 & 0.7 & 0.8 & 0.9 & 1.0 \\ 0.1 & 0.1 & 0.2 & 0.3 & 0.4 & 0.5 & 0.6 & 0.7 & 0.8 & 1.0 & 0.0 \\ 0.1 & 0.1 & 0.2 & 0.3 & 0.4 & 0.5 & 0.6 & 0.7 & 1.0 & 0.1 & 0.0 \\ 0.1 & 0.1 & 0.2 & 0.3 & 0.4 & 0.5 & 0.6 & 1.0 & 0.2 & 0.1 & 0.0 \\ 0.1 & 0.1 & 0.2 & 0.3 & 0.4 & 1.0 & 0.4 & 0.3 & 0.2 & 0.1 & 0.0 \\ 0.1 & 0.1 & 0.2 & 0.3 & 1.0 & 0.5 & 0.4 & 0.3 & 0.2 & 0.1 & 0.0 \\ 0.1 & 0.1 & 0.2 & 1.0 & 0.6 & 0.5 & 0.4 & 0.3 & 0.2 & 0.1 & 0.0 \\ 0.1 & 0.1 & 1.0 & 0.7 & 0.6 & 0.5 & 0.4 & 0.3 & 0.2 & 0.1 & 0.0 \\ 1.0 & 1.0 & 0.8 & 0.7 & 0.6 & 0.5 & 0.4 & 0.3 & 0.2 & 0.1 & 0.0 \\ 1.0 & 1.0 & 0.8 & 0.7 & 0.6 & 0.5 & 0.4 & 0.3 & 0.2 & 0.1 & 0.0 \end{vmatrix} \tag{7.23}$$

Assume now the current values of input parameters:

$$x_1^c = 55, \qquad x_2^c = 160$$

Then, using (7.14), (7.15), we obtain,

$\widetilde{X}_1^c \cong$ {0.5/0 + 0.6/1 + 0.7/2 + 0.8/3 + 0.9/4 + 1.0/5 + 0.9/6 +0.8/7 + +0.7/8 + 0.6/9 + 0.5/10}

$\widetilde{X}_2^c \cong \{0.3/10 + 0.4/1 + 0.5/2 + 0.6/3 + 0.7/4 + 0.8/5 + 0.9/6 + 1.0/7 +$
$+0.9/8 + 0.8/9 + 0.7/10\}$

Consequently,

$\bigcap_i \widetilde{X}_i^c \cong \{0.3/10 + 0.4/1 + 0.5/2 + 0.6/3 + 0.7/4 + 0.8/5 + 0.9/6 + 0.8/7 +$
$+0.7/8 + 0.6/9 + 0.5/10\}$

After normalization $\widetilde{x}_n^c$ takes the form,

$\widetilde{X}_n^c \cong \{0.3/0 + 0.4/1 + 0.5/2 + 0.7/3 + 0.8/4 + 0.9/5 + 1.0/6 + 0.9/7 +$
$+0.8/8 + 0.7/9 + 0.5/10\}$

Composing $\widetilde{X}_n^c$ with the matrix (7.23) we obtain,

$\widetilde{Y}_n^c \cong \{0.7/0 + 0.7/1 + 0.8/2 + 0.9/3 + 1.0/4 + 0.9/5 + 0.6/6 + 0.8/7 +$
$+0.7/8 + 0.5/9 + 0.4/10\}$

Consequently,

$$\propto(4) = 1 \Leftrightarrow v_i^* = 4$$

and

$$w^* = 0$$

Further, the current value of $\widetilde{y}_1$ is calculated,

$\widetilde{Y}_1^c \cong \text{hard defect} \cong \{1.0/0 + 0.5/1 + 0.0/2\}$

On the other hand, using the proposition P2:

P2= " IF $\widetilde{y}_1$ is ready for the next operation THEN $\widetilde{z}$ is big OR $\widetilde{z}$ is small"

Then, on the base of (7.14) and (7.17) we construct the fuzzy relations matrix:

$$R(A_2(y_1), A_2(z)) = \begin{bmatrix} 1.0 & 0.0 & 0.0 \\ 0.0 & 1.0 & 0.0 \\ 0.0 & 0.0 & 1.0 \end{bmatrix} \tag{7.24}$$

Composing the fuzzy set $\widetilde{Y}_1^c$ with the matrix (7.24) we obtain the current value of $\widetilde{Z}$,

$$\widetilde{Z}^c \cong \text{small} \cong \{1.0/0 + 0.5/1 + 0.0/2\}$$

Consequently,

$$w_Z^* = 0$$

Substitute the value of w_z^* in (7.22) and obtain,

$$\psi_i = \pi/2$$

The outlined approach was applied in the production of domestic refrigerators for the rejection of evaporators and may be recommended as an effective tool while constructing a fuzzy control system for various discrete processes.

7.4 Genetic Algorithms Based Fuzzy Regression Analysis and its Applications for Quality Evaluation

Complex systems, such as industry, economy, finance, marketing, and ecology function in the real world in an environment of uncertainty and imprecision. Such systems require decisions based on human expertise and judgment and involve human-machine interactions. In such environments we often cannot obtain exact numerical data about the system. Therefore, in such situations the classical regression analysis may not be applicable for modeling of complex systems [21,24,35]. The nature of information about complex systems with vagueness is frequently fuzzy and, as it is shown in [21,35] fuzzy regression seems intuitively more realistic.

In the pioneering work [43] in this field the authors use Zadeh's extension principle [52], α-level procedure, interval arithmetic, and linear programming techniques to develop a fuzzy linear regression analysis. A modification of this approach for nonlinear case is reported in [11]. A simple fuzzy regression model introducing the corresponding distances is considered in [17]. Minimization of these distances in fuzzy number space with respect to the unknown parameters of regression models leads to solving systems of equations. The generalization of this approach to general hyper planes and use of Hausdorff-like distances is proposed in [10,11]. The methodology for global approximation of fuzzy data is given in [12].

In [37] linear regression model fitting fuzzy-valued data by using a metric of fuzzy number space by integral of distance of every α-level set is proposed. [38] applies this approach to S-curve regression model and creating of this model is reduced to the solution of systems of linear equations. In [48] genetic algorithm is used to classify the input data for increasing the effectiveness of fuzzy regression analysis based on [21,43]. However, genetic algorithm is not applied directly for the determination of unknown fuzzy coefficients of fuzzy regression model.

The above approaches to fuzzy regression analysis are based on α-level sets and interval arithmetic, and require specifying and solving a set of linear programming problems or system of equations. Also, in these approaches fuzzy data and unknown fuzzy coefficients are frequently represented as fuzzy symmetrical triangular membership functions, whereas in reality they may be of arbitrary shape. This simplification significantly reduces the accuracy of solutions. The above approaches result in not-so-easy calculation problems. Therefore, there is a need to consider relatively more general regression equations and more effective approaches to estimate their parameters.

Another significant drawback of the above approaches is that the comparison of the calculated and the desired output is not considered in the process of determining fuzzy regression coefficients. [14] advocated the use of fuzzy chaos for fuzzy regression. In this approach the direct comparison of the desired and actual outputs is used. Other advantages of the approach proposed in [14] include the idea of extension of the method to the polynomial regression, and avoidance of using α-level procedures, the disadvantages of which have been mentioned above. The disadvantage of the proposed method is that the search for fuzzy regression coefficients is undirected.

In this section we propose nonlinear fuzzy regression analysis using genetic algorithms for determining unknown fuzzy coefficients of fuzzy regression models. The overall complexity of the proposed method is less than the above-mentioned approaches. The idea of using genetic algorithms for fuzzy regression analysis is proposed in [36,49]. Here we present a methodology for constructing fuzzy regression models on the basis of fuzzy arithmetic and genetic algorithms[8].

Assume, K fuzzy data expressed as fuzzy numbers are given for multiple input-single output system. The problem is to create a regression model for fitting this type of data. The following fuzzy regression model is considered:

$$\tilde{y}^M(\tilde{x}_1,\tilde{x}_2...,\tilde{x}_N)=\tilde{a}_0+\sum_{i=1}^{N}\tilde{a}_i\cdot\tilde{x}_i+\sum_{i=1}^{N}\tilde{b}_i\cdot\tilde{x}_i^2+\sum_{\substack{i=1\\i<j}}^{N}\sum_{j=1}^{N}\tilde{c}_{ij}\cdot\tilde{x}_i\cdot\tilde{x}_j \tag{7.25}$$

It is required to determine fuzzy coefficients (numbers) $\tilde{a}_i$, $\tilde{b}_i$, and $\tilde{c}_{ij}$, $i=\overline{0,N}$, $j=\overline{1,N}$ given fuzzy input data $\tilde{x}_{ik}$,$i=\overline{1,N}$, $k=\overline{1,K}$, and output data $\tilde{y}_{ik}$, $k=\overline{1,K}$, so that

$$\tilde{d}(\tilde{y}^M,\tilde{y})=\sum_{i=1}^{K}\left|\tilde{y}_i-\tilde{y}_i^M\right|\rightarrow\min \tag{7.26}$$

Here $\tilde{d}$ is Hamming fuzzy distance in general fuzzy number space between each fuzzy datum $\tilde{y}_i$, and the corresponding fuzzy number $\tilde{y}_i^M$ calculated by (7.25).

In order to define unknown parameters (shapes and centers of membership functions) $\tilde{a}_i$, $\tilde{b}_i$ and $\tilde{c}_{ij}$, $i=\overline{0,N}$, $j=\overline{1,N}$ the aggregated distance $\tilde{d}$ (7.26) is minimized. Use of least squares and gradient based methods for solving the problem (7.25-7.26) is not feasible, because $\tilde{d}$ in (7.26) is non-differentiable, fuzzy numbers are from general fuzzy number space, and the regression model is general (i.e. not restricted to linear form). We use genetic algorithms for solving this problem.

The only requirement for the fitness function is that it should adequately represent the objective function of the original problem. The fitness function does not have to be differentiable. This feature is particularly useful when the objective function of the problem at hand is not differentiable which is the case with fuzzy regression analysis.

The fitness function is defined by the following formula:

$$\tilde{F}=[1+\tilde{d}]^{-1}, \tag{7.27}$$

where $\tilde{d}$ is the aggregate distance between the desired and actual outputs of a regression model.

Input and output data and regression coefficients are represented as arbitrary fuzzy numbers in this work. The calculations are performed based on the fuzzy arithmetic operations [2,18]. Each fuzzy regression coefficient is represented by q strings of bits, where q is the number of parameters of a fuzzy number. Using pointwise representation may result in generating non-valid fuzzy numbers, i.e. one cannot guarantee that all the points for the parameters of fuzzy numbers would be in the correct order. Therefore, the parametric representation of fuzzy numbers is used to represent fuzzy coefficients. Combination of all strings for all regression coefficients is a chromosome, i.e. a potential specification for the parameters of a regression model.

Initially, a pool of binary chromosomes is randomly generated. Each chromosome is then translated into the set of fuzzy coefficients using pre-specified ranges for the peaks and parameters for those coefficients. Every coefficient can be an arbitrary fuzzy number represented in the following general form [27]:

$$\tilde{a}(x) = \begin{cases} 1, & \text{for } x \in [a_L, a_U] \\ l(x), & \text{for } x \in [-\infty, a_L] \\ r(x), & \text{for } x \in [a_U, +\infty] \end{cases}, \tag{7.28}$$

where a_L and a_U are the lower and upper bounds of the core of $\tilde{a}$ respectively; l is a monotonic, increasing, continuous from the right function from $(-\infty, a_L)$ to [0, 1], and such that $l(x) = 0$ for $x \in (-\infty, a_{SL})$; r is a monotonic, decreasing, continuous from the left function from (a_U, ∞) to [0, 1], and such that $r(x) = 0$ for $x \in (a_{SU}, \infty)$. Functions l and r are determined by m and n parameters respectively. Then, every fuzzy coefficient will be determined by m+n+1 (one extra parameter for the peak of fuzzy number) parameters if a = b in (7.28), and by m+n+2 parameters if a < b. In this case every fuzzy regression coefficient can be expressed as:

$$\left(\underbrace{\underbrace{01110}_{a_{ip}}\ \underbrace{01101}_{a_{il1}}\ \underbrace{01101}_{a_{il2}} \ldots \underbrace{11101}_{a_{ilm}}\ \underbrace{01011}_{a_{ir1}}\ \underbrace{11000}_{a_{ir2}} \ldots \underbrace{00001}_{a_{irn}}}_{\tilde{a}_i} \right)$$

Here, a_{ip} is the peak value of fuzzy coefficient $\tilde{a}_i$; a_{ilj}, $j = \overline{1,m}$ and a_{irk}, $k = \overline{1,n}$ are the jth and kth parameters of functions l and r respectively. In particular, assume that fuzzy regression coefficients have triangular form and are represented in L-R format:

$$\tilde{a} = (a_p, a_L, a_R)_{LR} \tag{7.29}$$

Then, the following chromosome encodes fuzzy coefficients of (7.25) represented in form (7.29):

$$\left(\underbrace{\underbrace{01100}_{a_{ip}}\,\underbrace{11101}_{a_{iL}}\,\underbrace{01000}_{a_{iR}}}_{\tilde{a}_i}\ \ldots\ \underbrace{\underbrace{11100}_{b_{ip}}\,\underbrace{10111}_{b_{iL}}\,\underbrace{11100}_{b_{iR}}}_{\tilde{b}_i}\ \ldots\ \ldots \underbrace{\underbrace{10010}_{c_{ijp}}\,\underbrace{11101}_{c_{ijL}}\,\underbrace{01100}_{c_{ijR}}}_{\tilde{c}_{ij}}\ \ldots\right)$$

Here, $\tilde{a}_i$, are fuzzy regression coefficients, and a_{ip}, a_{iL}, a_{iR} are the values for the peak and left and right deviations of a fuzzy triangular coefficient. In this example each parameter is encoded as a five-bit string.

The coefficients are plugged into the fuzzy model and the fuzzy outputs ($\tilde{y}_i^M$) are calculated for the set of given fuzzy inputs. These outputs are compared to the desired outputs ($\tilde{y}_i$) and the distance is calculated. Then, the value of fitness is determined using (7.27) for that chromosome.

After all the chromosomes in the population have been evaluated in terms of their fitness values, the selection process takes place. During selection the new generation is created by randomly choosing chromosomes from old population with the probabilities proportional to the fitness values of the chromosomes. We used elitist approach, where the best chromosome is preserved in the new population, i.e. the best solution always survives.

After selection, the newly generated population undergoes crossover and mutation operations. The probabilities of mutation and crossover control the level of exploration/exploitation in solution search process. The crossover operation randomly chooses two chromosomes with the specified probability and exchanges their parts. First, the split point is randomly selected. Then, two child chromosomes are generated each of them having a substring up to the split point from one parent, and another substring from the other parent chromosome. The following example demonstrates the crossover operator for two chromosomes representing two different sets of potential fuzzy regression coefficients.

Before crossover:

$$\left(\ldots\underbrace{01100\ 11101\ 01000}_{\tilde{a}_i}\ \ \underbrace{11100\ 101\overset{SP}{\ }|11\ 11100}_{\tilde{b}_i}\ \ldots\underbrace{10010\ 11101\ 01100}_{\tilde{c}_{ij}}\ldots\right)$$

$$\left(\ldots\underbrace{10100\ 00111\ 11010}_{\tilde{a}_i}\ldots\underbrace{01110\ 001\overset{SP}{\ }|01\ 00000}_{\tilde{b}_i}\ \ldots\underbrace{01111\ 00001\ 10101}_{\tilde{c}_{ij}}\ldots\right)$$

Here SP stands for ”split point“. After crossover the child chromosomes become:

$$\left(\ldots\underbrace{01100\,11101\,01000}_{\tilde{a}_i}\ldots\underbrace{11100\,101\overset{\mathrm{SP}}{\,|\,}01\ 00000}_{\tilde{b}_i''}\ldots\underbrace{01111\,00001\,10101}_{\tilde{c}_{ij}''}\ldots \right)$$

$$\left(\ldots\underbrace{10100\,00111\,11010}_{\tilde{a}_i'}\ldots\underbrace{01110001\overset{\mathrm{SP}}{\,|\,}1111100}_{\tilde{b}_i'''}\ldots\underbrace{10010\,11101\,01100}_{\tilde{c}_{ij}}\ldots \right)$$

As the result of crossover the child chromosome inherits part of fuzzy coefficients from one parent, and part from the other parent. Most probably one of the coefficients will not belong to either parents, because the split point may fall inside the substring for that coefficient as is shown in the example above.

Mutation chooses a bit in a chromosome with the specified probability of mutation and flips it to the opposite value. For example:

Before Mutation:

$$\left(\ldots\underbrace{0100\,1110101000}_{\tilde{a}_i}\ldots\underbrace{111001\,\overset{\mathrm{Mt}}{01}\,1111100}_{\tilde{b}_i}\ldots\underbrace{1001011101\,01100}_{\tilde{c}_{ij}}\ldots \right)$$

$$\left(\ldots\underbrace{011001110101000}_{\tilde{a}_i}\ \ldots\ \underbrace{111001\overset{\mathrm{Mt}}{1}11111100}_{\tilde{b}_i'}\ldots\underbrace{10010\,1110101100}_{\tilde{c}_{ij}}\ldots \right)$$

After mutation:

Here Mt marks the gene that undergoes mutation. In effect, a single mutation operation randomly changes one of the parameters of some fuzzy coefficient.

The evolution runs for the specified number of generations, and leads to quasi-optimal solutions, (i.e. close-to-minimum distance) solutions. The best chromosome defines the fuzzy regression coefficients.

The algorithm for performing fuzzy regression can be summarized as follows:

Step 1. Mapping solution space ($\tilde{a}_i$, $\tilde{b}_i$, $\tilde{c}_{ij}$) into genetic search space Ω (binary strings). Constructing fuzzy fitness function $\tilde{F}$ using fuzzy measure of distance $\tilde{d}$ (7.26) (objective function).

Step 2. Creating initial population (set of chromosomes) randomly, i.e. a population of fuzzy regression coefficients which are randomly specified.

Step 3. Evaluating each chromosome in the population in terms of fitness value using (7.27).

Step 4. If termination conditions are met go to step 8.

Step 5. Generating new population using selection operator. This operator randomly selects chromosomes from the current population with the probabilities proportional to the values of fitness of the chromosomes.

Step 6. Creating new chromosomes by mating randomly selected (with some specified probability called probability of crossover, P_c) chromosomes. The resulting offspring replaces the original parent chromosomes in the population.

Step 7. Mutating some randomly selected (with some specified probability called probability of mutation, P_m) chromosomes. Return to Step 3.

Step 8. Stop, return the best chromosome and translate it into the fuzzy regression coefficients.

The stop criterion may be, for example, maximum number of generations. This iterative process leads to the improved performance of candidate set of fuzzy coefficients.

The computer simulation showed the effectiveness of the proposed approach. We used GA-based fuzzy regression example given in [44]. For measuring the performance of fuzzy regression fuzzy error (7.26) was used. The fuzzy error is compared with the case of perfect match of the regression outputs with the desired outputs. Because of fuzzy subtraction operation used, the right and left deviation of fuzzy error is not expected to be crisp zero [13]. Therefore, to assure that learning is acceptable, it is necessary to define a set that would reflect the required accuracy [13]. Let the interval [y^d_{iL}, y^d_{iR}] be the support of $\tilde{y}^d_i$ represented as a triangle LR - fuzzy number $\tilde{y}^d_i = \left(y^d_{ip}, y^d_{iL}, y^d_{iR}\right)$, then left and right end points of support of $\tilde{y}^d_i$ can be easily calculated as:

$$y^d_{iSL} = y^d_{ip} - y^d_{iL}\,; \qquad y^d_{iSU} = y^d_{ip} + y^d_{iR}$$

If $\tilde{y}_i = \tilde{y}^d_i$, then the support of $\tilde{d}$ is [0, λ], where

$$\lambda = \sum_{i=1}^{K} \left| y^d_{iSU} - y^d_{iLU} \right| \tag{7.30}$$

If the fuzzy error derived from (7.26) is within the boundaries specified by (7.30) plus some user-specified acceptance level ε, then we accept the model. $\varepsilon > 0$ is a small acceptable positive number. It is chosen, like in classical case, with the consideration of the required accuracy for the problem at hand. If we want the error to be zero $\tilde{0} = (-0.5, 0, 0.5)$, ε is defined considering how close to zero the error should be, e.g. it should be in the interval [-0.5- ε ; 0.5+ ε].

Tanaka et al [44] used data with crisp input and fuzzy output numbers to fit a fuzzy linear regression model. This data was also used in [26,41]. Fuzzy regression model is represented in the following linear form:

$$\tilde{y}^M = \tilde{a}_0 + \tilde{a}_1\tilde{x} \tag{7.31}$$

where $\tilde{a}_0$ and $\tilde{a}_1$ are represented as fuzzy numbers with triangular membership functions. In [26] the following difference measure is used:

$$E = \frac{\int_{S_Y \cap S_{Y^M}} \left| \propto_Y(x) - \propto_{Y^M}(x) \right| dx}{\int_{S_Y} \propto_Y(x)dx}, \tag{7.32}$$

where $\propto_Y(x)$ and $\propto_{Y^M}(x)$ are membership functions for observed and estimated fuzzy numbers, and S_Y and S_{Y^M} are the corresponding supports. We have used the same data for the GA-based fuzzy regression. The parameters of GA used for this problem are listed in Table 7.3. The following estimates of fuzzy coefficients were obtained:

$$\tilde{a}_0 = (4.893; 1.827; 2.01);$$
$$\tilde{a}_1 = (1.621; 0.184; 0.095).$$

Table 7.3. GA Parameters

Population size	50
Chromosome length	50
Number of generations	500
Probability of crossover	0.7
Probability of mutation	0.01
Selection method	Elitist

More refined estimates can be obtained by further tuning the GA parameters. The pre-specified acceptance interval was set to [0; 30] (with $\lambda = 23.2$). The calculated fuzzy distance expressed point wise was (0; 5.6; 29.19), which falls within the acceptable region. We also calculated the fuzzy difference (7.32) as suggested in [26] for our model. Table 7.4 gives the comparison of errors using the methods described in [26], [44], [41] and GA-based fuzzy regression (GAFR). As one can see from the table the method proposed here (GAFR) outperforms the above methods in terms of total error as measured by fuzzy difference. Note, that our approach is not restricted to the case of crisp inputs as, for example, the one proposed in [26].

Table 7.4 Comparison of Errors

Method \ Obs.	1	2	3	4	5	Total Error
FLR [23]	1.86	1.3	0.58	0.86	1.0	5.6
FLSLR [13]	1.54	1.52	0.7	1.16	0.86	5.78
FMLS [14]	1.22	1.38	0.4	1.12	0.36	4.48
GAFR [8]	1.31	1.27	0.18	1.32	0.09	4.17

Let's consider some applications of the proposed method.

We use the proposed GA-based fuzzy regression method for intermediate quality evaluation in the manufacturing process of aluminum heaters. The quality is determined by human operators, hence, it is expressed by fuzzy numbers defining estimations on a linguistic scale. Therefore, we are given the sets of fuzzy inputs ($\tilde{x}_{1i}, \tilde{x}_{2i}$) and fuzzy outputs $\tilde{y}_i$, $i=\overline{1,10}$. The problem is to construct a fuzzy linear regression model for mapping the inputs to the outputs with an acceptable accuracy for predicting the quality of the semi-product of the aluminum plates.

Fuzzy-valued data expressed as fuzzy numbers are given in the Table 7.5. To fit these data the following fuzzy regression model is used:

$$\tilde{y} = \tilde{a} \cdot \tilde{x}_1^2 + \tilde{b} \cdot \tilde{x}_2^2 + \tilde{c} \cdot \tilde{x}_1 \cdot \tilde{x}_2, \tag{7.33}$$

where $\tilde{a}$, $\tilde{b}$, $\tilde{c}$ are unknown fuzzy parameters. For this example we set the accuracy for the acceptance interval as [0, 35] ($\lambda = 29.13$).

Table 7.5. Fuzzy data for quadratic fuzzy regression mode

Obs.	$\tilde{x}_1$	$\tilde{x}_2$	$\tilde{y}$
1	(0.1 0.1 0.1)	(0.1 0.1 0.1)	(0.037 0.120 0.129)
2	(0.3 0.2 0.1)	(0.1 0.1 0.2)	(-0.005 0.233 0.450)
3	(0.6 0.2 0.2)	(0.5 0.2 0.1)	(0.912 1.470 1.372)
4	(0.9 0.1 0.3)	(0.5 0.3 0.4)	(0.729 2.394 3.339)
5	(0.7 0.1 0.1)	(0.95 0.05 0.05)	(3.405 1.325 1.750)
6	(0.5 0.1 0.1)	(0.6 0.4 0.3)	(1.360 2.305 2.068)
7	(0.1 0.1 0.1)	(0.5 0.2 0.1)	(0.861 0.883 0.542)
8	(0.7 0.1 0.1)	(0.6 0.4 0.3)	(1.321 2.568 2.504)
9	(0.5 0.1 0.1)	(0.5 0.3 0.4)	(0.937 1.586 2.194)
10	(0.1 0.1 0.1)	(0.8 0.1 0.1)	(2.149 0.967 0.924)

Using the proposed method with GA (population size: 50; generations: 1,000; chromosome length: 120; crossover rate 0.7; mutation rate: 0.003) we obtain:

$\tilde{a} = (-0.878, 0.186, 1.148)$

$\tilde{b} = (3.1980, 0.276, 0.391)$

$\tilde{c} = (1.424, 0.563, 0.338)$

The total fuzzy error is (0; 0.01; 29.19) is within the desired accuracy and quite acceptable.

Estimation of Fuzzy Profit in an Oligopolistic Environment

We next consider a greatly simplified example of a firm's profit model in an oligopolistic environment where a number of firms compete in a market for a single product. In such industry the actions of one firm have an impact on the industry demand. The firm's management pursues profitability objectives. The variables that affect the firm's profit include: firm's price and advertising expenditures for a product (controllable), as well as the average price and advertising expenditures of the competitors (uncontrollable).

The firm's management needs to estimate competitors' average prices and advertising expenditures before making decision on firm's price and advertising. However, the information about competitor future actions is not known precisely. Competitor actions may be estimated in fuzzy terms. For example: "the average competitor price will be about $315". To cope with this kind of data we employ a fuzzy regression model for predicting the expected fuzzy profit. Table 7.6 lists the data for the profit estimation problem. The prespecified value of the aggregate fuzzy error is in the range (in million USD): [0; 3.0], ($\lambda = 2.4$). The calculated aggregate fuzzy error is: (0, 4.22, 2.84). Thus, the derived regression model is acceptable.

The advantage of the proposed method is that it can be applied to general fuzzy regression model and fuzzy number space. The simulation results show the effectiveness of the proposed method. The accuracy in terms of distance (7.31) by the suggested method is higher than by the existing methods. The application of the suggested approach to the real life manufacturing and business problems demonstrated its acceptability. The overall complexity of problem solving of fitting fuzzy valued data including computer simulations by the proposed method is less than that of the existing methods.

Table 7.6. Data for fuzzy profit problem

Avg. Price ($)	Avg.Advertising (in $1,000)	Firm Price ($)	Firm Advertising (in $1,000)	Profit (in $100,000)
(295, 4, 4)	(60, 5, 5)	(300, 4, 4)	(60, 5, 5)	(383.7, 60, 60)
(300, 4, 4)	(80, 5, 5)	(300, 4, 4)	(70, 5, 5)	(374.7, 60, 60)
(300, 4, 4)	(90, 5, 5)	(305, 4, 4)	(60, 5, 5)	(307.0, 60, 60)
(310, 4, 4)	(60, 5, 5)	(310, 4, 4)	(80, 5, 5)	(501.1, 60, 60)
(310, 4, 4)	(90, 5, 5)	(300, 4, 4)	(60, 5, 5)	(370.6, 60, 60)
(320, 4, 4)	(90, 5, 5)	(320, 4, 4)	(60, 5, 5)	(344.9, 60, 60)
(320, 4, 4)	(90, 5, 5)	(310, 4, 4)	(70, 5, 5)	(401.8, 60, 60)

(320, 4, 4)	(60, 5, 5)	(330, 4, 4)	(100, 5, 5)	(570.6, 60, 60)
(320, 4, 4)	(90, 5, 5)	(320, 4, 4)	(60, 5, 5)	(344.8, 60, 60)
(320, 4, 4)	(80, 5, 5)	(310, 4, 4)	(80, 5, 5)	(449.5, 60, 60)
(330, 4, 4)	(90, 5, 5)	(300, 4, 4)	(80, 5, 5)	(459.0, 60, 60)
(310, 4, 4)	(80, 5, 5)	(330, 4, 4)	(100, 5, 5)	(402.9, 60, 60)
(330, 4, 4)	(100, 5, 5)	(330, 4, 4)	(60, 5, 5)	(332.2, 60, 60)
(305, 4, 4)	(80, 5, 5)	(330, 4, 4)	(80, 5, 5)	(298.4, 60, 60)
(330, 4, 4)	(60, 5, 5)	(330, 4, 4)	(80, 5, 5)	(510.1, 60, 60)
(315, 4, 4)	(90, 5, 5)	(325, 4, 4)	(90, 5, 5)	(381.8, 60, 60)
(330, 4, 4)	(60, 5, 5)	(330, 4, 4)	(100, 5, 5)	(590.9, 60, 60)
(310, 4, 4)	(70, 5, 5)	(330, 4, 4)	(90, 5, 5)	(419.4, 60, 60)
(320, 4, 4)	(80, 5, 5)	(305, 4, 4)	(100, 5, 5)	(509.4, 60, 60)
(330, 4, 4)	(100, 5, 5)	(330, 4, 4)	(100, 5, 5)	(422.6, 60, 60)

7.5 An intelligent system for diagnosis of the oil-refinery plant

The absence of enough statistical data for establishing objective relationships between the values of features and the probabilities of diagnoses has led to a heuristic description of these relationships. And this is done as a set of rules defining the fuzzy probabilities of the diagnoses on the basis of knowledge and experience of experts-technologists. The separate local sets of rules describe various technological modules of the chemical complex and are integrated into the system's knowledge base. The diagnosis problem is formulated as follows[5]. Let $X_1,...,X_n$ is a series of variables on the basis of which a decision is made on the probability of the diagnoses from a predefined series of diagnoses $D_1,...,D_k$. Each of $X_i, i=\overline{1,n}$ takes a value from the set $\{x_{i1},..., x_{im_i},..., x_{ip_i}\}$. At the instant *t* the state of the object is described by the feature vector

$$\overline{X}^t = [X_1^t, X_2^t,..., X_n^t \mid X_i^t = x_{im_i}]^T,\ m_i \in [1, p_i],$$

where X_i^t - is the actual value of feature X_i at the instant *t*. It is required to make an estimate of the probability (a degree of possibility) of the diagnoses Pb^j:

$$\forall j \in [1, 10]: Pb^j = Pb\,(D_j / \overline{X}^t). \tag{7.34}$$

The second possible way to present the expert knowledge is a system of rules using the same notions as follows

$$\forall i,\ \forall j,\ \forall_{mi} : X_i = x_{im_i} \Rightarrow Pb\,(D_j|X_i) = Pb_s(D_j|x_{im_i}) \tag{7.35}$$

Either of the considered ways for presenting the expert knowledge has specific characteristics. The algorithms to process the information presented in either way are also specific. The most convenient form for presenting implicative knowledge to the human is the linguistic form, in which the expert operates with fuzzy categories such as: "If the value X_i is very big then the probability of D_j is small". That is why in creating the model the linguistic approach was used. In the suggested fuzzy diagnosis algorithm, the features $X_1,\ldots,X_n$ and probabilities Pb are represented as linguistic variables defined by the touples:

$$\langle X_i, T_i, V_i, G_i, M_i \rangle,\ i = \overline{1,n}, \tag{7.36}$$

$$\langle Pb, P, U, S, Q \rangle, \tag{7.37}$$

where X_i and Pb are designations for corresponding linguistic variables, T_i and P are term-sets of the variables X_i and Pb, respectively, i.e. the sets of their linguistic values representing the names for fuzzy variables A_{if_i} $(f_i = \overline{1, p_i}\ |p_i = \text{card}\ T_i)$ and $B_l (l = \overline{1, m}\ |\ m = \text{card}\ P)$ with the values from the universes of discourse V_i and U ; G_i and S are syntax rules generating the names A_{if_i} and B_l for the variables X_i and Pb; M_i and Q are semantic rules permitting to convert every new value of linguistic variable into a fuzzy variable.

The fuzzy variables A_{if_i} and B_l, in its turn, are represented by the following touples:

$$\langle A_{if_i}, V_i, \tilde{C}_{if_i} \rangle, i = \overline{1,4}, \tag{7.38}$$

$$\langle B_l, U, \tilde{E}_l \rangle, l = \overline{1,11}, \tag{7.39}$$

where A_{if_i} and B_i are names for fuzzy variables; V_i and U are the same as in (7.36) and (7.37); $\tilde{C}_{if_i} = \bigcup_{v \in V_i} \propto_{\tilde{C}}(v)/v$ and $\tilde{E}_i = \bigcup_{u \in U} \propto_{\tilde{E}}(u)/u$ are fuzzy subsets of sets V_i and U, describing the constraints on possible values of fuzzy variables A_{if_i} and B_i; $\propto_{\tilde{E}}(u)$ and $\propto_{\tilde{C}}(v)$ are membership functions, defined for $\tilde{E}_I$ and $\tilde{C}_{if_i}$.

Thus, the feature chosen for the reactor block according to the following linguistic variables:

$$\left\langle \text{fall in P2-1}, T_{1,} [-15^0, 8^0], G_1, M_1 \right\rangle;$$
$$\left\langle \text{fall in P2-2}, T_{2,} [-8^0, 8^0], G_2, M_2 \right\rangle;$$
$$\left\langle \text{fall in P2-3}, T_{3,} [-8^0, 8^0], G_3, M_3 \right\rangle;$$
$$\left\langle \text{fall in P2-4}, T_{4,} [-8^0, 8^0], G_4, M_4 \right\rangle.$$

The term-sets are defined as follows:

$T_1 = T_2 = T_3 = T_4 =$ {increased a bit, increased, increased significantly, decreased a bit, decreased, decreased significantly}. The estimation of the probability is represented by the appropriately named linguistic variable. The term-set P consists of the following linguistic values:

B_1 – definitely not, B_2 – almost impossible, B_3 – rather improbable, B_4 – hardly probable, B_5 – more or less probable, B_6 – probable, B_7 – rather probable, B_8 – rather probable, B_9 – very probable, B_{10} – almost certainly, B_{11} – definitely.

According to the designations introduced in (7.36)-(7.39), the fuzzy rules can be written in the following general form:

$$\forall i, \forall j: \text{IF } X_1 \text{ is } A_{1f_1}, \dots, X_n \text{ is } A_{nf},$$
$$\text{THEN } Pb(D_j/(X_1, \dots, X_n)) \text{ is } B_1; \qquad (7.40)$$
$$\forall i, \forall j, \forall f_i: \text{IF } X_i \text{ is } A_{if_i}, \text{THEN } Pb(D_j/(X_i) \text{ is } B_i$$

In (7.40) the rules are used which establish the relationships between all linguistic values for each feature considered separately and the value of the subjectively defined conditional probability for each diagnosis. The possibility of such a representation for expert knowledge is derived from crisp statistic approach. In particular, the widely used Bayesian formula for calculating probability of diagnoses yields the relationships:

$$\forall j=\overline{1,k}: P(D_j/(X_1,\ldots,X_n)) = F[P(D_j/X_1),\ldots,p(D_j/X_n)], \quad (7.41)$$

Formula (7.41) is extended for the case of independent features. Thus, in the type (7.40) diagnosis model, the condition of independence of the features is implied.

The analysis of the implemented technology has shown that it is possible to separate almost independent diagnostic features. This allows the use of type (7.40) models for the description of relationships between probabilities of diagnoses and fuzzy values of features. Moreover, in most practical problems, it is admissible to accept the assumption about the independence of the features of knowledge base. The rules like (7.37) can be represented as follows:

IF X_i is A_{if_i} THEN D_j is with the probability of B_l.

Table 7.7 illustrates a fragment of knowledge base represented as a set of linguistic rules for determining the estimates for the probability of diagnoses ΔT_4. The elements m_{jf} (j is the number of diagnosis from {D} and f is the number of linguistic value from the term-set T) are obtained on the basis of elements m^{γ}_{jf}, r=1,…,5, of all the 5 versions, filled out by 5 experts. The answers by experts represent the values B_l of linguistic variable Pb, the estimates of the probability.

The further processing of the knowledge base is reduced to the construction of fuzzy relational matrices $\tilde{R}^j_i$.

Figure 7.4 shows a chart of the sequential procedure of calculating the estimate of the probability of one of diagnoses. Values of technological parameters (features) enter module 1. This module serves as a linguistic interpreter, which determines features' linguistic values from the term-set. At the output we have fuzzy constraints $\tilde{G}_{if_i}$ for each feature, which then enter to module 2. This module is also fed by relational matrices $\tilde{R}^j_i$. For each X_i in module 2, a solving rule is used in form of conditional categorical syllogism or MODUS PONENS expressed by the formula

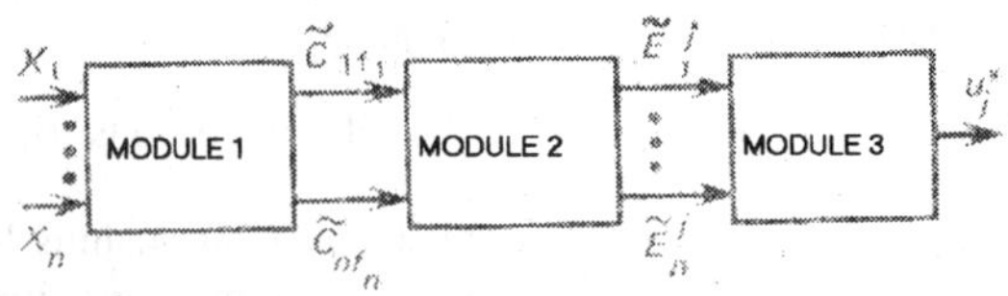

Figure 7.4. Structure of computing and reasoning system

Proposition 1: IF X_i is A_{if_i}, THEN Pb IS B_l

Proposition2: X_i is A_{if_i} (7.42)

Conclusion: Pb is B_l

The rule (7.42) is implemented using the compositional inference rule by L.Zadeh, saying that the solution of the assignment equation

$$R(v,\ u) = \tilde{R}_i^j \text{ and } R(v) = \tilde{C}_{if_i}$$

has the form

$$\forall i = \overline{1,\ n} : R(u) = \tilde{E}_{il}^j = \tilde{C}_{if_i} \circ \tilde{R}_i^j$$

$$\propto_{\tilde{E}}(u) = \max\ [\min\,(\propto_{\tilde{C}}(v),\ \propto_{\tilde{R}}(v,\ u))],$$

where $\circ$ is the composition designation.

The created this way fuzzy subsets $\tilde{E}_{il}^j = \bigcup_u \propto_{\tilde{E}}(u)/u$ define fuzzy values of linguistic variable Pb for the diagnosis D_j corresponding to the probability Pb (D_j/X_i), at current values of features

Table 7.7. A fragment of KB

IF ΔT_4		Increased a bit	Increased	Increased significantly	Decreased a bit	Decreased	Decreased significantly	Did not change
	the percentage of naftens in the raw is increased	Probably enough	Very probable	Almost certain	Definitely not	Definitely not	Definitely not	Definitely not
	the percentage of naftens in the raw is decreased	Definitely not	Definitely not	Definitely not	Probable enough	Very probable	Very probable	Definitely not
	the humidity of dry gas is increased	Definitely not	Definitely not	Definitely not	Probable	Probable	Rather probable	Almost impossible
	the humidity of dry gas is decreased	More or less probable	Probable	Probable	Definitely not	Definitely not	Definitely not	More or less probable
THEN the probability that	the consumption of dry gas is high	Definitely not	Definitely not	Definitely not	Rather improbable	More or less probable	Probable	Almost impossible
	the consumption of dry gas is low	More or less probable	Probably	Rather probable	Definitely not	Definitely not	Definitely not	Rather improbable
	More heavy raw is detected	Rather probable	Probably enough	Very probable	Definitely not	Definitely not	Definitely not	Hardly probable
	More light raw is detected	Definitely not	Definitely not	Definitely not	Hardly probable	Probable	Rather probable	Almost impossible
	the consumption of chlor-organic is high	Hardly probable	Probable	Probable	Definitely not	Definitely not	Definitely not	Almost impossible
	the consumption of chlor-organic is low	Definitely not	Definitely not	Definitely not	Hardly probable	Probable	Rather probable	Almost impossible

Let at the instant t, after the confirmation of the fact that the reason of the breach in stability of the technological process belongs to subset of non-controllable disturbances, the system gets the following information about the features:

X_1^t - "increased a bit", X_2^t - "increased", X_3^t - "increased a bit", X_4^t - "increased".

After applying the well-known operation of fuzzy inference, we get fuzzy subsets for the values of probability estimates such as diagnosis of D_1 – "percentage of naftens in the raw has increased". These values are as follows: $Pb(D_1/X_1^t) = B_9$ - "very probable"; $Pb(D_1/X_2^t) = B_{10}$ – "almost certain"; $Pb(D_1/X_3^t) = B_8$ - "probable enough" and $Pb(D_1/X_4^t) = B_9$ - "very probable". After applying the intersection operation on the obtained fuzzy subsets (Fig.7.5) we will have: $Pb(D_1/(X_1^t,...,X_n^t)) = u_1^* = 0.76$

Similarly the probability estimates for the other diagnoses are calculated. Then, a sequence of preferences is constructed – a monotonically decreasing series of diagnoses according to the estimates of their probabilities.

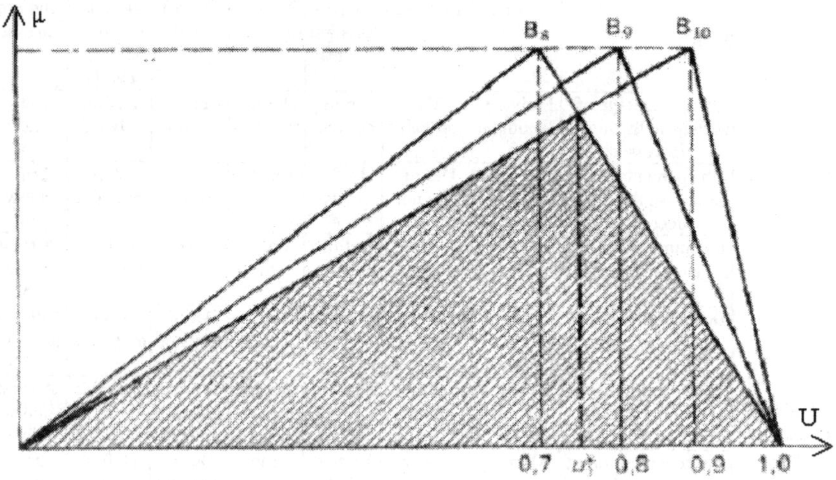

Figure 7.5. Membership functions of Pb

7.6 Neuro-Fuzzy Pattern Recognition in Manufacturing

Architecture and working principles of Neuro-fuzzy recognition system

Robotic manipulators are widely used in manufacturing processes. The manipulators are often used to classify semiproducts into different categories on the basis of their quality or type. These obtained quality or type usually determine the type of next processing the semiproducts should have on. Various measurement data are used to perform the classification. These data usually represent colors, sizes, dimension metrics, weights, voltage, etc.

In this section we consider the application of the Neuro-Fuzzy approach to the classification by using a robotic manipulator of class PUMA-560. Such a manipulator is used to classify figures (each representing a different sort of semiproducts) on the basis of the characteristics: shape, size and color. The manipulator is equipped with a camera to get the digital image of semiproduct to be classified. This image is processed by computer to extract the measurements on the basis of which the classification can be done. Fig. 7.6 shows the chart of the classification process.

First the camera scans the image over the manipulator work area. Then the image is coded according a supported digital image format. Then it is simplified and optimized to remove noise and unused background information. Then the image is converted to the format supported by Neuro-Fuzzy Recognition System. The next process is appropriate segmentation of image to separate different objects on the work area. Then each separate object is undergone further processing consisting in necessary scaling/orienting/shape-color-optimizing. The formatted and optimized object then enters to the Neuro-Fuzzy Recognition System (NFRS), which performs the actual recognition of the class the input object belongs to. The structure of the NFRS is shown in Fig. 7.7.

The fuzzy rules used are as follows:

IF $\mathit{ShapeClass}_i(\mathit{measurement\ of\ shape}) * \mathrm{Weight}_{(i,shape)}\&$
$\mathit{SizeClass}_i(\mathit{measurement\ of\ size}) * \mathrm{Weight}_{(i,size)}\&$
$\mathit{ColorClass}_i(\mathit{measurement\ of\ color}) * \mathrm{Weight}_{(i,color)}$

THEN $\mathit{Class}=\mathit{Class}_i$

where i=1,...,n; n is the number of rules (and classes); $\mathit{ShapeClass}_i(...)$, $\mathit{SizeClass}_i(...)$, and $\mathit{ColorClass}_i(...)$ are fuzzy values representing certain features of class i, *Weight(...)* represents the importance of particular feature for certain class.

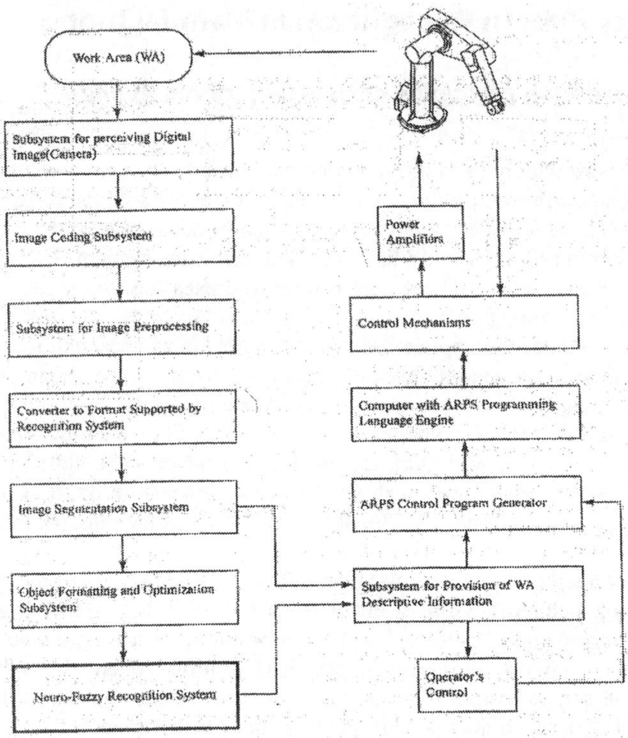

Fig. 7.6. The chart of the classification process

The two information flows, one from Image Segmentation System and the other from Neuro-Fuzzy Recognition System, allows preparation of formalized descriptive information about the state of the work area. This information is passed to ARPS Control Program Generator (ARPS is a specific programming language understood by PUMA-560). This program forces the control mechanisms of the manipulator to actuate all series of movements for required classification.

Below we will consider the Neuro-Fuzzy technology in more detail with application to manufacturing parts recognition process on an example of recognition of codes (figures) and symbols (words, signs).

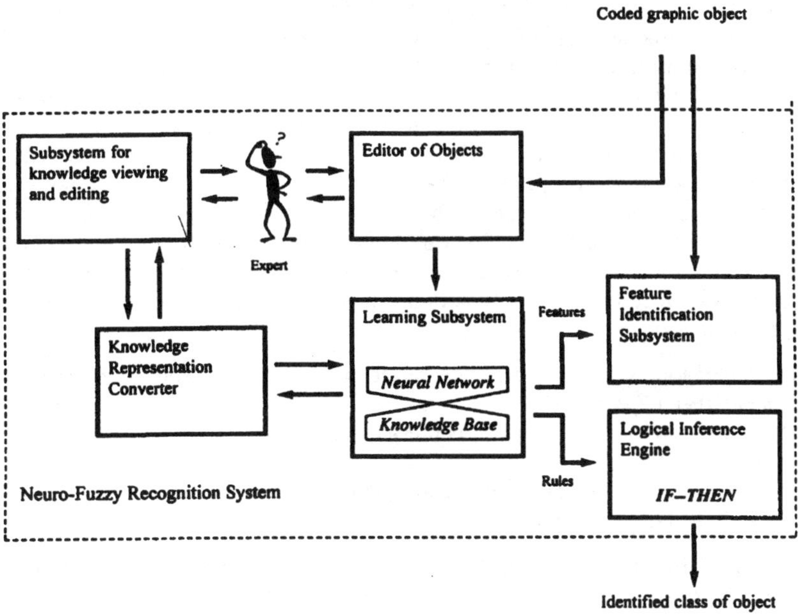

Fig. 7.7. Neuro-Fuzzy Recognition System

Methodology of Neuro-Fuzzy Image Recognition

Image recognition such as reading characters a task easily accomplished by human being presents significant difficulties when attempting to design artificial computer systems that can do the same work [15]. The problem becomes especially difficult when the recognition objects are fuzzy, in the case of manufacturing images. In this case the knowledge about the patterns cannot be expressed as an accumulation of bitmap images of classified figures, but in some way by rules containing fuzzy features describing the patterns under the consideration. Fuzzy rules give two advantages: first, they accord with the nature of such objects and so support their adequate description; and secondly, they are compact and allow human analysis and tuning. The human analysis can be required to exclude from rules, generated automatically by some method, unnecessary information that an artificial system may consider, e.g. features of a subjective working style. Here we try to apply the latest achievements of Soft Computing [8,15,16] in order to make a step toward the solution of the manufacturing pattern recognition problem.

As it is shown above the knowledge for the artificial recognition system can be described by fuzzy rules [7]. These rules describe the reaction of the system to an input object. The result of the inference process depends on whether the prescribed pattern features are found in the analyzed object. The feature set must contain all the pattern features sufficient for fuzzy description of each class of objects in the environment (i.e. all patterns involved in the recognition). Each class can be described by some subset of the feature set. Each feature is associated with two fuzzy terms: some relational fuzzy location [7] specifying the region in which this feature is mainly localized in a group of patterns; and some color value describing fuzzily the density of color in that fuzzy location. The condition part of a rule reveals features that the pattern corresponded to this rule should meet.

Let's give a simple example [7]. Assume that the considered environment includes the patterns known to us as characters *'U', 'O', 'C'*. In this case the fuzzy location variable may take values from the set: *"center-top"*, *"center-center"*, *"right-center"*, and the fuzzy color variable - from the set: *"white"* and *"black"*. The feature set can be produced as a subset of the above two sets relation. The features are combined by the *and* operation. Thus, the rules may look somewhat like the following:

1. If "center-top" is "white" and "center-center" is "white" and "right-center" is "black" then object is 'U'
2. If "center-top" is "black" and "center-center" is "white" and "right-center" is "black" then object is 'O'
3. If "center-top" is "black" and "center-center" is "white" and "right-center" is "white" then object is 'C'

It should be noted, that operation *"and"* used in these rules operates in a slightly different way than simply taking the minimum value as in traditional fuzzy logic. It should be noted as well that each feature may have different *"degree of importance"* for every rule. Say, feature *"center-top"* is of higher degree of importance for rule 1 than for others as it plays the principal role exclusively in recognizing character *'U'*. Feature *"center-center"* takes the same value for all the rules. It is of little importance as it does not make a considerable contribution to the recognition.

The final decision is made depending on the final membership degrees resulting from the rules. The rule with the highest final membership degree is said to have fired, and the class associated to this rule is the result of recognition [7].

$$\text{Class} = \text{Pattern}[\operatorname{Arg}\max_j M_j]$$

$$M = \frac{1}{N} \cdot \sum_i m_i w_i$$

$$m_i = \max_x \{\min\{\propto_{x_i}(x), \propto_{C_i}(c(x))\}\}$$

Here x_i is the fuzzy location of some feature i, C_i is its fuzzy color density and c(x) is the function of image color density the argument x of which is crisp location (two-dimensional).

Good performance of fuzzy rules depends on how well they have been tuned to match the environment. Here the use of neural learning is indispensable. We can easily transform our fuzzy model to neural representation. We consider neurons (a kind of RBFNN) the output signal f of which is calculated as follows:

$$f(X) = \sum_i w_i |x_i - \theta_i| \tag{7.43}$$

where x_i is one of neuron's inputs (X is vector of inputs); w_i and θ_i are parameters of the neuron;

This formula can act as the inference for the fuzzy rules. Indeed the summation plays the role of the *"and"* operation while the expression inside is inverted triangle-shaped membership functions in which w_i determines their slope and θ_i - the center-value.

Formula (7.43) can also be considered to be a function calculating a mismatch measure between an object and a class. Indeed x_i represents features of an input object while θ_i represents associated features of the candidate class. The formula calculates the distance measure between two vectors considering some weighting factors for associated elements of the vectors. Those weighting factors represent the *"degrees of importance"* mentioned above.

The parameters θ_i are estimated when collecting information about samples of the classes for a particular environment. Samples even belonging to the same class may still have noticeable difference in their appearance. Such classes often require use of several neurons to form the appropriate knowledge. Adjusting primarily implies tuning properly the parameters w_i. It is clear from the theory of recognition that when solving the classification problem the vectors representing different classes must be distanced from each other as far as possible. The adequate adjustment is the following:

$$\Delta w_{lki} = \frac{\gamma\left|\theta_{ki} - \theta_{li}\right|}{\left\{\sum_i w_{ki}\left|\theta_{ki} - \theta_{li}\right|\right\}} \tag{7.44}$$

$$\text{for} \quad k,l = 1,2,\ldots,Q$$

where Q is the number of classes in the environment.

The above formula can be derived from the gradient descent method if the total error for the neural network performance is assumed to be

$$E = \sum_{k,l,k\neq l}\left(\sum_i w_{ki}\left|\theta_{ki} - \theta_{li}\right|\right)^{-1} \quad \text{for} \quad k,l = 1,2,\ldots,Q \tag{7.45}$$

The method described above was used for software implementation of an experimental handwritten recognition system. The software system, after appropriate learning for necessary environment, allows to convert computer files containing handwritten graphics to text files.

Let's describe some points of the implementation in detail.

The size of the feature set depends on both the degree of fuzziness of the input objects (graphical images) and the number of classes in the environment. We found applicable a rectangle area to accept each input object. This area is divided to equal intervals in both (horizontal and vertical) directions so that we have equal regions that stand for features. Optimum number of the divisions is to be found experimentally (depending on the environment conditions). We found that 5x5 is quite enough for recognizing the English character set.

The features must allow some fuzziness in their location, so the borders of the regions are fuzzy. A displacement is allowable for a region until it covers no more than half of any of the neighboring regions. Therefore when processing an object, the system divides the area under the object to 10x10 subregions, and, combining 4 appropriate subregions, forms 25 fuzzy features each with 9 membership degrees. The membership degree for the middle region of each feature is intuitively set to 1. These membership degrees are then related to the values of color calculated by counting number of pixels of either color found inside each of 9 regions The upper images in Fig. 7.8 represent the fuzzy patterns reconstructed from the fuzzy rules generated after learning on the original images given below. Density of the black points reflects the value of membership degree.

Membership functions for color shade are set to be symmetrical triangle-shaped curves with their maximums in the crisp values of averaged color shade calculated separately for each region.

The final membership degrees obtained by combining those of terms in condition parts of the rules should not be so sensitive as to be affected considerably by mismatches found in one or a few features. Therefore the *"and"* operation is replaced by summation weighted with the degrees of importance.

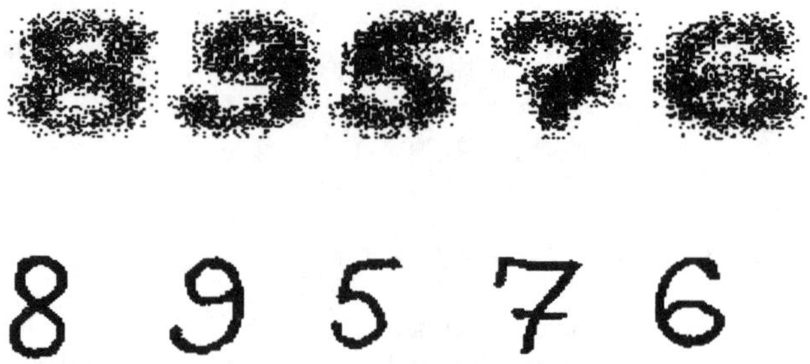

Fig. 7.8. Fuzzy patterns reconstructed from the fuzzy rules

One of the environments the system was tested on was the environment of figures drawn manually (Fig. 7.9).

The images in the upper two strings were used as learning samples. After the learning the knowledge describing the environment was generated. Information below is what, for example, the system had learned about the digit '8' in this experiment.

Fuzzy features of a pattern are given as 25 fuzzy values represented as sets each consisting of 9 elements (Table 7.8). Position (index) of each element inside the set determines the membership degree of the feature located in the corresponding position area on image. The value of the element determines the feature crisp color density value related to the maximum membership degree of the feature fuzzy color density.

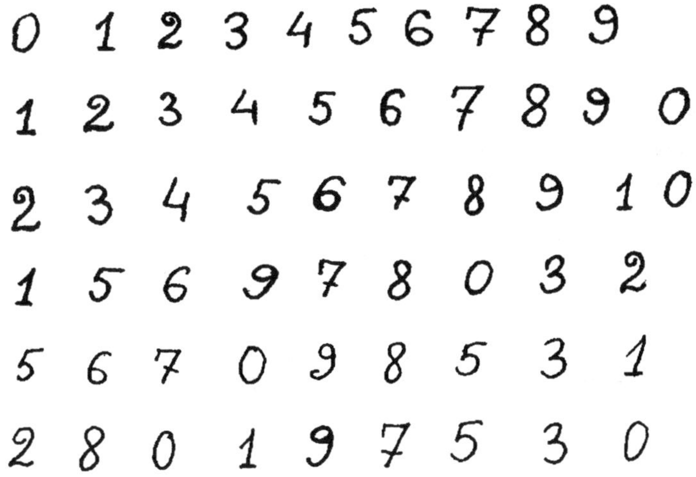

Fig. 7.9. Test example for recognition

The appropriate values of importance for all the 25 features are given in Table 7.9.

Table 7.8. Representation of pattern '8'

0	87	295	295	456	514	514	496	499	499	452	284	284	107	6
12	319	755	755	790	617	617	505	512	512	579	650	650	575	222
48	544	913	913	512	104	104	9	13	13	131	580	580	954	492
48	544	913	913	512	104	104	9	13	13	131	580	580	954	492
48	511	914	914	575	128	128	8	45	45	222	641	641	843	383
12	199	521	521	700	616	616	500	537	537	580	555	555	369	107
12	199	521	521	700	616	616	500	537	537	580	555	555	369	107
0	47	328	328	771	871	871	787	896	896	779	348	348	59	0
164	459	630	630	590	390	390	295	412	412	539	582	582	394	99
164	459	630	630	590	390	390	295	412	412	539	582	582	394	99
440	892	566	566	122	8	8	0	8	8	122	566	566	707	255
497	953	522	522	76	12	12	13	25	25	163	599	599	673	223
497	953	522	522	76	12	12	13	25	25	163	599	599	673	223
254	624	644	644	528	496	496	484	474	474	564	640	640	375	67
33	151	326	326	452	484	484	471	449	449	401	245	245	62	0

Given the file containing the whole graphics represented in Fig. 7.9 on its input the system recognized successfully all the digits.

The second test concerned the English capital letters. In this case the learning set first includes 50 samples. The system successfully recognized 75% of the test handwritten text. The rest of characters were rejected by the system as having too much mismatch error in features with all of the classes. After some decrease made in the admitted level for the mismatch error the system recognized correctly 90% of the characters. We had to extend the learning sample set up to 100 in order to get better performance. The time the recognition took was 1.5 times as much as in the first case, but the performance of the system became 98%.

Table 7.9. Values of importance for features of pattern '8'

176	187	76	111	167
194	219	140	176	356
211	355	249	242	230
342	114	97	174	239
194	105	120	95	172

Example on the recognition of parts

The use of the Neural and Fuzzy approaches allowed construction of an effective manufacturing recognition system capable of learning for a particular environment. A sample environment containing 7 types of different parts were considered. The parts were scattered randomly on a rectangular area. Then the area was photographed by a camera. The picture converted to a computer format (a TIFF file) was an input to the recognition system (Fig. 7.10).

The system processed the image to remove the noise and unnecessary details. Then the regions containing separate objects were identified. Then the regions were processed one by one to find to which part type the corresponding object belongs. The complexity of this case over the symbol recognition was that in this case a part can be freely rotated. That means that the system should consider not only displacement but also rotation angle. However, the simplicity of this case is that the number of recognition classes is usually smaller (only 7 in this case) and also the part size imprecision is neglectable. Hence we can use the physical area measure as an essential classification feature (which is rotation independent).

A number of experiments have been performed on the recognition system with various positioning of the parts. In the experimental software system most of the processor time is spent for graphical conversions and performing fuzzy inference on the base of a specific neural network. The recognition system without any error identified parts shown in Fig. 7.10 into 7 categories independent of parts location and orientation. The neural network even when implemented on a traditional computer offered the advantages of learning, compact knowledge base, and operative processing (as compared with the classic hard-model based computing systems). In addition it can be supposed that the use of appropriate fuzzy and neural chips will allow us to improve significantly the performance of the recognition system.

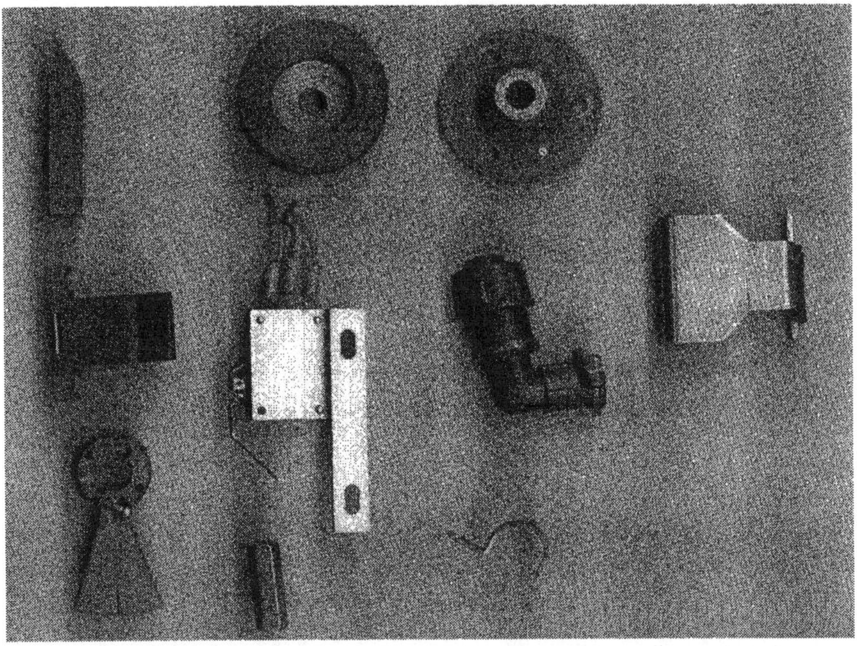

Fig. 7.10. Example of robot working environment.

7.7 Soft Computing based inventory control

Inventory is the stock of any item or resource used in an organization. An inventory system is the set of policies and control that monitors levels of inventory and determines what level should be maintained, when stock should be replenished, and how large the order should be [39].

Inventory control system provides firms to maintain independence of operations, to meet variation in product demand, to allow flexibility in production scheduling, to take advantage of economic purchase order size, etc. For each of preceding reasons, be aware that inventory is costly.

In making any decisions in inventory systems the following costs must be considered [39,46].

- □ Purchase cost. This is connected with purchases from an external seller. It is necessary to note that for large sales usually there is a discount depending on the volume. Such stepped prices will be considered hereafter in the representation of the purchase cost in the supply management model.
- □ Ordering cost. These costs refer to the managerial and clerical costs to prepare the purchase or production order.

- Holding cost. It includes the costs for storage facilities, handling, insurance, taxes, and the opportunity cost of capital. Obviously, high holding costs tend to favor low inventory levels and frequent replenishment.
- Shortage costs. When the stock of an item is depleted, an order for that item must either wait until the stock to satisfy demand and the costs resulting from stockout.

Very frequently when considering the statement and solution of the inventory control problem, the concerned parameters and the demand for each planning period appear to be both inexact and uncertain. That is why when constructing an inventory control model one needs to consider the uncertainty of these parameters. A more adequate uncertainty measure in this case is the fuzzy measure [30,42,46,50]. That is why in the considered model the information about the costs and demand will be represented as fuzzy numbers. The questions of fuzzy inventory control are considered in [19,46,50].

Let's give a formal statement of the fuzzy inventory control problem. Assume that for every i-th interval of plan period T, consisting of N intervals, the demand is D_i while the level of supply for the end of i-th interval is x_i. It is required to order such amount of z_i, $i=\overline{1,N}$ that would fully and timely satisfy the demand while keeping the overall expenses at a minimum level. Here the overall expenses are calculated as follows.

Holding cost is determined to be proportional to the ending inventory x_{i+1}. Considering the balance condition, we can write:

(the supply level at the end of i-th interval)=(the supply level at the beginning of i-th interval)+(amount of supply replenishment at the i-th interval)-(demand at the i-th interval),

or formally as in [46] can be written:

$$x_{i+1} = x_i + z_i - d_i \tag{7.46}$$

Therefore the holding cost will be determined as:

$$C_i^H = h_i(x_i + z_i - d_i) \tag{7.47}$$

If the ordering cost for the i-th period is K, then the total cost of obtaining z_i units at the beginning of i-th interval is determined as [46]

$$C_i^0(z_i) = \begin{cases} 0, & \text{if } z_i = 0 \\ K_i + C_i(z_i), & \text{if } z_i \geq 1 \end{cases} \tag{7.48}$$

where

$$C_i(z_i) = \begin{cases} az_i, & \text{if } 0 \leq z_i \leq d_i \\ a\tau_i + b(z_i - d_i), & \text{if } z_i \geq d_i \end{cases} \tag{7.49}$$

Here we assume that the supplier makes discounts based on the volume of order. Formula (7.48) illustrates the stepped function of price in such situations, when a is the price per unit for volumes below d_i and b is the unit price for volumes above d_i.

For solving the problem of optimal supply scheduling we need to consider the uniformity of the developed schedule. We will designates the measures "demand exceeds supply" and "supply exceeds demand" as $V(z_i)$ and $W(z_i)$. Considering the above described for the interval i the total amount of expenses C_i will be calculated as

$$C_i(z_i) = C_i^0(z) + h_i V(z_i) + p_i W(z_i)$$

Here h_i is the holding cost per unit for the i-th period, and p_i is the penalty cost due to shortages.

The total inventory cost over the N periods is

$$C(z_i) = \sum_{i=1}^{N} C_i^0(z) + h_i V(z_i) + p_i W(z_i) \tag{7.50}$$

Some constraints are applied to the values of the variables z_i and x_i. First the order volumes are integer values, i.e.

$$z_i = 0,1,2,3,... \tag{7.51}$$

Secondly, it would be better to keep the supply close to zero at the end of interval N:

$$x_{N+1} = 0 \tag{7.52}$$

Lastly, there is a condition of complete and timely satisfaction of the demand for every interval in the plan period T. This is the balance condition and can be rewritten on the basis of (7.46) as

$$x_i - x_{i-1} + z_i = d_i$$

Thus, the considered inventory control problem can be formulated as:

$$C(z_i) = \sum_{i=1}^{N} \left(C_i^0(z) + h_i V(z_i) + p_i W(z_i) \right) \to \min \tag{7.53}$$

subject to

$$x_i - x_{i+1} + z_i = d_i \tag{7.54}$$

$$x_{N+1} = 0 \tag{7.55}$$

In the model (7.46)-(7.55) d_i, x_i $(i \geq 2)$, h_i, p_i, K_i are fuzzy and represent fuzzy subsets of R.

In these conditions, the problem (7.53)-(7.55) is a typical problem of fuzzy dynamic programming [1,25,29,42,46,50]. However, we should note that the fuzzy dynamic approach is very computationally intensive for solution of a problem like (7.53)-(7.55). In [46] a genetic algorithm is used to solve this problem.

Let's consider an example of solving the problem of fuzzy inventory control considered in [46]. N=4, x_1 (initial inventory) is 15 units.

Fuzzy demand (d_i), fuzzy ordering cost (K_i), fuzzy holding cost (h_i) and fuzzy penalty cost (p_i) are shown in Table 7.10

Table 7.10. Data for fuzzy inventory problem

Period	d_i	K_i	h_i	p_i
1	(92,96,100,109)	(70,74,78,82)	(0.8,1,1,1.2)	(2.0,3.0,3.0,4.0)
2	(108,112,116,120)	(24,26,26,28)	(0.8,0.9,1.1,1.2)	(2.5,2.8,3.3,4.0)
3	(179,183,187,191)	(86,89,91,94)	(0.6,0.8,1.2,1.4)	(2.0,2.7,3.3,4.0)
4	(64,68,72,76)	(64,65,69,70)	(0.7,1,1,1.3)	(2.5,5.9,3.1,3.5)

The values of these fuzzy data are represented as trapezoidal fuzzy numbers.

According to [46] first an initial population $z = (z_1, z_2, z_3, z_4)$ for $z_i > 0$ integer is randomly generated. Then with the use of the selection operator fittest individuals providing the minimization of the objective function (7.53), are chosen. The next generation is formed through crossover and mutation operators [46,8]. In [46] the problem minimization of fuzzy objective function is changed into a multi-objective optimization problem. To check the feasibility of selection it is necessary to calculate

$$x_{i+1} = V_i(z_i), \quad i=\overline{1,3}$$

and finally $x_5(1)$. If the condition $x_5(1) = 0_f$, where 0_f is fuzzy zero, is satisfied, then $z = (z_1, z_2, z_3, z_4)$ is keeping.

Continuing this procedure some predefined number of times we choose among the keeping z such that provides the smallest value of the objective function. The solution on the basis of the data given in the table is as follows

$$z_1 = 59.9568, \; z_2 = 27.4523, \; z_3 = 71.7562, \; z_4 = 67.2431$$

7.8 Fuzzy Project Scheduling

Project management can be defined as planning, directing, and controlling resources (people, equipment, material) to meet the technical, cost, and time constraints of the project.

Let's give some definitions of Project Management [39]. A project starts out as a statement of work. A task is a further subdivision of a project. It is usually not longer than several months in duration and is performed by one group or organization. A subtask maybe used if needed to further subdivide the project into more meaningful pieces.

A work package is a group of activities combined to be assignable to a single organization unit. It still falls into the format of all project management; the package provides a description of what is to be done, when it is to be started and completed, the budget, measures of performance, and specific events to be reached at a points in time. The work breakdown structure (WBS) defines the hierarchy of project tasks, subtasks and work packages.

There are two basic classic methods to facilitate scheduling, planning, and controlling projects: PERT (Program Evaluation and Review Technique) and CPM

(Critical Path Method). In principle these two methods are quite similar and focus on finding the longest time – consuming path through a network of tasks as a basis for planning and controlling a project. Both PERT and CPM use nodes and arrows for display. Originally, the basic differences between PERT and CPM was that PERT used the arrow to represent an activity and CPM used the node. The other original difference was that PERT used three estimates – optimistic, pessimistic and best – of an activity's required time where as CPM used just the best estimate.

Usually the normal project scheduling includes the following phases (for the sake of simplicity in this case we consider CPM with single time estimate):

- identification of each activity to be done in the project;
- activity sequence determination and network reflecting the precedence relationships construction;
- determination of the critical path which reflects the longest sequence of connected activities through the network and is defined as the path with the zero slack time. Here the four time values for each activity have to be calculated: earliest start times (ES); earliest finish times (EF); latest finish times (LF), and latest start times (LS).

If a single estimate of the time, required to complete an activity, is not reliable, the best procedure is to use three time estimates. It should be noted that several assumptions need to0 be made to use project network and CPM or PERT analysis. First, there is a precise beginning and ending point for each activity. Secondly, project activity sequence relationship can be specified and networked. Finally, the activity times in PERT follow the beta distribution, with the variance of the project assumed to equal the sum of the variance along the critical path.

It should be noted that in Project Management, as a rule, every project in its nature is unique and the duration of activities in a project cannot be determined accurately. Ignoring the uncertainty of activity duration can lead to incorrect scheduling decisions.

To deal with uncertainty researchers introduced a probability estimate for completion time for the entire network. Experts formulate three estimates for each activity time completion: best (most likely estimate) t_b, maximum (pessimistic) t_p, and minimum (optimistic) to times. The estimated activity time is computed as weighted average of these times with generally used ratio: 4,1,1. In probabilistic PERT the estimated completion time of the network is computed using basic statistics. However, because the three times t_b, t_{max} and t_{min} for each activity come from experts who use their knowledge and experience, the nature of uncertainties in this problem is rather fuzzy than probabilistic.

Traditional project management tools such as PERT and CPM have been intended to deal with more real-fuzzy uncertainty issue [19,22,31]. The objective of this section is to develop a fuzzy scheduling methodology for management of project.

First we consider the extension of PERT for project management for fuzzy case [19].

Assume that each time activity (Fig. 7.11) is estimated by N experts, expressed as triangular members

$$t_i^A,\ t_i^B,\ t_i^C,\ t_i^D,\ t_i^E,\ t_i^F,\ t_i^G, \qquad i = \overline{1,3}$$

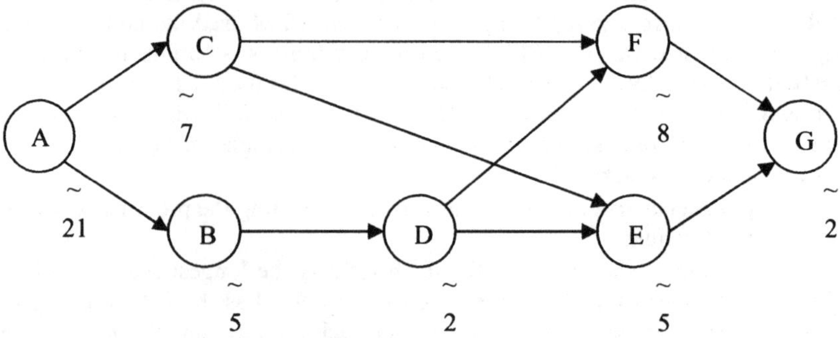

Fig. 7.11. Network of tasks

The aggregated expert opinions for completion of activity A is calculated as

$$T_{ave}^{A} = \left(\frac{\sum_{j=1}^{N} t_{jo}^{A}}{N}, \frac{\sum_{j=1}^{N} t_{jb}^{A}}{N}, \frac{\sum_{j=1}^{N} t_{jp}^{A}}{N} \right) \tag{7.56}$$

The fuzzy total time for project completion t is the time for completion of the activities on the critical path

$$t = t_{ave}^{A} + t_{ave}^{C} + t_{ave}^{F} + t_{ave}^{G} = (t_{ave\ o}, t_{ave\ b}, t_{ave\ p})$$

The project duration is obtained from defuzzification of t. Activity time duration and allocation of resources are in a close relationship. Now, let's consider the use of fuzzy PERT to schedule allocation of resources [19]. To this end we'll use time-cost model extensions of PERT and CPM. It is necessary first to create a CPM-type network diagram which includes for each activity: normal cost C_n; normal time t_n associated with each normal cost; crash time t_c – the shortest activity time; crash cost C_c associated with each crash time.

For the sake of simplicity here we will consider shortening activity time on the critical path. The normal time t_n for each activity is estimated as was shown above and is equal to t_b. C_n, t_c, and C_c for each activity could be determined similarly to the normal time t_n as it was done before (with the use of Fuzzy Delphi [19, 28]). After the defuzzification of these parameters, represented as triangular fuzzy numbers, we can calculate the cost slope. For example for activity A it is computed as:

$$K_A = (C_c^A - C_n^A) : (t_n^A - t_c^A)$$

The activities are ranked according to their cost slopes. Best candidates for shortening the length of the project are activities with smallest K, i.e. with smallest cost per unit of time to expedite each activity. By the other words, first we execute the activity ranked first, then we consider the second ranked activity and so forth.

A more effective model of project management at presence of uncertainty and imprecision of temporal information is fuzzy resource-constrained project scheduling models considered in [22]. Related problems are also considered in [20].

A formal statement of fuzzy resource-constrained project scheduling problem is as follows.

Assume that the considered project has ready-time t_r and deadline t_d. The execution of each activity requires certain amount of resources denoted as vector $R_i(r_{i1}, r_{i2}, ..., r_{in})$. The resource availability for the project is defined by R_i^a. First we need to define a criterion based on which we choose an optimal project schedule $S_{opt} \in S_c$ from the set of alternatives S_c.

It is common in practice of project management to use the minimum project lateness criterion, makespan or others [32, 33] as such a criterion. Because temporal information in project management is uncertain, then to predict the exact completion time of a project in advance is very difficult. That is why it is not purposeful to use the above-mentioned criteria for the optimization of project scheduling.

In [22] the schedule risk is used as a performance measure of project scheduling decisions in an uncertain project management environment. In [22] this performance measure (denoted as P) is formulated as follows:

$$\begin{aligned} P = \beta \sup_x \min[\propto_{D(s)}(x), \propto_{(t_d - t_r, \infty)}(x)] + \\ +(1-\beta) \inf_x \max[1 - \propto_{D(s)}(x), \propto_{(t_d - t_r, \infty)}(x)] \to \min \end{aligned} \tag{7.57}$$

where D(s) is the fuzzy project duration with respect to s, β is the optimistic-pessimistic index, (t_d-t_r, ∞) denotes the set of numbers necessary greater than (t_d-t_r). Here t_d and t_r are fuzzy numbers.

Let's identify the constraints for the project scheduling optimization problem.

For i-th activity's start time t_{s_i} the following constraint should be satisfied [22]

$$t_{s_i} \in [t_{es_i}, +\infty] \tag{7.58}$$

that means that for all activities s_i must be possibly greater than or equal to its earliest start time t_{es_i}.

Here

$$t_{es_i} = \begin{cases} \max_j \{t_{es_i}, t_{es_j} + t_j\}, & \text{if activity i is preceded by activity j,} \\ t_r, & \text{if activity i has no predecessors} \end{cases} \tag{7.59}$$

The late start time should satisfy the constraint

$$t_{s_i} \in (-\infty, t_{ls_i}) \tag{7.60}$$

that means that t_{s_i} should be smaller than or equal to its late start time t_{ls_i}; t_{ls_i} is equal to t_{lf_i} -t_i, where

$$t_{ls_i} = \begin{cases} \min\limits_j \{t_{ls_j}, t_{ls_j} - t_i\}, & \text{if activity } j \text{ is preceded by actiivity i,} \\ t_d, & \text{if activity i has no snecessors} \end{cases} \tag{7.61}$$

The finish time of i-th activity should be smaller than or equal to the start time of j-th activity

$$t_{s_j} - t_{s_i} \in [t_i, \infty) \tag{7.62}$$

The following are constraints for m-th type resources:

$$\sum_{i \in J} r_{im} \le r_m^a \tag{7.63}$$

Thus the problem (7.57)-(7.63) presents a project schedule optimization problem under uncertain environment. The use of direct optimization methods for solving this problem is not effective. Heuristic methods appear to be more useful. In [22] authors suggest an effective fuzzy beam search algorithm modified from the branch-and-bound method for solving problems like (7.57)-(7.63). Let's consider an example of fuzzy project management described in [22] for which the fuzzy beam search algorithm is used. In the example the project scheduling is determined for the design of an electronic product.

The considered project includes seven phases: system analysis, system definition, requirements allocation, preliminary design, detailed design, engineering model, and qualification tests [22].

On the whole the project includes 51 activities which are described in [22]. 4 resource types are required for the implementation of the project: r_1, r_2, r_3 (system, software, and hardware engineers, respectively), r_4 (supporting engineers).

In this problem, the uncertain data are represented as trapezoidal fuzzy numbers. The following source data are given: ready-time t_r=(0,0,0,0), deadline of the project t_d=(220,220,220,235), R^a(.)=(5,4,5,4), β=0.5.

In [22] the mentioned method is used to generate a project schedule with the makespan=(180,239,239,294) and the schedule risk=0.63. The table below illustrates the results of the considered problem solution with different makespans and the corresponding P under distinct resource availability.

Table 7.11. Fuzzy project makespans and schedule risks under distinct recourse availability

№	Additional recources	Resource availability	Project makespan	Schedule risk
1	0	(5,4,5,4)	(180,239,239,294)	0.63
2	1 unit of r_1	(6,4,5,4)	(178,235,235,289)	0.60
3	1 unit of r_2	(5,5,5,4)	(180,239,239,294)	0.63
4	1 unit of r_3	(5,4,6,4)	(172,227,227,280)	0.49
5	1 unit of r_4	(5,4,5,5)	(175,232,232,284)	0.56
6	2 unit of r_1	(7,4,5,4)	(177,233,233,286)	0.58
7	2 unit of r_2	(5,6,5,4)	(180,239,239,294)	0.63
8	2 unit of r_3	(5,4,7,4)	(168,223,223,275)	0.43
9	2 unit of r_4	(5,4,5,6)	(174,229,229,280)	0.52

10	1 unit of r_1 and r_2	(6,5,5,4)	(178,235,235,289)	0.60
11	1 unit of r_2 and r_3	(5,5,6,4)	(172,227,227,280)	0.49
12	1 unit of r_3 and r_4	(5,4,6,5)	(178,235,235,289)	0.60
13	1 unit of r_1 and r_4	(6,4,5,5)	(163,214,214,264)	0.34

Changing the resource availability, an acceptable schedule risk P can be found. For example, as can be seen from Table 7.11, adding another hardware engineer could probably reduce the fuzzy project duration to (172,227,227,280) with the schedule risk 0.49.

7.9 CW Based Decision Analysis on Risk Assessment of an Engineering Project

We are considering a project of building a power plant [47]. It is necessary to choose the best site among alternatives $a_i, i=\overline{1,n}$ for building a power plant based on criteria $C_j, j=\overline{1,m}$. Let's assume for simplicity that $i=\overline{1,2}$ and $j=\overline{1,3}$. Here a_1 is the first alternative (site), a_2 is the second alternative (site), C_1 is the influence on the environment, C_2 is the change in landscape, and C_3 is the complexity of construction works [47]. Decision Maker has to choose optimal on $C_j, j=\overline{1,3}$ an alternative from the predefined two alternatives a_1 and a_2. The choice of either alternative is connected with certain risk, which needs to be calculated. This risk is related to the uncertainty of the future events which more adequately can be described by fuzzy measure. To solve the above presented decision making problem, we will use the CW technology described in section 3.7.

1. Description of the problem domain in natural language

To describe the problem domain we will use conditional fuzzy granules:

$$\text{IF } x = u_i \text{ THEN } y \text{ is } G_i \quad (7.64)$$

where X and Y are linguistic variables and G_i is a fuzzy subset of some set V.

The description of the subject domain can be presented as a collection of granules (7.64). Thus the information is conveyed by fuzzy conditional granules of the IF-THEN form. The IDS includes the following propositions for the criterion C_1 - influence on the environment:

1. *For the description of the distance to the closest inhabited area:*

IF (plant is pretty close to inhabited area) with probability $Pb_{1,1}$ THEN influence is medium

IF (plant is close to inhabited area) with probability $Pb_{1,2}$ THEN influence is moderately high

IF (plant is very close to inhabited area) with probability $Pb_{1,3}$ THEN influence is high

2. For the description of the forecasted air pollution:

IF (air pollution is medium) with probability $Pb_{2,1}$ THEN influence is moderately high

IF (air pollution is not high) with probability $Pb_{2,2}$ THEN influence is low

3. For the description of the forecasted water pollution:

IF (water pollution is very low) with probability $Pb_{3,1}$ THEN influence is low

IF (water pollution is low) with probability $Pb_{3,2}$ THEN influence is medium

IF (water pollution is medium) with probability $Pb_{3,3}$ THEN influence is high

Analogously we prepare collection of fuzzy granules for the criteria C_2 and C_3.

2. Formation of a question

After the formation of a collection of fuzzy granules one can ask question, the answer to which can be determined from the IDS. Here the question is formed as: "What is the probability of (Y is Q)?", where Q is a fuzzy subset of V. The calculated probability is an interval value $[Pb_{min}, Pb_{max}]$, which is determined from the expected possibility (upper boundary) and the expected certainty (lower boundary) [47]. These values can be determined by the method suggested in [54].

3. Calculation of interval probabilities

It is possible to calculate what are probabilities of values that different criteria can take. For example, we can calculate the probability that the project corresponding to the first alternative will be of high construction complexity.

Let's assume that the results of reasoning have the form shown in Tables 7.12 and 7.13.

Table 7.12. Probabilities of different criteria values for the first alternative

Criterion \ Value	Low	Medium	High
Influence on the environment	[0.11; 0.43]	[0.34; 0.57]	[0.07; 0.24]
Changes in landscape	[0.15; 0.31]	[0.4; 0.74]	[0.1; 0.12]
Complexity of construction	[0.03; 0.18]	[0.52; 0.82]	[0.31; 0.93]

Table 7.13. Probabilities of different criteria values for the second alternative

Criterion \ Value	Low	Medium	High
Influence on the environment	[0.03; 0.12]	[0.13; 0.46]	[0.26; 0.78]
Changes in landscape	[0.1; 0.21]	[0.36; 0.82]	[0.43; 0.93]
Complexity of construction	[0.12; 0.36]	[0.34; 0.87]	[0.31; 0.52]

The main problem facing the DM is the risk evaluation. The risk assessment on the basis of calculated interval probabilities in [47] is suggested to do using generalized definition of entropy. That is why in [47] the entropy is generalized to the case of interval-valued probabilities. The procedure of determining the risk value in [47] can be summarized as follows. A set of systems is built that corresponds to each of the alternatives and entropy for each system is evaluated. Then the obtained values for separate systems are summed to get the overall entropy value. The latter can be considered as the evaluation of risk in terms of lacking information about an alternative.

It is obvious that in the considered problem we have three systems, because the number of criteria is 3. Assuming that every system has three possible values, namely, Low, Medium, and High, on the basis of data given in Tables 7.12 and 7.13, we can calculate entropies of systems that correspond to C_j, $j=\overline{1,3}$ (Table 7.14).

Table 7.14. Entropies of systems corresponding to different criteria

Alternative \ Criterion	Influence on the environment	Changes in landscape	Complexity of construction
The first alternative	[1.016; 2.162]	[0.835; 1.814]	[0.516; 2.21]
The second alternative	[0.746; 2.258]	[0.589; 2.212]	[0.867; 2.377]

Hence, the overall evaluation of the entropy for alternatives will be: for a_1 - [2.367; 6.186], for a_2 - [2.202; 6.847]. Because the second interval contains the first one, we conclude that the alternative a_2 is more uncertain than a_1, and thus more risky.

References

1. Aliev RA, Mamedova GA, Aliev RR (1993) Fuzzy Sets Theory and its Application. Tabriz: Tabriz University Press
2. Aliev R, Aliev F, Babaev M (1991) Fuzzy Process Control and Knowledge Engineering. Koln Verlag TUV Rheinland

3. Aliev RA (1994) Fuzzy Expert Systems. In Aminzadeh F and Jamshidi M (eds) Soft Computng: Fuzzy Logic, Neural Networks and Distributed Artificial Intelligence. NJ: PTR Prentice Hall
4. Aliev RA (1989) Fuzzy intelligent production systems. In: Bezdek JC(ed.) Thirth IFSA Congress. The coming of age of Fuzzy Logic. Seattle, Washington. pp 311-314
5. Aliev RA, Shakhnazarov MM, Abdikeev NM (1990) Intelligent control systems. Moscow: Radio I svyaz. (in Russian)
6. Aliev RA, Shachnazarov MM, Gulko DE (1988) Scheduling expert systems. (in English and Russian). J News of Academy of Sciences of USSR. Tech. Cybernetics 5: 118-128
7. Aliev RA, Guirimov BG (1997) Handwritten Image Recognition by Using Neural and Fuzzy Approaches. J Intelligent Control and Decision Making Systems. Baku. Publishing House of Azerb.SOA. pp.3-7
8. Aliev RA, Aliev RR (2001) Soft Computing and its Application. World Scientific. New Jersey. London. Singapore. Hong Kong
9. Allahverdiyev AA (2001) Fuzzy Decision Making for selection of vehicles. İn: First International Conference on Soft Computing and Computing with Words in System Analysis, Decision and Control. Turkey
10. Albrecht M (1992) Approximation of functional relationships to fuzzy observations. J Fuzzy Sets and Systems 49(3): 301-305
11. Bardossy A (1990) Note on Fuzzy Regression. J Fuzzy Sets and Systems 37: 338-353
12. Bandemer H, Nather W (1992) Fuzzy Data Analysis. Dordrecht: Kluwer Academic Publishers
13. Buckley JJ, Hayashi Y (1994) Fuzzy Neural Networks In: Yager R and Zadeh L(eds) J Fuzzy Sets, Neural Networks, and Soft Computing. 233-249
14. Buckley JJ, Hayashi Y (1998) Applications of Fuzzy Chaos to fuzzy simulation. Fuzzy Sets and Systems 99: 151-157
15. Bezdek JC, Pal SK (ed.) (1992) Fuzzy Models for Pattern Recognition, New York. IEEE Press
16. Bonfig KW, Guirimov BG (1996) Neuro-Fuzzy Based Generation of Fuzzy Knowledge and the Flexible Procedure of Logicai Inference for the Graphic Pattern Recognition. In: ICAFS'96. Siegen. Germany. pp 314-318
17. Diamond Ph (1988) Fuzzy Least Squares. J Information Sciences 46: 141-157
18. Dubois D, Prade H (1980) Fuzzy Sets and Systems. Theory and Applications. NY: Academic Press
19. George Bojadziev, Maria Bojadziev (1997) Fuzzy Logic for business, finance, and management. World Scientific
20. Hapke M, Slowinski R (1996) Fuzzy priority heuristics for project scheduling, J Fuzzy Sets and Systems 83 (3): 291-299
21. Ishlbuchi H, Tanaka H. (1992) Fuzzy regression analysis using neural networks. J Fuzzy Sets and Systems 50(3): 257-265
22. Juite Wang (2002) A fuzzy project scheduling approach to minimize schedule for product development. J Fuzzy Sets and Systems 127: 99-116
23. Krause B, Pozybill M (1996) Fuzzy Logik in der Verkehrsbeeinflussung. Straβenverkehrstechnik. vol 8. pp 356-362
24. Kacprzyk J, Fedrizzi M(eds) (1992) Fuzzy regression Analysis. Omnitech Press
25. Kacprzyk J (1983) Multistage Decision-Making under Fuzziness, Verlag TÜV Rheinland Köln, Germany

26. Kim B, Bishu RR (1998) Evaluation of Fuzyy Linear Regression Models by Comparing Membership Functions. J Fuzzy Sets and Systems 100: 343-352.
27. Klir G, Yuan B (1995) Fuzzy Sets and Fuzzy Logic.Theory and Applications. NJ: PTR Prentice Hall
28. Kuo RJ (1998) A decision support system for the stock market through integration of fuzzy neural networks and fuzzy Delphi. J Applied Artificial Intelligence. 12:501-520
29. Liu B, Esogbue AO (1999) Decision Criteria and Optimal Inventory Processes. Kluwer Academic Publishers. Norwell. Mass
30. Lee HM, Yao JS (1999) Economic Order Quantity in Fuzzy Sense for Inventory without Backorder Model, J Fuzzy Sets and Systems 105: 13-31
31. Loostma A (1989) Stochastic and fuzzy PERT. European J Oper. Res 43: 174-183
32. Morton TE, Pentico DW (1993) Heuristic Scheduling Systems. Wiley. New York
33. Ozdamar L, Ulusoy G (1995) A survey on the resource-constrained project scheduling problem. J IIE Trans 27 (5) :574
34. Park D, Kandel A, Laugholz G (1994) Genetic-based new fuzzy reasoning models with application to Fuzzy Control. J IEEE Trans. Systems, Man, and Cybernetics: 39-46
35. Pedrycz W (ed) (1996) Fuzzy modeling: paradigms and practice. Dordrecht:Kluwer Academic Publishers
36. Pedrycz W(ed) (1997) Fuzzy Evolutionary Computation. Dordrecht. Kluwer Academic Publishers
37. Ruoning XA (1991) Linear regression model in fuzzy environment. J Adv. Modeling Simulation 27: 31-40
38. Ruoning X (1997) S-Curve regression model in fuzzy environment. J Fuzzy Sets and Systems 90: 317-326
39. Richard BCh, Nicholas JA, F Robert Jacobs (1998) Production and Operations Management. Irwin/McGraw-Hill
40. Shortliffe E.H (1976) Computer-based medical consultations. MYCIN. American Elsevier. NY
41. Savic DA, Pedrycz W (1991) Evaluation of Fuzzy Linear Regression Models. J Fuzzy Sets and Systems 39: 51-63
42. Sommer G (1981) Fuzzy Inventory Scheduling in Applied Systems. In: GE Lasker (ed.). Applied Systems and Cybernetics. vol. VI. Pergamon Press. NY. pp 3052-3062.
43. Tanaka H, Uejima S, Asai K (1982) Linear regression analysis with fuzzy modeling. J IEEE Trans. Systems, Man, and Cybernetics. vol. SMC-12(6): 903-907
44. Tanaka H, Hayashi I, Watada J (1989) Possibilistic Linear regression Analysis for fuzzy Data. J European J. of Operations Research 40: 389-396
45. Tarko A , J Tsai, NM Rouphail (1993) Detection of Traffic Congestion Using Fuzzy Operator Logic in ADVANCE-like System. In: Proc. of the Intelligent Vehicles Symposium. pp 189-194
46. Thomas Feuring, James J Buckley (2000) Fuzzy Inventory Control. An Evolutionary Algorithm Solution. In:Fourth International Conference on Application of Fuzzy Systems and Soft Computing. ICAFS-2000. Siegen. Germany
47. Valishevsky A, Borisov A(2003) Fuzzy-Granule-Based Information Theoretic Approach towards Risk Analysis. In: Proc. of Second International Conference on Soft Computing, Computing with Words and Perceptions in System Analysis – ICSCCW-2003. Antalya. Turkey. pp. 275-281

48. Wang L, Zhang L, Hob H, Sekl H (1994) A fuzzy regression method based on genetic algorithm. In: Third Inter. Conf. on Fuzzy Logic. Neural Nets and Soft Computing. pp 471-472
49. Yabuchi Y, Watada J, Tatsumi K (1994) Fuzzy regression analysis of data with error. Japanese J. Fuzzy Theory and Systems 6(6): 673-685
50. Yao JS, Lee HM (1999) Fuzzy Inventory with or without Backorder Quantity with Trapezoidal Fuzzy Numbers. J Fuzzy Sets and Systems 105: 311-337
51. Zadeh LA (1975) The Concept of a Linguistic Variable and its Application to Approximate Reasoning. J Inform. Sci. 8: 199-249; 301-357
52. Zadeh LA (1965) Fuzzy Sets. J Information and Control 8: pp.338-353
53. Zimmerman HJ (2001)Approximate Reasoning in Logistics. J IEEE:2763-2766
54. Zadeh LA(1979) Fuzzy Sets and Information Granularity, Advances in Fuzzy Set Theory and Applications. In: MM Gupta, R.K. Ragade et al. (Eds.). North-Holland Pub lishing Company. pp. 3-18

8 Soft Computing in Finance

8.1 Soft Computing Based Stock Market Predicting System

The stock market is very attractive due to high expected profit. On the other hand it is very risky. This creates a need for intelligent stock-trading systems that are intended to help the investors make realistic prediction for taking optimal decisions. Conventional approaches address Regression and Time Series Analysis methods for stock market prediction [5,14]. These methods do not give expected results in situations when the data are influenced by subjective factors such as psychological, macro-economical, or political issues. Also we cannot ignore those factors at all, because technical indexes only are not capable of proper description of a complicated real-world environment. An effective stock trading system must use both qualitative and quantitative factors.

In [17] an intelligent stock-trading decision support system based on both quantitative and qualitative factors to assist investors in making the right decision is considered. Adopted from [17] below we describe the structure of an intelligent stock trading system on the base of Soft Computing technology (Fig. 8.1). The considered system is a multi-module system and includes 3 models. The first model is a quantitative model for predicting stock tendency, which develops the relationship between the independent variables $x_1, x_2, ..., x_k$, where k is the number of technical indexes, and the independent variable y_1, the stock market tendency, generated by quantitative data. It should be noted that this model provides only the general pattern of the stock market. The second model, a judgmental model, makes a relationship between judgmental information $x_{k+1}, x_{k+2}, ..., x_n$, say, political, financial, economic etc. inputs and dependent stock tendency outputs $y_2, y_3, ..., y_l$ generated by non-quantitative inputs. Note that the influence of these inputs continues during the lifetime. Therefore it is necessary to consider also a sequence of $(n-k)$ data for the time effect. Here n is the total number of inputs, $(n-k)$ is the number of judgmental inputs, l is the total number of outputs (stock tendency parameters), and $l-1$ is the number of outputs of the judgmental model.

Hence a total of $(2l-1)$ inputs should enter the third integration model to compute the resulting (total) stock market tendency y_{total} generated by both quantitative and non-quantitative data. Thus the information from the feed-forward neural network (quantitative model) and fuzzy neural network (or Fuzzy Delphi) are integrated with the time effect by using the Integration Fuzzy Neural Network to provide the final decision.

Below we describe the operational principles of the separate modules of the considered intelligent stock market system (Fig. 8.1.).

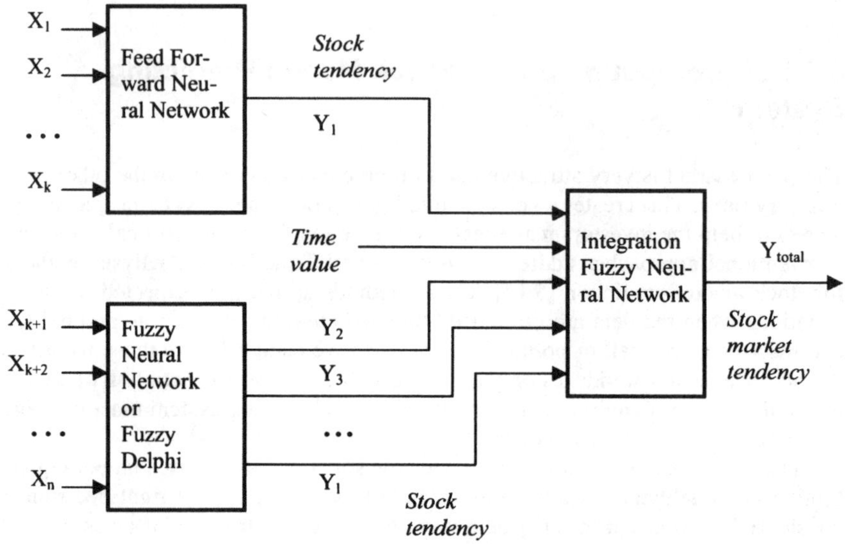

Fig. 8.1. The structure of the intelligent stock trading system

The feed-forward neural network (quantitative model)

The conventional approach usually uses a multiple regression model to set a relationship between the input factors and the stock index. However, the assumptions essential to the regression analysis as well as the constraints and complexity of the model determination have limited the application of these models. Similar problems appear when using the time series analysis technique ARMA [14].

Artificial Neural Networks (ANNs) have been used in stock market trading systems. ANNs are very attractive in creating intelligent systems due to such advantages as computational speed, learning by examples, and generalization ability. The advantage of neural models over conventional statistical models for predicting is obvious, which is well argued in [1,18,26]. In this case it is purposeful to use a feed-forward neural network to implement the qualitative model. The reason is a wide range of available learning algorithms for this type of neural network, and also that these networks are able to predict a continuous output. The structure of the neural network is defined under the following requirements. The network consists of multi-layer feed-forward structure with input, hidden, and output layers. The number of neuron in the input layer is determined by the number of independent variables, i.e. quantitative factors. The number of hidden layers and neurons is chosen based on the required accuracy of forecasting stock tendency. A frequently used method here is the trial and error method. However the best results can be obtained using genetic algorithms for the optimization of the structure of neural network. The output layer includes one neuron that provides the performed stock tendency.

Qualitative model

The input ($x_{k+1}, x_{k+2}, ..., x_n$) and output ($y_2, y_3, ..., y_l$) variables of this model are linguistic variables. As it was shown in section 4.3 good results in this case can be obtained using Fuzzy Delphi Method. In [17] the modified Fuzzy Delphi Method is used to construct a qualitative model for a DSS for the stock market. Interestingly enough the author of this paper states that this approach can be successfully used when the input judgmental information sources (all dimensions) are independent. However, in real life these dimensions are usually correlated. Hence, it would be more purposeful to use neural networks instead of Fuzzy Delphi. More advantageous is to use a fuzzy neural network type 3 (FNN_3) (see section 3.1 of the book for reference) in this case with a genetic algorithm based learning. In this case every dimension is represented as a linguistic variable ($x_{k+1}, x_{k+2}, ..., x_n$) and there are $l-1$ linguistic variables in the output of FNN representing the stock tendency.

Integration Fuzzy Neural Network

This FNN can be a feed-forward fuzzy neural network with input neurons, the number of which is determined as a total quantity of outputs from quantitative and qualitative models plus $(l-1)$ inputs for considering the time effect. The output layer of this FNN includes one fuzzy neuron, which represents the total stock market. In [17] the problem of constructing an intelligent trading system through an integration of fuzzy neural networks and Fuzzy Delphi on an example of Taiwan stock market is considered in detail. Both quantitative and qualitative factors are collected from various sources: company records, journals, newspapers, etc.

With the help of experts, a set of relevant factors were extracted including volume, average volume for 10 days, index in open, index in close, index fluctuation, financing buying and others. To account for time dynamics, previous (one day before) values of the factors, technical index are also included. One output was the forecast index which is used as the "buy-sell" signal. The computational results showed that neural network predicting (with factor input) is lower (MSE= 0.13477) than time series predicting (MSE=0.18706) of stock tendency. The authors, presuming that MSE is not suitable to adequately express the result of stock predicting, use the clarity of the buy-sell points and buy-sell performance. The value of the buy-sell signal allows investor who owns the stock to make decision:

IF signal is below the lower bound, THEN sell the stock
IF signal is over the upper bound, THEN keep the stock
IF signal is between the lower and upper bounds, THEN sell the stock

Similar rules can be derived for investor who does not own the stock. A clarity optimized neural network is chosen which has number of buying signals – 89, number of selling signals – 95, total number of days - 290, the clarity rate of buying points – 0.307, the clarity rate of selling points – 0.328, the clarity rate of selling-buying points – 0.634.

The buy-sell performance is determined on the basis of the above decision rules as the difference between the index value when the stock was bought and the index value when the stock was sold. The chosen neural network model has 13 transactions and performance equal to 3740.9 points. As a qualitative model in [17] the modified Fuzzy Delphi Model is used. This model uses 6 linguistic variables as inputs: political, financial, economic, message, technical, international. On the basis of this information obtained from newspapers, economic and business-related journals, and government reports, a first version questionnaire is compiled and handed to experts for ranking. After three iterations of the survey, the membership functions of the above mentioned linguistic variables converge with 90 percent confidence. The knowledge obtained from the fourth questionnaire is integrated with the results obtained from the quantitative model (neural network). The integrated model in [17] includes as its input one signal from the quantitative model and six linguistic data from the qualitative model. Because the influence of either of these dimensions is related to its longevity, the time levels are also included for each linguistic variable. The integration neural network has only one output (one node of FNN) – the total stock tendency.

MSE of integration FNN is 0.1175, which is significantly lower than MSE of the neural network (0.1810) for predicting only on the basis of technical index.

The net-performance of buying and selling for intelligent system, integrating the qualitative and quantitative models is equal to 3573.5 while the same index for the system with only quantitative model (without the consideration of judgmental information) is 2286.1. Thus it is obvious that the integration of quantitative and qualitative approaches really improves the performance.

8.2 Fuzzy Nonlinear Programming Approach to Portfolio Selection

Financial analysts in banks and insurance companies have devoted considerable attention to mathematical models that assist in managing portfolios of common stocks, and other securities. Inherent in such models is an assessment of a proposed portfolio's expected gain and the associated risk, or probable variation, in the actual gain [28].

The considered portfolio selection model is a fuzzy generalization of the Markowitz models [19,20] which laid the basis of Modern Portfolio Theory [22]. In this section we will show how rational investors can construct near optimal portfolios under condition of fuzzy uncertainty. The objective is to minimize the risk of the portfolio for a given level of return. Let's consider a formal statement of the problem of near optimal portfolio construction.

Let $\tilde{x}_j$ represent the fraction of the portfolio value to be allocated to asset j, and $\tilde{\alpha}_j$ associated expected gain at the end of the planning horizon. Let $\tilde{b}$ denote an acceptable lowest expected gain per dollar invested in the entire portfolio.

The account of the risk measure in the model is reflected by an objective function $\tilde{z}$ which is calculated as:

$$\tilde{z} = \sum_{i=1}^{N}\sum_{j=1}^{N} \tilde{\sigma}_{ij}\tilde{x}_i\tilde{x}_j, \quad i,j=\overline{1,N} \tag{8.1}$$

where $\tilde{\sigma}_{ij}$ is the fuzzy coefficients of the $N \cdot N$ matrix defined for asset i and asset j to be determined on the basis of $[\tilde{a}_i - \tilde{\alpha}_i]\cdot[\tilde{a}_j - \tilde{\alpha}_j]$.

The constraints of the model are:

$$\sum_{j=1}^{N} \tilde{\alpha}_j x_j > b \tag{8.2}$$

$$\sum_{j=1}^{N} \tilde{x}_j = \tilde{1} \text{ and } \tilde{x}_j \geq \tilde{0}, \quad j=\overline{1,N} \tag{8.3}$$

Taking into account (8.2)-(8.3), the problem of fuzzy optimization of portfolio selection can be reformulated as:

$$\min \tilde{z} = \sum_{i=1}^{N}\sum_{j=1}^{N} \tilde{\sigma}_{ij}\tilde{x}_i\tilde{x}_j \tag{8.4}$$

Subject to

$$\sum_{j=1}^{N} \tilde{\alpha}_j \tilde{x}_j \geq \tilde{b} \tag{8.5}$$

$$\sum_{j=1}^{N} \tilde{x}_j = \tilde{1} \tag{8.6}$$

$$\tilde{x}_j \geq \tilde{0}, \quad j=\overline{1,N} \tag{8.7}$$

The problem (8.4)-(8.7) is a fuzzy nonlinear programming problem. Varying the desired level of return $\tilde{\alpha}$ and repeatedly solving the fuzzy nonlinear programming problem (8.4)-(8.7), we can obtain the minimum risk portfolio for each value of $\tilde{\alpha}$.

In more sophisticated versions of the model, there can be additional constraints on the composition of the portfolio, several time periods, and measures of risk other than (8.1).

Let's now consider the procedure for solving the problem (8.4)-(8.7).

First we will consider two simplified modifications of the problem (8.4)-(8.7) and later suggest a fully fuzzified problem (8.4)-(8.7) [2].

Optimization over fuzzy constraints

The problem of optimization over fuzzy constraints is formulated as follows: find

$$\min\ z(x) \tag{8.8}$$

Subject to

$$\tilde{\varphi}_i(x) \subseteq \tilde{B}\ ,\quad i=\overline{1,\ m};\qquad x\in X, \tag{8.9}$$

where $X=(x_1,...,x_n)$, $X\subseteq R^n$, $f:R^n \to R$.

The method used for solving this problem is based on substitution of fuzzy sets by α – level fuzzy sets. Then for solving (8.8)-(8.9) it is sufficient to solve the problem

$$z(x) \to \min; \tag{8.10}$$

subject to

$$\varphi_i^\alpha(x) \le B_i^\alpha,\quad x\in X, \tag{8.11}$$

which is the classical non-linear mathematical programming problem in each α – level. So, the initial problem (8.10)-(8.11) is reduced to solve infinitely sets of classical optimization problem.

Let x_α be solution of the problem (8.10)-(8.11). Then decision of the problem (8.10)-(8.11) is defined as

$$x^*(x_\alpha)=\sup\{\alpha/z(x_\alpha)/=\min_{x_\alpha} z(x)\}$$

Optimization with fuzzy goal

The optimization with fuzzy goal is formulated as follows:

$$\tilde{z}(x) \to \min \tag{8.12}$$

subject to

$$\varphi(x)=0;\ \ x\in X \tag{8.13}$$

It is known that the goal function is an element of the set

$$Z(R^n)=\{z(x)/z(x):R^n\to R\}.$$

Fuzzy goal in the optimization problem is a usual fuzzy goal and it is a fuzzy set on $F(R^n)$ with membership functions

$$\propto_{\tilde{z}}(z):Z(R^n)\to[0,1].$$

For solving the problem (8.12)-(8.13) α – level set of the fuzzy goal function is used. Then each fuzzy set is a fuzzy preference relation between elements of admissible domain and, therefore, the solution of fuzzy non-linear programming problem is reduced to solving of multi-criteria decision making problem (the number of the criteria is equal to the numbers of levels α, $\alpha \in [0,1]$). This means that for the solving the problem (8.12)-(8.13) it is sufficient to solve

$$z^{\alpha}(x) \to \min$$

subject to

$$\varphi(x) = 0;\ x \in X,$$

where $\alpha \in [0,1]$, $\propto_{\tilde{z}}(z) \geq 1-\alpha$, which is a multi-criteria optimization problem.

Optimization over fuzzy goal and fuzzy constraints

Let $\propto_{\tilde{z}} \in [Z(R^n)]$ be a fuzzy goal and $\propto_{\tilde{\varphi}} \in Z(R^n)$ be fuzzy constraints:

$$\tilde{z}(x) \to \min \tag{8.14}$$

subject to

$$\tilde{\varphi}(x) \subset \tilde{B};\quad x \in X \tag{8.15}$$

For solving (8.14)-(8.15) we shall combine the two mentioned above cases, i.e. the optimization problem with fuzzy goal and non-fuzzy constraints and the optimization problem with non-fuzzy goal and fuzzy constraints and we get the following model:

$$z^{\alpha_i}(x) \to \min,\quad i = \overline{1.k},$$

$$\varphi^{\alpha_i}(x) \leq B^{\alpha_i},$$

$$x \in X,$$

where $\alpha_i \in [0,1]$,

$$\tilde{z} = \sum_{i=1}^{k} \propto_i / z^{\alpha_i},\quad \tilde{\varphi} = \sum_{i=1}^{k} \propto_i / \varphi^{\alpha_i},\quad \propto_i \geq 1-\alpha$$

8.3 Neuro-Fuzzy Approach to Modeling of Credit Risk in Trading Portfolios

Managing portfolio risk requires clear policy and objectives. Several methods (mainly Credit Metrics and Credit Risk $^{+}$) have been in operation for a number of years for market risk in trading portfolios. These methods are frequently based on deterministic, stochastic, and probabilistic models and have yielded limited success. As it is shown in [10] "two of the models, J.P. Morgan's Credit Metrics and Credit Suisse Financial Product's Credit Risk $^{+}$, have been released freely to the public since 1997 and have quickly become influential benchmarks". Models laid on the basis of these methods are statistical by nature and are mathematically similar. They differ in distributional assumptions, reliance on approximation formulae and functional forms.

As it is shown in [10] both models are highly sensitive to the probability distribution shape and volatility of default probabilities. However, in practice it is very difficult to determine expected default probabilities and their volatility. Capital decisions depend strongly on tail percentile values of the portfolio loss distribution, which are directly defined by higher moments of the distribution of default probabilities. Practitioners have difficulty with estimating these higher moments using available data. In these methods the set of outcomes is usually restricted to two states: default and non-default. However, within these extreme states there are other intermediate states estimated linguistically.

The latent variables in these methods are taken to be linear functions of risk factors and the idiosyncratic effects. This assumption is ill-argued.

Credit Metrics method uses Monte Carlo technique which is time consuming. Consequently, banks can not use these methods successfully in real-time applications.

Finally, in application of these statistical methods it is necessary to analyze given data to check if they satisfy general assumptions of probability theory.

The main weakness of statistical approaches to modeling market risk in trading portfolios is that they do not integrate judgmental and exogenous factors into models.

We need new tools, techniques and technologies to build systems to improve the judgment of practitioners and policy makers and optimize portfolio management.

Constructions of quality portfolios requires application of emerging technology tools, such as fuzzy logic, neural networks and other advanced technologies for trading in financial markets more intelligently.

There exist other methods (mainly based on classical regression analysis) for solving the above problems. In [6] a regression analysis procedure is used for constructing the relationship between risk factors (variables) and country's ability to service its debt. The main weakness of classical regression based risk predicting methods is that all known four assumptions of regression analysis need to be satisfied. These assumptions are very strict in practice.

General procedure for constructing of neuro-fuzzy paradigm based portfolios is described below.

1. True "probability distribution" for possible future outcomes is constructed (substituted) by convoluting together the expert knowledge and empirical knowledge. Representation of knowledge is based on fuzzy logic.
2. Choice of "risk factors", including political, economical, social and other factors, features obligors(linguistically described profile of obligors), credit quality, etc.
3. Construction of fuzzy mathematical models of the interaction between inputs (risk factors) and desired trading strategies(obligor defaults in the portfolio).It should be mentioned, that fuzzy logic, which is based on fuzzy measure(not on probability measure) is free of assumptions of probability theory; it is pos-

sible to create mathematical model of value-at-risk for market risk in trading portfolios by using appropriate number of fuzzy IF. ..THEN rules and shapes of membership functions.

4. Calculate obligor defaults in the portfolio given current data, using fuzzy IF... THEN model and fuzzy reasoning methods.
5. Integrate neural networks to implementation of fuzzy model, i.e. to use neuro-fuzzy technology in order to adapt portfolio constructing fuzzy model to quickly changing banking environment. Fuzzy inference and other fuzzy computing procedures are implemented by neuro-computing.

Use of neural networks in cooperation with fuzzy logic in portfolio construction gives us a chance to combine expert know - how, which is fuzzy, with finding from empirical time series, which are patterns that can be detected by neural networks.

Portfolio construction on base of neuro-fuzzy technology allows banks to use proposed methodology in real-time applications.

IF – THEN model of VAR

Assume that the following variables are used as risk factors:

1. Per capita income (PCI) (x_1);
2. Gross domestic product (GDP)growth (x_2);
3. Inflation (I) (x_3);
4. Fiscal balance (KB) (x_4);
5. External balance (EB) (x_5);
6. External debt (ED) (x_6);
7. Economic development (EcD) (x_7);
8. Default degree (DD) (x_8);
9. Credit quality (CQ) (x_9);

Relationship between these input variables and VAR is required. The Table 8.1. describes the input and output variables of the model, their units of measurement, ranges, and sources.

Table 8.1. Representation of input and output variables

Variable Name	Unit of Measurement	Range	Labels	Sources
VAR				
GDP				
.				
.				
.				

Relationship between input variables and VAR are represented by fuzzy model specified by necessary number of If-Then fuzzy rules. The following are the examples of fuzzy rules in knowledge base:

IF PCI is HIGH AND GDP is MEDIUM AND I is LOW AND ...

THEN VAR is MEDIUM

.

.

.

IF PCI is LOW AND GDP is HIGH AND I is LOW AND ...
THEN VAR is HIGH

Here, PCI, GDP, I, FB, EB, ED, EcD, DD, CQ and VAR are fuzzy variables, and HIGH,..., LOW are their linguistic values. For example, linguistic labels of a variable I is shown in Fig. 8.2.

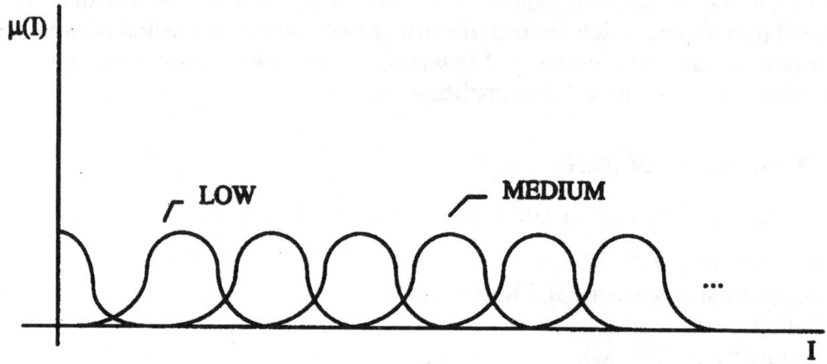

Fig. 8.2. Linguistic labels of a variable I

Given fuzzy (or non-fuzzy) input information, i.e. current values of PCI, ..., CQ, the current fuzzy value of VAR is calculated using fuzzy inference rules. Using the defuzzification procedure the crisp (exact) value of VAR is obtained.

Fuzzification and defuzzification procedures, fuzzy inference mechanism, and rule generation in fuzzy model representing relation between input variables and VAR are implemented on the basis of neural networks. Neural network part of the proposed Neuro-Fuzzy VAR predicting system has 9 input neurons (in accordance with the number of risk variables), which receive fuzzy or non-fuzzy information in on-line mode. It allows:

1. To adapt fuzzy model to changing banking environment by adjusting the shapes of the membership functions of the labels and the number of fuzzy rules in the knowledge base;
2. To calculate VAR in real-time mode.

The accuracy and effectiveness of the Neuro-Fuzzy system for predicting VAR were compared with the statistical methods using real banking environment. The results have shown the advantage of the Neuro-Fuzzy approach over the others.

8.4 A Fuzzy Approach to Constructing Credit Portfolio

The classic problem to optimize a portfolio has essential features in the case of credit portfolio. Credit portfolios are less dynamic as compared, for example, with securities portfolio on account of bank that is unable to change its portfolio profile immediately in any moment and have to evaluate it because of returns. Inflexibility in managing portfolio obliges an accuracy and an overall effectiveness of constructing credit portfolio.

Mainly, constructing portfolio is based on the YIELD-RISK curve (Fig.8.3) [24]. Optimizing a portfolio calls for determining the value of effective portfolios (Pareto-area).

Naturally, among the same quality portfolios (indistinguishable portfolios from quality point of view) the more effective will be the portfolio with greatest yield. Just because of this, real portfolios never consist only of high quality obligors' loans.

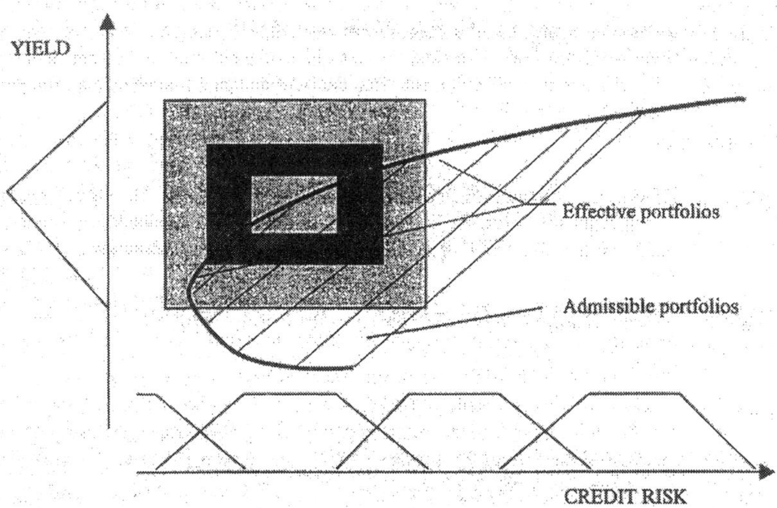

Fig. 8.3. An admissible portfolios area.

In [10] the range of credit quality allows representation by four credit quality distributions - "High", "Average", "Low" and "Very Low". The first three grades are created using data from internal Federal Reserve Board surveys of large banks. The "Average" is average distribution across the surveyed banks of total outstanding in each S&P grade. Corresponding, "High" and "Low" distributions are drawn from the higher and lower quality distributions among the banks in the sample. The "Very Low" distribution is not found in Federal Reserve sample, but is intended to represent a very weak large bank loan portfolio during a recession. Speculative grades (BB and below) loans account for half of outstanding in the "Average" portfolio.

Thus, a portfolio-constructing problem calls for the determination of mean and membership function of portfolio percentage of grades $\propto_{ni}$ by given means and membership functions of $\propto_{yi}$ yields and grades $\propto_{qi}$ qualities [21].

This is a multicriteria decision-making problem based on the following procedure [16].

$$R = \begin{array}{c} \\ T_1 \\ T_2 \end{array} \begin{array}{c} \begin{array}{cccccc} n_1 & n_2 & \dots & n_i & \dots \end{array} \\ \begin{bmatrix} r_{11} & r_{12} & \dots & r_{1i} & \dots \\ r_{21} & r_{22} & \dots & r_{2i} & \dots \end{bmatrix} \end{array} \tag{8.16}$$

We ignore distribution within grade as the credit risk is the same within a grade. Mainly, loan size and quantity within a grade depends on business served by bank, investing amount and so on. Of course, increasing the number of obligors leads to closeness of real result to analytical.

For restricted number of grades we have found intervals of existing membership functions for portfolios described above. These portfolio constructions are fuzzy decision-making solutions.

We have used trapezoid-shaped membership functions (Fig 8.4). Because of this, we just used four point of membership function (Table 8.2).

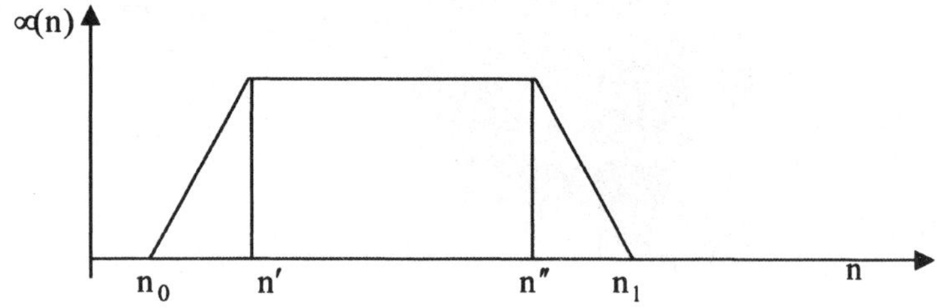

Fig. 8.4. Trapezoidal membership function

Thus, we have shown the possibility to solve a credit portfolio-constructing problem by fuzzy logic methodology.

Table 8.2. Points of membership functions

	High				Average				Low				Very low			
	n_0	n'	N''	n_1	n_0	n'	n''	n_1	n_0	n'	n''	n_1	n_0	n'	n''	n_1
AAA	2.5	3	4.5	5	1.5	1.9	3.2	4	0	0.6	1.8	2.5	0	0	1.2	1.5
AA	5	5.6	6.4	7	3	3.5	5.1	5.5	0.5	1.2	2.6	3.5	0	0	2.6	3
A	14	28	35	56	8	9.2	17.5	19	0	3.6	9.8	14	0	0	5.5	7
BBB	30	38	53	60	25	28.5	38.6	42	5	12	26.5	35	0	5.8	16.5	22
BB	0	8.5	15.6	25	5	9.2	19.5	32	25	26	39.8	41	30	38.6	45.6	52
B	0	2.5	8.9	12	2	8.6	19.8	27	20	23	39.7	42	35	42.5	43.5	52
CCC	0	0	1.6	2	1	3.8	4.3	6	4	5.8	8.6	10	6	9	11	14

Final portfolio choice depends, for sure, on bank's preferences. Usually, bank's inclination to risk is determined by finding an "ideal point" and creating lines of indifferent preferences.

In this section we don't touch on expert information analysis for estimating an inclination to risk. We just point out that portfolio-constructing problem is fuzzy by nature and not only because of bank's preferences for fuzziness.

Following the development of fuzzy modeling of an expert's information of bank's inclination to risk, fuzzy classification of obligor's grades, disposition to default, fuzzy estimating of awaiting yield allows creation of an adequate framework and software for accurate interactive constructing of credit portfolio.

8.5 Soft Computing Based TDSS Multi-Agent Systems in Finance

Billions of dollars are made and lost in global financial markets everyday. Consequently there is a strong demand for technology that can provide an "edge" in recognizing patterns and making trading decisions. Trading Decision Support Systems (TDSS) differ in terms of the "degree to which they make the decision". Some are query systems providing relevant information the user may choose to explore including real time data and simple data analysis. Others use descriptive and forecasting models and include "what-if' capabilities for generation and evaluation of alternatives. Still some others provide inference capabilities of intelligent systems. Recent developments in the area of soft computing have allowed additional enhancements to the above three categories of TDSS to improve search/retrieval, modeling, and inference capabilities. As it is shown above Neural network (NN) provide capabilities for building non-linear models and real time learning and update of models. Fuzzy Logic allows handling of fuzzy input and output variables as well as use fuzzy rules. Genetic algorithms (GA) increase the ability to optimize system parameters as well as perform specialized tasks as part of hybrid systems.

Generally most systems in use today are proprietary and very little has been published about their design and performance to prevent loss of competitive edge. However, dynamic and often volatile characteristics of markets indicate that no single system is ideal for all market conditions [13]. A contingency approach is necessary to fit the TDSS to the trading problem context. This paper proposes to build TDSS on the basis of Multi-Agent Distributed Intelligent Systems (MADIS) concepts (see 3.8) [4,8,23, 29]. This would allow a number of agents, each using different models, to simultaneously solve the problem and compete for the solutions. The following describes the stock trading problem, reviews the application of traditional and soft computing technologies for solving the trading problem, and proposes an architecture for trading decision support systems based on MADIS.

Although it is probably impossible to accurately predict the future price of a stock, forecasting systems are useful. The objective of investors and regulatory agencies is for stock markets to represent efficient markets where all information about the market and companies are equally available to the investing public and access to insider information is illegal. However, even under ideal market conditions with instantaneous availability of news of events that influence the stock price, the news does not have an instantaneous influence. It takes some time for the investors to agree on the consequence of the new information. Generally the new stock price does not immediately reflect the influence of new information which makes forecasting useful.

Stock analysts routinely predict the target price of stocks at some future point of time. They use this information to make buy-hold-sell recommendations. These predictions are based on the historical as well as current facts and predicted values of certain data items that are believed to influence the future value of a stock. The data items include both technical and fundamental indicators. Using TDSS should improve analyst predictions as manual methods and the human mind are not very good in handling and synthesizing large amount of information.

Predicting the future value of stocks is a very difficult task. There are usually a number of financial analysts who specialize in a given stock. However, in spite of having access to the same information, the analysts usually disagree and differ on their recommendations. Assuming access to the same information, differences in analyst performance in predicting the future value of stocks can be attributed to differences in know-how, skills, and TDSS used.

The prediction process involves weighing evidence obtained through analyses of vast amount of monetary, political and economic fundamentals, technical indicators, psychological factors, current events, etc. TDSS models provide the tool which helps predict the stock price more quickly and accurately. However, financial markets change and models must be updated frequently in terms of parameters as well as structure. Furthermore, there is no single model that one can rely on for both long term and short term predictions. The analyst must determine the appropriate forecasting horizon and select the right model. Long horizon may include consideration of many unpredictable events that may invalidate the forecast while short horizon may include much noise. Some evidence points to appropriate forecasting horizon in the range of several months. However, in turbulent markets one may only be able to predict the short term.

Stock market is an evolutionary, complex, non linear, dynamic system [9,12,30, 31]. Stock market is a complex system because many variables interact with each other in a non-linear way. Stock market is an evolutionary, not stationary, system where the underlying structure changes with time. Generally, the market has stable periods lasting of several months followed by a transition period before settling into another stable period. Thus, stock market is neither an ordered system explainable by simple rules nor a totally random system which is unpredictable. The rules for trading systems for trending markets are different than the rules for non-trending markets (low volatility). The rules are also different for each stable period. One must know when to switch from one set of rules to the other.

The following are major categories of Trading Decision Support Systems based on different modeling technologies:

- Rule-based systems.

Traditional rule-based systems use moving averages and other technical indicators and pattern descriptors to predict future prices of stock. Rule based system assumes that all factors are independent which may not be realistic. Rule based system may over simplify the behavior of the market. Where rules are common knowledge, the market will tend to adjust itself to such rules making forecasting less useful Rule-based systems do not have a good record of performance (less that 50% profitable trade with low profitability to risk ratio).

One could increase the number of rules by using multiple experts (in one example over 600 rules were assembled), however consistency and validity of such a rule base is suspect.

- Neural Network- based Systems.

In NN-based systems the financial analyst needs to specify what the neural networks needs to learn so that the network can develop the appropriate relationships between inputs and outputs. Inputs include capitalization, price earning ratio, price book value, expected earnings while the output is future price indicators. The forecast horizon is selected by the analyst.

The network learns the stock characteristics which in the past has led to under performance or over performance relative to the market. The network then predicts future price movements relative to market if such characteristics are present in the stock today. Neural Networks generalize and perform well against data the system has never seen before and are able to work in high dimensional spaces. However, some systems have shown that one can predict short term small movements and miss most of the large movements. It is necessary to retrain the network to adapt it to the changes in the market conditions by discarding the oldest data and including the newest data in the training set. Deer & Co. has developed and used a neural network-based system successfully, where the output is the predicted change in a stocks price (relative to the S&P average) for some future period.

- Fuzzy Rule based systems.

Fuzzy rule bases can be designed to include rules provided by traders or can be estimated from trading samples. Fuzzy rule based system avoids over reliance on quantitative data. The number of rules for fuzzy rule-based system is reduced to a very few compared to mechanical rule-based systems. A fuzzy rule-based system is more flexible than a mechanical rule-based system as it has fewer rules and can handle overlap and ambiguity between rules. Furthermore, fuzzy rule-base needs less frequent updates of the rule base than mechanical rule-base. This makes fuzzy rule-based systems easier to evaluate as it is difficult to evaluate the performance of a system that keeps changing over time.

In a fuzzy rule-based system the inputs such as trends and volatility measures are fuzzified, the association between inputs and outputs are expressed as fuzzy rules and fuzzy outputs are defuzzified into crisp recommendations. An advantage of fuzzifying a time series is the eliminations of noise.

It is appropriate to use fuzzy logic when several of the input variables are continuous and traditional rule-based systems are complex.

- Hybrid systems.

Neural networks based and fuzzy logic based systems are improvements over the traditional mechanical rule-based systems. However, they have strengths and weaknesses. Hybridization is a way to improve TDSS performance. The following are some hybridization schemes: NN plus GA:

GA is used to find a neural network that optimizes the performance objectives while observing constraints including risk limits. The objective may be average return, the constraints maximum fluctuations, and the adjustable parameters could be the neural network weights and the buy and sell signals.

The primary objective of combining GA with NN are discovering the new optimal topology of NN and finding the optimum weights. For NN, when the data are not too noisy, the optimal choice is the simpler models with few inputs. GA can help find the minimum number of inputs.

- NN plus Fuzzy logic.

It is difficult to include a priori information about financial markets into Neural Networks. Fuzzy logic can improve the performance of NN-based systems by incorporating knowledge of expert traders.

- Fuzzy Logic plus GA.

GA can be used to optimize the performance of fuzzy trading systems by minimizing complexity and optimizing the shape of membership functions.

The failure of TDSS stem from the changing support requirements for predicting a non-linear, complex, and dynamic market. A TDSS must be capable of:

- Adapting to market condition.

The system must be able to take new information into account and adapt to the new market condition by changing its own structure.

- Considering Political and Economic Events.

The system must be able to use qualitative information including judgment calls by the financial analyst.

- Matching to the task requirements.

The system must be appropriate for the task. The same system is unlikely to be useful for both short term and long term predictions.

The nature of the problem points at using MADIS where agents are built using different problem solving strategies, models, and technologies. Different agents use different models to make recommendations. Each agent solves the problem independently and makes recommendations. The financial analyst then uses his/her judgment to take the advice of one of the agents.

Multi-agent distributed intelligent systems with cooperation and competition among agents (MADIS) are composed of several agents working in parallel at any instance of time to solve a problem [4]. Only one of the competing candidate solutions is selected as the winning solution. Most decisions are repetitive and over time proposals of different agents are selected as the choice solution. In this sense the intelligent agents cooperate with each other in making a series of decisions.

The following describes the decision making process:

Intelligence:

Each Intelligent agent has access to all the available information (input data as well as objectives).

Design:

Each Intelligent Agent proposes solution(s) to the full problem (no partial solutions).

The Intelligent Agents work on the problem simultaneously and in parallel.

Choice:

The competing solutions of the intelligent agents form the candidate solutions to the problem. These alternative solutions are evaluated on the basis of the achievement of the stated objectives as well as the track record of the intelligent agents in solving similar problems in the contexts relevant to the problem. One of the solutions is selected as the winner solution.

The MADIS-based TDSS Architecture

Fig. 8.5. shows the architecture of the MADIS-based TDSS [4]. The components are described below:

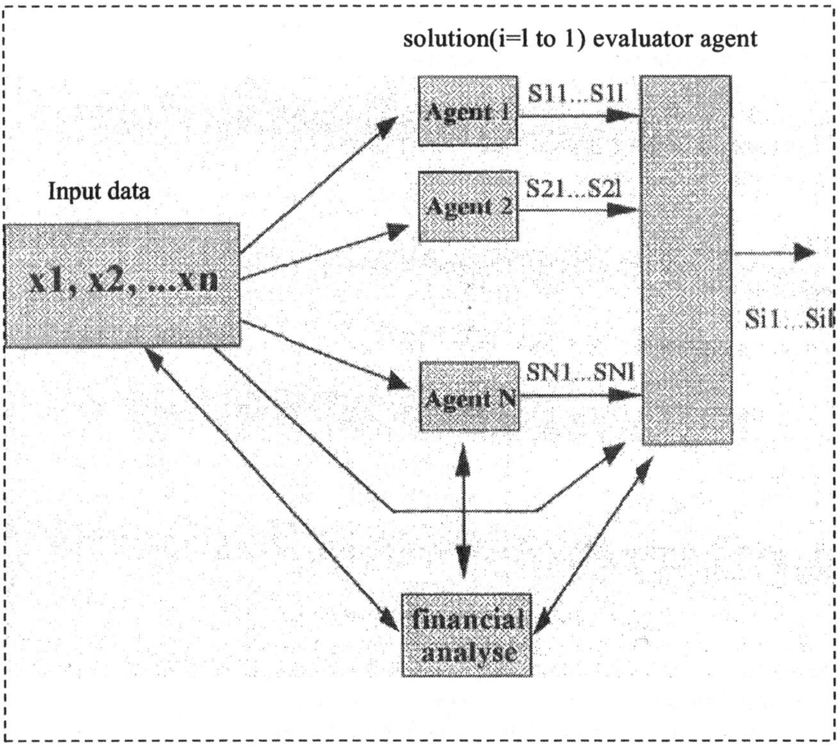

Fig. 8.5. Architecture of MADIS

Solution Agents

The solution agents are agents built using different models and technologies where each one generates an alternative solution.

Evaluator Agent
The evaluator agent selects the winning solution.

The three key issues relating to each agent are: measuring the on-going performance, determining when the model is reliable, and trusting the model. Evaluation criteria are profitability and consistence. Profitability can be interpreted as the difference between the predicted and actual- if we took the advice we would incur a gain or loss in a given time window. Consistence is the standard deviation of profit and losses- or sharp ratio.

There are certain conditions under which a model performs well and other conditions where it works poorly. The evaluation agent must be able to detect market changes and accurately identify the state change. There are also extraordinary market shaping forces, such as war, oil embargo, major scandals, and collapses of some securities markets that may make the prediction of some agents unreliable. Also there are periods when some agents show very poor performance. These periods arise unexpectedly, intermittently, and independently of networks. The Tokyo prediction system showed that the system fails in predicting large market movements resulting from significant political/ military conflicts and psychological factors. The evaluation agent with the aid of the decision maker must take these into consideration.

Knowledge of the characteristics attributed to the technologies used by agents helps the evaluator agent in selecting an agents solutions.

Mechanical rule-based systems use technical/statistical indicators such as moving averages, momentum stochastic, etc. to make buy/sell recommendations. Most rule-based systems are trend following systems. These systems do not have the ability to learn and adapt to new market conditions. This means these systems can fail when the market conditions change. This type of system will fail in no trending markets.

Neural nets are good at finding relationships in huge input/output space even if the patterns between inputs and outputs are ill defined. Neural Networks are good at recognizing patterns from noisy, complex data.

Market signals are the sum of trend and stochastic components. In choppy markets with an absence of trend, trend-following systems do poorly. When markets mature they start becoming choppy. Neural network models are more effective in choppy markets. Also non-linear dynamics methods work better in choppy (chaotic) markets.

Fuzzy logic does not have good learning capability. Fuzzy systems are useful for decision making where there is uncertainty and vagueness.

Another evaluation criterion is robustness dealing with the structure of the model. In TOKYO case, prediction rules change so frequently that networks which are trained can not be used over an extended period of time.

It is important for the evaluator agent to find the best prediction period for each agent. This can probably be best determined through approximation (trial & error).

In conclusion it is necessary to note that the stock market has identifiable patterns that can guide investors. Discovery of these patterns requires use of models on contingency basis. Systems based on machine learning techniques or financial models perform best when they match the context of the market. Success comes from knowing when to switch from one system to another. Here we presented an architecture that takes advantage of all relevant technologies and provides the means for switching from system to system.

8.6 Neural Nonlinear Modeling for Risk Management in Banking

In this section we will consider the market demand forecasting for the generation of business scenarios and will herein focus on one specific functional product, which has a high interest for many banks: the mortgage loan.

Predictions of mortgage market evolution are important not only for optimal asset and liability management but also in setting marketing strategies. There are many papers describing econometric models of mortgage markets. But it is still very difficult to formulate correct predictions for mortgage markets trends, as also for money markets in general. In this section non-linear modeling techniques will be used to fit the mortgage demand function, built in accordance with existing econometric works [27]. Most papers about mortgage markets use time series. Considerable number of econometric studies have been conducted in the post-war period.

Let's first consider the estimation of mortgage demand function.

In [27] the mortgage demand function is described by the following equation

$$\mathrm{Md} = \mathrm{g}(\mathrm{r}, \mathrm{P}^{\#}, \mathrm{DY}, \mathrm{P}, \mathrm{NOH}, \mathrm{NRH}) \qquad (8.17)$$

where r is the rate of interest of mortgages, H is housing services, P is general price of goods, DY is disposable income, NOH is the number of owner-occupied houses, NRH is the number of houses for rent.

We will show here that neuro-computing technology, using more data, permits to more effectively to fit the function Md (see (8.17))

The last few years have seen a considerable rise in the use of neural networks in financial and economic applications including predicting the ratings of corporate bonds, asset allocation in Bond markets, marketing, advertising, sales statistics and business forecasting [15]. The consensus is that neural networks perform at least as well as traditional techniques, and in many instances significantly outperform them [15].

In [15] interesting results are demonstrated comparing different forecasting techniques with the use of data from M-competition (Makpidakis et al.). Makpidakis et al. used 111 real-world time series and presented them to various groups of forecasters, with the most recent values held back. The competitors were asked to make forecasts over the withheld data and the results were compared. Although this competition pre-dates neural network time series modeling, the experiment has since been re-run using neural networks on a number of occasions.

Assuming that it would be interesting and useful for the reader, we offer some data on the results of comparison. Hull et al. using all 111 time series data found that neural forecasting is superior to classical models. Sharada and Patel using 75 of 111 time series data show that neuro-computing based forecasting is compatible to auto box. They also show (using the same 111 data) that neuro-computing outperforms Box-Jenkins. Lapedes and Farber using 3 chaotic time series proved that neural networks are better than regression models. Foster investigated all 111 time series and obtained results proving that neural networks are inferior to classical forecasting models.

To fit (8.17) we will use a feed-forward network. The architecture of this network is a three-layer feed-forward perception with back-propagation of error. Number of input neurons is the same as number of inputs values choice of which was made in accordance with economic theory and existing econometric models of mortgage market, with an empirical choice for the dynamic specification.

For the approximation (8.17) monthly observations were available for all the above mentioned variables, with the exception of DY for which only annual data is available. Monthly DY was estimated by linear interpolation. The experiments on neural forecasting of mortgage are described in [27].The variable to be predicted was the production of new mortgages in total value, MINDV. In these experiments the static version of the demand function has six explanatory variables, but a considerable number of lagged variables, and also dummies, have still to be introduced. At least one neuron for each input variable is necessary in the input layer. For the forecasting we use 21 input variables.

Sample set has to be divided in three parts: training set, test set and prediction set [27]. The prediction set is kept apart and is never used in training. The training samples are used to train the network; training phase is stopped when the error on test set has reached a minimum.

To evaluate the network performances conventional ordinary least squares was applied to the same data. The R squared was 0.7413, the standard error was 0.12, using only data up to December 1992. Regression on the whole data set, including 1993 produced an R squared of 0.7747 and a standard error of 0.11.

It is necessary to note that the increase of effectiveness of neuro-forecasting of mortgage demand in particular, and financial and economic processes in general is related significantly to the following points [27]: the interpretation and choice of the data set; the choice of the right filters and/or transformations for the variables used in the model; the optimal choice of structure of model, training parameters and model complexity.

8.7 Neuro-Fuzzy Loan Assessment System

Loan and borrower assessment is a very important problem for credit organizations that offer financial aid for large projects [25]. Normally when clients apply for a loan, they are given a questionnaire to fill out or are engaged in a several-stage interview. The organizations gather the information in order to evaluate the client (the potential borrower) in advance in terms of creditability. They also check if the activity of clients would agree with the functions or purposes of the organization. The ability to make right assessments is important to make right decisions and avoid losses — loan offering business is very risky and the cost for the wrong decision may be very high. Decision-making should be based on a well structured hierarchical several level approach for the analysis, evaluation and selection.

Having clearly understood that (1) the knowledge from the experts can be extracted in mostly linguistic form and iteratively and that (2) this information would contain ambiguities/contradictions and (3) it is being continually accumulated as the organization exists and operates, we decided to create a potential borrower assessment system on the basis of Neuro-Fuzzy approach. Besides preserving information consistency, this would allow relearning system's knowledge base as new data emerge or some conditions are no more the case—thereby keeping the system always operative and up-to-date.

In addition to the documentary evaluation of a client we do also visual and verbal evaluations of the client.

Table 8.3. shows a fragment from the list of input questions used by the system to make an evaluation

Table 8.3. A fragment from the list of input questions

1. Documentary evaluation of the client
   a) business is financially sound
   b) business has capable management
   c) business has necessary technical staff and know-how
   d) sales projections (including export if any) are realistic and achievable
   e) adequate supplies of raw materials, energy, water and others are achievable
   f) customer's contribution is sound and viable
   g) business conforms with all appropriate regulations and has received all necessary licenses
   h) project implementation schedule is realistic taking into account local availability of goods, services & labor
   i) proposed loan repayment schedule is feasible, and has the debt service coverage ratio of ___________
   j) proposed collateral and additional guarantees protect against lending risk
2. Client self-evaluation
   a) sales and marketing review questions
   b) business operations review questions
   c) financial review questions
   d) personnel related questions
   e) administrative management related questions
3. Visual Evaluation of the client
   a) outward appearance
      - age
      - physical conditions
      - outer clothes
      - hands
      - eyes

b) self-conduct
- waiting manner
- posture

4. Verbal evaluation of the client
a) Talking
- coherency and clearness
- talking manner (loudness, confidence)

b) conception delivering ability
- concreteness (no mixing allowed)
- lucidity

c) question answering manner
- exactness ("10 sheep", not "about 10")
- concreteness (no lyrical digressions)

To provide the system knowledge base the features/abilities of expert tractability, continuous self-tuning, automatic integrity reinforcement and flexible and reliable inference mechanism, we create the system's intelligent part on the basis of Neuro-Fuzzy approach [3,7].

Fuzzy rules used by the system are of the following form:

IF F_{11}/w_{11} AND F_{12}/w_{12} AND … AND F_{1n}/w_{1n} AND … AND F_{1N}/w_{1N}
THEN $D=d_1$

IF F_{21}/w_{21} AND F_{22}/w_{22} AND … AND F_{2n}/w_{2n} AND … AND F_{2N}/w_{2N}
THEN $D=d_2$

… … …

IF F_{r1}/w_{r1} AND F_{r2}/w_{r2} AND … AND F_{m}/w_{m} AND … AND F_{rN}/w_{rN}
THEN $D=d_r$

… … …

IF F_{R1}/w_{R1} AND F_{R2}/w_{R2} AND … AND F_{Rn}/w_{Rn} AND … AND F_{RN}/w_{RN}
THEN $D=d_R$

Where an element $F_m = \{f_n = v_m\}$ is called an attribute (or feature) and expresses the presence of some attribute (attribute or feature number n) in the input (object, event or situation) to be qualified for the output (object class, result event, or control), defined for rule r; f_n is a fuzzy variable taking a value from a set of input fuzzy constant (linguistic terms) $V = \{v_1, v_2, ..., v_M\}$; v_m is a value from set V used in rule r for variable f_n; N is total number of variables; R is the number of fuzzy rules; M is the dimension of fuzzy constant set V; D is the combined output variable; d_r is a fuzzy contribution to output D from rule r; w_m is the importance value of feature F_m.

Value of d_r in our case is fuzzy singleton, showing the confidence degree of output r.

A simplified rule looks like:

IF
{<Business is financially sound>="VERY"}/W=0.6 AND ...
{<Adequate supplies of row materials>="GOOD"}/W=0.5 AND ...
{<Debt service coverage ratio>=<Net income>/<Principal & Interest payments>="ABOUT N%"}/W=0.7 AND ...
THEN
<Loan Rating>="VERY GOOD"

Output, e.g., <Loans overall rating> is evaluated as combination of such values as <Loan Rating>, <Client Rating>, <Business relevance>, <Expected interest>, <Business Risk Level>, etc.

As can be seen we use non-traditional "weighted" fuzzy rules. This kind of rules are more effective than traditional ones in problems such as "fuzzy pattern recognition problem" in which features (or attributes) on which the decision is based are often very "subjective" and changeable in time. Also note that rules used a modified conjunction operator suggested in [7].

For providing learning and automatic integrity reinforcement the fuzzy inference is implemented by a Neuro-Fuzzy network.

The Neuro-Fuzzy network used is Radial Basis Function (RBF) neural network. The network consists of 3 layers of neurons. The input layer is fed by parameters specifying current customer attributes. The next layer performs normalization of these parameters values to fall into the certain range. Neurons in third layer calculate the distance measure:

$$y_j = \sum_i \left| F_{ji} - x_i \right| w_{ij}$$

Based on the values of distance measure, the final output, representing the rating value, is calculated.

Fig. 8.6. shows the generalized structure of the Neuro-Fuzzy Loan Assessment System (LAS). As can be seen the system consists of the following modules: Neural Network (which keeps the system knowledge in the weight parameters), Raw Data Store (linking to various sources of unprocessed data; filtering and conversion to the system database format), Learning Subsystem (machine based learning from database data), Fuzzy Rules Viewing/Editing Subsystem (human friendly tools for rule proving/editing), Knowledge Representation Converter (Updates user changes to the knowledge with applying appropriate changes to the network weights) and the others. To implement the Neural Network we used a modification of RBF neural networks, which is very useful for implementation of fuzzy inference mechanism in problems like the recognition problem. The node processors and their processing laws are designed so as to provide a convenient way to convert the knowledge base from fuzzy rules representation to neural representation and vice versa.

The importance values of features are evaluated for current state of system as:

$$w_{Li} = \sum_{K,L,K \neq L} \left\{ \frac{\left| F_{Li} - F_{Ki} \right|}{\sum_j \left| F_{Lj} - F_{Kj} \right|} \right\}$$

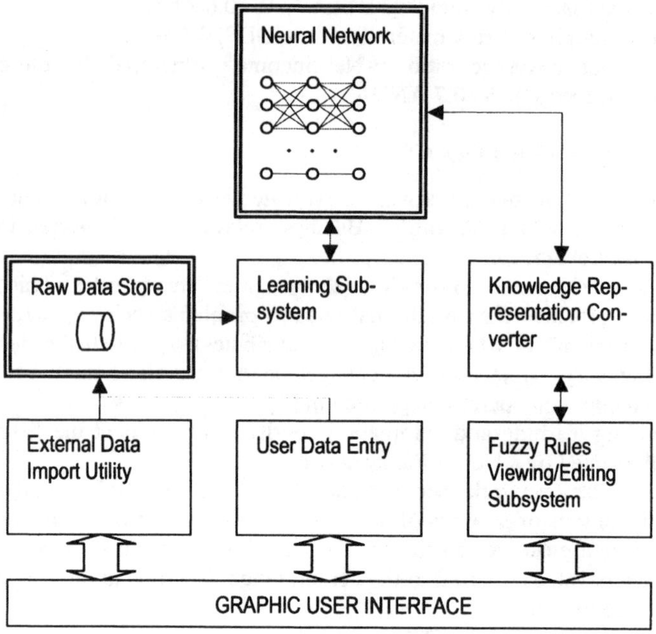

Fig. 8.6. Structure of the Neuro-Fuzzy LAS

An experimental version of system has been implemented in Borland C++. Algorithms of Neuro-Fuzzy inference and learning, written in C programming language, were adopted from [3] (Fig.8.7.).

The inputs to the systems can be supplied in different ways. Normal way is through the user interface provided with the system. The user enters values choosing one from 3 to 7 linguistic terms best answering the question. The inputs can be imported also from an Excel sheet with appropriately formatted data fields. The second form is more convenient when gathering data from old files or receiving replies by e-mails, or from experts having not installed the system on their machines (or having no desire to get it).

MS Office software is commonly accepted tool to use for business records and calculations by many organizations all over the world. The Excel documents have been used to collect information to learn the system. It was very convenient because most part of the documentation and records used by older paper-based system were based on Excel. New data can be entered to the system manually using the provided graphic user interface.

Trough the intuitive function menu the user can easily access the data, relearn (update) the system with newly entered data and evaluate the current customer in terms of creditability on the basis of all knowledge accumulated from complete historical data.

An experienced user (expert) can make modifications to automatically generated knowledge. The knowledge is available for editing in form of fuzzy IF-THEN rules.

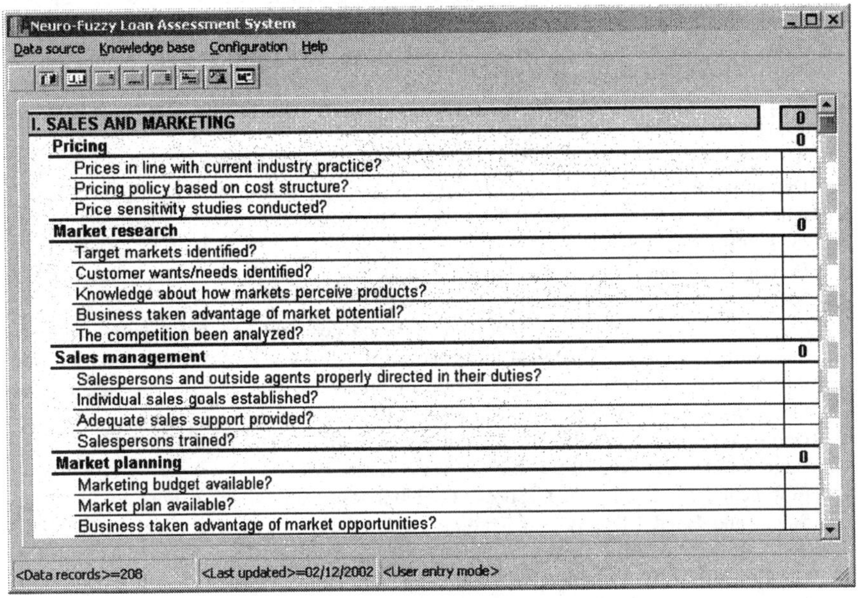

Fig.8.7. Neuro-Fuzzy Loan Assessment Application (screenshot)

After the user confirms the entry the system's knowledge base is reloaded. An interesting feature is that no source files (e.g. Excel documents) are changed or deleted – but only the system's internal neural network based knowledge modified. So rollback is always available.

To conclude, we just want to list here some of the advantages of the Neuro-Fuzzy Loan Assessment System:

- Less time for the user to learn how to work, can be used by any non-sophisticated (naïve) user
- Integration of information with commonly used office software and easy retrieval
- Mechanism to keep knowledge up-to-date
- Appropriate data archiving, no loss of information
- Better than traditional DBMS-based systems in integrity reinforcement and response time to get decision
- Self-optimizing characteristics and compact neural network based memory model
- Combined expert/automatic knowledge base tuning for quickest and correct decision

References

1. Aliev RA, Aliev RR. Soft Computing and its Application. World Scientific, New Jersey, London, Singapore, Hong Kong, 2001, 444 p
2. Aliev RA, Mamedova GA, Aliev RR Fuzzy Sets Theory and Its Application. Tabriz: Tabriz University Press, 1993, p.244
3. Aliev RA, Vaguidov RM, Neural networks. Theory and practice, Tabriz University, Tabriz, Iran, 1993
4. Aliev, R, Fazlollahi, B, Vahidov, R „Soft Computing Based Multi-Agent Marketing Decision Support System", The Journal of Intelligent and Fuzzy Systems. 1998
5. Allen Webster. (1992) Applied Statistics for Business and Economics. Richard Instructor's Edition. D. Irvin. Inc. 1992
6. Berenguer M and Davies P"Credit Risk: Modeling To Manage". Balance sheet, vol. 7, No 1
7. Bonfig KW, Guirimov BG "Neuro-Fuzzy based generation of fuzzy knowledge and the flexible procedure of logical inference for the graphic pattern recognition, in proc. International Conference on Application of Fuzzy Systems ICAFS-96, Siegen, Germany, pp. 314-318, 1996
8. Deboeck, G, Cader, M „Pre-and Post-processing of financial Data", in Trading on the Edge, Guido Deboeck, Editor, John Wiley & Sons, 1994
9. Embrechts, M, Cader, M, Deboeck, G „Nonlinear Dimensions of Foreign Exchange, Stock, and Bond Markets", in Trading on the Edge. Guido Deboeck, Editor, John Wiley & Sons, 1994
10. Gordy MB "A Comparative Anatomy of Credit Risk Models" December 8, 1998
11. Guirimov BG "Algorithms and methods for recognition of fuzzy images", Ph.D. Dissertation, Baku State University, 2000
12. Hall, J „Adaptive Selection of U.S. Stocks with neural Nets", in Trading on the Edge. Guido Deboeck, Editor, John Wiley & Sons, 1994
13. Jang, G, Lai, F „Intelligent Trading of an Emerging Market", in Trading on the Edge. Guido Deboeck, Editor, John Wiley & Sons, 1994
14. Kendall, SM, and K. Ord. 1990. Time series. New York: Oxford University Press. 3rd ed
15. Kingdon J. Intelligent Systems and Financial Forecasting, Springer,1998
16. Klir GJ, U.StClair, Bo Yuan. Fuzzy Set Theory. Prentice Hall
17. Kuo RJ A Decision Support System for the Stock Market through integration of fuzzy neural networks and Fuzzy Delphi. Applied Artificial Intelligence 12:501-520, 1998
18. Lippmann, RP 1987. An introduction to computing with neural nets. IEEE Magazine 3(4):4-22
19. Markowitz HM (1952): "Portfolio Selection:, Journal of Finance, 7, p.77-91
20. Markowitz HM (1959): "Portfolio Selection: Efficient Diversification of Investments", Wiley, New York, NY
21. Mehta D, Aliev RA, F.Sh.Bagirov. A fuzzy model of credit risk, In: Proceedings of ICAFS 2000, 2000, pp. 128-132
22. Mitra G, Kyriakis T, Lucas C, Pirbhay M (2002): "A Review of Portfolio Planning: Models and Systems", In Advances in Portfolio Construction and Implementation, Satchel S.E., and Scowcroft A.E. (Eds.), Butterworth and Heinemann, Oxford

23. Nwana, H, Azarmi, N, (Ed..), Software Agents and soft Computing: Towards Enhancing Machine Intelligence - concepts and applications. Springer, Berlin, 1998
24. Richard A Brealey, Stewart C. Myers. Principles of Corporate Finance.-USA, 1991.4-1
25. Thomas LC The survey of credit and behavioral scoring: forecasting financial risk of lending to customers, International Journal of Forecasting, Vol. 16, 2000, pp. 149-172
26. Trading on the edge. Neural, Genetic and Fuzzy Systems for Chaotic Financial Markets. Guido J. Dedoeck, John Wiley & Sons, Inc., New-York, Chichester, Brisbane, Toronto, Singapore
27. Virilli F. Neural Network Time Series Models for Financial Risk Management; Berlin: Wiss. Verl. Berlin, 2001
28. Wagner HM (1975): "Principles of Operations Research", Prentice-Hall, Inc., Englewood Cliiffs, New Jersey
29. Wooldridge, M, Jennings, N. Intelligent agents: theory and practice", The Knowledge Engineering Review. Vol. 10:2, 1995, ppl 15-152
30. Ye, Z, Gu, L „ A Fuzzy System for Trading the Shanghai Stock Market", in Trading on the Edge, Guido Deboeck, Editor, John Wiley & Sons, 1994
31. Yoda, M. Predicting the Tokyo Stock Market", in Trading on the Edge. Guido Deboeck, Editor, John Wiley & Sons, 1994

9 Soft Computing in Electronic Business

9.1 A Multi-Agent system for E-Commerce Decisions

Introduction

In an era of electronic commerce businesses and consumers have access to vast amounts of potentially important information. While this is beneficial for making informed decisions, the decision makers could easily become overwhelmed by the information, much of which may not be relevant. Thus, there is a pressing need for decision aiding tools that would effectively process, filter, and deliver the right information to the organizational and individual decision makers. In this work we argue that combination of DSS and agent technologies could prove a very powerful tool for rendering decision support in e-commerce application. We propose an architecture for a multi-agent DSS for e-commerce, and describe a prototype system for making on-line investment decisions.

Electronic Commerce is revolutionizing the way business processes are conducted [34]. Businesses and individual consumers are gaining access to vast amount of information available on the web. This information, if properly used can facilitate improved decision making both in business-to-business as well as business-to-customer segments of E-Commerce.

However, although the amount and accessibility of potentially useful information has increased, the nature of unaided human decision making remains "boundedly" rational. The ability of human decision makers to process large amounts of data, filter out irrelevant information, and guard against harmful cognitive biases in making purchasing/selling decisions is largely limited.

Decision Support Systems (DSS) can help improve consumer decision-making. Although the field of DSS is quite mature, it's applications in e-commerce deserves special attention of researchers and developers. Empirical studies have shown that use of DSS promotes effective decision making [33]. Also adapting the DSS to the decision maker, decision task, and decision context, improves decision support effectiveness [20]. In the e-commerce application of DSS, the customer is the decision maker making a purchase decision in a context that provides fast and easy access to vast amount of information outside the boundaries of DSS. Unlike the pre-web era, when DSS was primarily used for managerial support in an "island" mode, with a weak link to the environment, in e-commerce applications the DSS can be an integral part of the digital environment and directly support the actions of the involved parties.

Relatively young software agent technology can help decision makers to cope with problems of information overload when utilized as a component of DSS. Through delegation of routine time-consuming decision-related tasks to software assistants human decision makers can better concentrate on higher-level cognitive and analytical aspects of decision making.

This section seeks to propose a method for integrating DSS and software agents within the context of e-commerce. We believe that this combination will lead to a very powerful synergy that would promote effective decision making by the parties involved in e-commerce activities.

Related Work

The customer decision-making process in the e-commerce environment is generally information intensive [15]. The age of web-enabled technologies and e-commerce brings about the abundance of information accessible to decision makers. If not properly filtered and pre-processed, this amplitude of information can easily "overload" the decision maker (e.g. [32,39]). As stated in [34], "Web has made the need for decision support ever more pronounced". Currently, there is a pressing need for tools that facilitate search, preprocessing, filtering of information, automation of routine tasks, and providing decision makers with valuable options and insights [14]. The early applications of DSS in e-commerce tend to belong to data-based DSS category [1]. The use of model-based category of DSS is increasing and examples reported in the literature include General Electric Plastics using simulation models and United Technologies Corporation using models that support design and selection of heating, ventilation, and air-conditioning units for buildings [28].

Because of differences among customers, their needs, and context for decision-making purchase, not all of the potentially available information and analysis are relevant in any given situation. The danger of "information overload" and "Negative affective reactions" requires enabling customers discover exactly what information they want [5] and become selective in use of information and analysis. Empirical evidence suggests that consumer satisfaction with the decision-making process was significantly increased when the models used by DSS fit the knowledge level of the customer [29]. Increasing satisfaction with the shopping experience may lead customers to make purchases and not just browse, as well as consider product attributes apart from just the price.

Software agent technology is arguably well suited for e-commerce applications [23]. Generally defined in terms of their properties of proactiveness, autonomy, social ability, and reactivity [42], these robots of cyberspace come handy in automating many information processing tasks. Well-known price comparison agents BargainFinder and Jango are by now classical examples of agent-assisted e-commerce.

Some researchers are looking to automate much of e-commerce related activities using agent technologies [3, 9]. For example, Chen et al. [9] describes multi-agent system for largely automating the stages of Consumer Buying Behavior model. Barbuceanu & Lo propose an agent-based approach for automating decision making/negotiation activities using optimization and multi-attribute utility functions [3].

While automation of well-structured tasks in e-commerce is an important direction of research, it must be noted that many ill-structured tasks are not subject to automation [31]. These tasks naturally require DSS based approach with proper combination of human and machine capabilities. Some deployment of web-based DSS in the context of e-commerce has been reported in the literature [28, 4]. However, the combination of agent-based technologies with DSS in this respect can prove particularly powerful [11]. In particular, an agent-enabled DSS can facilitate an intelligent interface for the decision makers [29]. For example, Conway & Koehler have demonstrated how an interface agent incorporating optimization tools can significantly reduce rationally bounded suboptimal decision making by humans.

There have been attempts to utilize agents as decision-aiding tools for e-commerce applications. We have already mentioned simple price-comparison agents earlier. Sen & Hernandez proposed an architecture for a system that informs the users about implicit constraints and preferences in search for appropriate products [32]. They describe a prototype for apartment location problem. Collins et al. describe an agent-based mixed-initiative DSS for semi-automated contracting, where agents recommend task-bid assignment decisions based on bids submitted by the suppliers [10]. Arguing that fully autonomous behavior is impractical, the authors allow the decision makers to override agent-generated proposals.

There is a fairly common realization among researchers in the field that personalization of information and suggestion is necessary when providing decision support [2,25,27]. Ardissono et al. describe a multi-agent system that maintains user profile and stereotypical information to guide the buyer in "rational" exploration of a catalog, and take initiative in suggesting products. Muller & Pischel describe agents that interact with user and back office and provide decision support to the users in choosing license type for an information product (e.g. electronic journal, virtual journal, etc.). Nunes-Suarez et al describe a multi-agent system that assist the user in making travel plans. The agents propose the plans in accordance with the maintained user profile.

While these developments in agent-based decision support for e-commerce applications are very encouraging, they do not take into account fundamental models and principles of human problem solving/decision making field. For example, the well-known Simon's model of problem solving (Intelligence-Design-Choice), extensively used in DSS literature, as well as divergence-convergence principles known from the problem solving literature are not incorporated in the above approaches. Indeed, the "recommendation" and "suggestion" - centric approaches are quite susceptible to well - known human biases of anchoring and adjustment, and premature convergence. Furthermore, the information presented by parties with vested interest in selling products is likely to be biased. Such abusive treatment of customers may result in loss of trust. In this paper we propose an architecture for multi-agent Decision Support System for e-commerce that incorporates the above principles of human problem solving.

Relation between Simon's decision making and consumer buying behavior models

The classical concepts of decision support advocates support for all stages of the decision process as well as tailoring the support to fit the user, task, and the decision context [20]. The classical buyer decision-making process consists of five stages: problem recognition, information search, evaluation of alternatives, purchase decision, and post-purchase behavior [13,16,26]. Problem recognition occurs when a consumer senses a disparity between his or her actual state and his or her desired state [7]. Problem recognition can be activated when a consumer is exposed to external stimuli, such as the smell of food. This causes the consumer to become aware of an unmet need such as hunger.

Aware of an unmet need, the consumer is motivated to gather information on finding a way to satisfy the need. Information gathering helps consumers become aware of competing brands or products and their features. If the consumer has low involvement with the products that can satisfy the perceived need, little effort will be exerted in the information search. For example, the consumer may process information through peripheral routes, relying more heavily on cues as opposed to detailed and elaborate product specific information [8].

The information search results in the identification of a set of alternative products that will reduce or eliminate a felt need [16]. The consumer will then identify product features/characteristics to determine the likely consequences of purchasing from the considered set of alternative products [30]. Using available internal, form memory, and external information, the consumer may process information by attributes or by products to develop a set of beliefs, or "product image", about a product with regard to its ability to deliver the desired benefits [6].

After evaluating alternatives, consumers typically form product preferences and perhaps the intention to purchase what they perceive as the most desirable. Several sub-decisions may be involved including brand choice, vendor selection, quantity to purchase, timing of purchase, payment method, etc.

In the final post-purchase behavior stage the consumer experiences some level of satisfaction or dissatisfaction. Marketers strive to satisfy the customers in order to enhance their relationship with the consumers [37]. Functionalities of an efficient and effective web site that may provide relevant information at all phases of consumer decision-making include:

- Problem recognition: The web selling design of Amanzon.com is a good example. When a user searches for a book, he/she gets access to the links similar customers used for other books. Users are considered similar if they searched for the same book. Customers are provided with these links because he or she may potentially need those books.
- Information search: To help the users make informed decisions, the DSS may provide more information on the different features of a series of products and the appropriate situations in which to use or apply those various products. A valuable aspect of decision support would also be identification of a small number of attractive diverse alternatives if the product search space is large.

- Evaluation of alternatives: During this phase of the decision making process, the DSS may provide decisional guidance (e.g. through critiquing) on forming the criteria for making a choice.
- Purchase decision: in this phase, the system can help finalize and execute the purchase decision.
- Post purchase behavior: a system can help review the implementation and consequences of a decision.

Miles and Howes indicated the analogy between some phases of Consumer Buying Behavior (CBB) and Simon's model [24]. Specifically, in their view the intelligence-design-choice triad corresponds to managing search criteria, product search, and comparison of products phases respectively. In our opinion this view is somewhat narrow, since important CBB phases are omitted from the model. We have summarized our perception of the link between CBB and Simon's model in Table 9.1.

Table 9. 1. Comparison of CBB and Simon's models

CBB	Simon's model	Comments
Need Recognition	Intelligence	Opportunity discovery
Information Search		Collection of relevant information
	Design	Filtering information. Reducing information space (e.g. product & merchant brokering)
Evaluation of Purchase	Choice	Comparisons of purchasing options. Making purchasing decision.
Purchase		
Post-Purchase	Intelligence	Reviewing the decision

Clearly, the need recognition and initial information search for certain products can be adequately referred to as intelligence phase. Here the decision maker is primarily engaged in initial exploration of the market. This exploration can help identify new categories of products that the buyer was unaware of, as well as general information about known products, e.g. price ranges, general features, sellers, etc. When the information search becomes more focused, shifting from explorative to exploitative mode, the decision maker effectively enters the design phase. Here the list of products and merchants is effectively reduced to a few viable alternatives using some formal or informal criteria (e.g. low-cost alternatives) for more in-depth analysis. For example, the Active Buyer's Guide (http://activebuyersguide.com/) features an active sales assistant that elicits buyer's preferences and generates a list of alternative recommendations for further analysis by the buyer. This process is still in the realm of the information search phase of the CBB.

Evaluation and subsequent purchase correspond to the choice phase of the Simon's model, the outcome of which is the final decision. One possibility here is to perform comparison of the options identified at the earlier phase. There are many sites that facilitate easy comparisons of such options. Another more advanced possibility (if applicable) is to make use of models. For example, in the already mentioned work [28] the authors discuss a case of using simulation models on the web for making purchasing decisions.

Finally, the after-purchase evaluation is similar to the '"reviewing" phase of the decision making model. This last phase was introduced by Simon at a later time [35] than the original publication [36]. However, as he originally indicated that reviewing is akin to the intelligence phase, which is why we put it under "intelligence" to stress the cyclical nature of the model.

There are no clear-cut boundaries between the phases, and in fact the iterative processes are common. Nevertheless, since Simon's model has been used extensively in DSS literature, our mapping gives us an opportunity to design DSS features for making e-commerce decisions.

Architecture of Multi-Agent DSS for E-Commerce

This section describes the architecture for Multi-Agent DSS for E-Commerce (MADEC). Fig. 9.1 represents the general architecture of MADEC. The architecture consists of two parts: the objects (passive entities) and the agents (active doers). The major components of the architecture include: Interface; Data & Models (Objects); Intelligence Team (Agents); Design Team (Agents); and Choice Team (Agents). The suggested naming reflects the phases of decision making (by Simon) that the corresponding agents support the most.

The Interface provides interaction between the user and MADEC. The Data and Models contain the data and models relevant to decision making. The user can access the data and models just as in the traditional DSS to extract information from the database, do useful calculations and what-if analysis using models, and perform other related tasks. For example, in purchasing airline tickets, the customer may choose to use a model for planning a round trip journey or a model for planning multiple segments separately. In MADEC, the Data & Models also provide information to different agents if required.

The Intelligence Team contains one or more information agents, the purpose of which is to obtain relevant information. The examples of these agents include serach agents, information filtering agents, collaborative filtering agents, and other information agents. Since there has been extensive research in this area, we will not elaborate on design principles for these agents. One interesting architecture for this type of agent-supported information retrieval is Retsina proposed by Sycara et al [38].

The Design Team incorporates agents for generating alternative suggestions. Agents in this group propose alternatives that should be weighed by the decision maker in making economic decisions. The name for a team corresponds to the design phase of problem solving model that deals primarily with generation of alternative courses of actions. Each proposing agent in this team is using its own unique model in recommending a decision. Inclusion of multiple proposing agents, therefore, reflects the fact that there is more than one viable alternative (e.g. product) to be considered. Our approach, thus departs from other "suggestion" type of approaches which assumes that an "optimal" "customized" solution for each customer can be found.

Furthermore, our agents not only suggest different alternatives, but also suggest ones that are *diverse.* In other words, we do not employ an approach where there is a single global criterion but a set of alternatives that satisfy that single criterion best is presented to the user. We are indeed diversifying the criteria. In fact, divergence is an essential element in problem solving. One could diversify in respect to the desired objectives of a decision, as well as to the way these objectives are achieved. In other words, one can identify different values in making a decision (e.g. maximize quality, minimize cost, etc.) as well as different views on how to achieve a given objective (e.g. how to choose a quality product).

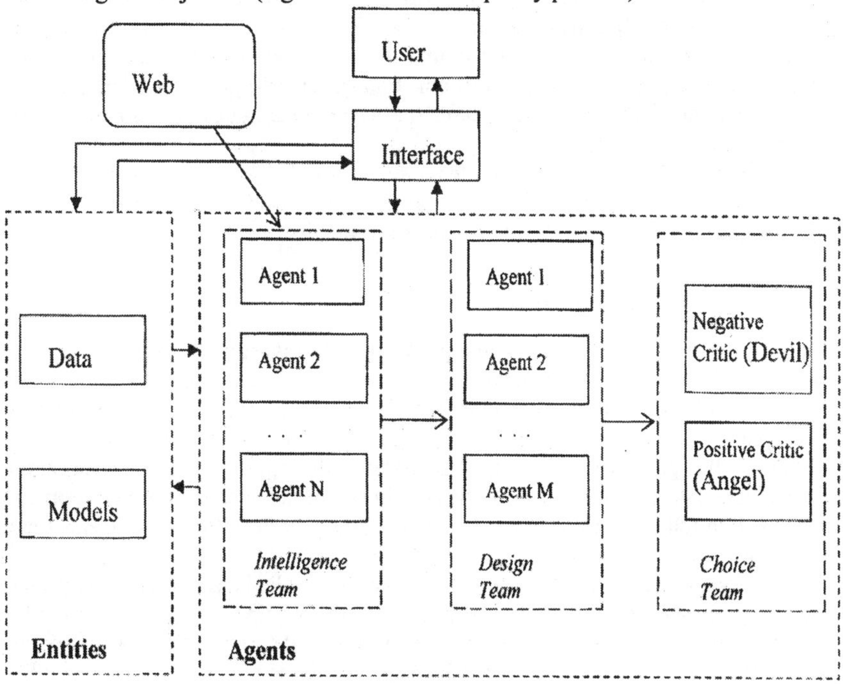

Fig. 9. 1. An architecture of MADEC

The agents of the design team adopt different values & views while making their recommendations. While we recognize that value dimension is more appropriate for diversification of alternatives in e-commerce, our example in the next section will incorporate both dimensions. Examples of value-diversified suggestions in a book/airline ticket buying context would be:

- Agent suggesting the lowest-price book/ticket;
- Agent suggesting the book with the best delivery time/shortest flight time;
- Agent suggesting the best quality (e.g. hardcover, paper type) book/non stop flight;
- Agent suggesting the best trusted source(book retailer/airline company).
- Some combinations of the above.

Hence, our proposal generates a set of "good" (in a Pareto-optimum sense) and diverse alternatives, countering human biases of premature convergence, and anchoring and adjusting, and promoting divergence in alternative search.

The Choice Team consists of two types of agents: the positive and negative critique agents, also referred here to as the "angel" and the "devil" respectively. These agents critique the proposals generated by the proposing team, as well as those modified/specified by the user. The design of such critiquing agents has been elaborated elsewhere [40]. The critique is organized around the objectives, preferences, and soft constraints of the user. For example, in a book buying scenario, one possible critique could be that the proposed book is somewhat expensive considering buyer's budget (soft constraint). By critiquing the available alternatives generated by the design team, the system facilitates convergence in the choice phase of decision making.

In nutshell, the MADEC generates a set of good diverse alternative products with pros and cons outlined. Next section describes application of MADEC architecture to develop a prototype for on-line investments.

MADEC Prototype for Investments

We have developed MADEC prototype for on-line investments. Fig. 9. 2 shows a sample screenshot of user interface (actual names of stocks are coded). Such an enhancement could potentially be included in an e-commerce site supporting online investments. Since our interest is primarily in the design and choice teams of MADEC, the prototype currently incorporates the agents of the corresponding teams. The data for different securities had been pre-downloaded from electronic resources. In a full system this would be accomplished by the information agents (intelligence team).

The Design Team consists of two assisting agents, called analysts and four proposing agents (proposers). The proposers and the analysts help to identify a small subset of perspective securities, from which the proposers determine their portfolios and submit them to the user. The two analysts are designed so that they incorporate different sets of views, namely, two well known different schools of thought in security analysis: Fundamental Analysis and Technical Analysis. Therefore, analysts are the sources of expertise in the two corresponding schools. The analysts differ on the view dimension of the prototype. The four proposes determine the portfolio based on the analysts' recommendations and the values (sets of objectives) they employ. The value dimension of the proposes adopted here is their attitude towards risk, since this is one of the major concerns in selecting adequate portfolio. Each analyst advises two proposes. One of these proposes in each of two groups is risk taker, while the other is risk averse. Therefore, the alternative investment scenarios reflect four distinct attitudes:

- technical analysis/risk-taking;
- technical analysis/risk-averse;
- fundamental analysis/risk-taking; and
- fundamental analysis/risk-averse.

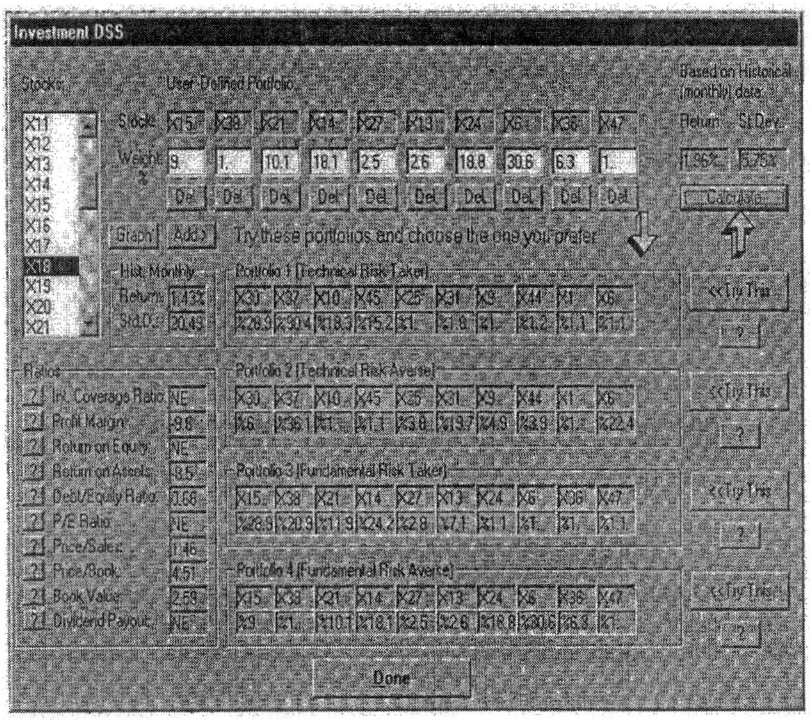

Fig. 9.2. MADEC prototype screenshot

The agents of the design team use rule-based methods as well as heuristic search techniques to generate proposals. Note that the agent tries to find an "optimal" solution within the adopted value-view.

The Choice team consists of two critiquing agents: the Devil and the Angel. These agents analyze and critique the proposed portfolios. There is a number of important factors taken into consideration while making investment decisions. These factors include: expected return and risk, user preferences and the source of alternative portfolios. Generally, one seeks to maximize return while keeping risk at a minimum level. This knowledge can be used for providing objective-related critique. If the user chooses a portfolio with the average expected return being high, the "angel" would detect that and tell user that "the expected return of this portfolio is high". The "devil", on the other hand, may detect high risk associated with the proposed allocation and prompt it to the user.

The preference related critique examines the solution from the point of view of the user's preferences. If the user is a risk taker high risk would trigger the positive critique commenting on the appropriateness of the portfolio to the user's risk preference. On the other hand, if the user is risk averse, the negative critique would identify the inadequate risk level.

The user can also come up with his/her own solutions, or modify the solutions generated by the agents. Our initial investigation suggests that the users of MADEC make better decisions in terms of the return on investment, and have higher level satisfaction with the system than the users of the DSS without agents.

9.2 Soft Computing and Personalization of Electronic Commerce

As the Internet expands and as online sale grows and become widely used, the problem of personalizing the interaction between the Web-store and the customer and of adequate product presentation is becoming more and more important.

The following are examples of questions that appear when implementing e-commerce systems:

- How to pick the right advertisement for a particular customer?
- How to design the content of an advertisement to better serve a customer?
- Which product to recommend to a particular customer?
- How to determine a good price for a particular customer?
- How to determine the right consumer segment to target for a given product?

Personalization is the mechanism and method to target customers individually so that their needs can be better served [18]. The required system should offer personalized interactions to customers and be able to represent the data from one source in various forms suitable for different users depending on their domain expertise, preferences and needs. Also the system must assist the users in selection of the products which best match their needs. This is considerably more than what is available from traditional filtering systems.

While presenting the information to the users, the system should monitor their actions and collect their preferences and needs in order to use the mined data for self-adapting. The adapted system should then anticipate the next session user actions presenting them more comfortable and friendly interface, and changing the contents of pages and links to reflect most likely required information. This requires many pages to be generated “on the fly” depending on the users short-term goals. Also the system should effectively maintain and constantly update the users/customers and store databases.

In order to create a customizable architecture of a Web store that can be tailored to the user needs and preferences in generating hypertextual pages of the store catalog, the system should be based on specialized agents working in parallel and communicating with each other that only can perform such a complicated task. A monolithic architecture, composed of a single agent cannot manage the problem in a satisfactory way. Each agent in the multi-agent architecture should offer a specific set of services. In this system the domain-dependent knowledge concerns information about products and customer declaratively represented features while the domain-independent knowledge consists of components, which represent the core of the store.

The architecture of the Web-Store system may consist of the following agents (Fig. 9.3):

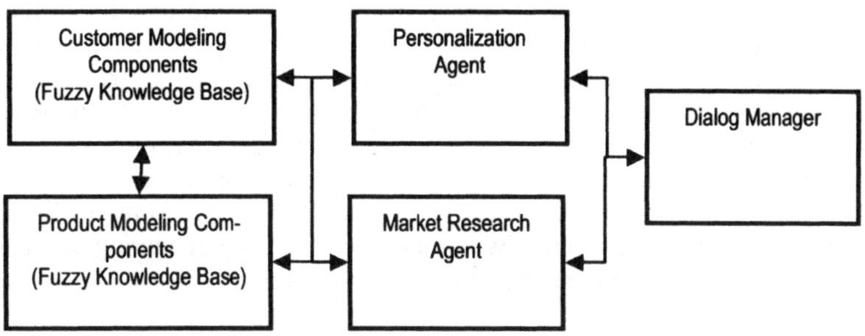

Fig. 9.3. Structure of Personalized Web Store system

1. Dialog manager (handles the interaction with the user and maintains contextual information; identifies and monitors the user actions for dynamic customer modeling purposes).
2. Customer Modeling Component – CMC (actually creates and maintains customer profiles and stereotype KB; contains customer descriptive information and preferences in form of linguistic values with probabilistic evaluations: [Name], [Importance], [Values] <Linguistic value, Probability>; and generates predictive information).
3. Product Modeling Component – PMC (dynamically builds a product taxonomy view to offer the user a personalized, constrained information); assists the user in selection of products
4. Market research Agent (mining new rules and targeting customers based on market segmentation).
5. Personalization Agent (builds code for personalized hypertextual Web-pages with necessary layout and contents).

Let's give some definitions and shortly describe on how fuzzy sets based approaches can be used in the components of a Web-Store system mentioned above.

A *customer model* is a repository of data/information about a particular customer. It stores preferences, attributes, interests, histories, etc. Explicitly enquiring the customers of these attributes and preferences is inadmissible. Neural network based unsupervised and reinforcement learning algorithms widely described in literature [21] can be successfully used for these purposes. Most customer attributes are inherently fuzzy. Thus, fuzzy sets can play a major role in customer profile representation. Fuzzy clustering method can be used to build adaptive consumer profile.

A *product model* is a repository of all product-related information. The behavior of a customer related to a particular product category depends on different attributes. The problem is to find which attributes can be used for best targeting specific products. Because the descriptions of product evaluation/categorization as well as customer perception about the product are fuzzy and dynamic, an appropriate way to describe relations between product categories and qualitative attributes is using fuzzy sets and membership functions. For example, "a good car" is a car that has *good* mileage, *reasonable* price, *satisfactory* reliability etc. All qualifiers such as good or reasonable are described by membership functions such that all products possess all attributes to certain degrees. Fuzzy sets based qualifiers provide best targeting of products whereas preserving their fine differentiation.

Market research implies conducting polls and surveys to collect important information about customer preferences of products, brands, ideas etc. Surveys provide ways for targeting customers based on market segmentation. Fuzzy approaches are used to summarize the survey results in the form of linguistic knowledge that can be interpreted by humans easily. Fuzzy clustering approaches are used to segment customer response to surveys [17].

9.3 Risk Analysis in Electronic Commerce Using Fuzzy Weighted Average

Risk analysis has been widely used in context of financial and engineering problems. A new way of conducting business over the Internet, called Electronic Commerce (EC) is growing exponentially. The use of the Internet for commerce has radically altered the way of doing business. More benefits can be achieved by EC and many companies have started doing business this way. However, the risk is also increasing. To avoid significant losses it is very important to carry out a thorough risk analysis before implementation an EC project [41].

Many managers believe that because of the growing dependency of business and technology, it is very important before implementation of any EC project to carefully plan and measure the associated risk [12,19]. However little research has been done to identify and assess the overall risk associated to overall risk of implementing an EC development. In this section we will briefly consider a model of risk analysis using fuzzy set in which the overall risk of an EC project is assessed on the basis of the fuzzy weighted average [22].

Risk analysis can be performed using the probability theory. EC is a relatively new strategy and less information is available regarding risk associated with EC to create an adequate statistical model for risk analysis for EC. Considering also that risk analysis is highly subjective and related to inexact and vague information, the application of fuzzy sets theory seems very appropriate to solve this problem.

Technically the risk is defined as the value of possible loss multiplied by the degree of probability at which this loss is expected. For example, if the risk of a $100,000 car crash is estimated with the probability of 0.001, then the risk rate is 1000. In more complicated cases such as electronic commerce development, the risk can be related with a number of criteria.

The first problem in development of EC is the identification of risk, i.e. to find potential risk factors. Based on these factors we create the qualitative or quantitative criteria. The other problem is how to aggregate the different criteria and perform a linguistic approximation to derive a final assessment.

Identification of risk

Understanding the taxonomy of risk factors is very important to an organization for managing the risk. The overall risks of an organization can be divided into three categories: technical, organizational, and environmental. The technical factors include operational (maintainability, reliability etc.), security (client-server relationships, data and transactions etc.), and requirement (design, stability etc.) factors. The organizational factors include business, resource (missed schedule, lack of budget, personnel loss etc.), managerial etc. The environment related factors consists of legal, cultural, performance and other factors. The common factors that are used in identifying risk have reliability, privacy, security, legal, technical, and trust nature. First of all, before doing risk assessment we need to determine a set of qualitative and quantitative criteria that identify the risk: $R_i, i = \overline{1,n}$. The criteria R_i are variables whose values are linguistic terms.

Calculation of risk factors and linguistic approximation

Usually the criteria R_i are found by averaging probabilities or likelihoods of different fuzzy risk factors given by different experts in questionnaires. Then it is very important to note that different risk factors may not be equally important. So it is a good idea to assign a weight to each risk factor W_i , $i = \overline{1,n}$. Note that these weights reflect the severity of risk factors and are also subjective and fuzzy. They are also obtained by averaging experts' opinion.

One of the most widely used approaches is to evaluate the risk level using the well known formula of weighted average [22]:

$$R = \sum_{i=1}^{n} W_i \cdot R_i \Big/ \sum_{i=1}^{n} W_i \tag{9.1}$$

To make calculations with fuzzy numbers, the α -cut method is used. When using triangle-shaped membership functions, it is enough to consider intervals with α =0, 0.5, and 1. After we have an output (fuzzy) value of R, we can use one of well-known defuzzification algorithms to make a final risk assessment. We can also do linguistic approximation which means translating the output fuzzy value back into linguistic term. To choose the most appropriate term, the Euclidean distance measure can be used.

Paper [41] demonstrates an example of calculating the risk assessment. The chosen risk factors are: x_1 (lack of budget), x_2 (key person loss), x_3 (miss schedule). Risk probabilities are evaluated by fuzzy values {"not likely", "possible", "probable", "likely", "very likely"}, and the severity is evaluated by values from the set {"minimal", "low", "moderate", "high", "critical"} (Fig. 9.4).

Values from the both sets are described by triangle-shaped membership functions. The overall decision $y = (w_1x_1 + w_2x_2 + w_3x_3)/(w_1 + w_2 + w_3)$. For α =0, the intervals for x_i are [0.5, 1], [0, 0.5], [0.25, 0.75], for w_i are [0.75, 1], [0.25, 0.75], [0.25, 0.75], and for y is [0.25, 0.875]. For α =0.5 y is [0.4044, 0.7143] and for α =1.0 y is [0.5625, 0.5625].

Then the resultant fuzzy set is mapped to one of fuzzy labels (Fig.9.4): "minimal", "low", "moderate", "high", and "critical". The Euclidean distance measure used for the linguistic approximation gave the minimal distance for the label "moderate":

D(*x*, "moderate")=0.1684

The above result means that the risk of the organizational resources in this case is moderate.

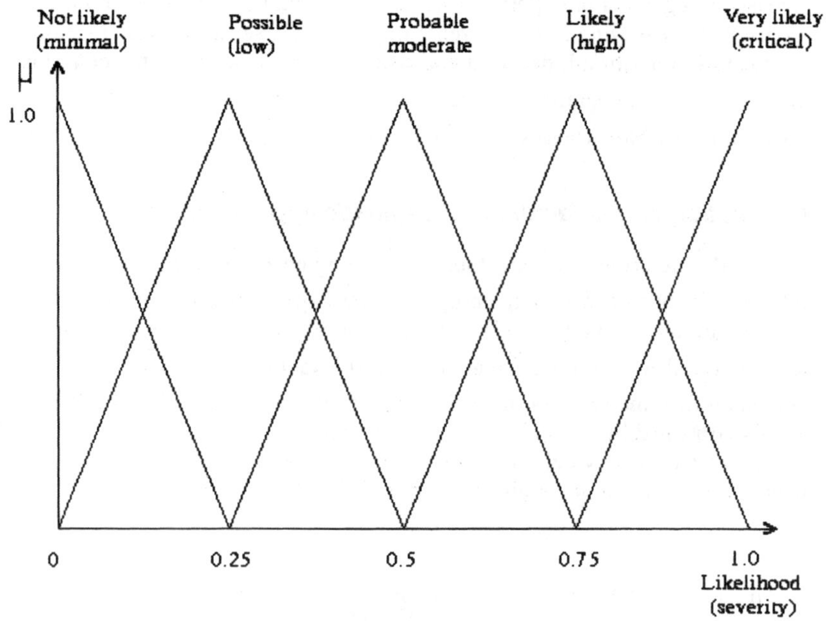

Fig.9. 4. Membership functions of linguistic terms

References

1. Alter S (1980) Decision Support Systems. Current Practice and Continuing Challenges. Reading, Mass. Addison-Wesley
2. Ardissono L, Barbero C, Goy A, Petrone G(1999) An Agent Architecture for Personalized Web Stores. In: Proc. Of Third International Conference on Autonomous Agents. Seattle. Washington
3. Barbuceanu M, Lo WK (2000) A Multi-Attribute Utility Theoretic Negotiation Architecture for Electronic Commerce. In: Proc. of Fourth International Conference on Autonomous Agents. Barcelona. Catalonia
4. Berndt D J (2001) Consumer Decision Support Systems. A Health Care Case Study. In: Proc. of the 34th Hawaii International Conference on System Sciences. Hawaii
5. Bertman S (1998) Hyper culture: The human cost of speed. Westport. CT: Praeger publishing
6. Bettman JR (1979)An Information Processing Theory of Consumer Choice. Reading. MA: Addison-Wesley
7. Bruner GC (1987) The effect of Problem Recognition Style on Information Seeking. J Academy of Marketing Science. vol. 15: 3341
8. Cacioppo J, Petty R, Kao C (1986) Central and Periferial Routes to Persuasion: An Individual Difference Perspective J Journal of Personality and Social Psychology-vol. 51: 1032-1043.
9. Chen Q, Hsu M, Dayal U, Griss M, (2000) Multi-Agent Cooperation. Dynamic Workflow and XML for E-Cornmerce. In: Proc. Of Fourth International Conference on Autonomous Agents. Barcelona. Catalonia
10. Collins J, Bilot C, Gini M (2000) Mixed-Initiative Decision Support in Agent-Based Automated Contracting In: Proc. Of Fourth International Conference on Autonomous Agents. Barcelona. Catalonia
11. Conway DG, Koehler GJ (2000) Interface agents: Caveat mercator in electronic commerce. J Decision Support Systems. vol. 27:355-366
12. Davis J (1999) E-Commerce still a risky business. E-Commerce Times
13. Engel JF, Kollat DT, Blackwell RL (1973) Consumer Behavior. 21 ed. New York. Holt, Rinehalt & Winston
14. Fazlollahi B (2001) Electronic Commerce: Determining B2C Web site Functions In: Information Resources Management Association International Conference. Toronto. Ontario Canada
15. Glazer R (1991) Marketing in an Information-Intensive Environment: Strategic Implications of Knowledge as an Asset. J Marketing. vol. 55: 1-9
16. Howard J, Sheth J (1972) The Theory of Buyer Behavior. New York. Free Press
17. Hsu T, Chu KM, Chan H (2000) The fuzzy clustering on market segment. In: the Ninth IEEE Intl. Conf. on Fuzzy Systems. vol. 2: pp 621-626
18. Jain V, Krishnapuram R(2001) Application of Fuzzy Sets in Personalization for e-Commerce, 0-7803-7078-3/01. IEEE. pp. 263-266
19. Jaleshgari RP (1999) Project maps minimize risk. J Information week. Issues 755: 138-143
20. Keen PG, Morton MS (1978) DSS: An Organizational Perspective. Reading: Addison-Wesley

21. Lawrence J (1994) Introduction to Neural Networks. California Scientific Software Press
22. Lee DH, Park D (1997) An efficient algorithm for fuzzy weighted average. J Fuzzy Sets and Systems. vol. 87: 39-45
23. Ma M (1999) Agents in E-Commerce. J Communications of ACM. vol. 42: 78-80
24. Miles GE, Howes A (2000) A Framework for Understanding Human Factors in Web-Based Electronic Commerce. J International Journal of Human-Computer Studies, vol. 52:131-163
25. Muller JP, Pischel M (1999) Doing Business in the Information Marketplace: A Case Study. In: Proc. Of Third International Conference on Autonomous Agents. Seattle. Washington
26. Nicosia FN (1982) Consumer Decision Processes: A Futuristic View. J Advances in Consumer Research, vol. 9: 17-19
27. Nunes-Suarez J, O'Sullivan D, Brouchoud H, Cros P, Moore C , Byrne C (2000) Experiences in the Use of FIPA Agent Technologies for the Development of a Personal Travel Application. In: Proc. Of Fourth International Conference on Autonomous Agents. Barcelona. Catalonia
28. O'Keefe RM , McEachern T (1998) Web-Based Customer Decision Support Systems. J Communications of the ACM, vol. 41: 71-78
29. Pereira RE (2000) Optimizing Human-Computer Interaction for the Electronic Commerce Environment. J Electronic Commerce Research, vol. 1
30. Phillips, Olson, Baumgartner (1995) Agent's role on the presence and Advertising effectiveness
31. Scarl A Gebauer J (1999) Between Flexibility and Automation. J Computer-Mediated Communication. vol. 5.
32. Sen S, Hernandez K (2000) A Buyer's Agent. In: Proc. Of Fourth International Conference on Autonomous Agents. Barcelona. Catalonia
33. Sharda R, Barr SH, McDonnell JC (1988) Decision Support System Effectiveness. A Review and an Empirical Test. J Management Science. vol. 34:139-159
34. Shaw M, Gardner JD, Thomas H (1997) Research Opportunities in Electronic Commerce. Decision Support Systems. vol. 21: 149-156
35. Simon HA (1977) The New Science of Management Decision. Englewood Cliffs. N J. Prentice-Hall
36. Simon HA (1960) The New Science of Management Decision. New York. Harper
37. Spreng RS, MacKenzie, Olshavsky HR (1996) A reexamination of the Determinants of Consumer Satisfaction J Marketing vol. 60: 1532.
38. Sycara KP, Zeng D (1996) Multi-Agent Integration of Information Gathering and Decision Support. In: European Conference on Artificial Intelligence
39. Tsvetovat M, Sycara K (2000) Customer Colaitions in the Electronic Marketplace In: Proc. Of Fourth International Conference on Autonomous Agents. Barcelona. Catalonia.
40. Vahidov R, Elrod R (1999) Incorporating Critique and Argumentation in DSS. J Decision Support Systems. vol. 26: 249-258
41. Wat FKT,Ngai EWT (2001) Risk analysis in electronic commerce development using fuzzy set. 0-7803-7078-3/01 In: IEEE. pp 807-811
42. Wooldridge M, Jennings N (1995) Intelligent Agents. Theory and Practice. Knowledge Engineering Review. vol. 10: 115-152

Printing: Strauss GmbH, Mörlenbach
Binding: Schäffer, Grünstadt